The Peripheral Actions of 5-Hydroxytryptamine

The Peripheral Actions of 5-Hydroxytryptamine

EDITED BY

JOHN R. FOZARD
Hypertension Research Project
Sandoz Ltd

OXFORD NEW YORK TOKYO
OXFORD UNIVERSITY PRESS
1989

Oxford University Press, Walton Street, Oxford OX2 6DP
Oxford New York Toronto
Delhi Bombay Calcutta Madras Karachi
Petaling Jaya Singapore Hong Kong Tokyo
Nairobi Dar es Salaam Capte Town
Melbourne Auckland
and associated companies in
Berlin Ibadan

Oxford is a trade mark of Oxford University Press

Published in the United States
by Oxford University Press, New York

British Library Cataloguing in Publication Data
The peripheral actions of 5-
hydroxytryptamine.
1. Medicine. Drug therapy. Serotonin &
tryptophan
I. Fozard, John R.
615'.36
ISBN 0–19–261683–8

Library of Congress Cataloging in Publication Data
The peripheral actions of 5-hydroxytryptamine/edited by John R. Fozard.
Includes bibliographies and index.
1. Serotonin—Physiological effect. 2. Nerves,
Peripheral—Physiology. I. Fozard, John R.
[DNLM:1. Serotonin—physiology. QV 126 P445]
QP801.S4P47 1989 615'.71—dc19 88–31303
ISBN 0–19–261683–8

Set by Footnote Graphics
Warminster, Wiltshire
Printed in Great Britain by
Courier International Ltd
Tiptree Essex

To my family—for thankfully putting in many commas but never a full stop

Preface

The principal condition for accepting the not inconsiderable task of editing this book was that a number of hand-picked colleagues and friends agreed that (a) it would be a worthwhile venture, and (b) they would help me with the project. *The Peripheral Actions of 5-Hydroxytryptamine* is testimony to their collective and generally enthusiastic response to my initial approach.

In a sense, this book is a follow-up to the earlier volume *The Neuropharmacology of Serotonin*, edited by Richard Green, which dealt mainly with the central actions of 5-HT. My idea was simple; to describe and put into context the more significant of the plethora of important new developments in the field of 5-HT in the periphery. Whilst, inevitably, a substantial part of thc book dcals with advances in the evergreen cardiovascular and gastro-intestinal areas, equally, several subjects have been covered (e.g. allosteric 5-HT_2 receptor modulation—chapter 3; secretion mechanisms—chapter 12; cardiopulmonary and carotid body reflexes—chapter 13) which to my knowledge have not previously been reviewed in detail. A recurrent theme, culminating in the final two chapters which deal specifically with the subject, has been the evaluation of the therapeutic significance of these developments.

The book unashamedly reflects a number of personal foibles. Abbreviations, for example, have been kept to a minimum and mainly restricted to compound names; the list appears at the front of the book. '5-HT' terminology has been used in preference to 'serotonin'. The classification of functional 5-HT receptors of Bradley *et al.* (*Neuropharmacology* **25**, 563–76, 1986), despite its limitations, has been adopted throughout the book. The core of this classification and its nomenclature is summarized in Table 1 of chapter 5. However, in recognition of the heterogeneity of '5-HT_1-like' receptors, the terminology derived from radioligand binding studies has also been used as appropriate, and, I hope, prudently. The key features of the subtypes of 5-HT recognition sites are summarized in section 2 of chapter 4.

My gratitude goes to all the contributors who have written so effectively on their allotted subjects, to the editorial staff at Oxford University Press for encouragement and guidance, and to Jacqueline Boucris and Sylvia Buld for secretarial assistance. Finally, I acknowledge the friendship and collaboration of all those colleagues past and present who have helped me indulge a magnificent obsession.

31 May, 1988 J.R.F.

Contents

16 Pathophysiological significance of 5-hydroxytryptamine in the periphery 377

DONALD S. HOUSTON AND PAUL M. VANHOUTTE

Contributors

Theresa A. Branchek Department of Anatomy and Cell Biology, Columbia University, New York, NY 10032, USA

Marlene L. Cohen Lilly Research Laboratories, Eli Lilly and Company, Lilly Corporate Center, Indianapolis, IA 46285, USA

Richard A. Cohen Peripheral Vascular Section, Robert Dawson Evans Department of Clinical Research, Boston University Medical Center, Boston, MA 02118, USA

I. S. De La Lande Department of Clinical and Experimental Pharmacology, University of Adelaide, Box 498, Adelaide, South Australia 5001, Australia

W. Feniuk Pharmacology Division, Glaxo Group Research Ltd., Ware, Hertfordshire SG12 ODJ, UK

J. R. Fozard Preclinical Research Department, Sandoz Ltd., CH-4002, Basel, Switzerland

Brian L. Furman Department of Physiology and Pharmacology, University of Strathclyde, Glasgow G1 1XW, UK

Michael D. Gershon Department of Anatomy and Cell Biology, Columbia University, New York, NY 10032, USA

Donald S. Houston Department of Medicine, Health Sciences Centre, University of Manitoba, Winnipeg, Manitoba, Canada

D. Hoyer Preclinical Research Department, Sandoz Ltd., CH-4002, Basel, Switzerland

P. P. A. Humphrey Pharmacology Division, Glaxo Group Research Ltd., Ware, Hertfordshire SG12 ODJ, UK

Alberto J. Kaumann Department of Pharmacology, Smith Kline and French Research Ltd., The Frythe, Welwyn, Hertfordshire AL6 9AR, UK

Paul Leff Analytical Pharmacology Group, Department of Pharmacology I, Wellcome Research Laboratories, Beckenham, Kent BR3 3BS, UK

D. S. McQueen Department of Pharmacology, University of Edinburgh Medical School, Edinburgh EH8 9JZ, UK

Graeme R. Martin Analytical Pharmacology Group, Department of Pharmacology I, Wellcome Research Laboratories, Beckenham, Kent BR3 3BS, UK

Gary M. Mawe Department of Anatomy and Cell Biology, Columbia University, New York, NY 10032, USA

A. K. Mir Preclinical Research Department, Sandoz Ltd., CH-4002, Basel, Switzerland

Ewan J. Mylecharane Department of Pharmacology, University of Sydney, NSW 2006, Australia

Christine A. Phillips Department of Pharmacology, University of Sydney, NSW 2006, Australia

Mila B. Turla Department of Physiology, University of Michigan Medical School, Ann Arbor, Michigan 48109–0622, USA

Paul M. Vanhoutte Department of Physiology and Biophysics, Mayo Clinic and Foundation, Rochester, Minnesota 55905, USA

Tony J. Verbeuren Fondax, Groupe de Recherche Servier 7, Rue Ampère, F-92800 Puteaux, France

D. I. Wallis Department of Physiology, University of Wales College of Cardiff CF1 1SS, UK

N. Graham Waton Department of Physiology and Pharmacology, University of Strathclyde, Glasgow G1 1XW, UK

R. Clinton Webb Department of Physiology, University of Michigan Medical School, Ann Arbor, Michigan 48109–0622, USA

Abbreviations

In addition to the following, several abbreviations specifically related to techniques of drug receptor classification are used in Chapters 2 and 3 and are defined there.

21–009	4-(3-ter-butyl-amino-2-hydroxypropoxy) indol-2-carbonic acid isopropylester
5-HTP-DP	*N*-acetyl-5-hydroxytryptophyl-5-hydroxytryptophan amide
8-OH-DPAT	8-hydroxy-2-(di-n-propylamino)tetralin
A23187	calcium ionophore of microbial origin
ACTH	adrenocorticotrophic hormone
ADP	adenosine 5′-diphosphate
ATP	adenosine 5′-triphosphate
AH25086	3-(2-aminoethyl)-1H-indole-5-(*N*-methyl)acetamide
BEA 1654	1-(3-acetamidophenyl)piperazine
BRL 24924	(+)-[endo]-4-amino-5-chloro-2-methoxy-*N*-(1-azabicyclo-[3,3,1]-non-4-ylbenzamide
BRL 43964	[endo]*N*-(9-methyl-9-azabicyclo-[3,3,1]-non-3-yl)-1-methylindazole-3-carboxamide
BW 501C67	(α-anilino-*N*-2-n-chloro)phenoxypropylacetamidine
CGS 12066B	7-trifluoromethyl-4(4-methyl-1-piperazinyl-pyrolo (1,2-a) quinoxaline
cyclic AMP	adenosine 3′,5′ cyclic monophosphate
cyclic GMP	guanosine 3′,5′ cyclic monophosphate
DNA	deoxyribonucleic acid
cDNA	complementary deoxyribonucleic acid
DOB	4-bromo-2,5-dimethoxyphenylisopropylamine
EDRF	endothelium-derived relaxing factor(s)
EDTA	ethylenediaminetetraacetate
EGTA	ethyleneglycol bis(aminoethylether)tetraacetate
GABA	γ-aminobutyric acid
GR38032	1,2,3,9-tetrahydro-9-methyl-3[(2-methyl-1H-imidazol-1-yl) methyl]-4-H-carbazol-4-one
GR43175	3-[2-dimethylamino]ethyl-*N*-methyl-1H-indole 5 methane sulphonamide

G (or N) proteins	guanine nucleotide regulatory binding proteins
GTP	guanosine 5′-triphosphate
H-7	1-(5-isoquinolinesulphonyl)-2-methyl-piperazine
ICI 89406	2-[2-hydroxy-3-[2-(3-phenylureido)ethylamino] propoxy]benzonitrile
ICI 118, 551	erythro-DL-1(7-methylindan-4-yloxy)-3-isopropyl-aminobutan-2-ol
ICI 169, 369	2-(2-dimethylaminoethylthio)-3-phenylquinolone
ICS 205–930	(3α-tropanyl)1-H-indole-3-carboxylic acid ester
I-HEAT	2-[β-(iodo,4-hydroxyphenyl)-ethyl-aminomethyl] tetralone
INPEA	1-(4-nitrophenyl)-2-(isopropylamino)-ethanol
LSD	D-lysergic acid diethylamide
LY53857	4-isopropyl-7-methyl-9-(hydroxy-1-methylpropxycarboxyl) 4,6,6A,7,8,9,10,10A-octahydroindole-(4,3-FG) quinoline maleate
mCPP	1-(*m*-chlorophenyl)piperazine
MDL 72832	8-[4-(1,4-benzodioxan-2-yl-methylamino)butyl]-8-azaspiro [4,5]-decane-7,9-dione
MDL 72222	1αH,3α,5αH-tropan-3-yl-3,5-dichlorobenzoate
MK 212	6-chloro-2-(1-piperazinyl)pyrazine
PAPP	1-[2-(4-aminophenyl)ethyl]-4-(3-trifluoromethyl-phenyl) piperazine
mRNA	messenger ribonucleic acid
pA_2	negative logarithm of the molar concentration of antagonist that causes a 2-fold rightward shift in the dose-response curve of an agonist
pK_A	negative logarithm of the agonist equilibrium dissociation constant
pK_B	negative logarithm of the antagonist equilibrium dissociation constant
pK_D	negative logarithm of the equilibrium dissociation constant determined by radioligand binding
pK_I	negative logarithm of the antagonist equilibrium dissociation constant determined using the Cheng and Prussoff equation
RU 24969	5-methoxy-3(1,2,3,6-tetrahydropyridin-4-yl)1H-indole
SCH 23390	(+)8-chloro-2,3,4,5-tetrahydro-3-methyl-5-phenyl-1H-3-benzazepine-7-ol hemimaleate

SDZ 206-830	(3α-homotropanyl)-1-methyl-5-fluoro-indole-3-carboxylic acid ester
SDZ 206–792	(*N*-desmethyl-3-α-homotropanyl)-1H-indole-3-carboxylic acid ester
TFMPP	1-(3-trifluoromethylphenyl)piperazine
U466191	11,9-epoxy-methano-prostaglandin H_2
WAL 1307	3-amino-6-methyl-1,13b-dihydro-dibenz(b,f)imidazo (1,5-d)-(1,4)-oxazepine
WB 4101	(2,b-dimethoxyphenoxyethyl)aminomethyl-1,4 benzodioxane
WY 26392	*N*-(1,3,4,6,7,11b α-hexahydro-2H-benzo-[a]-quinolizin-2β-yl), *N*-methyl-n-propyl-sulphonamide

1

Synthesis, storage, release, and metabolism of 5-hydroxytryptamine in peripheral tissues

TONY J. VERBEUREN

1.1 Introduction

5-HT is present in both plants and animals. In mammals, significant levels of 5-HT are associated with the enterochromaffin cells of the gastro-intestinal mucosa, the brain, the pineal gland, and the blood platelets. In addition, 5-HT has also been detected in peripheral nerves of the gut, in the heart, the lung, the kidney, the spleen, the thyroid, in blood vessels, and in mast cells of certain species. Although a role for 5-HT has been recognized in the brain (neurotransmitter) and the pineal gland (precursor of melatonin), the significance of its presence in other tissues is not yet understood.

The aim of the present review is to summarize the biosynthesis, uptake, storage, release, and metabolism of 5-HT in peripheral tissues. The complex interactions between 5-HT and adrenergically innervated tissues will receive special attention.

1.2 Biochemical pathways of 5-HT: synthesis and metabolism

1.2.1 Synthesis of 5-HT

5-HT is synthesized in most of the tissues in which it is stored. An important exception is the blood platelet: the significant pool of 5-HT found in the platelets has its origin in the enterochromaffin cells of the gastro-intestinal tract (Erspamer and Testini 1959). The biosynthesis of 5-HT is schematically represented in Fig. 1.1. The precursor of 5-HT is the essential amino acid L-tryptophan. The first step in the synthetic pathway is the hydroxylation of tryptophan; this step, catalysed by the enzyme tryptophan hydroxylase, is the rate limiting step. The 5-hydroxytryptophan thus formed is then transformed to 5-hydroxytryptamine by the non-specific enzyme aromatic L-amino acid decarboxylase. Only 2 per cent of the ingested tryptophan is utilized for the daily synthesis of about 10 mg of

5-HT. A detailed overview of the activities and the regulation of the enzymes involved in the biosynthesis of 5-HT has recently been presented by Tyce (1985).

1.2.1.1 *Tryptophan hydroxylase*

In almost all the tissues which contain meaningful amounts of 5-HT, tryptophan hydroxylase has been detected. Activity of the enzyme has been found in the enterochromaffin cells of the gastro-intestinal mucosa (Legay *et al.* 1983b), in the pineal gland (Deguchi 1977), in mouse mast cells (Lovenberg *et al.* 1965), in 5-HT neurones of the gastro-intestinal tract (Gershon *et al.* 1977), and in carcinoid tumour cells (Lovenberg *et al.* 1967). No tryptophan hydroxylase activity has been detected in platelets (Morrissey *et al.* 1977). Inside the cell, the enzyme activity may be localized in the cytosol (Lovenberg *et al.* 1967) or associated with subcellular organelles (Green and Sawyer 1966; Joh *et al.* 1975).

1.2.1.2 *Aromatic L-amino acid decarboxylase*

Aromatic L-amino acid decarboxylase is distributed in many body tissues. The enzyme catalyses the decarboxylation of 5-hydroxytryptophan to 5-hydroxytryptamine but is non-specific; it also catalyses the decarboxylation of tryptophan, tyrosine, phenylalanine, and dihydroxyphenylalanine (see Tyce 1985).

1.2.2 Metabolism of 5-HT

The predominant metabolic pathway of 5-HT is oxidative deamination by monoamine oxidase with the resulting formation of 5-hydroxyindoleacetic acid. A schematic representation of the major pathways of metabolism of 5-HT is shown in Fig. 1.1. Besides deamination, followed by oxidation (5-hydroxyindoleacetic acid) or deamination followed by reduction (5-hydroxytryptophol), conjugation with sulphate (5-HT-*O*-sulphate) or glucuronide (5-HT-*O*-glucuronide) and *N*-acetylation (*N*-acetyl-5-HT) represent minor metabolic pathways of 5-HT. *O*-Methylation and *N*-methylation of 5-HT have also been described. In the pineal gland, *N*-acetylation of 5-HT followed by *O*-methylation leads to the formation of melatonin (see Axelrod 1974; Tyce 1985).

1.2.2.1 *Oxidative deamination*

Monoamine oxidase is the principal enzyme involved in the inactivation of 5-HT. The enzyme is present both intra- and extraneuronally and has a particularly high activity in the liver. Its activity can be attributed to the effects of two isoenzymes, designated type A and type B (Johnston 1968; Knoll and Magyar 1972; see Verbeuren and Vanhoutte 1982). Monoamine oxidase will catalyse the oxidative deamination of a wide variety of biogenic

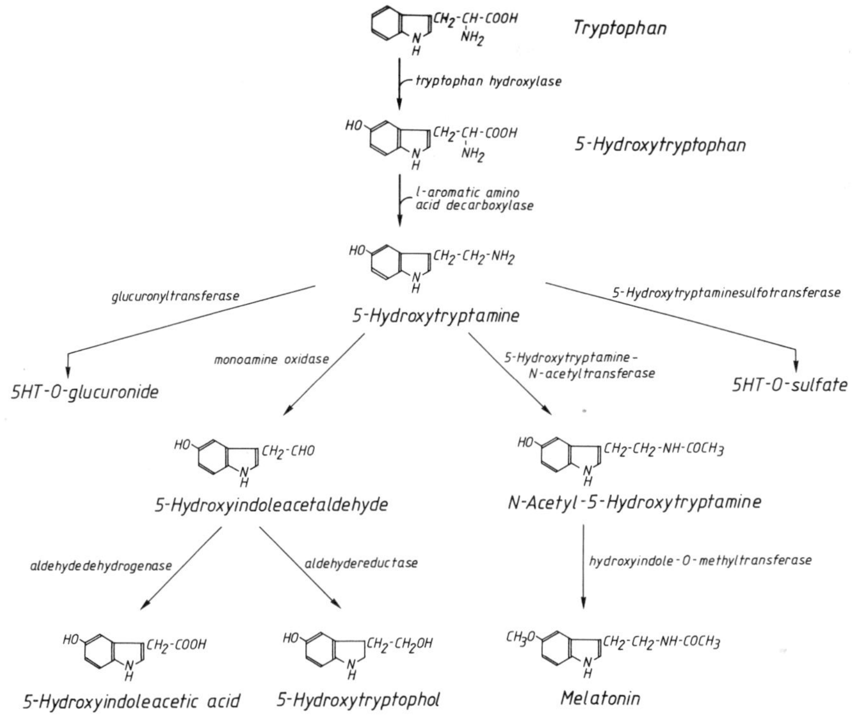

Fig. 1.1 Biosynthesis and metabolism of 5-HT

amines such as dopamine, noradrenaline, tyramine, phenylethylamine, tryptamine, and 5-HT. The B form of the enzyme, which is predominantly, if not exclusively, localized extraneuronally, deaminates preferentially phenylethylamine; the A form is localized both intra- and extraneuronally and is responsible for the oxidative deamination of 5-HT and noradrenaline (see Verbeuren and Vanhoutte 1982; Tyce 1985).

The 5-hydroxyindoleacetaldehyde, formed by oxidative deamination of 5-HT, is rapidly oxidized to 5-hydroxyindoleacetic acid by aldehyde dehydrogenase, which is present in many body tissues (see Tyce 1985). 5-Hydroxyindoleacetic acid is present in many 5-HT-containing tissues (e.g. Legay *et al.* 1983a and b). 5-Hydroxyindoleacetaldehyde can also be reduced to the alcohol, 5-hydroxytryptophol, in a reaction catalysed by aldehyde reductase. Although, under normal conditions, this pathway is less important than the oxidation to 5-hydroxyindoleacetic acid, 5-hydroxytryptophol has been detected in peripheral tissues (Tyce *et al.* 1968).

1.2.2.2 *Sulphation and glucuronidation*

The enzyme 5-HT-sulphotransferase, which inactivates 5-HT and forms

5-HT-*O*-sulphate, is found in brain, liver, and lung (Hidaka *et al.* 1969). *O*-Sulphation of 5-HT results in a loss of biological activity; however, in many tissues arylsulphatase enzymes can regenerate free 5-HT by cleavage of 5-HT-*O*-sulphate (Kishimoto *et al.* 1961). 5-HT can be conjugated with glucuronic acid by a reaction catalysed by the enzyme uridine diphosphate glucuronyltransferase. This enzyme is found in many tissues but the liver has the highest activity; in the isolated perfused rat liver, 5-HT-*O*-glucuronide is a major metabolite of 5-HT (Tyce *et al.* 1968).

1.2.2.3 N-*acetylation,* O-*methylation,* N-*methylation*

In the pineal gland, *N*-acetylation of 5-HT is the first step in the biosynthesis of the hormone melatonin (see Fig. 1.1). The enzyme that catalyses this reaction is 5-HT-*N*-acetyltransferase; this enzyme is also present in the liver (Tyce *et al.* 1968). The activity of 5-HT-*N*-acetyltransferase in the pineal body regulates the diurnal rhythm of the synthesis of melatonin (Klein and Weller 1970). *N*-Acetyl-5-HT may be further *O*-methylated in the pineal body by the enzyme hydroxyindole-*O*-methyltransferase (Fig. 1.1), resulting in the formation of melatonin (see Tyce 1985). In the brain 5-HT itself can be *O*-methylated, resulting in the formation of 5-methoxytryptamine (Green *et al.* 1973). In the lung and in the brain, *N*-methylation of 5-HT has been described (Axelrod 1962; Morgan and Mandell 1969).

1.3 5-HT in the gastro-intestinal tract

In mammals, most of the 5-HT is associated with the enterochromaffin cells of the gastro-intestinal tract (Erspamer 1954 and 1966) which probably accounts for the bulk of the 5-HT which reaches the blood. However, in the intestine, 5-HT is not only contained in enterochromaffin cells; increasing evidence suggests that the monoamine is also present in 5-HT neurones (Gershon *et al.* 1977; Costa *et al.* 1982; see also Chapter 11).

1.3.1 5-HT in enterochromaffin cells

As long ago as 1906 it was observed that certain cells of the gut mucosa and adrenal medulla stain similarly after dichromate fixation; the gut cells were named 'enterochromaffin cells' (Ciaccio 1906). Because of their metabolic properties, the mucosal enterochromaffin cells are considered to be members of the APUD (Amine Precursor Uptake and Decarboxylation) systems (Pearse 1968; Legay *et al.* 1983a). Further studies concerning the structure and function of these cells led also to their being designated as paraneurons, in recognition of their common features with neurones (Fujita and Kobayashi 1977). The enterochromaffin cells, which are dispersed in the gut epithelium of animals and humans, contain numerous large, electron dense neuro-secretin-like granules, in which 5-HT is stored (Polak *et al.* 1976; Mourre *et al.* 1981; Ahlman and Dahlström 1982; Nemoto *et al.* 1982). Such 5-HT-containing enterochromaffin cells are not only found in the intestine, but appear also to be present in other body tissues—for instance, the pancreas, the lung, the thymus, the thyroid, and the urogenital tract (Polak *et al.* 1982).

In intestinal mucosa, high levels of endogenous 5-HT are present (over 7 μg/g of tissue, about 20 times the content of the whole brain) and this 5-HT is mostly associated with the enterochromaffin cells (Forsberg and Miller 1983; Legay *et al.* 1983a).

The presence of tryptophan hydroxylase in the enterochromaffin cells has been shown in two different ways: [^{3}H]5-HT can be formed from [^{3}H]tryptophan in rat caecum mucosal cells (Legay *et al.* 1983a), and parachlorophenylalanine, an inhibitor of tryptophan hydroxylase, decreases both the 5-HT content and formaldehyde-induced fluorescence due to 5-HT in enterochromaffin cells (Grübe 1976; Legay *et al.* 1983a). Only very low levels of 5-hydroxytryptophan were detected in rat mucosa, suggesting that the hydroxylation of tryptophan, rather than the decarboxylation of 5-hydroxytryptophan, is the rate-limiting step in the synthesis of 5-HT (Legay *et al.* 1983a).

The enterochromaffin cells also possess a high affinity uptake mechanism for 5-HT with a K_m value (about 5×10^{-8} M) comparable to that found in the brain (Mourre *et al.* 1981; Ternaux *et al.* 1981; Legay *et al.* 1983a).

Early observations by Bülbring and her colleagues demonstrated that 5-HT can be released into the gut lumen as a result of pressure exerted on the mucosa (Bülbring *et al.* 1958; Bülbring and Crema 1959). Subsequently, several *in vivo* studies showed that intraluminal sucrose solutions (O'Hara *et al.* 1959), a low luminal pH (Resnick and Gray 1962), and vagal nerve stimulation (Ahlman *et al.* 1981a and b; Ahlman and Dahlström 1982) increase the levels of 5-HT within the intestinal lumen.

Since in the enterochromaffin cells the storage granules are mostly located near the base of the cells, a release of 5-HT across the basolateral membrane is most likely to occur; the amine thus released may then escape into the portal circulation (Ahlman and Dahlström 1982). Thus, instillation of hypertonic glucose into the intestine (Drapanas *et al.* 1962), intra-arterial administration of nicotine, morphine, catecholamines, and acetylcholine, scratching the serosal surface (Burks and Long 1966), acid perfusion of the duodenum (Kellum and Jaffe 1976), vagal nerve stimulation (Pettersson *et al.* 1979) and splanchnic nerve stimulation (Larsson *et al.* 1979) have all been shown to increase 5-HT levels in the portal blood. The effect of vagal nerve stimulation appears to be mediated by a β-adrenoceptor mechanism, since it can be partially antagonized by propranolol (Pettersson *et al.* 1979; Ahlman and Dahlström 1982).

More recently, *in vitro* studies have provided more direct information about the release of 5-HT from the gut enterochromaffin cells. In gut specimens of the rat, the β-adrenoceptor agonists, adrenaline and isoprenaline, reduced the 5-HT content of the enterochromaffin cells; similarly, noradrenaline and acetylcholine were able to decrease slightly the 5-HT content (Petersson *et al.* 1978). The effect of adrenaline appears to be mediated by β_2-adrenoceptors (Pettersson *et al.* 1978). Field stimulation also decreases the 5-HT content of the gut strips and this effect can be prevented by propranolol and 6-hydroxydopamine, indicating that a neurogenic mechanism mediated by β-adrenoceptors is involved (Pettersson *et al.* 1979; see Ahlman and Dahlström 1982). Cholera toxin decreases 5-HT levels in enterochromaffin cells of the feline jejunum and is the most powerful 5-HT-releasing agent tested so far in enterochromaffin cells (Ahlman and Dahlström 1982).

In 1982, Forsberg and Miller, using an *in vitro* model of the rabbit duodenal mucosa, demonstrated that cholinergic agonists, when added to the serosal side of the mucosa, stimulated the release of 5-HT to the vascular side of the enterochromaffin cells. This effect was blocked by atropine but not by hexamethonium and tetrodotoxin, demonstrating that stimulation of muscarine receptors, presumably located on the surface of the enterochromaffin cells postsynaptically to the cholinergic neurones, causes vectorial release of 5-HT (Forsberg and Miller 1982). A basal release of 5-HT to both serosal and mucosal sides of the enterochromaffin cells, has also been described; this basal release was not blocked by atropine (Forsberg and Miller 1982 and 1983). Adrenaline, isoprenaline, bombesin, substance P, neurotensin, and vasoactive intestinal peptide, when added to the serosal side, were not able to evoke a mucosal or serosal release of 5-HT (Forsberg and Miller 1982). Since 8-bromo-cyclic AMP and phosphodiesterase inhibitors induce 5-HT release from the mucosa (serosal side), it is suggested that cyclic AMP is important in mediating 5-HT release from enterochromaffin cells (Forsberg and Miller 1983). Evidence has also appeared implicating an important role for Ca^{2+}; thus removal of Ca^{2+} and addition of several di- and trivalent cations inhibited the 5-HT release evoked by carbachol, while Ca^{2+}-ionophores caused release of 5-HT (Forsberg and Miller 1983). Finally, somatostatin, which inhibits the release of most gastrointestinal hormones, also inhibits the release of 5-HT from the duodenal enterochromaffin cells (Forsberg and Miller 1983).

Concerning the metabolism of 5-HT by the enterochromaffin cells, only minimal information is available. In a recent study, it was shown that a very low level of 5-hydroxyindoleacetic acid was present in rat enterochromaffin cells; the low ratio of 5-hydroxyindoleacetic acid to 5-HT (0.03) indicated a slow utilization of the amine when compared with, for example, the brain (Legay *et al.* 1983a). These experiments demonstrate, however, the presence of monoamine oxidase in the enterochromaffin cells.

1.3.2 5-HT in enteric neurones

Since the original suggestion by Gershon *et al.* (1965), that 5-HT might be stored in enteric neurones where it could act as a neurotransmitter, considerable evidence has accumulated indicating that 5-HT indeed fulfills the criteria needed for an enteric neurotransmitter. Thus 5-HT is synthesized by myenteric neurones where high concentrations of the amine are subsequently stored. 5-HT can be released from these neurones which possess a high affinity uptake mechanism for the amine and which contain the enzymes necessary for the inactivation of 5-HT. Our current knowledge about the interaction of 5-HT and the enteric neurones is not further discussed here since it is reviewed in detail in Chapter 11.

1.4 5-HT in platelets

Platelets may be the most important source of the 5-HT which interacts with the cardiovascular system. Since several recent reviews have dealt in detail with the interaction of 5-HT with platelets (for example, Holmsen 1985; Stolz 1985; Zucker and Nachmias 1985), the subject will be discussed only briefly in this chapter.

The platelets of all mammals are rich in 5-HT; almost all the circulating 5-HT is contained within the platelets. It is generally believed that platelets do not possess the enzymes required for synthesis of 5-HT. Under conditions in which maximal activity of tryptophan hydroxylase in pineal glands and brain could be detected, no enzyme activity was measured in platelets; moreover, with systems adequate to determine 5-HT synthesis in pineal gland cells, no 5-HT synthesis in human or rat platelets could be detected (Morrissey *et al.* 1977). More recent investigations, however, seem to indicate that low level activities of the enzymes necessary for 5-HT synthesis can be detected in platelets (Stolz 1985).

The bulk of the 5-HT associated with the platelets originates in the enterochromaffin cells of the intestinal mucosa (Erspamer 1954; Pletscher 1968; Stolz 1985). 5-HT is taken up by the platelets in the circulating blood and subsequently stored within the dense granules. This storage of 5-HT is not a simple process since it involves at least three steps: transport through the platelet membrane, transport through the storage granule membrane, and formation of complexes within the storage granules (Stolz 1985).

The mechanism for storage of 5-HT in the granules is not completely clear. Inside the dense granule, the amine is sequestered with large aggregates of ADP, ATP, and divalent metal ions (Ca^{+} in humans) (see Holmsen 1985). Binding proteins, such as seronectin, contribute to the storage and/or the uptake of 5-HT into various cells including platelets (Stolz 1985).

When platelets are activated, they respond with shape change, aggregation, and secretion of their storage granules. The 5-HT stored within the platelets can be released by exocytosis, by pharmacological displacement, by inhibition of the granular transport carrier, or by aminophores and protonophores which form complexes with the amine.

Platelets contain monoamine oxidase but under normal conditions, the amounts of 5-HT metabolized to 5-hydroxyindoleacetic acid and 5-hydroxytryptophol are very low (Pletscher 1968), because the amine is protected in the dense granules (Da Prada and Picotti 1979; Holmsen 1985).

1.5 5-HT in the cardiovascular system

Several studies have shown that substantial amounts of 5-HT can be detected in the heart and in blood vessels of mammals including humans (Beauvallet *et al.* 1968; Madan *et al.* 1970; Votavova *et al.* 1971; Berkowitz *et al.* 1974; Jarrott *et al.* 1975; Sole *et al.* 1979; Kalsner and Richards 1984; Niwa *et al.* 1984; Edvinsson *et al.* 1984; Ozaki *et al.* 1986; Levitt and Duckles 1986). In this section, attention will be focused on the storage, release, and metabolism of 5-HT in the different compartments of the cardiovascular system, although in some instances only sparse information is available.

1.5.1 5-HT in the heart

Normal hearts of different mammals, including humans, contain significant amounts of 5-HT. Between 0.3 and 0.5 μg/g of tissue is the norm and this concentration is about six times lower than that of noradrenaline (Beauvallet *et al.* 1968; Madan *et al.* 1970; Berkowitz *et al.* 1974; Jarrott *et al.* 1975; Sole *et al.* 1979; Niwa *et al.* 1984). After inhibition of tryptophan hydroxylase with α-propyldopacetamide, cardiac 5-HT levels were significantly reduced, indicating actual synthesis of the indoleamine within the heart (Sole *et al.* 1979). In hamster and rat hearts, a concentration of 5-hydroxyindoleacetic acid, about 10 to 20 times lower than that of 5-HT, was detected (Sole *et al.* 1979; Niwa *et al.* 1984). When monoamine oxidase was inhibited with pargyline, an increased cardiac 5-HT content was detected and the levels of 5-hydroxyindoleacetic acid decreased markedly (Berkowitz *et al.* 1974; Sole *et al.* 1979). These experiments indicate that metabolism by monoamine oxidase occurs in the mammalian heart and that 5-HT is synthesized within the heart. The available evidence indicates that the cardiac 5-HT is not associated with platelets, sympathetic or 5-HT-containing nerves, or mast cells (Jarrott *et al.* 1975; Sole *et al.* 1979).

The physiological significance of the cardiac 5-HT remains speculative; the indoleamine has profound effects on the mammalian heart and it has been suggested that 5-HT might be implicated in the aetiology of certain cardiac diseases (see Sole *et al.* 1979).

1.5.2 5-HT in the blood vessel wall

Isolated arteries and veins of different mammals, including humans, contain 5-HT (Berkowitz *et al.* 1974; Jarrott *et al.* 1975; Edvinsson *et al.* 1984; Kalsner and Richards 1984; Ozaki *et al.* 1986; Levitt and Duckles 1986). In mesenteric arteries of rats, pre-treatment with pargyline results in an increased 5-HT content and a decreased 5-hydroxyindoleacetic acid content, thus demonstrating the metabolism of 5-HT by monoamine oxidase in these blood vessels. Concerning the storage, release, and metabolism of 5-HT, several studies have now been performed showing an interaction of 5-HT with different cellular components of the vasculature; interactions between 5-HT and endothelial cells, smooth muscle cells and sympathetic nerves, have all been investigated (Figs 1.2 and 1.3).

1.5.2.1 *5-HT in endothelial cells*

Endothelial cells line the heart and the large blood vessels and they are the principal cellular components of the microvascular system. They thus are the first cells with which circulatory substances come into contact and

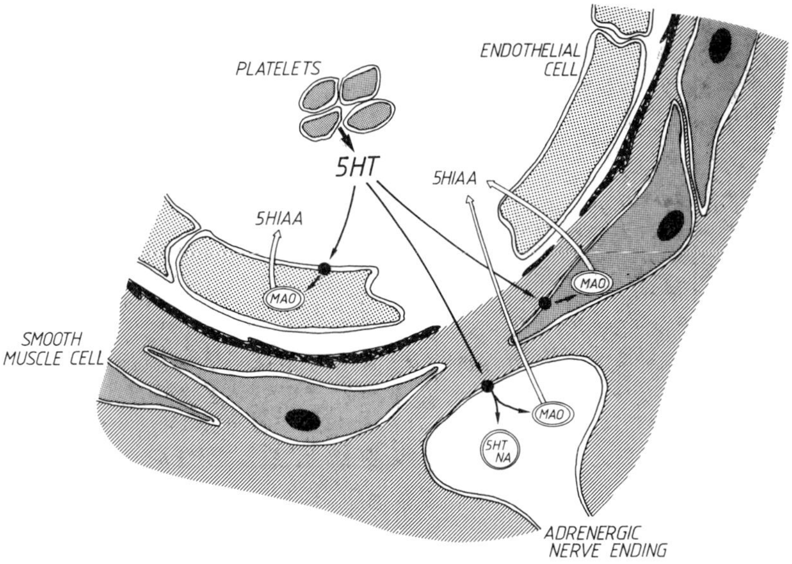

Fig. 1.2 Uptake and metabolism of 5-HT by the blood vessel wall. At sites of endothelial denudiation, 5-HT can penetrate the vessel wall and be taken up by vascular smooth muscle cells and sympathetic nerve terminals. 5HIAA = 5-hydroxyindoleacetic acid; MAO = monoamine oxidase; NA = noradrenaline

interact (D'Amore and Shepro 1982). Since under normal conditions, 90 per cent of the 5-HT detected in the blood is present in a functionally inactive form, stored in the dense granules of the platelets, plasma levels of 5-HT are too low (generally lower than 20 ng/ml) to produce measurable cardiovascular effects (see Crawford 1963; Tranzer *et al.* 1966). Two processes will tend to limit any increase in the concentration of free 5-HT in the blood: (a) enzymatic degradation by monoamine oxidase, particularly in the river; and (b) rapid removal of 5-HT by the endothelial cells, especially in the lungs (see Gillis 1985). These processes are important regulators of the blood levels of 5-HT which may increase in conditions such as carcinoid syndrome (Brown 1977), experimental pulmonary embolization (Rosoff *et al.* 1971), or endotoxin shock (Stein and Thomas 1967). The free 5-HT which escapes deamination in the liver is transported to the lungs where it is rapidly cleared from the circulation by an avid uptake process into the endothelial cells; this uptake is followed by oxidation catalysed by the A-form of monoamine oxidase.

The ability of the lung to remove 5-HT was first observed by Starling and Verney (1925), who reported that it was impossible to maintain adequate circulation through an isolated kidney using defibrinated blood, unless the lungs were included in the perfusion circuit. Later, Thomas and Vane (1967) reported that lungs of anaesthetized dogs could remove up to 90 per cent of an intravenous infusion of 5-HT. It is now accepted that the primary morphological sites for the removal of 5-HT are the capillary and small vessel endothelial cells. The removal of 5-HT is considered to be an important non-ventilating function of the lung (Junod 1975; Gillis and Pitt 1982; Gillis 1985).

1.5.2.1.1 *Uptake of 5-HT by endothelial cells* The transport of 5-HT into the endothelial cells, which is the rate-limiting step in the extraction of 5-HT from the blood, exhibits the characteristics of a carrier-mediated process in being Na^+-dependent, saturable, and temperature sensitive (Junod 1975; Gillis 1980 and 1985). The apparent K_m value, determined by several techniques, is 2 μM (see Gillis 1985).

The removal of 5-HT is inhibited by hypothermia, high K^+ or low Na^+, cocaine, imipramine, and cardiac glycosides (see Junod 1975; Gillis 1980 and 1985). Although these properties resemble those governing the uptake of noradrenaline (or 5-HT) into sympathetic nerves (Bosin and Lahr 1981; Verbeuren *et al.* 1983), experiments with 6-hydroxydopamine to induce sympathetic denervation have revealed that the removal of 5-HT from the circulation does not depend on the presence of sympathetic nerves in the pulmonary blood vessels (Iwasawa and Gillis 1974).

The uptake of 5-HT by endothelial cells is not restricted to the endothelium of the lung. It has been demonstrated that endothelial cells from bovine aorta and human umbilical vein (Small *et al.* 1976) and from porcine aorta and pulmonary artery (Junod and Ody 1977) also take up 5-HT by a process very similar to that described for the lungs. Similarly, capillaries of the brain have the capacity to take up 5-HT by mechanisms comparable with those in the lung, which may be of importance regarding the passage of 5-HT from the blood to the brain (Spatz *et al.* 1981). In a recent study, Fukuda *et al.* (1986a) reported that little uptake of 5-HT takes place into the endothelial cells of the rat aorta, while those of the rat mesenteric artery avidly accumulate the amine; this indicates that the functional role of the endothelium, with respect to the uptake of 5-HT, may vary in different vascular beds. It thus seems likely that 5-HT is removed from the circulation by endothelial cells throughout the vasculature. However, due to the large surface area of the pulmonary vasculature and because the entire cardiac output passes through the lungs, this organ seems to be uniquely situated to regulate the arterial 5-HT concentration (Bosin and Lahr 1981; Gillis 1985).

1.5.2.1.2 *Metabolism of 5-HT by the endothelium* After its uptake by the lung endothelium, 5-HT is deaminated by monoamine oxidase of the A type (Roth and Gillis 1975; Gillis 1985). The primary metabolite formed is 5-hydroxyindoleacetic acid although small amounts of 5-hydroxytryptophol have also been detected in the pulmonary effluent (Crooks *et al.* 1979). The enzymatic deamination of 5-HT by isolated lungs is a very rapid process (Junod 1975). No methylation of the indoleamine could be detected even when S-adenosyl methionine was provided as a methyl donor (Crooks *et al.* 1979). Similarly in brain capillaries, the uptake of 5-HT is followed by extensive metabolism to 5-hydroxyindoleacetic acid (Spatz *et al.* 1981).

Recent studies from our laboratory illustrate that intraluminal application of [^{3}H]5-HT to perfused dog saphenous veins and coronary arteries results in an accumulation of the amine by the blood vessels. Similarly, the [^{3}H]catecholamines, noradrenaline and adrenaline, are taken up and stored when applied to the luminal surface of the saphenous veins (Verbeuren *et al.* 1988). The most likely storage sites for all three amines in the blood vessels are the sympathetic nerves (Verbeuren *et al.* 1983 and 1988). Removal of the endothelium from the veins and arteries causes a marked elevation of the tissue accumulation of [^{3}H]5-HT, [^{3}H]noradrenaline and [^{3}H]adrenaline. To what extent endothelial monoamine oxidase acts as a barrier for these amines in the isolated perfused blood vessels, remains to be elucidated; preliminary data obtained with the monoamine oxidase inhibitor pargyline indicate that the endothelial monoamine oxidase contributes to the prevention of the accumulation of noradrenaline but not (or to only a limited extent) to that of 5-HT (Verbeuren *et al.* 1988). Our data, however, clearly indicate a potent barrier function of both venous and arterial endothelium toward the penetration of luminally applied 5-HT in canine blood vessels.

It should be stated that the *in vivo* administration of 5-HT can cause an accumulation of significant amounts of unchanged 5-HT in mouse or rat lungs (Bosin and Lahr 1981). Thus, in addition to the endothelium, the lung may contain other storage sites for 5-HT. Whether sympathetic nerves (see Verbeuren *et al.* 1983 and 1988) or enterochromaffin cells (Polak *et al.* 1982) contribute to this 5-HT accumulation in the lungs remains to be determined.

1.5.2.2 *5-HT in smooth muscle cells*

Vascular smooth muscle cells possess the capacity for taking up and metabolizing 5-HT. Denervated blood vessels or vessels treated with neuronal uptake inhibitors such as cocaine or desipramine are able to accumulate significant amounts of [^{3}H]5-HT (Verbeuren *et al.* 1983 and 1988; Paiva *et al.* 1984; Cohen 1985; Levitt and Duckles 1986). Similar to the extraneuronal uptake of catecholamines, the accumulation of 5-HT by the smooth muscle of the dog saphenous vein is only partially inhibited by hydrocortisone and it thus appears that the extraneuronal uptake has a corticosteroid-sensitive and a corticosteroid-resistant component (Paiva *et al.* 1984). In the rat aorta, a cocaine-sensitive extraneuronal uptake of 5-HT, different from that of noradrenaline, has been described; this uptake appears to occur primarily in the smooth muscle layers close to the lumen (Fukuda *et al.* 1986a and b). Following extraneuronal uptake into the smooth muscle cells, extensive metabolism of 5-HT takes place and the available evidence indicates that 5-HT is mainly degraded by monoamine oxidase of the A type (Paiva *et al.* 1984).

1.5.2.3 *5-HT in sympathetic nerves*

Experiments performed with sympathetically innervated, non-vascular preparations, such as the pineal gland and the vas deferens, have established that 5-HT can be taken up by, be stored in, and be released from, sympathetic nerves (Thoa *et al.* 1969; Jaim-Etcheverry and Zieher 1971 and 1980). A particularly intriguing observation is that, in the pineal gland of mammals, 5-HT is stored both in the pinealocytes (as a precursor of melatonin) and in the sympathetic nerve fibres supplying the gland. In the nerve vesicles of these sympathetic fibres, 5-HT is stored together with noradrenaline (Owman 1964; Jaim-Etcheverry and Zieher 1971; Axelrod 1974) and this observation was probably the first demonstration of nerve cells which contain at least two neurotransmitters. The experimental data indicate that 5-HT is formed in the pinealocytes and is subsequently taken up by the sympathetic nerves where it is stored in the nerve vesicles (Jaim-Etcheverry and Zieher 1971).

Indirect evidence obtained in vascular tissues suggests that 5-HT can also be taken up by the sympathetic nerves of arteries and veins: thus, relatively high concentrations (above 10^{-6}M) of 5-HT can cause displacement of the neurotransmitter noradrenaline from its storage granules (McGrath 1977; Starke and Weitzell 1978). Moreover, the fact that the contractile responses to threshold concentrations of 5-HT are markedly potentiated in the presence of inhibitors of the neuronal uptake of noradrenaline (McGrath 1977; Curro *et al.* 1978; see Vanhoutte *et al.* 1981), also indicates that low concentrations of 5-HT may enter the sympathetic nerve terminals.

On the basis of these observations, Verbeuren *et al.* (1983) performed a systematic study to examine the uptake, accumulation and release of 5-HT in isolated canine blood vessels (Fig. 1.3).

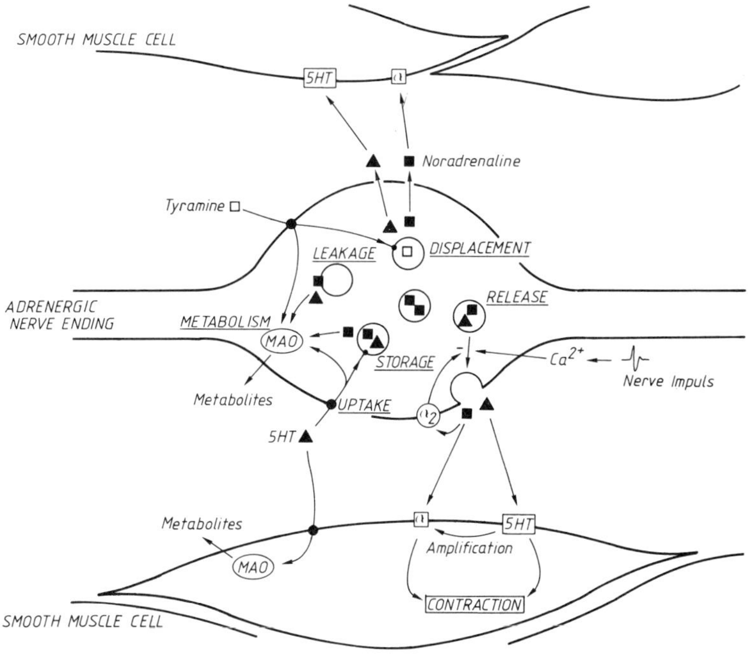

Fig. 1.3 Uptake, metabolism, storage, and release of 5-HT by the vascular sympathetic nerves. ▲ = 5-HT; ■ = noradrenaline; MAO = monoamine oxidase

1.5.2.3.1 *Uptake of 5-HT into sympathetic nerves* When strips of isolated canine saphenous veins and basilar arteries are incubated with [^{3}H]5-HT (even with concentrations as low as 10^{-8}M), significant amounts of the amine accumulate in the tissues. Treatment of the vessel segments with cocaine or denervation with 6-hydroxydopamine prior to incubation, causes a marked reduction in the accumulation of [^{3}H]5-HT, suggesting that most of the accumulation occurs into the sympathetic nerves present in the vessel wall (Verbeuren *et al.* 1983). Uptake of radioactive 5-HT by vascular sympathetic nerves has since been confirmed in saphenous veins and basilar and coronary arteries of the dog, in rabbit basilar arteries, and in the mesenteric vascular bed of the rat (Kawasaki and Takasaki 1984 and 1986; Paiva *et al.* 1984; Cohen 1985; Levitt and Duckles 1986; Verbeuren *et al.* 1986). In canine saphenous veins, the cocaine-sensitive accumulation of $[^3H]$5-HT is markedly augmented after monoamine oxidase inhibition with pargyline, suggesting that rapid intraneuronal deamination of the amine normally takes place (Verbeuren *et al.* 1988). $[^3H]$5-HT, previously incorporated into platelets, can also enter the blood vessel wall of dog saphenous veins and be taken up by the sympathetic nerves; this transfer of 5-HT from platelets to sympathetic nerve terminals is markedly augmented when the endothelium is removed (Verbeuren *et al.* 1988).

The situation is somewhat more complicated with respect to cerebral arteries where evidence is available suggesting an innervation by 5-HT-containing nerves (Chan Palay 1976; Griffith *et al.* 1982). Since, however, the accumulation of $[^3H]$5-HT by dog basilar arteries is decreased by cocaine and 6-hydroxydopamine treatment, the available evidence indicates that in this artery most of the $[^3H]$5-HT is taken up by sympathetic nerve terminals (Verbeuren *et al.* 1983 and 1986). For rabbit basilar arteries, a similar conclusion was recently reached on the basis that desipramine and 6-hydroxydopamine treatment reduced the accumulation of $[^3H]$5-HT, whilst the specific blocker of 5-HT-uptake, fluoxetine, was without effect (Levitt and Duckles 1986). Thus, in some cerebral blood vessels, 5-HT is not (or only to a minor extent) taken up into specific 5-HT-containing nerve terminals.

1.5.2.3.2 *Release of 5-HT from sympathetic nerves* The radioactive 5-HT, previously incorporated into the sympathetic nerves of the isolated guinea-pig vas deferens, is slowly and spontaneously released from the tissues; this spontaneous release can be enhanced by noradrenaline and tyramine, which displace 5-HT from the storage granules (Thoa *et al.* 1969). Electrical stimulation *in vivo* causes release of both noradrenaline and 5-HT from the sympathetic nerves of the pineal body (Jaim-Etcheverry and Zieher 1980). Thus, in non-vascular preparations, 5-HT present in sympathetic nerves can be released from these nerves spontaneously, by displacement and by discharging stimuli.

The $[^3H]$5-HT which accumulates in the nerves of isolated blood vessels slowly and spontaneously leaks from the storage sites and overflows into the superfusion medium; the radioactivity detected in this basal efflux consists of over 80 per cent of 5-hydroxyindoleacetic acid (Verbeuren *et al.* 1983; Cohen 1985). The spontaneous ^3H release is not influenced by drugs such as tetrodotoxin, phentolamine, dihydroergocristine, and methiothepin; it is, however, dependent upon the ^3H content of the tissues, is decreased in tissues pre-treated (before the incubation) with cocaine or 6-hydroxydopamine, and enhanced by tyramine and by high concentrations of the alkaloid, raubasine (Verbeuren *et al.* 1983, 1986 and 1988). These observations indicate that a large part of the basal ^{3}H-release is due to [^{3}H]5-HT leaking from the storage granules into the cytoplasm of the sympathetic nerves where it is deaminated by the intraneuronal monoamine oxidase before it reaches the synaptic cleft (Verbeuren *et al.* 1983; Cohen 1985). In canine basilar arteries, labelled with [^{3}H]5-HT, a greater spontaneous ^{3}H release is observed as compared with tissues incubated with [^{3}H]noradrenaline; this observation indicates that [^{3}H]5-HT leaks more easily from the intravesicular storage sites than [^{3}H]noradrenaline, probably because the indoleamine is less firmly bound inside these storage granules (Verbeuren *et al.* 1986).

In isolated blood vessels labelled with [^{3}H]5-HT, electrical stimulation causes release of tritium (Verbeuren *et al.* 1983, 1986 and 1988; Paiva *et al.* 1984; Kawasaki and Takasaki 1984 and 1986; Cohen 1985), and chromatographic analysis shows that a large fraction of the tritium released consists of intact [^{3}H]5-HT (Verbeuren *et al.* 1983; Cohen 1985). The pattern and the amounts of the [^{3}H]5-HT released by electrical stimulation from basilar arteries are comparable with those of [^{3}H]noradrenaline released from arteries previously labelled with the [^{3}H]catecholamine (Verbeuren *et al.* 1986). The release of 5-HT is frequency-dependent (Verbeuren *et al.* 1983), is blocked by tetrodotoxin (Verbeuren *et al.* 1983; Cohen 1985), is absent in blood vessels pre-treated (before labelling) with cocaine (Verbeuren *et al.* 1983; Paiva *et al.* 1984), is absent in denervated tissues (Verbeuren *et al.* 1983), and is abolished in Ca^{2+}-free solution (Verbeuren *et al.* 1983; Kawasaki and Takasaki 1984). Elevation of the external K^+-concentration evokes the release of [^{3}H]5-HT from the isolated dog saphenous vein and this release also is absent in denervated veins or in veins pre-treated with cocaine (Verbeuren *et al.* 1983). All these experiments support the concept that the [^{3}H]5-HT, which is released from the isolated vascular preparations, has its origin in the sympathetic nerves.

To test whether the [^{3}H]5-HT release is sensitive to pre-junctional receptor modulation, the effect of four α-adrenoceptor antagonists has been studied. Low concentrations of the highly selective α_1-adrenoceptor antagonist, prazosin, do not alter the stimulation-induced release of 5-HT (Kawasaki and Takasaki 1986); in contrast, phentolamine, yohimbine, and raubasine augment the release of 5-HT induced by electrical stimulation by inhibiting the pre-junctional α_2-adrenoceptors (Verbeuren *et al.* 1983 and 1986; Kawasaki and Takasaki 1986). In contrast, dihydroergocristine, which stimulates α_2-adrenoceptors, decreases the release of [^{3}H]5-HT as it does that of [^{3}H]noradrenaline (Verbeuren *et al.* 1986). From these results it can be concluded that the release of [^{3}H]5-HT from the isolated blood vessels is controlled by the same pre-synaptic mechanisms which regulate the release of the endogenous transmitter, thus reinforcing the concept that the ^{3}H-indoleamine is released together with noradrenaline from the sympathetic nerves. The selective 5-HT_2 receptor antagonist, LY 53857 (Kawasaki and Takasaki 1986), and the non-selective 5-HT receptor antagonist, methiothepin (Verbeuren *et al.* 1983), do not alter the stimulation-induced 5-HT overflow in vascular tissues suggesting that 5-HT receptors do not participate in the control of 5-HT release from the blood vessel wall.

Whether the accumulation of 5-HT and its subsequent release from sympathetic nerves plays a physiological role in health or disease remains to be determined. However, recent studies have revealed that 5-HT released from the sympathetic nerves of vascular tissues can influence or alter the contractile response of the blood vessels. Thus, in the isolated perfused rat mesenteric artery, pre-treatment with 5-HT causes a marked, specific augmentation of the vasoconstrictor response to peri-arterial nerve stimulation (Kawasaki and Takasaki 1984 and 1986). Moreover, in canine coronary arteries, 5-HT derived from aggregating platelets and present in sympathetic nerves, can be released by electrical impulses and convert the normal sympathetic vasodilator response into a vasoconstrictor response mediated by 5-HT receptors (Cohen 1985).

1.6 5-HT in the liver, the pineal gland, and mast cells

1.6.1 5-HT in the liver

5-HT, released from the enterochromaffin granules or by carcinoid tumours of the gastro-intestinal tract, overflows into the portal blood which then passes through the liver. In the liver a substantial clearance of the amine occurs. The liver metabolises injected 5-HT by multiple pathways and releases the products into bile and plasma (Tyce *et al.* 1968). The removal of 5-HT by the liver, which can reach as much as 80 per cent, is largely a reflection of metabolism by monoamine oxidase A, resulting in the formation of 5-hydroxyindoleacetic acid (Gillis 1985). Indeed, the rat liver perfused at a normal flow rate clears 5-HT at roughly the rate expected for catabolism by monoamine oxidase. This contrasts with the clearance of 5-HT by the rat lung where the rate is over three times that predicted from the kinetics of the lung monoamine oxidase (Wiersma and Roth 1980; Gillis 1985). This difference is most likely due to the high rate of 5-HT removal into the pulmonary endothelium; the absence of such a process in the liver may be attributed to the observation that in this organ the endothelial lining of the vasculature is discontinuous, allowing a more direct access of 5-HT to the hepatic enzymes (see Gillis 1985). Besides 5-hydroxyindoleacetic acid, the liver forms significant amounts of 5-hydroxytryptophol, indicating that reduction of 5-hydroxyindoleacetaldehyde also takes place (Tyce *et al.* 1968; Tyce 1985).

In the liver 5-HT can also be inactivated by 5-HT-sulphotransferase with the resulting formation of 5-HT-*O*-sulphate and by 5-HT-*N*-acetyltransferase which forms *N*-acetyl-5-HT. The amine can be conjugated with glucuronic acid by the enzyme uridine diphosphate glucuronyltransferase, resulting in the production of 5-HT-*O*-glucuronide which is a major metabolite of 5-HT in the rat liver (Tyce *et al.* 1968; Tyce 1985). Inhibition of monoamine oxidase in the liver causes an increase in the formation of 5-HT-*O*-glucuronide and the glucuronidation pathway can easily compensate for the blocked monoamine oxidase activity (Tyce *et al.* 1968). It seems apparent that under normal circumstances, the liver can actively control the blood concentration of 5-HT liberated into the portal circulation by the intestinal enterochromaffin granules.

1.6.2 5-HT in the pineal gland

The pineal gland is rich in 5-HT which is contained both in its parenchymal cells, the pinealocytes, and in the sympathetic nerve fibres suppling the gland; in the pinealocytes, 5-HT is synthesized from tryptophan and this 5-HT serves as the precursor for the hormone melatonin (see Jaim-Etcheverry and Zieher 1971; Axelrod 1974). As shown in Fig. 1.1, 5-HT is first *N*-acetylated by the enzyme 5-HT-*N*-acetyltransferase using acetyl-coenzyme A. The *N*-acetyl-5-HT thus formed is then *O*-methylated, using S-adenosylmethionine as the methyldonor, in a reaction catalysed by the enzyme hydroxyindole-*O*-methyltransferase; the resulting compound is melatonin (Klein and Weller 1970; Axelrod 1974; Tyce 1985). In the pineal body, 5-HT is also metabolised to 5-hydroxyindoleacetic acid by monoamine oxidase and 5-HT which leaves the pineal cells can be taken up by and be stored in the sympathetic nerve terminals together with noradrenaline (Jaim-Etcheverry and Zieher 1971; Axelrod 1974, see above). The striking diurnal rhythm of the melatonin concentration in the pineal body is due to marked variations in the activity of the enzyme 5-HT-*N*-acetyltransferase; the activity of this enzyme in the pineal gland is 50 times greater during the night than during the day (Klein and Weller 1970; Axelrod 1974). The driving force behind this circadian rhythm is a biological clock arising in or transmitted via the suprachiasmatic nucleus in the hypothalamus and is generated in the pineal body by the diurnal release of the neurotransmitter noradrenaline (Axelrod 1974).

1.6.3 5-HT in mast cells

Mast cells of rodents can take up and store 5-HT (Uvnas 1978) and it is believed that the mast cells store 5-HT and histamine together, most likely in their large secretory granules (Morrisson *et al.* 1974). Both histamine and 5-HT can be secreted simultaneously from mast cells by stimuli that cause exocytosis, such as compound 48/80 and several peptides. Some evidence has recently been presented indicating that under certain conditions, a differential release of 5-HT and histamine occurs (Theoharides *et al.* 1982; Tamir *et al.* 1982; Carraway *et al.* 1984).

Storage of 5-HT in mast cells may contribute to the content of 5-HT observed in certain organs such as the lung (Gillis 1985), the heart and blood vessels (Berkowitz *et al.* 1974; Jarrott *et al.* 1975). In the rat kidney treated with formaldehyde vapour, small fluorescent cells have been found around the main renal artery and the interlobular and arcuate arteries; the fluorescence appears to be attributable to 5-HT associated with mast cells (Petrovic and Bell 1984). In humans, mast cells of nasal mucosa and nasal polyps contained significant amounts of 5-HT (Bumsted *et al.* 1979).

1.7 Concluding remarks

5-HT occurs in a wide variety of tissues throughout the body of most mammalian species. In addition to its function as a neurotransmitter in the brain, hypotheses have been presented indicating or suggesting that 5-HT may be implicated in several diseases such as hypertension, atherosclerosis, vasospasm, Raynaud's disease, carcinoid syndrome, eclampsia, migraine, acute respiratory failure, and shock (see Chapter 16).

As potent, specific 5-HT-receptor antagonists have become available during the past decade, extensive studies have been performed not only into the possible mechanisms of action of 5-HT in disease, but also to determine its function in the normal physiology of many body organs. The fact that the indoleamine can be synthesized, taken up, stored, released, and metabolised in various cell types distributed throughout the body certainly increases the potential of 5-HT as a candidate for an important role in health and disease.

Acknowledgements

I wish to express my appreciation to Professor Dr A. Herman for the many discussions and comments and for his valuable support. I am indebted to Mrs L. Van den Eynde for her careful secretarial assistance.

Studies from our laboratory reported in this chapter were supported by grants from the Belgian Foundation of Medical Research (Fonds voor Geneeskundig Wetenschappelijk Onderzoek; FGWO), from the Queen Elizabeth Medical Foundation (Koningin Elizabeth Stichting), and from the University of Antwerp (UIA).

References

Ahlman, H. and Dahlström A. (1982). Storage and release of 5-hydroxytryptamine in enterochromaffin cells of the small intestine. In *5-Hydroxytryptamine in peripheral reactions* (ed. F. De Clerck and P.M. Vanhoutte) pp. 1–21. Raven Press, New York.

Ahlman, H., Bhargava, H.N., Dahlström, A., Larsson, I., Newson, B., and Pettersson, G. (1981a). On the presence of serotonin in the gut lumen and possible release mechanisms. *Acta Physiologica Scandinavica* **112,** 263–9.

Ahlman, H., De Magistris, L., Zinner, M., and Jaffe, B.M. (1981b). Release of immunoreactive serotonin into the gut lumen of the feline gut in response to vagal nerve stimulation. *Science* **213,** 1254–5.

Axelrod, J. (1962). The enzymatic N-methylation of serotonin and other amines. *Journal of Pharmacology and Experimental Therapeutics* **138,** 28–33.

Axelrod, J. (1974). The pineal gland: a neurochemical transducer. *Science* **184,** 1341–8.

Beauvallet, M., Godefroy, F., and Weil-Fugazza, J. (1968). Modification de la teneur en 5-hydroxytryptamine du coeur au cours d'une surcharge du regime en chlorure de sodium. *Comptes Rendus des Séances de la Société de Biologie et de ses Filiales (Paris)* **162,** 2085–8.

Berkowitz, B.A., Lee, C.H., and Spector, S. (1974). Disposition of serotonin in the rat blood vessels and heart. *Clinical and Experimental Pharmacology and Physiology* **1,** 397–400.

Bosin, T.R., and Lahr, P.D. (1981). Mechanisms influencing the disposition of serotonin in mouse lung. *Biochemical Pharmacology* **30,** 3187–93.

Brown, H. (1977). Serotonin-producing tumors. In *Serotonin in health and disease*, Vol. 4 (ed. W.B. Essman) pp. 393–423. Spectrum Publications, New York.

Bülbring, E. and Crema, A. (1959). The release of 5-hydroxytryptamine in relation to pressure exerted on the intestinal mucosa. *Journal of Physiology (London)* **146,** 18–28.

Bülbring, E.R., Lin, C.Y., and Schofield, G.C. (1958). An investigation of the peristaltic reflex in relation to anatomical observations. *Quarterly Journal of Experimental Physiology* **43,** 26–37.

Bumsted, R.M., El-Ackad, T., Smith, J.M., and Brody, M.J. (1979). Histamine, norepinephrine and serotonin content of nasal polyps. *The Laryngoscope* **89,** 832–43.

Burks, T.F. and Long, J.P. (1966). Catecholamine induced release of 5-hydroxytryptamine from perfused vasculature of isolated dog intestine. *Journal of Pharmaceutical Science* **55,** 1383–6.

Carraway, R.E., Cochrane, D.E., Granier, C., Kitabgi, P., Leeman, E., and Singer, E.A. (1984). Parallel secretion of endogenous 5-hydroxytryptamine and histamine from mast cells stimulated by vasoactive peptides and compound 48/80. *British Journal of Pharmacology* **81,** 227–9.

Chan-Palay, V. (1976). Serotonin axons in supra- and subependymal plexuses and in the leptomeninges; their roles in local alterations of cerebrospinal fluid and vasomotor activity. *Brain Research* **102,** 103–30.

Ciaccio, C. (1906). Sur une nouvelle espèce cellulaire dans les glands de Lieberkühn. *Comptes Rendus des Séances de la Société de Biologie et de ses Filiales (Paris)* **1,** 76–9.

Cohen, R.A. (1985). Platelet-induced neurogenic coronary contractions due to accumulation of the false neurotransmitter, 5-hydroxytryptamine. *Journal of Clinical Investigation* **75,** 286–92.

Costa, M., Furness, J. B., Cuello, A. C., Verhofstad, A. A. J., Steinbusch, H. W. J., and Elde, R. P. (1982). Neurons with 5-hydroxytryptamine-like immunoreactivity in the enteric nervous system: their visualization and reactions to drug treatment. *Neuroscience* **7,** 351–64.

Crawford, N. (1963). Plasma free serotonin (5-hydroxytryptamine). *Clinica Chimica Acta* **8,** 39–45.

Crooks, P. A., Dreyer, R. N., Sulens, C. H., Gillis, C. N., and Coward, J. K. (1979). Deamination of 5-hydroxytryptamine metabolites in isolated perfused rabbit lung by high-performance liquid chromatography. *Analytical Biochemistry* **93,** 143–52.

Curro, F. A., Greenberg, S., Verbeuren, T. J., and Vanhoutte, P. M. (1978). Interaction between alpha-adrenergic and serotonergic activation of canine saphenous veins. *Journal of Pharmacology and Experimental Therapeutics* **207,** 936–49.

D'Amore, P. and Shepro, D. (1982). Captation of 5-hydroxytryptamine and effects on endothelial cell metabolism. In *5-Hydroxytryptamine in peripheral reactions* (ed. F. De Clerck and P. M. Vanhoutte) pp. 37–47. Raven Press, New York.

Da Prada, M. and Picotti, G. B. (1979). Content and subcellular localization of catecholamines and 5-hydroxytryptamine in human and animal blood platelets: monoamine distribution between platelets and plasma. *British Journal of Pharmacology* **65,** 653–62.

Deguchi, T. (1977). Tryptophan hydroxylase in pineal gland of rat: postsynaptic localization and absence of circadian change. *Journal of Neurochemistry* **28,** 667–8.

Drapanas, T., MacDonald, J. C., and Stewart, J. D. (1962). Serotonin release following instillation of hypertonic glucose into the proximal intestine. *Annals of Surgery* **156,** 528–36.

Edvinsson, L., *et al.* (1984). Indoleaminergic mechanisms in brain vessels: localization, concentration, uptake and in vitro responses of 5-hydroxytryptamine. *Acta Physiologica Scandinavica* **121,** 291–9.

Erspamer, V. (1954). Pharmacology of indole alkylamines. *Pharmacological Reviews* **6,** 425–87.

Erspamer, V. (1966). Peripheral physiological and pharmacological actions of indole alkylamines. In *Handbook of experimental pharmacology: 5-Hydroxytryptamine and related indolealkylamines*, Vol. 14 (ed. V. Erspamer) pp. 245–359. Springer-Verlag, Berlin.

Erspamer, V. and Testini, A. (1959). Observations of the release and turnover rate of 5-hydroxytryptamine in the gastrointestinal tract. *Journal of Pharmacy and Pharmacology* **11,** 618–23.

Forsberg, E. J. and Miller, R. J. (1982). Cholinergic agonists induce vectorial release of serotonin from duodenal enterochromaffin cells. *Science* **217,** 355–6.

Forsberg, E. J. and Miller, R. J. (1983). Regulation of serotonin release from rabbit intestinal enterochromaffin cells. *The Journal of Pharmacology and Experimental Therapeutics* **227**, 755–66.

Fujita, T. and Kobayashi, S. (1977). Structure and function of gut endocrine cells. *International Review of Cytology* supplement 6, 187–233.

Fukuda, S., Su, C., and Lee, T. J.-F. (1986a). Extraneuronal serotonin accumulation in peripheral arteries of the rat. *Experientia* **42,** 1244–5.

Fukuda, S., Su, C., and Lee, T. J.-F. (1986b). Mechanisms of extraneuronal serotonin uptake in the rat aorta. *The Journal of Pharmacology and Experimental Therapeutics* **239,** 264–9.

Gershon, M. D., Drakontides, A. B., and Ross, L. L. (1965). Serotonin: synthesis and release from the myenteric plexus of the mouse intestine. *Science* **149,** 197–9.

Gershon, M. D., Dreyfus, C. F., Pickel, V. M., Joh, T. H., and Reis, D. J. (1977). Serotonergic neuron in the peripheral nervous system: identification in gut by immunohistochemical localization of tryptophan hydroxylase. *Proceedings of the National Academy of Sciences of the U.S.A.* **74,** 3086–9.

Gillis, C. N. (1980). Metabolism of vasoactive hormones by pulmonary vascular endothelium: possible functional significance. In *Vascular neuroeffector mechanisms* (ed. J. A. Bevan, T. Godfraind, R. A. Maxwell, and P. M. Vanhoutte) pp. 304–15. Raven Press, New York.

Gillis, C. N. (1985). Peripheral metabolism of serotonin. In *Serotonin and the cardiovascular system* (ed. P. M. Vanhoutte) pp. 27–36. Raven Press, New York.

Gillis, C. N. and Pitt, B. R. (1982). The fate of circulating amines within the pulmonary circulation. *Annual Review of Physiology* **44,** 269–81.

Green, A. R., Koslow, S. H., and Costa, E. (1973). Identification and quantitation of a new indolethylamine in rat hypothalamus. *Brain Research* **51,** 371–4.

Green, H. and Sawyer, J. L. (1966). Demonstration, characterization, and assay procedure of tryptophan hydroxylase in rat brain. *Analytical Biochemistry* **15,** 53–64.

Griffith, S. G., Lincoln, J., and Burnstock, G. (1982). Serotonin as a neurotransmitter in cerebral arteries. *Brain Research* **247,** 388–92.

Grübe, D. (1976). Metabolism of biogenic monoamines in entero-endocrine cells. In *Chromaffin, enterochromaffin and related cells* (ed. R. E. Couplano and T. Fujita) pp. 285–292. Elsevier, Amsterdam.

Hidaka, H., Nagatsu, T., and Yagi, K. (1969). Occurrence of a serotonin sulphotransferase in the brain. *The Journal of Neurochemistry* **16,** 783–5.

Holmsen, H. (1985). Platelet activation and serotonin. In *Serotonin and the cardiovascular system* (ed. P. M. Vanhoutte) pp. 75–86. Raven Press, New York.

Iwasawa, Y. and Gillis, C. N. (1974). Pharmacological analysis of norepinephrine and 5-hydroxytryptamine removal from the pulmonary circulation: differentiation of uptake sites for each amine. *The Journal of Pharmacology and Experimental Therapeutics* **188,** 386–93.

Jaim-Etcheverry, G. and Zieher, L. M. (1971). Ultrastructural cytochemistry and pharmacology of 5-hydroxytryptamine in adrenergic nerve endings. III. Selective increase of norepinephrine in the rat pineal gland consecutive to depletion of neuronal 5-hydroxytryptamine. *The Journal of Pharmacology and Experimental Therapeutics* **178,** 42–8.

Jaim-Etcheverry, G. and Zieher, L. M. (1980). Stimulation-depletion of serotonin and noradrenaline from vesicles of sympathetic nerves in the pineal gland of the rat. *Cell Tissue Research* **207,** 13–20.

Jarrott, B., McQueen, A., Graf, L., and Louis, W. J. (1975). Serotonin levels in vascular tissue and the effects of a serotonin synthesis inhibitor on blood pressure in hypertensive rats. *Clinical and Experimental Pharmacology and Physiology* supplement 2, 201–05.

Joh, T. H., Shikimi, T., Pickel, V. M., amd Reis, D. J. (1975). Brain tryptophan hydroxylase: purification of, production of antibodies to, and cellular and ultrastructural localization in serotonergic neurons of rat midbrain. *Proceedings of the National Academy of Sciences of the USA* **72,** 3575–9.

Johnston, J. P. (1968). Some observations upon a new inhibitor of monoamine oxidase in brain tissue. *Biochemical Pharmacology* **17,** 1285–97.

Junod, A. F. (1975). Metabolism, production and release of hormones and mediators in the lung. *American Review of Respiratory Diseases* **112,** 93–108.

Junod, A. F. and Ody, C. (1977). Amine uptake and metabolism by endothelium of pig pulmonary artery and aorta. *American Journal of Physiology* **232,** C88–C94.

Kalsner, S. and Richards, R. (1984). Coronary arteries of cardiac patients are hyperreactive and contain stores of amines: a mechanism for coronary spasm. *Science* **223,** 1435–7.

Kawasaki, H. and Takasaki, K. (1984). Vasoconstrictor response induced by 5-hydroxytryptamine released from vascular adrenergic nerves by periarterial nerve stimulation. *The Journal of Pharmacology and Experimental Therapeutics* **229,** 816–22.

Kawasaki, H. and Takasaki, K. (1986). Pharmacological characterization of presynaptic-adrenoceptors in the modulation of the 5-hydroxytryptamine release from vascular adrenergic nerves in the rat. *Japanese Journal of Pharmacology* **42,** 561–70.

Kellum, J. M. and Jaffe, B. M. (1976). Release of immunoreactive serotonin following acid perfusion of the duodenum. *Annals of Surgery* **184,** 633–8.

Kishimoto, Y., Takahashi, N., and Egami, F. (1961). Synthesis and properties of serotonin O-sulfate. *Journal of Biochemistry* **49,** 436–40.

Klein, D. C. and Weller, J. L. (1970). Indole metabolism in the pineal gland: an arcadian rhythm in N-acetyltransferase. *Science* **169,** 1093–5.

Knoll, J. and Magyar, K. (1972). Some puzzling pharmacological effects of monoamine oxidase inhibitors. *Advances in Biochemical Psychopharmacology* **5,** 393–408.

Larsson, I., Ahlman, H., Bhargava, H. N., Dahlström, A., Pettersson, G., and Kewenter, J. (1979). The effects of splanchnic nerve stimulation on the plasma levels of serotonin and substance P in the portal vein of cat. *Journal of Neuronal Transmission* **47,** 89–98.

Legay, C., Faudon, M., Hery, F., and Ternaux, J. P. (1983a). 5-HT metabolism in the intestinal wall of the rat. I. The mucosa. *Neurochemistry International* **5,** 721–7.

Legay, C., Faudon, M., Hery, F., and Ternaux, J. P. (1983b). 5-HT metabolism in the intestinal wall of the rat. II. The nerve plexuses—interactions between 5-HT containing cells. *Neurochemistry International* **5,** 571–7.

Levitt, B. and Duckles, S. P. (1986). Evidence against serotonin as a vasoconstrictor neurotransmitter in the rabbit basilar artery. *The Journal of Pharmacology and Experimental Therapeutics* **238,** 880–5.

Lovenberg, W., Jequier, E., and Sjoerdsma, A. (1967). Tryptophan hydroxylation in pineal gland, brainstem and carcinoid tumor. *Science* **155,** 217–9.

Lovenberg, W., Levine, R. J., and Sjoerdsma, A. (1965). A tryptophan hydroxylase in cell-free extracts of malignant mouse mast cells. *Biochemical Pharmacology* **14,** 887–9.

Madan, B. R., Khanna, N. K., Godhwani, J. L., and Pendse, V. K. (1970). Changes in the 5-hydroxytryptamine content of the heart during ectopicventricular arrhythmias and consequent to its reversion by quinidine. *Indian Journal of Medical Research* **58,** 130–4.

McGrath, M. A. (1977). 5-Hydroxytryptamine and neurotransmitter release in canine blood vessels. *Circulation Research* **41,** 428–35.

Morgan, M. and Mandell, A. J. (1969). Indole (ethyl) amine N-methyltransferase in the brain. *Science* **165,** 492–3.

Morrissey, J. J., Walker, M. N., and Lovenberg, W. (1977). The absence of tryptophan hydroxylase activity in blood platelets. *Proceedings of the Society for Experimental Biology and Medicine* **154,** 496–9.

Morrison, D. C., Roser, J. F., Henson, P. M., and Cochrane, C. G. (1974). Activation of rat mast cells by low molecular weight stimuli. *Journal of Immunology* **112,** 573–82.

Mourre, C., Calas, A., and Gonella, J. (1981). Radioautographic study of uptake and storage of indole amines in the rabbit enterochromaffin cells. *Neuroscience Letters* **23,** 251–6.

Nemoto, N., Kawaoi, A., and Shikata, T. (1982). Immunohistochemical demonstration of serotonin (5HT)-containing cells in the human and rat intestines, *Biomedical Research* **3,** 181–7.

Niwa, M., Kunisada, K., Himeno, A., Kawaguchi, A., and Ozaki, M. (1984). 5-Hydroxytryptamine content in the rat heart: quantitation by high-performance liquid chromatographic electrochemical detection. *Japanese Journal of Pharmacology* **34,** 264–7.

O'Hara, R. S., Fox, R. O., and Cole, J. W. (1959). Serotonin release mediated by intraluminal sucrose solutions. *Surgical Forum* **10,** 215–8.

Owman, C. (1964). Sympathetic nerves probably storing two types of monoamines in the rat pineal gland. *International Journal of Neuropharmacology* **2,** 105–12.

Ozaki, M., Himeno, A., Uchida, S., Ohta, H., and Niwa, M. (1986). Accelerated uptake of serotonin in mesenteric arteries of young spontaneously hypertensive rats. *Journal of Hypertension* **4,** S227–8.

Paiva, M. Q., Caramona, M., and Osswald, W. (1984). Intra- and extraneuronal metabolism of 5-hydroxytryptamine in the isolated saphenous vein of the dog. *Naunyn-Schmiedeberg's Archives of Pharmacology* **325,** 62–8.

Pearse, A. G. E. (1968). Common cytochemical and ultrastructural characteristics of cells producing hormones (the APUD series) and their relevance to thyroid and ultimo-brachial C-cells and calcitonin. *Proceedings of the Royal Society London (Biology)* **170,** 71–80.

Petrovic, T. and Bell, C. (1984). Small fluorescent cells in the rat kidney contain 5-hydroxytryptamine not a catecholamine. *Journal of the Autonomic Nervous System* **11,** 329–38.

Pettersson, G., Ahlman, H., Bhargava, H. N., Dahlström, A., Kewenter, J., Larsson, I., and Siepler, J. K. (1979). The effect of propranolol on the serotonin concentration in the portal plasma after vagal nerve stimulation in the cat. *Acta Physiologica Scandinavica* **107,** 327–31.

Pettersson, G., Dahlström, A., Larsson, I., Lundberg, J. M., Ahlman, H., and Kewenter, J. (1978). The release of serotonin from rat duodenal enterochromaffin cells by adrenoceptor agonists studied in vitro. *Acta Physiologica Scandinavica* **103,** 219–24.

Pletscher, A. (1968). Metabolism, transfer and storage of 5-hydroxytryptamine in blood platelets. *British Journal of Pharmacology and Chemotherapy* **32,** 1–16.

Polak, J. M., Heitz, P., and Pearse, A. G. E. (1976). Differential localization of substance P and motilin. *Scandinavian Journal of Gastroenterology* supplement 39, 39–42.

Polak, J. M., De Mey, J., and Bloom, S. R. (1982). 5-Hydroxytryptamine in mucosal endocrine cells of the gut and lung. In *5-Hydroxytryptamine in peripheral reactions* (ed. F. De Clerck and P. M. Vanhoutte) pp. 23–35. Raven Press, New York.

Resnick, R. H. and Gray, S. J. (1962). Chemical and histological demonstration of hydrochloric acid-induced release of serotonin from intestinal mucosa. *Gastroenterology* **42,** 48–55.

Rosoff, C. B., Salzman, E. W., and Gurewich (1971). Reduction of platelet serotonin and the response to pulmonary emboli. *Surgery* **70,** 12–9.

Roth, R. A. and Gillis, C. N. (1975). Multiple forms of amine oxidase in perfused rabbit lung. *The Journal of Pharmacology and Experimental Therapeutics* **194,** 537–44.

Small, R., Macarak, E., and Fisher, A. B. (1976). Production of 5-hydroxyindoleacetic acid from serotonin by cultured endothelial cells. *Journal of Cell Physiology* **90,** 225–32.

Sole, M. J., Shum, A., and Van Loon, G. R. (1979). Serotonin: metabolism in the normal and failing heart. *Circulation Research* **45,** 629–34.

Spatz, M., Maruki, C., Abe, T., Rausch, W. D., Abe, K., and Merkel, N. (1981). The uptake and fate of the radiolabeled 5-hydroxytryptamine in isolated cerebral microvessels. *Brain Research* **220,** 214–9.

Starling, E. H. and Verney, E. G. (1925). The secretion of urine as studied on the isolated kidney. *Proceedings of the Royal Society of London* **97,** 321–63.

Starke, K. and Weitzell, R. (1978). Is histamine involved in the sympathomimetic effect of nicotine? *Naunyn-Schmiedeberg's Archives of Pharmacology* **304,** 237–48.

Stein, M. and Thomas, D. P. (1967). Role of platelets in the acute pulmonary response to endotoxin. *Journal of Applied Physiology* **23,** 47–52.

Stolz, J. F. (1985). Uptake and storage of serotonin by platelets. In *Serotonin and the cardiovascular system* (ed. P. M. Vanhoutte) pp. 37–42. Raven Press, New York.

Tamir, H., Theoharides, T. C., Gershon, M. D., and Askenase, P. W. (1982). Serotonin storage pools in basophil leukemia and mast cells: characterization of two types of serotonin binding protein and radioautographic analysis of the intracellular distribution of [^{3}H] serotonin. *The Journal of Cell Biology* **93,** 638–47.

Ternaux, J. P., Gonella, J., Legay, C., Faudon, M., Barrit, M. C., and Héry, F. (1981). 5-HT metabolism in the intestinal wall of the rabbit. *Journal of Physiology, Paris* **77,** 319–26.

Theoharides, T. C., Bondy, P. K., Tsakalos, N. D., and Askenase, P. W. (1982). Differential release of serotonin and histamine from mast cells. *Nature (London)* **297,** 229–31.

Thoa, N. B., Eccleston, D., and Axelrod, J. (1969). The accumulation of C^{14}-

serotonin in the guinea-pig vas deferens. *The Journal of Pharmacology and Experimental Therapeutics* **169,** 68–73.

Thomas, D. P. and Vane, J. R. (1967). 5-Hydroxytryptamine in the circulation of the dog. *Nature (London)* **216,** 335–8.

Tranzer, J. P., Da Prada, M., and Pletscher, A. (1966). Ultra-structural localization of 5-hydroxytryptamine in blood platelets. *Nature (London)* **212,** 1574–5.

Tyce, G. M. (1985). Biochemistry of serotonin. In *Serotonin and the cardiovascular system* (ed. P. M. Vanhoutte) pp. 1–14. Raven Press, New York.

Tyce, G. M., Flock, E. V., and Owen, C. A. Jr. (1968). Uptake and metabolism of 5-hydroxytryptamine by the isolated perfused rat liver. *American Journal of Physiology* **215,** 611–9.

Uvnas, B. (1978). Chemistry and storage function of mast cell granules. *Dermatology* **71,** 76–80.

Vanhoutte, P. M., Verbeuren, T. J., and Webb, R. C. (1981). Local modulation of the adrenergic neuroeffector interaction in the blood vessel wall. *Physiological Reviews* **61,** 151–247.

Verbeuren, T. J. and Vanhoutte, P. M. (1982). Deamination of released ^{3}H-norepinephrine in the canine saphenous vein. *Naunyn-Schmiedeberg's Archives of Pharmacology* **318,** 148–57.

Verbeuren, T. J., Jordaens, F. H., and Herman, A. G. (1983). Accumulation and release of [^{3}H]5-hydroxytryptamine in saphenous veins and cerebral arteries of the dog. *The Journal of Pharmacology and Experimental Therapeutics* **226,** 579–88.

Verbeuren, T. J., Zonnekeyn, L. L., Jordaens, F. H., and Herman, A. G. (1986). Effects of iskedyl and its two constituents raubasine and dihydroergocristine on the release of [^{3}H]noradrenaline and [^{3}H]serotonin in canine basilar arteries. *European Journal of Pharmacology* **125,** 1–10.

Verbeuren, T. J., Jordaens, F. H., Bult, H., and Herman, A. G. (1988). The endothelium inhibits the penetration of serotonin and norepinephrine in the isolated canine saphenous vein. *The Journal of Pharmacology and Experimental Therapeutics* **244,** 276–82.

Votavova, M., Boullin, D. J., and Costa, E. (1971). Specificity of action of 6-hydroxydopamine in peripheral cat tissues: depletion of noradrenaline without depletion of 5-hydroxytryptamine. *Life Sciences* **10,** 87–91.

Wiersma, D. A. and Roth, R. A. (1980). Clearance of 5-hydroxytryptamine by rat lung and liver: the importance of relative perfusion and intrinsic clearance. *The Journal of Pharmacology and Experimental Therapeutics* **212,** 97–102.

Zucker, M. B. and Nachmias, V. T. (1985). Platelet activation. *Arteriosclerosis* **5,** 2–18.

2

Quantification of the actions of 5-hydroxytryptamine receptor agonists and antagonists

PAUL LEFF AND GRAEME R. MARTIN

2.1 Introduction

Quantitative analysis of agonist and antagonist action is the basis of receptor classification and drug design. Attention to quantitative rigour is especially important in the case of 5-HT receptor research due to the complex variety of pharmacological and implied physiological and pathophysiological actions that 5-HT exhibits. The purpose of this chapter is to consider the extent to which reliable quantitative information can be obtained from studies, in isolated tissues, of agonist and antagonist interactions with 5-HT receptors. Initially, we exemplify the nature of the information required and the conditions needed to provide it. Then we examine existing and potential problems that can affect the generation or interpretation of this information.

2.2 The quantitative information required of agonists and antagonists

2.2.1 Antagonists

In principle, the interaction of an antagonist with a receptor can be characterized by an affinity constant. The Schild equation (Arunlakshana and Schild 1959) is conventionally used for this purpose. However, estimation of antagonist affinity is only valid when there is evidence for simple competition between the antagonist and the agonist for the receptor. A set of analytical criteria must be satisfied (Schild 1973; Black *et al.* 1983a). First, the fractional increase in agonist concentration required to overcome the effects of the antagonist should be independent of agonist concentration. Therefore, the antagonist should produce parallel displacement of agonist concentration-effect curves plotted semilogarithmically. Secondly, the affinity of the antagonist should be independent of concentration, meaning that the fractional increase in agonist concentration should be proportional to antagonist concentration. Therefore, the gradient of a Schild plot should be unity. Finally, the estimated antagonist affinity should be independent of the agonist used.

An example where each of these criteria is satisfied is illustrated in Fig. 2.1 (Leff and Martin 1986). Trazodone applied in the concentration range 0.1–10 μM produced effectively parallel rightward displacements of 5-HT concentration-effect curves in rings of rabbit thoracic aorta. The gradient of the Schild plot was 1.00 (±0.13, SEM) and the estimated affinity constant, the pK_B, was 7.23 (±0.19, SEM). When the same experiment was performed using the α-methyl derivative of 5-HT as agonist, the results were virtually identical: the Schild plot slope was 1.09 (±0.09, SEM) and

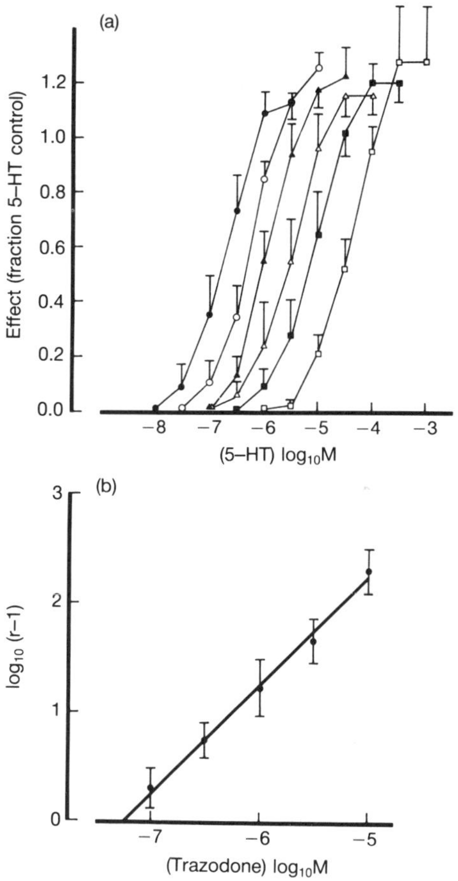

Fig. 2.1 Antagonist effects of trazodone on 5-HT-elicited responses in rabbit aorta. (*a*) 5-HT concentration–effect curves in the presence of zero (●), 0.1 μM(○), 0.3 μM (▲), 1.0 μM (△), 3.0 μM (■), and 10 μM (□) trazodone. (*b*) Schild plot analysis of trazodone-induced curve displacements. (Redrawn from Leff and Martin 1986 with permission)

the pK_B estimate was 7.39 (±0.11, SEM). Analytically, therefore, the antagonistic action of trazodone represents an ideal case.

2.2.2 Agonists

In principle, the interaction of an agonist with a receptor may be characterized by two quantities: affinity and efficacy. These quantities can be estimated by traditional null methods (Stephenson 1956; Furchgott 1966; Mackay 1966; Barlow *et al.* 1967) or by direct model-fitting procedures (Black *et al.* 1985; Leff 1987). The conditions and criteria for valid estimation of agonist affinity and efficacy are not so well established as those for estimation of antagonist affinity. However, recent theoretical modelling methods have shown that analysis of concentration-effect curve shape can assist in developing such criteria (Leff 1987). Briefly, using the operational model of agonism (Black and Leff 1983) predictions can be made for the relation between partial and full agonist curves and for the relation between agonist curves produced in the presence and absence of irreversible receptor antagonism. Testing accordance with these predictions allows confidence to be attached to the affinity and efficacy estimates made.

Figure 2.2 illustrates an example where this approach has been taken. The diagram illustrates the typical effect of irreversible receptor occlusion, using phenoxybenzamine, on 5-HT-elicited responses in the rabbit isolated aorta (Black *et al.* 1985). The operational model of agonism was used to make certain predictions about these effects. First, on the basis that the concentration-effect curve for 5-HT in this system is steep compared to a rectangular hyperbola (its mid-point gradient is greater than 0.576), it was predicted that the curve should become inherently flatter upon reduction of efficacy. It was also predicted for such 'steep' cases that the location of the curve produced upon receptor occlusion should be to the right on the concentration axis of the estimated agonist affinity constant, the pK_A (negative logarithm of the agonist dissociation constant). Both predictions were borne out by the data as illustrated by Fig. 2.2. Indeed, the model could be directly fitted to the data giving the lines shown in the diagram and providing estimates of affinity and efficacy. From nine replicate experiments, the pK_A was 6.95 (±0.04, SEM). The efficacy estimate, i.e. the value of τ (see legend to Fig. 2.2), obtained in the absence of irreversible antagonism was 1.96 (±0.16, SEM). In addition to according with theoretical expectations, the effects of phenoxybenzamine also fulfilled a practical criterion for valid estimation of agonist affinity and efficacy (Ariëns *et al.* 1964; Furchgott 1966) in that they could be protected against by concomitant administration of a simple competitive antagonist methysergide (Barrett *et al.* 1986). Analytically, therefore, these data can be regarded as ideal in that they allow reliable agonist affinity and efficacy estimates to be made.

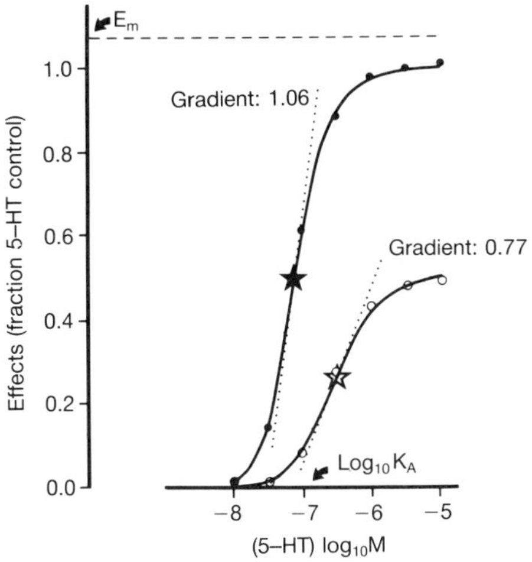

Fig. 2.2 Analysis of 5-HT agonism in rabbit aorta by irreversible receptor occlusion. Two successive 5-HT curves were obtained in a single tissue before (●) and after (○) treatment with phenoxybenzamine (0.1 μM, 30 min). The operational model of agonism (Black and Leff 1983) was fitted directly to the data in order to estimate K_A (the agonist dissociation constant), τ (the efficacy of the agonist in the system), E_m (the maximal effect possible in the system), and G (the gradient parameter: 0.576 for a rectangular hyperbola). The star symbols indicate the $[A_{50}]$ values for the curves, showing that the curve obtained after phenoxybenzamine treatment is to the right (on the $\log_{10}$ M axis) of the $\log_{10} K_A$ estimate, a result which is expected because the concentration–effect curve is initially steeper than a rectangular hyperbola. The change in gradient is also predicted under these conditions (see Black *et al.* 1985 for details). (Redrawn from Black *et al.* 1985 with permission)

2.3 Experimental conditions for quantifying agonism and antagonism

In order to obtain reliable estimates of antagonist affinity and of agonist affinity and efficacy, a number of experimental conditions must be satisfied. Furchgott (1968) reviewed the factors that should be considered. They include: alterations in the sensitivity of the assay system to the agonist; desensitization induced by the agonist; multiple receptor interactions; indirectly mediated agonist effects; post-receptor actions of antagonists; influence of agonist-removal mechanisms; lack of attainment of agonist or antagonist equilibrium when the effect is measured; and loss of agonist or antagonist from the assay medium.

Among these are factors which have affected or may affect quantification of 5-HT receptor agonists and antagonists. For example, it has been shown that under-estimation of antagonist affinity results in a tissue, such as the rat superior cervical ganglion, which accumulates 5-HT via a high affinity, saturable uptake mechanism (Ireland *et al.* 1987). Even in the absence of a specific uptake process reliable estimation of antagonist affinity requires inhibition of the metabolic disposal of less polar analogues of 5-HT, e.g. tryptamine (Bradley *et al.* 1985), since these amines gain access to intracellular monoamine oxidase by lipophilic penetration of cell membranes. Failure to take account of such agonist disposal mechanisms previously led to suggestions of a distinct tryptamine receptor in rat caudal artery (Hicks and Langer 1983).

The heterogeneous receptor actions of 5-HT and its analogues also appear to be a problem. In the rabbit aorta, tryptamine—unlike 5-HT—activates α_1-adrenoceptors in the same concentration range required to elicit contractions via 5-HT_2 receptors. Stollak and Furchgott (1983) showed that, in the absence of α_1-adrenoceptor blockade, affinity estimates obtained for each of several 5-HT_2 receptor antagonists differed by up to 26-fold depending upon whether 5-HT or tryptamine was the agonist used. These apparent agonist-dependent differences were abolished following α_1-adrenoceptor inactivation.

Multiplicity of 5-HT receptors in particular tissues might also pose a similar problem. The identity of the 5-HT receptor in the canine basilar artery has been claimed by different workers to be 5-HT_2 (Müller-Schweinitzer and Engel 1983) or '5-HT_1-like' (Peroutka *et al.* 1983; Taylor *et al.* 1986). It may be that a mixture of the two receptor types occurs in the tissue. If so, judicious use of a range of agonists and antagonists should demonstrate this in terms of complex concentration-effect curves and non-linear Schild plots (Kenakin 1984). Conditions for quantifying interactions at one or other of the receptors present should then become clear.

More generally, the tendency for a mixture of two receptors to resemble a third, as judged by the effects of agonists and antagonists (Kenakin 1984) emphasizes the need to establish conditions which are sufficiently rigorous for affinity and efficacy estimates to be obtained. These comments may be apposite in the case of 5-HT_3 receptors, where three different receptor subtypes have been claimed to exist in different tissues (Richardson *et al.* 1985), although steps have not been taken to exclude the possibility that only two types exist but in different relative densities in the various tissues studied.

2.4 Analysis of antagonism

5-HT antagonist studies of the kind illustrated in Fig. 2.1, which fulfil the necessary analytical criteria for simple competition, are not commonly reported in the literature. Multiplicity of receptors in particular tissues may play a part in this, but perhaps a more important problem is the variable dynamics that many 5-HT receptor antagonists exhibit, i.e. the tendency for ligands to behave as surmountable antagonists in some systems and as non-surmountable antagonists in others. Furthermore, when antagonists conform to the criteria for simple competition in different tissues problems in reproducing affinity estimates are encountered. This section will consider these problems in a little more detail.

2.4.1 Non-surmountable antagonism

Non-surmountable action is common amongst the antagonists conventionally used to classify 5-HT receptors. 5-HT_2 receptor-mediated agonist effects in bovine coronary arteries (Kaumann and Frenken 1985), in rat jugular vein and in rat caudal artery (Leff and Martin 1986) are antagonized non-surmountably by methysergide. Ketanserin demonstrates this action in the rat caudal artery (Leff and Martin 1986), in rat renal vasculature (Charlton *et al.* 1984) and in guinea-pig trachea (Lemoine and Kaumann 1986). '5-HT_1-like' receptor-mediated effects on presynaptic sympathetic neurones (Charlton *et al.* 1986) and on vascular endothelium (Van Nueten *et al.* 1984) are antagonized in a non-surmountable fashion by methiothepin. 5-HT_3 receptor-mediated effects in the rat vagus (Ireland and Tyers 1987), in rabbit superior cervical ganglion (Round and Wallis 1987), and in rabbit nodose ganglion (Round and Wallis 1987) are antagonized non-surmountably by MDL 72222. ICS 205–930 produces similar effects in each of these tissues (Ireland and Tyers 1987; Round and Wallis 1986). Fig. 2.3 illustrates some examples of non-surmountable actions.

Non-surmountable antagonism presents a problem in quantification because of the indefinite nature of the mechanism(s) responsible for the phenomenon. In the case of competitive antagonism, the Law of Mass Action provides a mechanistic basis for quantifying antagonist affinity accompanied by analytical criteria for assessing the validity of the affinity estimate. This is not the case with non-surmountable antagonism. The first, and obvious, problem is that antagonist action of this kind fails one of the criteria for simple competition so that affinity estimation using the Schild plot approach is precluded. The second problem is that no single alternative mechanism can be proposed. Where models have been suggested it is unclear how the antagonist action is to be quantified in a meaningful way and how analytical criteria can be developed.

A key factor which has emerged from the few studies into the nature of non-surmountable antagonism is the ability for simple competitive antagonists to protect against the action of non-competitive agents. Black *et al.* (1983b) showed that methysergide, which behaves competitively in the rabbit aorta preparation, protects against the non-surmountable antagonism produced by LSD. In systems where methysergide behaves as

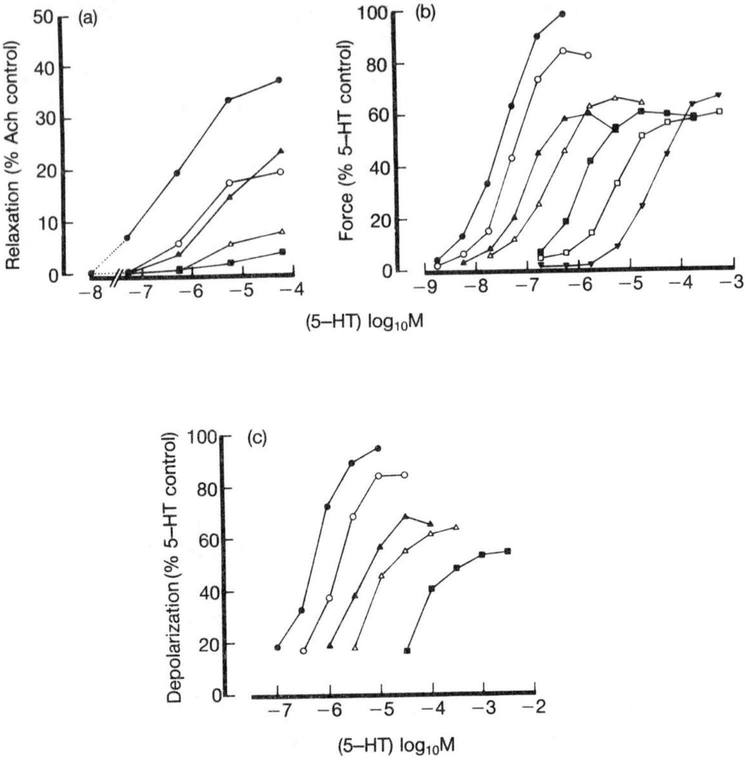

Fig. 2.3 Examples of non-surmountable antagonism. (*a*) Relaxational effects of 5-HT in canine coronary arteries with intact endothelium. Antagonism by methiothepin at zero (●), 5.3nM (○), 21nM (▲), 85nM (△), and 340nM (■). (Redrawn from Van Nueten *et al.* 1984 with permission). (*b*) Constrictor effects of 5-HT in guinea-pig trachea. Antagonism by ketanserin at zero (●), 0.3nM (○), 1nM (▲), 3nM (△), 10nM (■), 30nM (□) and 100nM (▼). (Redrawn from Lemoine and Kaumann 1986 with permission). (*c*) Depolarizing effects of 5-HT in rat vagus nerve. Antagonism by MDL 72222 at zero (●), 30nM (○), 100nM (▲), 300nM (△) and 1μM (■). (Redrawn from Ireland and Tyers 1987 with permission)

a non-surmountable antagonist, its actions can be prevented by agents which still behave competitively. Ketanserin can effect this protection in bovine coronary arteries (Kaumann and Frenken 1985); α, α-dimethyltryptamine, which behaves as a simple competitor in the rat jugular vein (Leff *et al.* 1986), protects against the non-surmountable action of methysergide in this tissue (Leff and Martin unpublished). The simplest interpretation for such observations is that the non-surmountable antagonist, the competitive antagonist, and the agonist interact in a mutually exclusive way with the receptor. The non-surmountable action could be explained on the basis of slow dissociation kinetics on the part of the antagonist producing this effect. Agonist- and tissue-dependence of non-surmountable antagonism can also be accommodated by this kinetic argument (Leff and Martin unpublished). Indeed, we favour this 'hemi-equilibrium' explanation. However, this explanation was rejected by De Chaffoy de Courcelles *et al.* (1986) in their interpretation of the effects of LSD, ketanserin, and methysergide, among others, on 5-HT-induced phosphatidic acid formation in platelets. These workers suggested an allosteric mechanism whereby non-surmountable antagonists induce either a change in affinity of the receptor for the agonist or a reduction of binding capacity. Kaumann and co-workers have suggested an allosteric mechanism for the non-surmountable effects of ketanserin and methysergide in a variety of tissues (Kaumann and Frenken 1985; Frenken and Kaumann 1985; Lemoine and Kaumann 1986; Kaumann *et al.* 1986). Their hypothesis, which is considered in detail in Chapter 3 also accounts for protection by competitive antagonists, albeit in a more complex way than the kinetic argument.

In addition to these explanations, it may be necessary to consider resultant actions of the antagonists involving post-receptor effects. For example, cyproheptadine, usually classified as a 5-HT_2 receptor antagonist, has been reported to possess calcium entry blocking properties (Winquist *et al.* 1984). This combination of properties also occurs in verapamil (Auguet *et al.* 1984). In such cases, it is theoretically possible to resolve the competitive and post-receptor actions of the agent by applying a recently developed concentration-ratios analysis (Hughes and Mackay 1985; Black *et al.* 1986), as has been demonstrated in the case of verapamil by Leff and Morse (1987). Quantification of the post-receptor component cannot be analysed meaningfully without knowledge of its mechanism.

Evidently, there are a number of mechanisms which could explain non-surmountable antagonism. Also, there is no guarantee that the same mechanism applies in each case. In addition to the problems of quantification caused by these circumstances, the agonist- and tissue-dependence of the phenomenon makes even an empirical approach to quantification difficult. The literature cited previously indicates that methysergide, for example, can behave as an apparently simple competitive antagonist in some tissues but non-surmountably in others. Furthermore, when this compound demonstrates non-surmountable antagonism the degree of depression of the agonist concentration-effect curve varies in relation to the rightward displacement of the curve. Thus, as shown in Fig. 2.3, for the guinea-pig trachea, substantial rightward displacement of the 5-HT concentration-effect curve accompanies the asymptote depression caused by methysergide. In contrast, in the bovine coronary artery (Kaumann and Frenken 1985) the 5-HT curve is depressed by methysergide with virtually no location change. Descriptions of the action of compounds like methysergide in terms of a single parameter such as a pD'_2 (negative logarithm of the concentration of antagonist required to cause 50 per cent depression of the agonist-concentration effect curve asymptote) is inappropriate under these conditions. This parameter does not describe the action and it is not a useful index of antagonist activity because it does not remain constant under different conditions, despite the same receptor being involved. On the other hand it is interesting that, in particular systems, pD'_2 values for a series of non-surmountable antagonists appear to correlate with binding affinities (Müller-Schweinitzer and Engel 1983), a result which is expected if a 'hemi-equilibrium' mechanism operated (Leff and Martin unpublished).

2.4.2 Variable antagonist affinity

The previous section dealt with the problems of quantification resulting from a situation in which the analytical criteria for simple competitive antagonism are not fulfilled. Here, we consider the situation where the criteria are met but in which problems of quantification or interpretation still occur.

Ideally, a particular receptor type would be characterized by a pK_B value for an antagonist (or a series of values for a set of antagonists) which varied within only narrow limits when different agonists or tissues containing that receptor type were used. In such circumstances, anomalous behaviour would be easily detected and, for example, new receptors identified with confidence. However, antagonist affinity does not appear to be a very stable parameter in the case of the ligands which are conventionally used for the purposes of 5-HT receptor identification. Estimates for the affinity of ketanserin, for example, vary by over two orders of magnitude between tissues which, by the accepted classification criteria (Bradley *et al.* 1986), all contain the same 5-HT_2 class of receptor (Bradley *et al.* 1983; Kaumann 1983; Lemberger *et al.* 1984; Mecca and Webb 1984). In each case, experimental conditions were appropriately controlled and the analytical criteria for simple competitive antagonism were reported to have been fulfilled. Furthermore, variation in the affinities of ketanserin and of other antagonists has been demonstrated under conditions where particular attention was paid to practical and analytical criteria (Leff and Martin 1986). A notable example was trazodone which, as illustrated in Fig. 2.1, behaved competitively with a pK_B value of 7.23 in the rabbit aorta but when studied in the rat caudal artery afforded a pK_B of 8.32.

A number of interpretations are available for such variability. It may be that 5-HT receptors, in this case, are a heterogeneous population. However, this conclusion does not help to bring order to an already challenging area of receptor research. It may be that the antagonists express properties additional to and independent of the receptor which superimpose on expression of simple competition. Alternatively, the variability could be viewed as an extension of the non-surmountable mechanism discussed in the previous section. In either of these latter cases, the utility of the antagonists in receptor classification would have to be called into question. A possibility that we have considered is that variable affinity is the consequence of antagonist binding to sites in receptors which are not utilized by the natural agonist, 5-HT (Leff and Martin 1986; Leff *et al.* 1986). Whereas the 5-HT attachment sites are conserved in different tissues and in different species these accessory sites may not be. Although this is evidently not an issue as regards the process of quantifying antagonism—despite variable pK_B estimates, the criteria for simple competition are fulfilled—it may be an important consideration as regards the application to which the pK_B estimates are put. If the accessory binding site explanation is valid, then discriminating between 5-HT receptors, with the emphasis on '5-HT', using compounds which attach to the accessory sites is illogical if the actions of 5-HT itself are subserved in the same way at each receptor. If the issue is not receptor classification but rather the quantification of potential drug actions, then variability in pK_B estimates could be looked upon as favourable from the point of view of drug selectivity.

2.5 Analysis of agonism

Whereas antagonists have been widely studied in 5-HT receptor research due to their important place in receptor classification and because of their potential as drugs, the same attention has not been given to agonists. This section illustrates the use of agonists in receptor classification and then considers existing and potential problems in obtaining the affinity and efficacy estimates required for proper quantification of their action.

2.5.1 Agonists in receptor classification

Figs 2.4 and 2.5 illustrate how affinity and efficacy estimates for agonists have assisted in confirming the similarity between the receptor populations in different tissues (Fig. 2.4) and in discriminating between different classes in different tissues (Fig. 2.5). In the first case, each in a series of four agonists displays the same affinity and relative efficacy in two tissues which purportedly contain the same 5-HT_2 receptor class but in which conventional antagonists exhibited variable dynamics and affinities (Leff and Martin 1986; Leff *et al.* 1986). In the second, another series of agonists simultaneously discriminate the receptors mediating endothelium-dependent and -independent vasorelaxation from each other and from the 5-HT_2 receptor class (Leff *et al.* 1987; Martin *et al.* 1987). In this case the use of agonists avoids the possible complications caused by variable antagonist dynamics.

These examples illustrate how quantitative information on agonists can

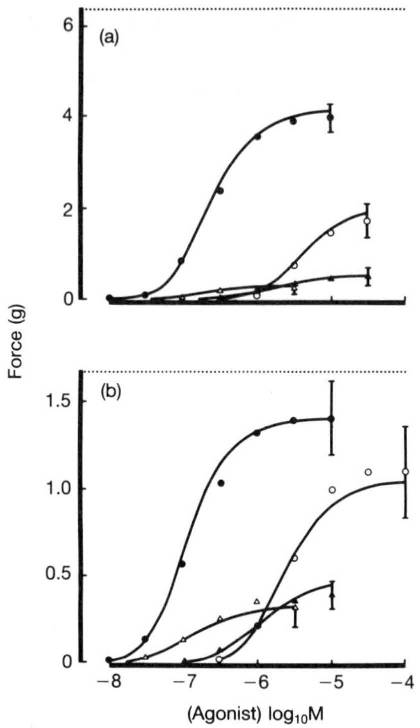

Fig. 2.4 Similarity of 5-HT receptors in different tissues demonstrated by agonist affinity and efficacy estimates. Concentration–effect curves for 5-HT (●), 5-cyanotryptamine (○), N, N-dimethyltryptamine (▲), and N-benzyl-5-methoxytryptamine (△) in rabbit aorta (panel (a)) and in rat jugular vein (panel (b)). (Redrawn from Leff *et al.* 1986 with permission). The lines drawn through the data were obtained by fitting the operational model of agonism (Black and Leff 1983). The dotted line indicates the E_m value in each case. This procedure provided the following affinity (pK_A) and efficacy (τ) estimates indicating that the receptors in the two tissues are of the same class (for details see Leff *et al.* 1986)

	Rabbit aorta		Rat jugular vein	
	pK_A	τ	pK_A	τ
5-HT	6.88	1.20	6.88	1.93
5-cyanotryptamine	5.80	0.77	5.80	1.24
N,N-dimethyltryptamine	6.34	0.44	6.34	0.71
N-benzyl-5-methoxytryptamine	7.30	0.36	7.30	0.58

For each agonist the value of τ was 1.61 higher in the jugular vein compared with the rabbit aorta, but the relative efficacies were the same

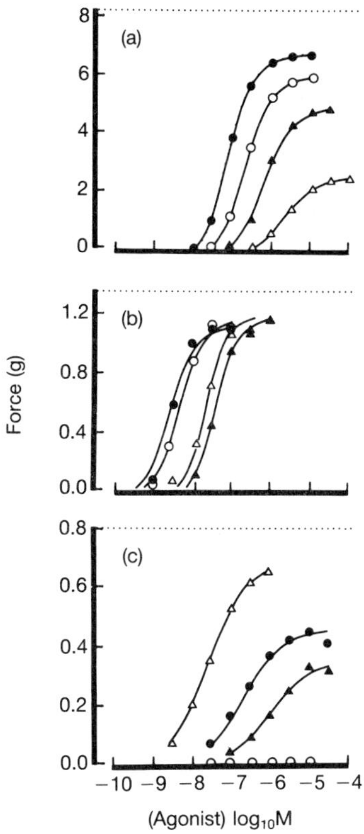

Fig. 2.5 Discrimination between 5-HT receptors in different tissues using agonist affinity and efficacy estimates. Concentration–effect curves for 5-HT (●), α-methyl-5-HT (○), 5-methyltryptamine (▲) and 5-carboxamidotryptamine (△) in rabbit aorta (contractions; panel (a)), rabbit jugular vein with intact endothelium (relaxations; panel (b)) and rabbit jugular vein with denuded endothelium (relaxations; panel (c)). The lines drawn through the data were obtained by operational model fitting (Black and Leff 1983). The dotted line indicates the E_m value in each case. The affinity and efficacy estimates were as follows

	Rabbit aorta		Rabbit jugular vein (+ endothelium)		Rabbit jugular vein (− endothelium)	
	pK_A	τ	pK_A	τ	pK_A	τ
5-HT	7.1	1.8	8.6	1.8	6.2	1.4
α-methyl-5-HT	6.7	1.5	8.3	1.9	<4.5	—
5-methyltryptamine	6.4	1.2	7.3	2.1	5.6	0.7
5-carboxamidotryptamine	6.0	0.7	7.5	2.2	6.6	6.5

These estimates indicated that the receptor classes in the three tissues can be simultaneously discriminated by this series of tryptamine analogues resolve some issues in receptor classification and extend our knowledge base with regard to receptor definition. Importantly, from the point of view of potential drug design, the affinity and efficacy estimates obtained can be rationalized in terms of chemical structure as demonstrated for the 5-HT_2 receptor (Clancy and Maayani 1985; Leff *et al.* 1986) and for the comparison between the 5-HT_2 receptor and the endothelial receptor (Leff *et al.* 1987).

2.5.2 Problems in agonist quantification

The relative scarcity of quantitative studies on 5-HT receptor agonists means that problems which influence their quantification may remain to be exposed. However, some features particularly associated with 5-HT receptor-mediated agonism are already known which may affect the quality of affinity and efficacy estimates obtained.

For example, the responses of many tissues to 5-HT and other agonists are generally transient, i.e. they rise and fade during continued exposure to the agonist, implying the existence of acute desensitization mechanisms. In our laboratory we have observed such responses in association with '5-HT_1-like' receptors on the endothelium of the rabbit jugular vein and rat stomach, with 5-HT_2 receptors in rat caudal artery and in guinea-pig trachea, and with 5-HT_3 receptors in the rat vagus. Under such conditions it is possible that response measurements are necessarily made at a time when equilibrium has not been established between the agonist and the receptor. Theoretical modelling of this situation indicates that agonist-concentration effect curves may be shifted rightwards on the concentration axis, leading to under-estimation of agonist affinity when, for example, the method of irreversible receptor inactivation is applied (Leff 1986). Moreover, this can occur without causing any changes in curve shape which may otherwise signify an anomaly.

Limited studies show that this may be a problem in certain tissues containing 5-HT_2 receptors. Fig. 2.6 illustrates representative recordings of the responses to single additions of 5-HT in guinea-pig trachea and in rabbit aorta (Leff and Martin 1988). Evidently, at equivalent levels, the responses are more transient in the former tissue than in the latter. When the method of irreversible receptor alkylation was used to estimate the dissociation constant for 5-HT in each tissue it was found that the value obtained in the trachea ($pK_A = 6.42$) was significantly less than that obtained in the aorta ($pK_A = 6.88$). The association between response transience and affinity estimation may be coincidental but it is maintained in at least two further tissues. In the rat jugular vein, responses are relatively well sustained and the same affinity estimate was made for 5-HT, as well as for other agonists, as in the rabbit aorta (see Fig. 2.4). In the rat caudal artery, where responses are more transient, a pK_A estimate of 6.20 for 5-HT was made. Nevertheless it would be desirable to obtain independ-

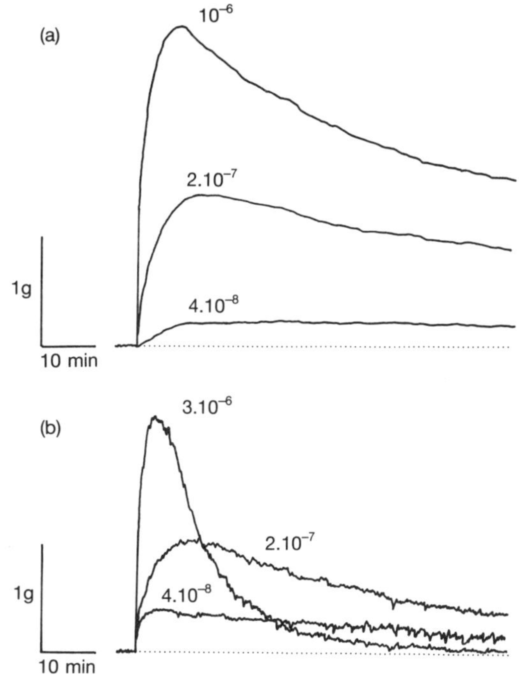

Fig. 2.6 Effect–time profiles for 5-HT obtained at concentrations which produce the same equivalent response in rabbit aorta (panel (a)) and guinea-pig trachea (panel (b)). At each equivalent concentration the responses in the trachea demonstrated more 'fade' than those in the aorta, the discrepancy being most pronounced at the high dose level (Leff and Martin, 1988)

ent estimates of agonist affinity in order to support or deny the argument. A general method for estimating agonist affinity which, in principle, avoids problems associated with the expression of agonism itself was suggested by Furchgott and Bursztyn (1968). It involves the effective conversion of the agonist in question into a competitive antagonist using irreversible receptor inactivation, then estimating its dissociation constant as an antagonist. This method requires the availability of an agonist with sufficiently higher efficacy than the test agonist for it to be active after receptor inactivation. Unfortunately this presents a problem in quantification in the case of 5-HT_2 receptors at least, because there appear to be no agonist mimetics of 5-HT with substantially higher efficacy than the natural agonist.

Another problem which may affect quantification of 5-HT receptor-mediated agonism is the existence of multiple coupling mechanisms subserving single receptor types. In the rabbit aorta, for example, 5-HT_2 receptor-mediated responses appear to be subserved only by plasmalemmal Ca^{2+} channels (Ratz and Flaim 1984; Barrett *et al.* 1986), whereas in the rat aorta (Nakaki *et al.* 1985) and in bovine coronary artery (Ratz and Flaim 1984) the responses are both extra- and intra-cellular Ca^{2+} dependent. Theoretically the operation of multiple transducers can lead to erroneous estimation of agonist affinity and efficacy (Leff, unpublished), meaning that information on post-receptor mechanisms should be considered when assessing the reliability of the estimates made.

2.6 Concluding remarks

Quantification of agonists and antagonists is ideally achieved by obtaining estimates of affinity and efficacy. This process is subject to the fulfilment of certain practical and theoretical criteria. In the case of 5-HT receptor agonists and antagonists a number of particular features of their action additionally influence or may influence their quantification.

The examples considered in this chapter emphasize the need for rigorous application of the analytical criteria but they also illustrate the necessary but not sufficient nature of those criteria. Thus the variable affinities of agonists and antagonists between tissues purportedly containing the same receptor type and under conditions when the analytical criteria are fulfilled throws up the challenge of understanding how receptor interactions behave simply when they are in fact complex. These kinds of problems are likely to require development of further chemical probes, pharmacological methods for detecting and evaluating resultant actions, and extensions to pharmacological receptor theory for their solution.

References

Ariëns, E. J., Simonis, A. M., and van Rossum, J. M. (1964). Drug-receptor interactions: interaction of one or more drugs with one receptor system. In *Molecular Pharmacology* (ed. E. J. Ariens) pp. 119–286. Academic Press, New York.

Arunlakshana, O. and Schild, H. O. (1959). Some quantitative uses of drug antagonists. *British Journal of Pharmacology and Chemotherapy* **14,** 45–58.

Auguet, M., Defeudis, F. V., and Clostre, F. (1984). Competitive inhibition of 5-HT receptors in rabbit isolated aorta by the Ca^{2+}-antagonist methoxyverapamil (D600). *Neurochemistry International* **6,** 701–10.

Barlow, R. B., Scott, N. C., and Stephenson, R. P. (1967). The affinity and efficacy of onium salts on frog rectus abdominis. *British Journal of Pharmacology and Chemotherapy* **31,** 188–96.

Barrett, V. J., Leff, P., Martin, G. R., and Richardson, P. J. (1986). Pharmacological analysis of the interaction between Bay K 8644 and 5-HT in the rabbit aorta. *British Journal of Pharmacology* **87,** 487–94.

Black, J. W. and Leff, P. (1983). Operational models of pharmacological agonism. *Proceedings of the Royal Society London, Series B* **220,** 141–62.

Black, J. W., Gerskowitch, V. P., and Leff, P. (1983a). Reflections on the classification of histamine receptors. In *Pharmacology of histamine receptors* (ed. C. R. Ganellin and M. E. Parsons) pp. 1–9, John Wright, Bristol.

Black, J. W., Brazenor, R. M., Gerskowitch, V. P., and Leff, P. (1983b). The problem of insurmountable antagonism in 5-hydroxytryptamine receptor classification. *British Journal of Pharmacology* **80,** 607P.

Black, J. W., Leff, P., Shankley, N. P., and Wood, J. (1985). An operational model of agonism: the effect of E/[A] curve shape on agonist dissociation constant estimation. *British Journal of Pharmacology* **84,** 561–71.

Black, J. W., Gerskowitch, V. P., Leff, P., and Shankley, N. P. (1986). Analysis of competitive antagonism when this property occurs as part of a pharmacological resultant. *British Journal of Pharmacology* **89,** 547–55.

Bradley, P. B., Humphrey, P. P. A., and Williams, R. H. (1983). Are vascular 'D' and '5-HT_2' receptors for 5-hydroxytryptamine the same? *British Journal of Pharmacology* **79,** 295 P.

Bradley, P. B., Humphrey, P. P. A., and Williams, R. H. (1985). Tryptamine-induced vasoconstrictor responses in rat caudal arteries are mediated predominantly via 5-hydroxytryptamine receptors. *British Journal of Pharmacology* **84,** 919–25.

Bradley, P. B., Engel, G., Feniuk, W., Fozard, J. R., Humphrey, P. P. A., Middlemiss, D. N., Mylecharane, E. J., Richardson, B. P., and Saxena, P. R. (1986). Proposals for the classification and nomenclature of functional receptors for 5-hydroxytryptamine. *Neuropharmacology* **25,** 563–76.

Charlton, K. G., Johnson, T. D., and Clarke, D. E. (1984). Vasoconstrictor and norepinephrine potentiating action of 5-hydroxykynuramine in the isolated perfused rat kidney: involvement of serotonin receptors and α_1-adrenoceptors. *Naunyn-Schmiedeberg's Archives of Pharmacology* **328,** 154–9.

Charlton, K. G., Bond, R. A., and Clarke, D. E. (1986). An inhibitory prejunctional 5-HT_1-like receptor in the isolated perfused rat kidney. Apparent distinction from the 5-HT_{1A}, 5-HT_2 and 5-HT_{1C} subtypes. *Naunyn-Schmiedeberg's Archives of Pharmacology* **332,** 8–18.

Clancy, B. M. and Maayani, S. (1985). 5-Hydroxytryptamine receptor in isolated rabbit aorta: characterisation with tryptamine analogs. *Journal of Pharmacology and Experimental Therapeutics* **233,** 761–9.

De Chaffoy de Courcelles, D., Leysen, J. E., Roevens, P., and van Belle, H. (1986). The serotonin – S_2 receptor: a receptor-transducer coupling model to explain insurmountable antagonist effects. *Drug Development Research* **8,** 173–8.

Frenken, M. and Kaumann, A. J. (1985). Ketanserin causes surmountable antagonism of 5-hydroxytryptamine-induced contractions of large coronary arteries of dog. *Naunyn-Schmiedeberg's Archives of Pharmacology* **328,** 301–3.

Furchgott, R. F. (1966). The use of β-haloalkylamines in the differentiation of receptors and determination of dissociation constants of receptor-agonist complexes. *Advances in Drug Research* **3,** 21–55.

Furchgott, R. F. (1968). A critical appraisal of the use of isolated organ systems for the assessment of drug action at the receptor level. In *Importance of fundamental principles of drug evaluation* (eds. D. H. Tedeschi and R. E. Tedeschi) pp. 277–97, Raven Press, New York.

Furchgott, R. F. and Bursztyn, P. (1968). Comparison of dissociation constants and of relative efficacies of selected agonists acting on parasympathetic receptors. *Annals of the New York Academy of Sciences* **144,** 882–98.

Hicks, P. and Langer, S. Z. (1983). Antagonism by tetra-hydro-β-carboline of the vasoconstrictor responses of tryptamine in rat tail arteries. *European Journal of Pharmacology* **96,** 145–9.

Hughes, I. E. and Mackay, D. (1985). Quantification of the characteristics of antagonists exhibiting both competitive antagonism and functional interaction. *British Journal of Pharmacology* **85,** 271–5.

Ireland, S. J. and Tyers, M. B. (1987). Pharmacological characterisation of 5-hydroxytryptamine-induced depolarisation of the rat isolated vagus nerve. *British Journal of Pharmacology* **90,** 229–38.

Ireland, S. J., Straughan, D. W., and Tyers, M. B. (1987). Influence of 5-hydroxytryptamine uptake on the apparent 5-hydroxytryptamine antagonist potency of metoclopramide in the rat isolated superior cervical ganglion. *British Journal of Pharmacology* **90,** 151–60.

Kaumann, A. J. (1983). Yohimbine and rauwolscine inhibit 5-hydroxytryptamine contraction of large coronary arteries of calf through blockade of 5-HT_2 receptors. *Naunyn-Schmiedeberg's Archives of Pharmacology* **323,** 149–54.

Kaumann, A. J. and Frenken, M. (1985). A paradox: the 5-HT-receptor antagonist ketanserin restores the 5-HT-induced contraction depressed by methysergide in large coronary arteries of calf. Allosteric regulation of 5-HT_2-receptors. *Naunyn-Schmiedeberg's Archives of Pharmacology* **328,** 295–300.

Kaumann, A. J., Frenken, M., and Lemoine, H. (1986). Allosteric regulation of 5-HT_2-receptors in smooth muscle. *Naunyn-Schmiedeberg's Archives of Pharmacology* **330,** R64.

Kenakin, T. P. (1984). The classification of drugs and drug receptors in isolated tissues. *Pharmacological Reviews* **36,** 165–282.

Leff, P. (1986). Potential errors in agonist dissociation constant estimation caused by desensitisation. *Journal of Theoretical Biology* **121,** 221–32.

Leff, P. (1987). Can operational models of agonism provide a framework for classifying hormone receptors? In *Perspectives on receptor classification* (Receptor Biochemistry and Methodology, Vol. 6, eds. J. W. Black, D. H. Jenkinson and V. P. Gerskowitch) pp. 157–67. Alan R. Liss, New York.

Leff, P. and Martin, G. R. (1986). Peripheral 5-HT_2-like receptors. Can they be classified with the available antagonists? *British Journal of Pharmacology* **88,** 585–93.

Leff, P. and Martin, G. R. (1988). Differences in agonist dissociation estimates for 5-HT at 5-HT_2 receptors: a problem of acute desensitisation? *British Journal of Pharmacology* **95,** 569–77.

Leff, P. and Morse, J. M. (1987). Resultant pharmacological actions of verapamil: quantification of competitive 5-hydroxytryptamine antagonism in combination with calcium antagonism. *Journal of Pharmacology and Experimental Therapeutics* **240,** 284–7.

Leff, P., Martin, G. R., and Morse, J. M. (1986). The classification of peripheral 5-HT_2-like receptors using tryptamine agonist and antagonist analogues. *British Journal of Pharmacology* **89,** 493–9.

Leff, P., Martin, G. R., and Morse, J. M. (1987). Differential classification of vascular smooth muscle and endothelial cell 5-HT receptors using tryptamine analogues. *British Journal of Pharmacology* **91,** 321–31.

Lemberger, H. F., Mason, N., and Cohen, M. L. (1984). 5-HT_2-like receptors in the rat portal vein: desensitisation following cumulative serotonin addition. *Life Sciences* **35,** 71–7.

Lemoine, H. and Kaumann, A. J. (1986). Allosteric properties of 5-HT_2-receptors in tracheal smooth muscle. *Naunyn-Schmiedeberg's Archives of Pharmacology* **333,** 91–7.

Mackay, D. (1966). A new method for the analysis of drug-receptor interactions. *Advances in Drug Research* **3,** 1–19.

Martin, G. R., Leff, P., Cambridge, D., and Barrett, V. J. (1987). Comparative analysis of two types of 5-hydroxytryptamine receptor mediating vasorelaxation: differential classification using tryptamines. *Naunyn-Schmiedeberg's Archives of Pharmacology* **336,** 365–73.

Mecca, T. E. and Webb, R. C. (1984). Vascular responses to serotonin in steroid hypertensive rats. *Hypertension* **6,** 887–92.

Müller-Schweinitzer, E. and Engel, G. (1983). Evidence for mediation by 5-HT_2 receptors of 5-hydroxytryptamine-induced contraction of canine basilar artery. *Naunyn-Schmiedeberg's Archives of Pharmacology* **324,** 287–92.

Nakaki, T., Roth, B. L., Chuang, D.-M., and Costa, E. (1985). Phasic and tonic components in 5-HT_2 receptor-mediated rat aorta contraction: participation of Ca^{++} channels and phospholipase C. *Journal of Pharmacology and Experimental Therapeutics* **234,** 442–6.

Peroutka, S. J., Noguchi, M., Tolner, D. J., and Allen, G. S. (1983). Serotonin-induced contraction of canine basilar artery: mediation by 5-HT_1 receptors. *Brain Research* **259,** 327–30.

Ratz, P. H. and Flaim, S. F. (1984). Mechanism of 5-HT contraction in isolated bovine ventricular coronary artery. *Circulation Research* **54,** 135–43.

Richardson, B. P., Engel, G., Donatsch, P., and Stadler, P. A. (1985). Identification of serotonin M-receptor subtypes and their specific blockade by a new class of drugs. *Nature* **316,** 126–31.

Round, A. and Wallis, D. I. (1986). The depolarising action of 5-hydroxytryptamine on rabbit vagal afferent and sympathetic neurones *in vitro* and its selective blockade by ICS 205–930. *British Journal of Pharmacology* **88,** 485–94.

Round, A. and Wallis, D. I. (1987). Further studies on the blockade of 5-HT depolarisations of rabbit vagal afferent and sympathetic ganglion cells by MDL 72222 and other antagonists. *Neuropharmacology* **26,** 39–48.

Schild, H. O. (1973). Receptor classification with special reference to β-adrenergic receptors. In *Drug Receptors* (ed. H. P. Rang) pp. 29–36. Macmillan, London.

Stephenson, R. P. (1956). A modification of receptor theory. *British Journal of Pharmacology and Chemotherapy* **11,** 379–93.

Stollak, J. S. and Furchgott, R. F. (1983). Use of selective antagonists for determining the types of receptors mediating the action of 5-hydroxytryptamine and tryptamine in the isolated rabbit aorta. *Journal of Pharmacology and Experimental Therapeutics* **224,** 215–21.

Taylor, E. W., Piper Duckles, S., and Nelson, D. L. (1986). Dissociation constants of serotonin agonists in the canine basilar artery correlate to K_i values at the

5-HT_{1A} binding site. *Journal of Pharmacology and Experimental Therapeutics* **236,** 118–25.

Van Nueten, J. M., Leysen, J. E., de Clerck, F., and Vanhoutte, P. M. (1984). Serotonergic receptor subtypes and vascular reactivity. *Journal of Cardiovascular Pharmacology* **6,** S564–S614.

Winquist, R. J., Siegl, P. K. S., Baskin, E. P., Bohn, D. L., Morgan, G., and Wallace, A. A. (1984). Calcium entry blocker activity of cyproheptadine in isolated cardiovascular preparations. *Journal of Pharmacology and Experimental Therapeutics* **230,** 103–9.

3
The Allosteric 5-HT_2 receptor system

ALBERTO J. KAUMANN

3.1 History of 5-HT_2 receptors

'In the absence of precise knowledge of the mechanisms involved it is best to use words which do not depend on any one theory, and it is for this reason that the word unsurmountable is used to describe antagonisms which cannot be overcome by increasing the concentration of the active drug.'

Gaddum *et al.* (1955).

Despite this cautious statement, crucial findings, ideas, and methods about the mode of action of 5-HT, still valid currently, came first from Gaddum and his colleagues. Both 5-HT and LSD contain an indole moiety. Based on this structural analogy, Gaddum (1953a) used LSD and found it to block the contractions caused by 5-HT in uterine muscle. It was soon noticed that the blockade by LSD could not be surmounted by 5-HT (Gaddum *et al.* 1955). Similar findings were made with 5-HT and LSD in arterial muscle (Gaddum and Hameed 1954). On the other hand, Gaddum *et al.* (1955) also found a class of indole-containing antagonists, such as methylgramine, that caused surmountable blockade of the effects of 5-HT. The pioneering work of Gaddum and his colleagues culminated in the postulation of the existence of two types of tryptamine receptor (Gaddum and Picarelli 1957). One receptor, designated D, was located on smooth muscle and blocked by dibenzyline (phenoxybenzamine). The other receptor, labelled M, was located on nerve endings and the effects of 5-HT mediated through this site were blocked by morphine. This receptor classification and the recognition that antagonists can cause either surmountable or insurmountable blockade are interrelated and have fundamentally stimulated related research for over 30 years.

The next impetus towards 5-HT receptor classification came from attempts to radiolabel the receptors. Using [^{3}H]5-HT, [^{3}H]spiperone and [^{3}H]LSD and rat brain membranes, Peroutka and Snyder (1979) found that 5-HT bound in nanomolar concentrations to a population of saturable sites and in micromolar concentrations to a second population of saturable sites. They proposed the name 5-HT_1 receptors for the high affinity sites and 5-HT_2 receptors for the low affinity sites. [^{3}H]LSD exhibited high affinity for both 5-HT_1 and 5-HT_2 sites, but [^{3}H]spiperone had high affinity only for 5-HT_2. Another high affinity ligand for the brain 5-HT_2 site, [^{3}H]ketanserin, was introduced subsequently (Leysen *et al.* 1982). Both spiperone and ketanserin antagonized the effects of 5-HT often, although not always, in a surmountable and competitive manner on smooth muscle of arteries (spiperone: Maayani *et al.* 1984; Leff and martin 1986; Leff and Morse 1987; ketanserin: Kaumann 1983; Frenken and Kaumann 1984a), and trachea (Cohen *et al.* 1985; Lemoine and Kaumann 1986). There is a rough agreement between the affinities of spiperone and ketanserin for some smooth muscle 5-HT receptors, estimated from competitive antagonism, and the affinities of the corresponding radioligands for brain 5-HT sites. It has become increasingly clear that peripheral D-receptors are in large part 5-HT_2 receptors (see for example Humphrey 1984). An important exception are the D-receptors of rat stomach fundus which are not 5-HT_2 receptors because they cannot be blocked by spiperone or ketanserin (Leysen and Tollenaere 1982; Cohen and Wittenauer 1985; Cohen *et al.* 1985).

3.2 Surmountable and insurmountable antagonists

Since Gaddum's finding that LSD antagonized 5-HT-induced contractions of smooth muscle, related compounds have appeared that are also antagonists of the effects of 5-HT. One of these agents is methysergide which was found to be a surmountable antagonist in some smooth muscle systems and an insurmountable antagonist in others (Müller-Schweinitzer and Weidmann 1978; Feniuk, 1984). Insurmountable blockade has created diverse reactions among investigators. Feniuk (1984) felt that the insurmountable blockade caused by methysergide could be due to some interference beyond the receptors in the pathway leading to the contractile response of 5-HT. Leff and his colleagues have been making painstaking efforts to exclude insurmountable antagonists (including methysergide) from receptor classification and to use only surmountable antagonists (Leff and Martin 1986) or partial agonists (Leff *et al.* 1986) for that purpose. On the other hand, in order to classify vascular 5-HT receptors, Peroutka (1984) has deliberately associated surmountable antagonists with 5-HT_2 receptors and insurmountable antagonists with 5-HT_1 receptors. With regard to the contractile response to 5-HT, Peroutka (1984) divided blood vessels into two groups. In the first group, 5-HT is a potent agonist ($EC_{50} < 100$ nM) but antagonists such as ketanserin and methysergide are weak and insurmountable blockers. In the second group, 5-HT is weak ($EC_{50} > 100$ nM) but ketanserin and methysergide are potent and surmountable blockers.

None of these views seems wholly justified. The concentrations of methysergide that can cause half maximal depression of the response to 5-HT (i.e. the IC_{50} values of methysergide) are often subnanomolar (Müller-Schweinitzer and Weidmann 1978; Kaumann and Frenken 1985) and, as will be discussed below, are likely to be related to a site of action within the 5-HT_2 receptor system. It is therefore not necessary to suggest an interference by methysergide in the pathway beyond the 5-HT_2 receptor system (Feniuk 1984).

Leff and his colleagues aim at characterizing surmountable antagonists with reproducible affinity among tissues as tools for receptor classification. They excluded methysergide because it often blocks in an insurmountable manner and ketanserin because its affinity estimates vary greatly and also because on occasions it blocks in a partially insurmountable manner (Leff and Martin 1986). They attributed ketanserin's properties to a lack of chemical resemblance to 5-HT and therefore introduced several compounds closely related to tryptamine that caused surmountable blockade. However, although affinity estimates agreed among tissues, these agents showed at least 100-fold lower affinities for 5-HT_2 receptors than ketanserin (Leff *et al.* 1986). Furthermore, as will be described later, tryptamine itself can act as a partially insurmountable antagonist (Fig. 3.6); clearly more work needs to be done with the tryptamine analogue approach of Leff *et al.* (1986) in different tissues to support the generality of the mode of action of these agents as tools for 5-HT receptor classification. The argument about chemical structure lacks a certain conviction, given that it was the chemical resemblance of LSD to 5-HT that led Gaddum (1953a) to discover its blocking properties. Because LSD is usually, and methysergide often, a rather potent insurmountable antagonist, their indole moieties actually suggest an interaction with the 5-HT_2 receptor system. As will be detailed below, it is my intention to try to understand insurmountability in order to learn more about the 5-HT_2 receptor system.

Peroutka's (1984) proposal to use surmountable and insurmountable antagonism for vascular 5-HT receptor classification is not supported by several pieces of evidence. For example, Müller-Schweinitzer and Engel (1983) found that ketanserin and spiperone blocked insurmountably the effects of 5-HT on canine basilar artery, with IC_{50} values of 1 and 2 nM which resemble the corresponding affinities for 5-HT_2 receptors. This is incompatible with Peroutka's (1984) view that the 5-HT receptors of canine basilar artery are 5-HT_1. However, it could be argued that because the IC_{50} of an insurmountable antagonist depends on the 5-HT concentration, the conclusion of Müller-Schweinitzer and Engel (1983) stands on shaky ground. Furthermore, there is no theoretical basis to relate IC_{50} values for the depression of 5-HT-induced contraction to the equilibrium dissociation binding constant of the antagonist as these authors did. Frenken and Kaumann (1986) added some support for a participation of 5-HT_2 receptors by finding that ketanserin caused nearly surmountable antagonism of the effects of 5-HT in canine basilar artery, and that the calculated affinity agreed with that expected for 5-HT_2 receptors. However, Frenken and Kaumann (1986) also found a component of the 5-HT-induced contraction in this tissue that was resistant to blockade by ketanserin and which is thus not inconsistent with Peroutka's (1984) view that a 5-HT_1 receptor is involved. It can be concluded that 5-HT mediates its contractile response in canine basilar artery through a combination of 5-HT_2 receptors and receptors that are perhaps of the '5-HT_1-like' subtype. These latter receptors appear to be of the 5-HT_{1A} subtype (Peroutka *et al.* 1986). A similar co-existence of ketanserin-sensitive and ketanserin-resistant 5-HT receptors (i.e. 5-HT_2 and perhaps '5-HT_1-like') has been reported by Frenken and Kaumann (1985a) in canine coronary artery. Another observation that disagrees with Peroutka's (1984) classification of vascular 5-HT receptors is the finding that subnanomolar or nanomolar concentrations of methysergide can cause insurmountable blockade of 5-HT-induced contractions in calf coronary arteries (Kaumann and Frenken 1985; Kaumann 1985), rat tail artery (Frenken and Kaumann 1987a) and trachea of calf and guinea-pig (Lemoine and Kaumann 1986). Because, as will be discussed later, methysergide interacts with ketanserin in each of these systems, it is suggested that methysergide blocks 5-HT_2 receptors allosterically. Even Peroutka's (1984) generalization that the EC_{50} values for vascular 5-HT_2 receptors are greater than 100 nM does not always hold (see for example Fig. 3.7 for calf coronary artery, $EC_{50} = 30$ nM; Fig. 4 of Frenken and Kaumann 1987a, for rat tail artery, $EC_{50} = 25$ nM).

In a recent attempt to classify functional 5-HT receptors, virtually nothing was discussed about the meaning and implications of insurmountable antagonism (Bradley *et al.* 1986). While surmountable and competitive antagonism clearly follows first principles and needs no further comment, insurmountable antagonism has either been cautiously bypassed or dealt with in a somewhat bewildering and empirical way. In this chapter I shall discuss and present information from my laboratory aimed at understanding insurmountable antagonism in 5-HT_2 receptor systems. I shall not deal in detail with cases of insurmountable antagonism not due to a direct action of the antagonist on the 5-HT_2 receptor system, such as a non-specific depressant effect, blockade of calcium channels, or the use of insufficiently high 5-HT concentrations to achieve surmountability of blockade. As an example of the latter, the insurmountable blockade by ketanserin observed by Brazenor and Angus (1982) on canine coronary artery became surmountable when sufficiently high 5-HT concentrations were administered by Frenken and Kaumann (1985a).

3.3 Fast and slow 5-HT-induced contractions

In 1981 I observed, but did not at the time understand, that 5-HT-induced contractions developed considerably more slowly in calf coronary arteries that had been exposed to phenoxybenzamine (followed by washout) or that were obtained in the presence of methysergide. Both phenoxybenzamine and methysergide depress the maximum response to 5-HT, but while the effect of the former is essentially irreversible (Fig. 3.1) (Frenken and Kaumann 1987b), that of the latter is completely reversible (Frenken and Kaumann 1987a) after washing out the drug. Residual responses to high 5-HT concentrations in the presence of nanomolar concentrations of methysergide or after phenoxybenzamine pre-treatment (followed by washout) have a half time five to seven times longer than the half time of matching responses to 5-HT in the absence of drug (Fig. 3.2). The slow contractions become manifest at considerably higher 5-HT concentrations than the fast contractions and it is possible to observe both contractions during a single cumulative concentration-effect curve for 5-HT (Fig. 3.1). Slow contractions have also been observed with 5-HT in the presence of methysergide in rat tail artery (Frenken and Kaumann 1987a). Slow contractions are unaffected by 1000 nM prazosin in calf coronary artery and pulmonary artery (unpublished experiments by K. Sampford in the author's laboratory), suggesting that α_1-adrenoceptors are not involved. Slow contractions are usually not elicited by 5-HT in the absence of specific drugs. Surmountable, competitive antagonists, such as ketanserin, yohimbine, and rauwolscine (Kaumann 1983), do not uncover slow contractions regardless of the concentration used and affinity for the 5-HT_2 receptor. Thus, an increase of 5-HT concentrations up to 10 000 fold required to surmount the blockade caused by ketanserin does not result in the appearance of slow contractions (Kaumann and Frenken 1985). The simplest interpretation of these experiments is that both phenoxybenzamine and methysergide uncover slow 5-HT-induced contractions and that these are mediated through a receptor insensitive to the two insurmountable antagonists but sensitive to blockade by the surmountable antagonists. However, this simple interpretation was found to be untenable due to a chance discovery.

3.4 A paradox

An experimental condition has to be detailed briefly to understand some unexpected findings. Preparations of calf coronary artery with its endothelium rubbed off, contract isometrically at 32.5°C upon administration of 5-HT in Krebs solution containing 6 μM cocaine, ascorbate and the organic substrates glucose 10 mM, L-glutamate 5 mM, pyruvate 5 mM, and fumarate 5 mM (Kaumann 1983). Under these conditions there is a tendency for the contractions to increase upon successive exposures to 5-HT. In 1983 Michael Frenken, working in my laboratory, tried to estimate the affinity of ketanserin for coronary 5-HT_2 receptors. After having determined a first concentration-effect curve for 5-HT, he unexpectedly ran out of organic substrates. He made up a new stock solution of substrates from new

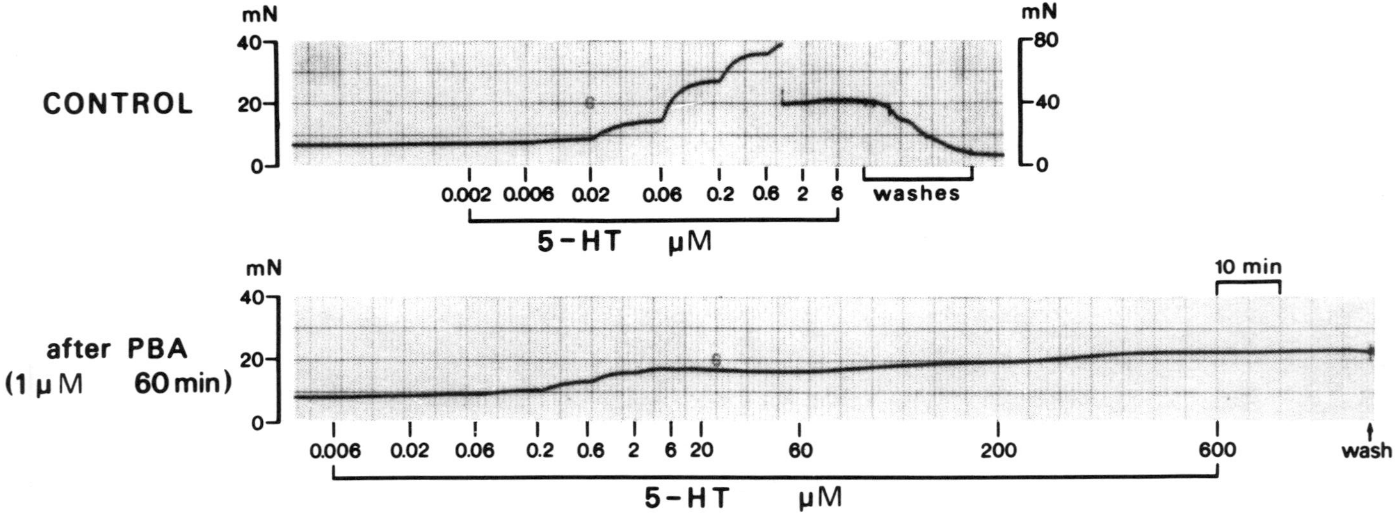

Fig. 3.1 Fast and slow contractions induced by 5-HT. Original recordings from a strip of calf coronary artery with the endothelium rubbed off. The tissue contracted isometrically at 32.5°C in modified Krebs solution containing 6 µM cocaine and 200 µM ascorbate as previously described (Kaumann 1983). After determining a first concentration–effect curve for 5-HT followed by washes (upper panel), the tissue was incubated for 1 hour in 1 µM phenoxybenzamine, washed, and a second curve for 5-HT (lower panel) determined. The amplification had to be halved in the upper panel. While the bottom tracing consists of fast contractions (up to 2 µM) and slow contractions (60 and 200 µM), the top tracing consists only of fast contractions. (From Frenken and Kaumann (1987b) with permission)

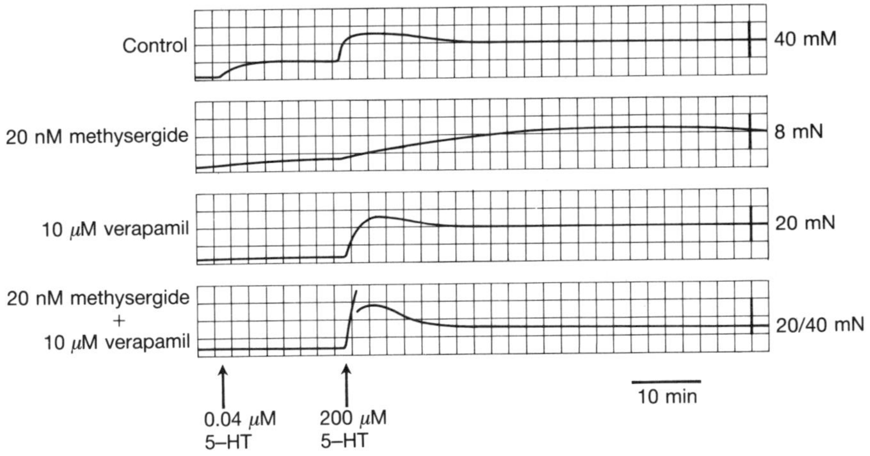

Fig. 3.2 Effects of methysergide, verapamil, and a combination of both, on the contractions evoked by a low and high concentration of 5-HT. Each tracing is from a different strip dissected from the same calf coronary artery. Verapamil and methysergide were incubated for 4 hours. The amplification of the bottom tracing had to be halved during the response to 5-HT. Verapamil abolished the response to the low 5-HT concentration. On the other hand, verapamil converted the slow contraction of 5-HT, induced by methysergide, into a fast and augmented response. Similar results were observed in another four strips from the same artery

batches. After incubating the tissues with the new substrates some of the tissues were also incubated with ketanserin while other preparations served as controls. Surprisingly, and unlike my own experiments (Kaumann 1983), Frenken noticed that the responses to 5-HT were depressed in the absence of ketanserin but not in its presence. Furthermore, the contractions elicited by 5-HT were slow in the absence and fast in the presence of ketanserin. It occurred to me that the new organic substrates contained some contaminant that acted in a methysergide-like fashion, i.e. it both caused partially insurmountable antagonism and induced slow 5-HT-induced contractions. Moreover, the effect was not seen in the presence of ketanserin. If this were true, ketanserin should also restore the fast 5-HT-induced contractions that had been depressed by methysergide. It was indeed found (in Krebs solution containing organic substrates that did not depress the response to 5-HT) that ketanserin was able to prevent both the depression of the 5-HT-induced contraction by methysergide and the appearance of slow contractions (Kaumann and Frenken 1985). High concentrations of methysergide were able to overcome, in a concentration-dependent manner, the protective effects of ketanserin without changing the sensitivity of the tissues to 5-HT. These experiments suggest that ketanserin and methysergide competed for a common site near or at the

5-HT_2 receptor that was not recognized by 5-HT. The paradox that a surmountable antagonist could restore the response to 5-HT depressed by an insurmountable antagonist led to the working hypothesis which is discussed in the next section.

The nature of the contaminant of the stock solution containing the organic substrates that depressed 5-HT-induced contractions is still unknown. A recent analysis with reverse phase high pressure liquid chromatography, carried out by N. Holland at ICI, and using tryptamine as a standard, ruled out tryptamine as the contaminant. However, comparison of stock solutions with normal organic substrates, with solutions containing the batch of substrates found to cause depression of the response to 5-HT, revealed that only the latter solution contained a compound whose nature is currently under investigation.

3.5 A two-state hypothesis

The observations that under certain conditions fast and slow contractions, elicited by 5-HT, are interconvertible, led to the idea that the 5-HT_2 receptor may exist in two interconvertible states $R \rightleftharpoons R'$ (Frenken and Kaumann 1984b; Kaumann and Frenken 1985). It is assumed that the 5-HT_2 receptor can exist in a highly active state R and a low active state R′ and that fast and slow contractions are mediated through R and R′ respectively (see Fig. 3.3). By competing with methysergide for an allosteric site, ketanserin favours R while methysergide facilitates the shift to R′. 5-HT and ketanserin, but not methysergide, also compete for a site on the 5-HT_2 receptor in the R-state. 5-HT has an equilibrium dissociation constant $K_{5\text{-HT}}$ of approximately 0.2 μM for R, as estimated by receptor occlusion with phenoxybenzamine by Frenken and Kaumann (1987b) for 5-HT_2 receptors of calf coronary (see also Fig. 3.1) and pulmonary artery. The affinity of 5-HT (i.e. the reciprocal of $K_{5\text{-HT}}$) for the R-state of the 5-HT_2 receptors of calf arteries agrees with similar affinity estimates for the 5-HT_2 receptors of rabbit aorta (Clancy and Maayani 1985; Barrett *et al.* 1986).

The affinity of 5-HT for R′ is unknown, but probably at least 100-fold lower than for R because only concentrations greater than 20 μM 5-HT elicit slow contractions. It could be argued that there is an inactive state, instead of a low active state, of the 5-HT_2 receptor and that the slow contractions evoked by high 5-HT concentrations reflect merely a time-dependent interconversion of the receptor from an active into an inactive state. However, this mechanism can be ruled out for kinetic reasons. Isomerization into an inactive receptor state would be expected to cause slow contractions with 5-HT of consistently smaller magnitude than fast contractions, which was not observed. For example, after fast contractions, caused by low 5-HT concentrations, had been depressed by phenoxybenzamine pre-treatment, high 5-HT concentrations evoked slow contractions of greater force than the depressed fast contractions (Fig. 3.1). Similarly, after 0.1 or 0.3 nM methysergide had considerably depressed fast contractions caused by low 5-HT concentrations, high 5-HT concentrations again caused slow contractions that were of greater strength than the fast contractions (e.g. Fig. 3 of Kaumann and Frenken 1985).

Do the R and R′ states pre-exist in the absence of drugs? Two observations are consistent with the pre-existence, and not induction, of the $R \rightleftharpoons R'$ states: (1) The slow contractions, induced by 5-HT, become conspicuous together with residual fast contractions when about 85 per cent of the receptors in the R-state have been made unavailable to 5-HT by a treatment with phenoxybenzamine followed by washout of this haloalkylamine (Frenken and Kaumann 1987b). (2) On occasions, when a cumulative concentration-effect curve for 5-HT is determined on calf coronary artery, higher than maximally effective concentrations (>6 μM) may cause some depression, but very high concentrations (>60 μM) again increase force (Kaumann unpublished experiments). These observations, made with 5-HT in the absence of other drugs, rule out the possibility that drugs cause 5-HT to induce slow contractions by inducing some sort of fit with a 5-HT receptor—e.g. induced-fit (Koshland 1970).

It might be argued that 5-HT elicits fast and slow contractions through distinct, non-interconvertible receptors. However, a selective blocking agent for the slow contractions has not been found. Instead, all agents studied so far, that cause the slow contractions to disappear also restore fast 5-HT-induced contractions. These agents have been designated activators (Kaumann 1987a) because it is assumed that they facilitate the R-state of the 5-HT_2 receptor which is assumed to be associated with the generation of fast 5-HT-induced contractions (for examples of activators see Tables 3.1 and 3.3).

A variety of observations suggest that the R′-state can slowly interconvert into the R-state. For example, in calf coronary artery, but not pulmonary artery, residual fast contractions induced by 5-HT can become more prominent when measured at 2 hourly intervals after phenoxybenzamine washout, suggesting an increased availability of receptors in the R-state due to interconversion from the R′-state (Frenken and Kaumann 1987b). Repetitive challenges with high 5-HT concentrations in the presence of methysergide, that had initially caused only slow contractions, may at later stages produce small fast contractions followed by slow contractions (unpublished observations in collaboration with Sarah Slee). These findings suggest spontaneous transitions between R and R′. The transition from R′ to R can probably be facilitated by ketanserin, even when administered after a phenoxybenzamine-induced receptor occlusion, because ketanserin hastens the recovery of fast 5-HT-induced contractions in calf coronary artery (Frenken and Kaumann 1987b). Although a spontaneous reappearance of fast contractions has not been observed in calf pulmonary artery after phenoxybenzamine treatment, ketanserin can partially restore fast contractions to 5-HT in this tissue (Frenken and Kaumann 1987b). The mechanism of this effect of ketanserin is not receptor protection because the drug was actively enhancing reponses to 5-HT after, and not before, phenoxybenzamine incubation and washout. Instead, the effect of ketanserin suggests that the 5-HT_2 receptor can only be irreversibly alkylated in its R-state but not in the R′-state (Fig. 3.3). Indeed, when calf arteries were pre-incubated with methysergide followed by incubation with phenoxybenzamine and washout of both drugs, the contractions caused by 5-HT were depressed; ketanserin, administered after methysergide and 5-HT had been washed out, completely restored the effects of high 5-HT concentrations (Frenken and Kaumann 1987b). Thus, the deliberate conversion by methysergide of most 5-HT_2 receptors into the R′-state prevented phenoxybenzamine from alkylating the receptors. Ketanserin merely reconverted most receptors from R′ to R, thereby allowing the production of maximum fast contractions with 5-HT.

These observations are consistent with the pre-existence of both the R-state and R′-state, and with the hypothesis that phenoxybenzamine can alkylate the 5-HT_2 receptor only in its R-state. None of these interpretations should be regarded as being established beyond doubt. However, the

Table 3.1 Activators of 5-HT_2 receptors in various tissues

Ligand	Tissue	Source
Ketanserin	Calf coronary artery	(1,2,3,4,5,6,7) Figs. 3.5, 3.6, 3.7
	Calf and guinea-pig trachea	(8)
	Rat tail artery	(4)
	Dog basilar artery	(9)
ICI 169,369 (15)	Calf coronary artery	(10) Fig. 3.4
	Rat tail artery	(10)
Spiperone	Calf coronary artery	(9)
	Rat tail artery	(11)
Phentolamine	Calf coronary artery	(5)
	Calf pulmonary artery	(9)
Trazodone	Rabbit aorta	(12)
Methysergide	Rabbit aorta	(12)
Cinanserin	Human platelets	(13)
Verapamil	Calf coronary artery	(14) Fig. 3.2

(1) Frenken and Kaumann 1984a; (2) Kaumann and Frenken 1985; (3) Kaumann 1985; (4) Frenken and Kaumann 1987a; (5) Kaumann 1988; (6) Frenken and Kaumann 1987b; (7) Frenken and Kaumann 1988; (8) Lemoine and Kaumann 1986; (9) Frenken in the author's laboratory (unpublished); (10) Frenken and Kaumann 1987d; (11) Teng and Lemoine in the author's laboratory (unpublished); (12) Black *et al.* 1983; (13) Bevan and Heptinstall 1986; (14) Sarah Slee in the author's laboratory (unpublished); (15) Blackburn *et al.* 1987

Table 3.2 Deactivators of $5\text{-}HT_2$ receptors in various tissues

Ligand	Tissue	Source
Tryptamine	Calf coronary artery	(1,2) Fig. 3.6
8-OH-DPAT	Calf coronary artery	(3)
Methysergide	Calf coronary artery	(3,4,5,6) Figs. 3.2, 3.4
	Calf pulmonary artery	(4,7)
	Rat tail artery	(8)
	Calf and guinea-pig trachea	(9)
Ritanserin	Calf coronary artery	(8)
	Rat tail artery	(8)
	Rat portal vein	(10)
LY53857	Calf coronary artery	Fig. 3.5
Cyproheptadine	Calf coronary artery	(11)
	Dog basilar artery	(11)
LSD	Calf coronary artery	Fig. 3.7
	Rabbit aorta	(12)
Ketanserin	Human platelets	(13)
I-HEAT	Calf coronary artery	(6)
Pizotifen	Calf coronary artery	(14)

(1) Frenken and Kaumann 1985b; (2) Frenken and Kaumann 1988c; (3) Frenken and Kaumann 1984b; (4) Kaumann and Frenken 1985; (5) Kaumann 1985; (6) Kaumann 1988; (7) Frenken and Kaumann 1987b; (8) Frenken and Kaumann 1987a; (9) Lemoine and Kaumann 1986; (10) Kaumann and Groszmann (unpublished); (11) Frenken in the author's laboratory (unpublished); (12) Black *et al.* 1983; (13) Bevan and Heptinstall 1986; (14) Kaumann (unpublished)

Table 3.3 Comparison of equilibrium dissociation constants K (−log, M) of activators for the R-state and for the allosteric site of the $5\text{-}HT_2$ receptor system of calf coronary artery

Activator ligand L	R state $-\log K = \log(EC_{50}$ ratio-1)-log[L] agonist	Allosteric site $-\log K = \log(IC_{50}$ ratio-1)-log[L] deactivator	Source
Ketanserin	9.5 5-HT	10.0 Methysergide	(1)
		10.0 Methysergide	(2)
		9.9 Methysergide	(3)
		9.4 Ritanserin	(3)
	9.9 Tryptamine	9.9 Tryptamine	(4) Fig. 3.6
	9.5 LSD	9.4* LSD	Fig. 3.7
ICI 169,369	9.1 5-HT	9.7 Methysergide	(5) Fig. 3.4
Phentolamine	7.4 5-HT	7.4 Methysergide	(6)

(1) Kaumann and Frenken 1985; (2) Kaumann 1985; (3) Frenken and Kaumann 1987a; (4) Frenken and Kaumann 1988; (5) Frenken and Kaumann 1987c; (6) Kaumann 1988 and additional experiments.
*K was calculated with IC_{30} ratio of LSD

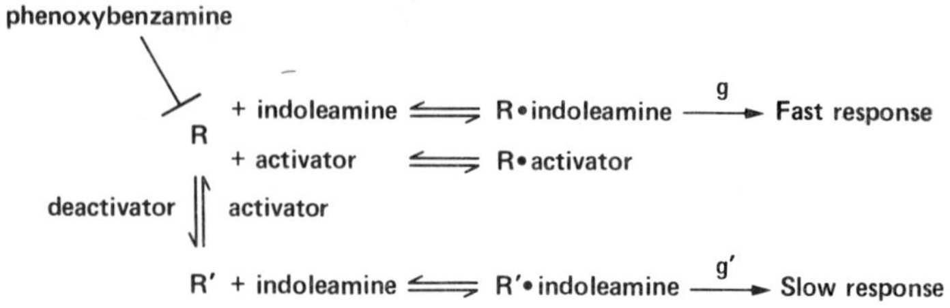

Fig. 3.3 Working scheme for the allosteric 5-HT_2 receptor system. The receptor interconverts between a high affinity state R and a low affinity state R′ for indoleamines. Through R indoleamines cause fast contractions and through R′ slow contractions. By competing for an allosteric site activators favour the R-state and deactivators the R′-state. Activators also compete with indoleamines for the receptor in the R-state, thereby antagonizing competitively fast concentrations. g and g′ are transfer functions relating receptor occupancy to effect

two-state model is attractive and a valuable working hypothesis for current and future research based on the discovery that 5-HT elicits interconvertible fast and slow contractions.

The two-state scheme of the 5-HT_2 receptor envisaged here is quite different from the two-state scheme proposed by Del Castillo and Katz (1957) for the nicotine receptor at the motor end-plate. Acetylcholine was assumed to bind first to an inactive state which then interconverts into an activatable, depolarizing, state. Katz and Thesleff (1957) also proposed that the activatable state had low affinity and an inactive (desensitized) state high affinity for acetylcholine. Low affinity and activatable states and a high affinity inactive state, have been verified with direct binding of cholinergic ligands (Sine and Taylor 1982). Unlike the nicotine receptor, both 5-HT_2 receptor states are activatable, and the more activatable state R is thought to have higher affinity than the less activatable state R′.

3.6 Activators and deactivators

Ligands that allosterically facilitate the R-state or R′-state of the 5-HT_2 receptor have been called activators or deactivators respectively (Kaumann 1987a). The characteristics of a deactivator are that it causes insurmountable antagonism of the effects of 5-HT and that insurmountability can be prevented or reversed by an activator. Activators often, but not always, cause surmountable antagonism of the effects of 5-HT. Some deactivators cause stimulant effects mediated through 5-HT_2 receptors. Activators are listed in Tables 3.1 and 3.3, deactivators in Table 3.2.

3.7 Activators as surmountable antagonists

A variety of activators shift concentration-effect curves for 5-HT towards higher concentrations in a parallel manner. Equilibrium dissociation constants for these ligands are independent of the concentration used to block the 5-HT-induced effects, which is equivalent to having Schild plots (Arunlakshana and Schild 1959) with slopes of one. Although this is consistent with unimolecular competition between the activator and 5-HT for one 5-HT_2 receptor, more complicated mechanisms also can account for a slope of unity of the Schild plot. These mechanisms include multiple binding sites with co-operative interactions (Colquhoun 1973; Thorn 1973) or receptor subtypes (Lemoine and Kaumann 1983).

Results with an activator, ICI 169,369 (see Table 3.1), are shown in Fig. 3.4. ICI 169,369 100 nM shifted in parallel manner the concentration-effect curve for the fast contractions of 5-HT towards about 100 times greater 5-HT concentrations. ICI 169,369 100 nM also completely prevented the

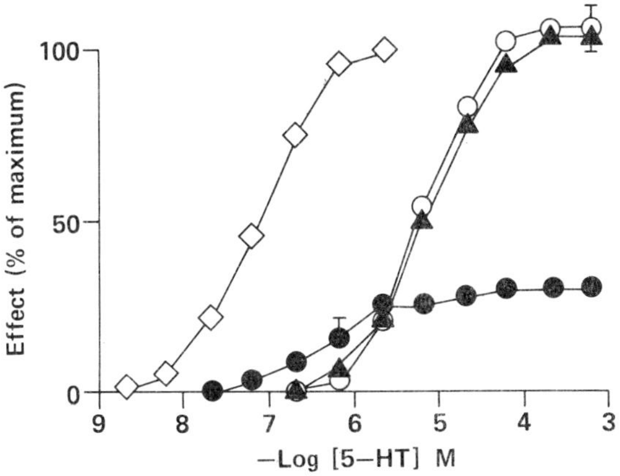

Fig. 3.4 Surmountable antagonism of the effects of 5-HT by ICI 169,369 and prevention by ICI 169,369 of the methysergide-induced depression of the maximum response to 5-HT. Two successive concentration–effect curves for 5-HT were determined on strips of coronary artery from two calves. Shown are the second curves either in the absence of drugs (◇ n = 4), in the presence of 20 nM methysergide (● n = 4), in the presence of 100 nM ICI 169,369 (▲), or in the presence of both 20 nM methysergide plus 100 nM ICI 169,369 (○ n = 4). Contractions induced by 5-HT in the presence of methysergide were fast up to 2 μM and slow with 20 μM and higher concentrations. All contractions induced by 5-HT in the absence of drugs and presence of ICI 169,369 plus methysergide were fast. Methysergide was incubated for 2h and ICI 169,369 for 3 hours. Symbols are mean ± SEM (SEM not shown if smaller than size of symbol)

depressant effect of methysergide 20 nM. The slow contractions observed with high 5-HT concentrations under the influence of methysergide were converted into fast concentrations in the presence of ICI 169,369.

Another remarkable feature of the interaction of activators and deactivators is that low concentrations of the latter do not modify the degree of surmountable blockade caused by the former. For example, the concentration–effect curve for 5-HT in the presence of ICI 169,369 100 nM is not modified by methysergide 20 nM (Fig. 3.4). Similar results have been found with methysergide in the presence of phentolamine (Kaumann 1987b) or ketanserin (Kaumann and Frenken 1985) in calf coronary artery and other tissues (Tables 3.1 and 3.3). This pattern does not preclude the possibility that methysergide may also compete with 5-HT for the receptor state (i.e. R-state) that causes fast contractions with 5-HT. The additional receptor occupancy produced by methysergide in the presence of an activator concentration that causes already high occupancy, would not result in a measurable increase in blockade. However, the methysergide concentrations that block 5-HT_2 receptors can be considerably higher ($-\log K_B = 8.25$, Leff and Martin 1986) than those that cause deactivation ($-\log IC_{50} = 9.80$, Kaumann and Frenken 1985), supporting the hypothesis for two different mechanisms.

One way of interpreting the opposite effects of activators and deactivators on the 5-HT-response pattern is to assume that they compete for a common site at or near the 5-HT_2 receptor system that allosterically modulates the $R \rightleftharpoons R'$ interconversion. The hypothesis for a common site received some support when it was shown that the concentration-effect curves of the deactivator, methysergide, as a depressant of the responses to 5-HT, could be shifted by the activator, ketanserin, towards higher methysergide concentrations (Kaumann and Frenken 1985; Kaumann 1985; Frenken and Kaumann 1987a). Thus, the protective effect of an activator can be surmounted by the deactivator. The ratio of equieffective depressant concentrations of the deactivator in the presence and absence of the activator has been used to obtain some sort of affinity estimate for the interaction of the activator with the hypothetical allosteric site. This affinity estimate is, of course, not unambiguous because activators also antagonize surmountably the effects of 5-HT and because residual contractions elicited by 5-HT in the presence of the deactivator can be fast at low 5-HT concentrations and slow at high 5-HT concentrations. The affinity estimate of ketanserin, ICI 169,369, and phentolamine for the allosteric site appears to be similar to the corresponding affinity estimates for the R-state, estimates from surmountable antagonism (Table 3.3). This similarity could mean that the hypothetical allosteric site is similar to the 5-HT receptor in the R-state, but caution is warranted with this interpretation because of the ambiguities mentioned above.

An allosteric site, to which methysergide binds with high affinity, may also modulate the receptor for 5-HT that mediates contraction of the stomach fundus of the rat. Frankhuizen and Bonta (1974) found that 1 nM methysergide caused insurmountable antagonism for the 5-HT-induced contractions of the fundus. The competitive surmountable antagonist piperoxan completely prevented the methysergide-induced depression of the responses to 5-HT. The receptors of the fundus have been classified as '5-HT_1-like' and appear to be different from 5-HT_2 receptors because they are not blocked by ketanserin (Bradley *et al.* 1986). The observations of Frankhuizen and Bonta (1974) suggest that the fundus 5-HT receptor is modulated allosterically in a similar way to the 5-HT_2 receptor. Furthermore, they suggested that the site of action of methysergide and piperoxan is the same. Because methysergide is as potent a deactivator in fundus as it is in 5-HT_2 receptor systems, the intriguing possiblity should be considered that the allosteric site has common features in the two 5-HT-receptor systems.

3.8 Activators as partially insurmountable antagonists

These are ligands that: (a) cause partially insurmountable antagonism; (b) block according to the Schild equation (Arunlakshana and Schild 1959) or equivalent equations; and (c) partially restore 5-HT-induced contractions depressed by a deactivator. Two cases have to be distinguished: first, a ligand such as ketanserin that can be both a weak deactivator and a strong activator; and second, a ligand such as verapamil that can depress the maximum response to 5-HT by a mechanism unrelated to an interaction with the 5-HT_2 receptor system.

Lemoine and Kaumann (1986) observed that even subnanomolar concentrations of ketanserin depressed by up to 40 per cent the maximum contractile response to 5-HT in smooth muscle of guinea-pig trachea. The dependence on the ketanserin concentration of the depression of the response to 5-HT could be fitted with the receptor occupancy curve by using the equilibrium dissociation constant K of ketanserin ($K = 0.3$ nM) estimated from the blockade of the effects of 5-HT (these experiments are also shown in Fig 2.3b, page 32). The experiments of Lemoine and Kaumann are interpreted as a slight shift of the 5-HT_2 receptors into the R′-state by ketanserin. Methysergide depressed the responses to 5-HT in guinea-pig tracheal muscle more than ketanserin, but ketanserin restored them up to 60 per cent of maximum. These results are consistent with a greater isomerization of R into R′ with methysergide than with ketanserin in guinea-pig tracheal muscle, and with some reversal from R′ to R when the two drugs were combined. Some activators may be more effective than ketanserin in causing a more complete isomerization of the receptor into the R-state, thereby restoring the response to 5-HT that had been depressed by ketanserin. It is relevant here to point out that in human platelets even ketanserin appears to be a strong deactivator (Bevan and Heptinstall 1986). Furthermore, methysergide appears to be an activator in platelets (Bevan and Heptinstall 1986) and perhaps also in rabbit aorta (Black *et al.* 1983). For reasons not yet understood, the direction and extent of the interconversion $R \rightleftharpoons R'$ by a drug appears to depend on tissue and species.

The different kinetics of 5-HT-induced contractions attributed to R (fast) and R′ (slow) suggested different mechanisms by which intracellular Ca^{2+} is increased. 5-HT 6 μM, a concentration causing fast contractions, can cause a transient Ca^{2+} influx in bovine coronary arteries (Ratz and Flaim 1984), perhaps through 5-HT_2 receptor regulated channels. Nothing is known about possible increases of intracellular Ca^{2+} during slow contractions induced by high 5-HT concentrations. We have started using verapamil, a Ca^{2+} channel blocking agent, as a putative tool to influence differentially fast and slow contractions (in collaboration with Sarah Slee). Interestingly, 10 μM verapamil abolished contractions evoked by low 5-HT concentrations, but converted slow contractions, caused by high 5-HT concentrations in the presence of methysergide, into fast contractions (Fig. 3.2). Although even a low concentration of verapamil (0.3 μM) was able to depress fast contractions it did not convert slow into fast contractions (not shown). These experiments suggest that verapamil has three actions—blockade of a 5-HT-sensitive Ca^{2+} channel at low concentrations, blockade of 5-HT_2 receptors, and isomerization of R′ into R at high concentrations. A dual action, consisting of blockade of both Ca^{2+}-channels and 5-HT_2 receptors, has already been proposed by Auguet *et al.* (1984) and analysed by Leff and Morse (1987). These latter authors estimated an equilibrium dissociation constant of 2 μM for the verapamil 5-HT_2 receptor complex in rabbit aorta. The relatively low affinity of verapamil for the 5-HT_2 receptor is in line with our observation that the drug is an activator at 10 μM but not at 0.3 μM.

3.9 Deactivators without stimulant effects

No stimulant effects have been observed with the deactivators ritanserin, cyproheptadine, I-HEAT, and LY53857. Methysergide does not contract tracheal muscle of calf and guinea-pig and only a marginal contraction of calf coronary artery has been observed with 10 μM methysergide (amounting to less than 4 per cent of the maximum response to 5-HT).

My experiments with LY53857 on calf coronary artery (Fig. 3.5) appear to be at variance with those reported by Cohen *et al.* (1983) on rat jugular vein. These authors reported an equilibrium dissociation constant of 0.05 nM for LY53857 but unfortunately did not mention whether the antagon-

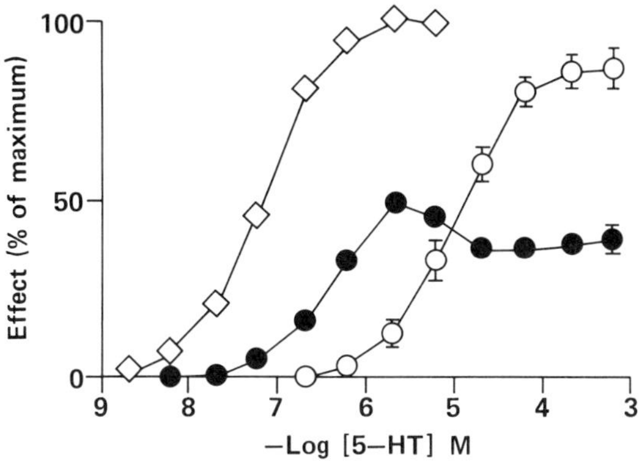

Fig. 3.5 Insurmountable antagonism by LY53857 and reversal by ketanserin of the effects of 5-HT. Two successive concentration–effect curves for 5-HT were determined on strips of coronary artery from one calf. Shown are the second curves either in the absence of drugs (◇ n = 4), in the presence of 2 nM LY 53857 (● n = 4), or in the presence of both 2 nM LY53857 plus 100 nM ketanserin (○ n = 4). Contractions induced by 5-HT in the presence of LY53857 were fast up to 6 μM and slow with 60, 200 and 600 μM. All contractions induced by 5-HT in the absence of drugs and presence of ketanserin plus LY53857 were fast. LY53857 was incubated for 3 hours and ketanserin for 3.5 hours. Symbols represent mean ± SEM (SEM not shown if smaller than size of symbol)

ism of the effects of 5-HT was surmountable or insurmountable in jugular vein. On the other hand, Cohen *et al.* (1983) showed that LY53857 caused insurmountable antagonism of the pressor effects of 5-HT in the pithed rat, which agrees with my own experiments with 2 nM (Fig. 3.5) and other concentrations (0.3–100 nM results not shown) of LY53857 that also caused insurmountable antagonism in calf coronary artery. LY53857 resembles both methysergide and LSD (see Fig. 3.7) in inducing slow contractions with high 5-HT concentrations that can be converted to fast contractions with ketanserin. The similarity of the interactions of these three drugs with 5-HT and ketanserin on the 5-HT_2 receptor system is not surprising in view of their close chemical kinship.

3.10 Deactivators with stimulant effects

Methysergide can cause marked contractions of canine basilar arteries which has led Frenken (see Table 3.2) to use cyproheptadine as a deactivator devoid of stimulant effects in this tissue. The depression of 5-HT-induced contractions by cyproheptadine in canine basilar artery is partially prevented by ketanserin, also suggesting an $R \rightleftharpoons R'$ interconversion for the 5-HT_2 receptors of this artery.

The deactivators, tryptamine and LSD, can, however, be used as tools to study the allosteric 5-HT_2 receptor system because the stimulant effects of low concentrations of the former fade (Fig. 3.6) and those of the latter never exceed 20 per cent of maximum response to 5-HT (Fig. 3.7).

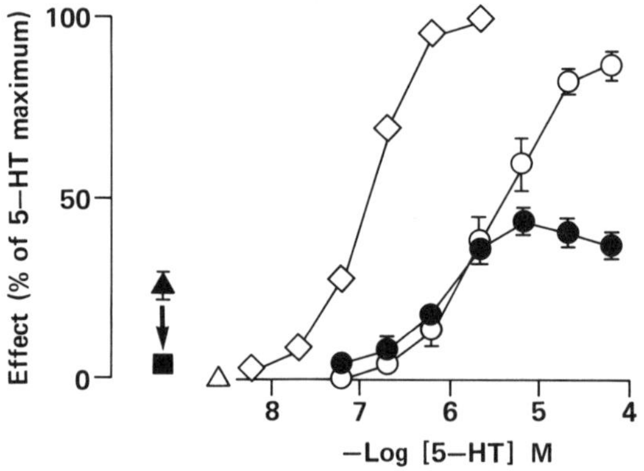

Fig. 3.6 Insurmountable antagonism by tryptamine and reversal by ketanserin of the effects of 5-HT. To inhibit monoamine oxidase irreversibly, the tissues were pretreated for 30 min with 100 μM pargyline followed by washout. Two successive concentration–effect curves for 5-HT were determined on strips of coronary artery from two calves. Shown are second curves either in the absence of drugs (◇ n = 4), in the presence of 1 μM tryptamine (● n = 4), or in the presence of both 1 μM tryptamine plus 10 nM ketanserin (○ n = 4). After the washout of 5-HT of the first curve followed by recovery of basal tone, the administration of 1 μM tryptamine caused a contraction (▲) which faded 2 hours later (■). Ketanserin 10 nM incubated 1 hour before the administration of tryptamine prevented its contraction (△). Symbols are mean ± SEM (SEM not shown if smaller than size of symbol)

The contractions caused by low tryptamine concentrations are blocked by ketanserin (Fig. 3.6) with an affinity expected for the 5-HT_2 receptor (Frenken and Kaumann 1988). The transient effects of tryptamine followed by fade are interpreted as occurring initially through R which slowly isomerizes into R′. Tryptamine also causes insurmountable blockade of the effects of 5-HT. The desensitization of the 5-HT response has previously been observed by Gaddum (1953b) who proposed that the two indoleamines interacted with a common receptor. Ketanserin antagonized the depressant effects of tryptamine on 5-HT contractions with the same potency as its affinity for 5-HT_2 receptors (Table 3.3).

Finally, high concentrations of tryptamine (mM) can also cause slow contractions which are facilitated by methysergide (Frenken and Kaumann

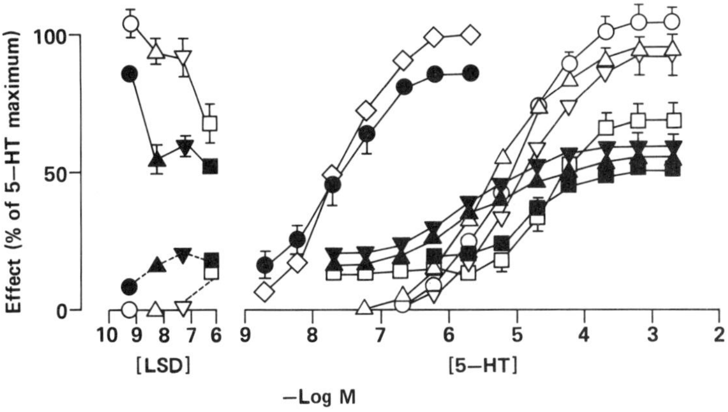

Fig. 3.7 Interactions of LSD, 5-HT, and ketanserin with the 5-HT_2 receptor system. Results are from coronary arteries of four calves. The LSD concentration used were (nM): 0.5 ●○, 5 ▲△, 50 ▼▽ and 500 ■□. Closed and open symbols represent concentrations of LSD in the absence and presence of 100 nM ketanserin, respectively. Two successive concentration–effect curves for 5-HT were determined, the first in the absence and the second in the presence of an LSD concentration or LSD + ketanserin. LSD was incubated for 2 hours and ketanserin for 2.5 hours before the second curve for 5-HT was begun. The right panel depicts second concentration–effect curves for 5-HT in the absence of drugs (◇, n = 36 strips) or presence of drugs (n = 4 strips for each curve). The left lower panel shows curves for the stimulant effects of LSD (broken lines). The left upper panel represents maximum responses to 5-HT, depressed by LSD, and the prevention of depression by ketanserin. Symbols represent mean ± SEM (SEM not shown if smaller than size of symbol)

1988), presumably by a synergistic interaction favouring the R′-state. The slow contractions of tryptamine resemble those of 5-HT and it is plausible that they are due to a low-affinity complex of the receptor in the R′-state.

The pattern of the effects of LSD (Fig. 3.7) is similar to that of tryptamine. Ketanserin antagonized with similar potency the stimulant effects of LSD, the depressant effects of LSD on the maximum response to 5-HT, and the fast contractions elicited by 5-HT (Fig. 3.7; Table 3.3). It therefore appears likely that the stimulant effects of LSD are evoked through the R-state of the receptor, while the depression of the 5-HT response is caused through an allosteric site; ketanserin appears to compete for both loci.

3.11 On endogenous allosteric modulators

When 5-HT, released from aggregating platelets, contracts an artery, an exogenous or endogenous allosteric modulator could, depending on the prevalence of the R or R′ state of the 5-HT_2 receptor, facilitate or reduce 5-HT-induced contractions. For example, it has recently been found that in calf coronary artery, the activator ketanserin can restore contractions (induced by 5-HT released from platelets) that had been depressed by the deactivators tryptamine or methysergide (Kaumann *et al.* 1987). An endogenous allosteric modulator should be active at concentrations compatible with its endogenous concentration. Tryptamine has been detected in brain cells (Juorio *et al.* 1987) at concentrations ($\simeq$ 100 nM) slightly lower than those that allosterically depress 5-HT responses in calf coronary artery (Fig. 3.6) (Frenken and Kaumann 1988). Nevertheless, higher than 100 nM concentrations could conceivably occur in the neighbourhood of the 5-HT_2 receptor system because the turnover of tryptamine is high (Warsh *et al.* 1979). Another candidate for an endogenous modulator has recently been reported by Barbaccia and Costa (1986), who extracted a heat-stable factor from brain which could be inactivated with pronase and that selectively inhibited the binding of ^{3}H-ketanserin to membranes of rat brain. Because it can be inferred from the binding features of [^{3}H]ketanserin that it may also label an allosteric site, the factor of Barbaccia and Costa may well be an endogenous activator or deactivator of the allosteric 5-HT_2 receptor system.

3.12 Biochemical considerations

The proposed two state scheme and its implications require biochemical verification. Unlike successful radiolabelling of brain 5-HT_2 receptors (Peroutka and Snyder 1979; Leysen *et al.* 1982), attempts to radiolabel 5-HT_2 receptors on smooth muscle have not been successful, perhaps because the density is low and because the radioligands available are not appropriate. When adequate binding conditions have been found, experiments should be designed to test the mode of action of activators and deactivators outlined in this chapter. Allosteric 5-HT_2 receptors would be those in which the binding of 5-HT would be affected by binding of activators or deactivators on sites distinct from the 5-HT binding sites. It should then be possible to measure the two interconvertible states $R \rightleftharpoons R'$, to estimate the postulated low affinity of 5-HT for R′ and to confirm the affinity of 5-HT for R (estimated in intact tissues, $K_{5\text{-HT}} \simeq 0.2\ \mu M$). The relative amounts of R and R′ should also be measured under several conditions. It would be expected that deactivators would merely enhance the allosteric constant $L = R'/R$, while activators would decrease it. (L is analogous to the equilibrium constant for the transition between two states of an enzyme protein—see Monod *et al.* 1965.) The assumption that 5-HT binds to both states $R \rightleftharpoons R'$ should be tested with deactivators and activators. In the presence of a deactivator, there would come a point where there would be so much R′-state present that as much 5-HT binds to it as to the small amount of receptors left in the R-state. When 5-HT binds to the R-state to form R.5-HT, the concentration of free R-state would be decreased and some receptors in the R′-state would be converted into the R-state to maintain equilibrium. The equilibrium would be shifted towards R by an activator and towards R′ by a deactivator.

Evidence in rat brain obtained by Leysen *et al.* (1982) is consistent with the interpretation that the activator ketanserin binds to more than one class of sites of the 5-HT_2 receptor system (Kaumann 1985). Leysen *et al.* (1982) observed that nanomolar concentrations of unlabelled ketanserin removed more [^{3}H]ketanserin from its binding sites than the deactivator methysergide. [^{3}H]Ketanserin therefore labels a site for which methysergide competes, and this could be the allosteric site for which both activators and deactivators may compete. In addition, [^{3}H]ketanserin also labels a site for which 5-HT, but not low methysergide concentrations ($\simeq$ 1 nM), may compete. Although [^{3}H]ketanserin can also label other receptors (e.g. α_1-adrenoceptors—Hoyer *et al.* 1987), its affinity for α_1-adrenoceptors is at least 20 times lower than for 5-HT_2 receptors (Leysen *et al.* 1981—binding to rat brain membranes; Frenken and Kaumann 1987b—antagonism of the effects of noradrenaline in calf pulmonary artery), and it is unlikely that [^{3}H]ketanserin labelled many α_1-adrenoceptors in the study of Leysen *et al.* (1982) because the density of α_1-adrenoceptors is relatively low in rat frontal cortex (Hoyer *et al.* 1987). It is, therefore, tempting to speculate that the methysergide-resistant binding of ketanserin could be to the 5-HT_2 receptor.

Binding experiments at different temperatures could perhaps give a clue as to whether a simple isomerization into R′ by deactivators or reisomerization into R by activators can merely be explained by a conformational receptor change (small temperature dependence) or by a more complicated biochemical modification of the receptor (large temperature dependence). Such experiments would probably only be meaningful with a 5-HT_2 receptor system in the test tube because changes in temperature in intact tissues would also cause modifications of fast and slow contractions due to receptor unrelated processes.

Chronic treatment of rats with drugs such as ritanserin (Leysen *et al.* 1985) causes a decrease (downregulation) of brain 5-HT_2 receptors. Because drugs that cause downregulation of 5-HT_2 receptors include several that often have deactivator properties—e.g. ritanserin, pizotifen, cyproheptadine (see Leysen *et al.* 1985 for further literature)—it is tempting to suggest that interconversion into the R′-state may be one mechanism that could account for the decrease in 5-HT_2 receptors accessible to radioligands.

The products of the phospholipase C mediated breakdown of phosphatidyl inositol 4,5-biphosphate, inositol 1,4,5 triphosphate, and 1,2-diacylglycerol (Berridge 1984), could be involved in the generation of fast and slow contractions elicited by 5-HT. Inositol 1,4,5 triphosphate has been found to enhance intracellular Ca^{2+} and cause contraction in vascular smooth muscle (Somlyo *et al.* 1985). Of considerable interest is the fact that Nakaki *et al.* (1985) have observed that in rat aorta 5-HT can cause phasic and tonic contractions. The phasic contractions, but not the tonic contractions, were abolished by Ca^{2+} channel blockade with nifedipine. On the other hand, the tonic contractions were inhibited by an inhibitor of phospholipase which also prevented phosphoinositide hydrolysis. Tonic contractions were mimicked with phorbol esters which, like 1,2-diacylglycerol, activate protein kinase C. Smooth muscle cells derived from rat aorta have 5-HT_2 receptors that mediate an increase in inositol phosphates and mobilize calcium (Doyle *et al.* 1986). It is possible that this effect plays a role in the fast contractions. Preliminary observations by Lemoine *et al.* (1988), in the author's laboratory, suggest that 5-HT increases inositol phosphates through the R-state but not through the R′-state of the 5-HT_2 receptor in tracheal smooth muscle. The phasic and tonic contractions in rat aorta resemble the fast and slow contractions caused by 5-HT in calf coronary artery and other tissues, suggesting the involvement of similar biochemical mechanisms. Future experiments will tell.

3.13 Concluding remarks

For several smooth muscle systems pharmacological evidence suggests that the 5-HT_2 receptor exists in two interconvertible states $R \rightleftharpoons R'$. Indolamines appear to have high affinity for R and low affinity for R′. 5-HT causes fast contractions through R and slow contractions through R′. By presumably acting through an allosteric site, activators shift the equilibrium towards R and deactivators towards R′. Activators usually cause surmountable antagonism of the effects of indoleamines.

Deactivators cause insurmountable but reversible antagonism. The receptor in the R-state, but not in the R′-state, can be irreversibly blocked by phenoxybenzamine. Interconversion of the receptor into the R′-state by a deactivator prevents the blockade by phenoxybenzamine. Some deactivators (tryptamine, LSD) cause effects through the R-state but also isomerize the receptor into the R′-state. The structure of the deactivators LSD, LY53857, and methysergide contains tryptamine which is itself a deactivator, suggesting that the isomerization of 5-HT_2 receptor into the R′-state occurs preferentially with indoleamines. The 5-HT_2 receptor system could be regulated by endogenous allosteric modulators.

Acknowledgements

Of the observations discussed in this chapter, those which originated in my laboratory owe much to my colleagues, whose names appear in the text. I wish to thank Sheila Hancox and Larraine Horwood for typing this manuscript.

References

Arunlakshana, O. and Schild, H. O. (1959). Some quantitative uses of drug agonists. *British Journal of Pharmacology and Chemotherapy* **14,** 48–58.

Auguet, M., De Feudis, F. V., and Clostre, F. (1984). Competitive inhibition of 5-HT receptors in rabbit isolated aorta by the Ca^{2+}-antagonist methoxyverpamil (D600). *Neurochemistry International* **6,** 701–10.

Barbaccia, M. L. and Costa, E. (1986). Endogenous ligands for ^{3}H-imipramine and ^{3}H-ketanserin recognition sites. *Clinical Neuropharmacology* **9,** Suppl 4, 223–5.

Barrett, V. J., Leff, P., Martin, G. R., and Richardson, P. J. (1986). Pharmacological analysis of the interaction between Bay K 8644 and 5-HT in rabbit aorta. *British Journal of Pharmacology* **87,** 487–94.

Berridge, M. J. (1984). Inositol triphosphate and diacylglycerol as second messengers. *Biochemistry Journal* **212,** 849–58.

Bevan, J. and Heptinstall, S. (1986). Effects of combinations of 5-hydroxytryptamine receptor antagonists on 5-HT-induced human platelet aggregation. *Naunyn-Schmiedeberg's Archives of Pharmacology* **334,** 341–5.

Black, J. W., Brazenor, R. M., Gerskowitch, V. P., and Leff, P. (1983). The problem of insurmountable antagonism in 5-hydroxytryptamine receptor classification. *British Journal of Pharmacology* **80,** 607P.

Blackburn, T. P., Thornber, C. W., Pearce, R. J., and Cox, B. (1988). In vitro studies with ICI 169, 369, a chemically novel 5-HT antagonist. *European Journal of Pharmacology* **150,** 247–56.

Bradley, P. B., Engel, G., Feniuk, W., Fozard, J. R., Humphrey, P. P. A., Middlemiss, D. N., Mylecharane, E. J., Richardson, B. P., and Saxena, P. R. (1986). Proposals for the classification of nomenclature of functional receptors for 5-hydroxytryptamine. *Neuropharmacology* **25,** 563–76.

Brazenor, R. M. and Angus, J. A. (1982). Actions of serotonin antagonists on dog coronary artery. *European Journal of Pharmacology* **81,** 569–76.

Clancy, M. and Maayani, S. (1985). 5-Hydroxytryptamine receptor in isolated rabbit aorta. Characterization with tryptamine analogs. *Journal of Pharmacology and Experimental Therapeutics* **233,** 761–9.

Cohen, M. L. and Wittenauer, L. A. (1985). Relationship between serotonin and tryptamine receptors in the rat stomach fundus. *Journal of Pharmacology and Experimental Therapeutics* **232,** 75–9.

Cohen, M. L., Fuller, R. W., and Kurz, K. D. (1983). LY53857, a selective and potent serotonergic ($5\text{-}HT_2$) receptor antagonist does not lower blood pressure in the spontaneously hypertensive rat. *Journal of Pharmacology and Experimental Therapeutics* **227,** 327–32.

Cohen, M. L., Schenck, K. W., Colbert, W., and Wittenauer, L. (1985). Role of 5-HT$_2$ receptors in serotonin-induced contractions of nonvascular smooth muscle. *Journal of Pharmacology and Experimental Therapeutics* **232,** 770–4.

Colquhoun, D. (1973). The relation between classical and cooperative models of drug action. In *Drug receptors* (ed. H. P. Rang) pp. 149–82. Macmillan, London.

Del Castillo, J. and Katz, B. (1957). Interaction at end-plate receptors between different choline derivatives. *Proceedings of the Royal Society London (Biology)* **146,** 369–81.

Doyle, V. M., Creba, J. A., Ruegg, U. T., and Hoyer, D. (1986). Serotonin increases the production of inositol phosphates and mobilises calcium via the 5-HT$_2$ receptor in A_7r_5 smooth muscle cells. *Naunyn-Schmiedeberg's Archives of Pharmacology* **333,** 98–103.

Feniuk, W. (1984). An analysis of 5-hydroxytryptamine receptors mediating contraction of isolated smooth muscle. *Neuropharmacology* **23,** 1467–72.

Frankhuizen, A. and Bonta, I. L. (1974). Receptors involved in the action of 5-HT and tryptamine on the isolated rat stomach fundus preparation. *European Journal of Pharmacology* **26,** 220–30.

Frenken, M. and Kaumann, A. J. (1984a). Interaction of ketanserin and its metabolite ketanserinol with 5-HT$_2$ receptors in pulmonary and coronary arteries in calf. *Naunyn-Schmiedeberg's Archives of Pharmacology* **326,** 334–9.

Frenken, M. and Kaumann, A. J. (1984b). Interconversion between two 5-HT$_2$ receptor states in large coronary artery in calf. *Proceedings of the 9th International Congress of Pharmacology.* London.

Frenken, M. and Kaumann, A. J. (1985a). Ketanserin and surmountable antagonism of 5-hydroxytryptamine-induced contraction on large coronary arteries of dog. *Naunyn-Schmiedeberg's Archives of Pharmacology* **328,** 301–3.

Frenken, M. and Kaumann, A. J. (1985b). Tryptamine interacts with both the 5-HT$_2$ receptor and an allosteric site in bovine large coronary artery. *Pflügers Archives* **403,** R32.

Frenken, M. and Kaumann, A. J. (1986). Surmountable antagonism by ketanserin of 5-hydroxytryptamine-induced contractions in dog basilar artery. *British Journal of Pharmacology* **89,** 550P.

Frenken, M. and Kaumann, A. J. (1987a). Allosteric properties of the 5-HT$_2$ receptor system of the rat tail artery. Ritanserin and methysergide are not competitive 5-HT$_2$ receptor antagonists but allosteric modulators. *Naunyn-Schmiedeberg's Archives of Pharmacology* **335,** 359–66.

Frenken, M. and Kaumann, A. J. (1987b). Interconversion into a low active state protects vascular 5-HT$_2$ receptors against irreversible antagonism by phenoxybenzamine. *Naunyn-Schmiedeberg's Archives of Pharmacology* **335,** 481–90.

Frenken, M. and Kaumann, A. J. (1988). Effects of tryptamine mediated through 2 states of the 5-HT$_2$ receptor in calf coronary artery. *Naunyn-Schmiedeberg's Archives of Pharmacology* **337,** 484–92.

Frenken, M. and Kaumann, A. J. (1987d). ICI 169,369 is both a competitive antagonist and an allosteric modulator of arterial 5-HT$_2$ receptors. *British Journal of Pharmacology* **91,** 350P.

Gaddum, J. H. (1953a). Antagonism between lysergic acid diethylamide and 5-hydroxytryptamine. *Journal of Physiology* **121,** 15P.

Gaddum, J. H. (1953b). Tryptamine receptors. *Journal of Physiology (London)* **119,** 363–8.

Gaddum, J. H. and Hameed, K. A. (1954). Drugs which antagonize 5-hydroxytryptamine. *British Journal of Pharmacology* **9,** 240–8.

Gaddum, J. H. and Picarelli, Z. P. (1957). Two types of tryptamine receptors. *British Journal of Pharmacology* **12,** 323–8.

Gaddum, J. H., Hameed, K. A., Hathway, D. E., and Stephens, F. F. (1955). Quantitative studies of antagonists for 5-hydroxytryptamine. *Quarterly Journal of Experimental Physiology* **40,** 49–74.

Heuring, R. A. and Peroutka, A. J. (1987). Characterization of a novel ^{3}H-5-hydroxytryptamine binding site in bovine brain membranes. *Journal of Neuroscience* **7,** 894–903.

Hoyer, D., Vos, P., Closse, A., Pazos, A., Palacios, S. M., and Davies, H. (1987). [^{3}H]Ketanserin lavels 5-HT_2 receptors and α_1-adrenoceptors in human and pig brain membranes. *Naunyn-Schmiedeberg's Archives of Pharmacology* **335,** 226–30.

Humphrey, P. P. A. (1984). Peripheral 5-hydroxytryptamine receptors and their classification. *Neuropharmacology* **23,** 1503–10.

Juorio, A. V., Greenshaw, A. J., and Nguyen, T. V. (1987). Effect of intranigral administration of 6-hydroxytryptamine and 5,7-dihydroxytryptamine on rat brain tryptamine. *Journal of Neurochemistry* **48,** 1346–50.

Katz, B. and Thesleff, S. (1957). A study of the desensitization produced by acetylcholine at the motor end-plate. *Journal of Physiology* **138,** 63–80.

Kaumann, A. J. (1983). Yohimbine and rauwolscine inhibit 5-hydroxytryptamine-induced contraction of large coronary arteries in calf through blockade of 5-HT_2 receptors. *Naunyn-Schmiedeberg's Archives of Pharmacology* **323,** 149–54.

Kaumann, A. J. (1985). Multiple receptor subtypes and responses in autonomic effectors. In *Trends in autonomic pharmacology* (ed. S. Kalsner) vol. 3, pp. 51–60. Urban and Schwarzenberg, Baltimore.

Kaumann, A. J. (1987a). A two-state model for the 5-HT_2 receptor. *Hoppe-Seylers Zeitschrift für Physiologische Chemie* **368,** 1131–2.

Kaumann, A. J. (1988). A two-state model for the 5-HT_2 receptor. Effects of α-adrenoceptor ligands. *Journal of Cardiovascular Pharmacology* 11 (Suppl. 1) S88–92.

Kaumann, A. J. and Frenken, M. (1985). A paradox: the 5-HT_2 receptor antagonist ketanserin restores the 5-HT-induced contraction depressed by methysergide in large coronary arteries of calf. Allosteric regulation of 5-HT_2 receptors. *Naunyn-Schmiedeberg's Archives of Pharmacology* **328,** 295–300.

Kaumann, A. J., Lemoine, H., and Frenken, M. (1987). Ketanserin restores tryptamine-depressed contractions elicited by 5-hydroxytryptamine (5-HT) released from platelets in calf coronary arteries. *Naunyn-Schmiedeberg's Archives of Pharmacology* **335,** R89.

Koshland, D. E. (1970). A molecular model for the regulatory behaviour of enzymes. *Harvey Lectures*, **64,** 33–57.

Leff, P. and Martin, F. R. (1986). Peripheral 5-HT_2-like receptors. Can they be classified with available antagonists? *British Journal of Pharmacology* **88,** 585–93.

Leff, P. and Morse, J. M. (1987). Resultant pharmacological actions of verapamil.

Quantification of competitive 5-hydroxytryptamine antagonism in combination with calcium antagonism. *Journal of Pharmacology and Experimental Therapeutics* **240,** 284–7.

Leff, P., Martin, G. R., and Morse, J. M. (1986). The classification of peripheral 5-HT_2-like receptors using tryptamine agonist and antagonist analogues. *British Journal of Pharmacology* **89,** 493–9.

Lemoine, H. and Kaumann, A. J. (1983). A model for the interaction of competitive antagonists with two receptor-subtypes characterized by a Schild plot with apparent slope unity. Agonist-dependent enantiomeric affinity ratios for bupranolol in tracheal but not right atria of guinea-pig. *Naunyn-Schmiedeberg's Archives of Pharmacology* **322,** 111–20.

Lemoine, H. and Kaumann, A. J. (1986). Allosteric properties of 5-HT_2 receptors in tracheal smooth muscle. *Naunyn-Schmiedeberg's Archives of Pharmacology* **333,** 91–7.

Lemoine, H., Pohl, V., and Teng, K. J. (1988). Serotonin (5-HT) stimulates phosphotidyl inositol (Pi) hydrolysis only through the R-state of allosterically regulated 5-HT_2 receptors in calf tracheal smooth muscle. *Naunyn-Schmiedeberg's Archives of Pharmacology* **337,** R 103.

Leysen, J. E. and Tollenaere, J. P. (1982). Biochemical models for serotonin receptors. *Annual Reports of Medicinal Chemistry* **17,** 1–10.

Leysen, J. E., Awouters, F., Kennis, L., Laduron, P. M., Vandenberk, J., and Janssen, P. A. J. (1981). Receptor binding profile of R41468, a novel antagonist at 5-HT_2-receptors. *Life Sciences* **28,** 1015–22.

Leysen, J. E., Niemegeers, J. E., van Nueten, J. M., and Laduron, P. M. (1982). [^{3}H]Ketanserin (R41468), a selective ^{3}H ligand for serotonin$_2$ receptor binding sites. Binding properties, brain distribution, and functional role. *Molecular Pharmacology* **21,** 301–14.

Leysen, J. E., Van Gompel, P., Gommeren, W., Woestenborghs, R., and Janssen, P. A. J. (1985). Down regulation of serotonin-S_2 receptor sites in rat brain by chronic treatment with the serotonin-S_2 antagonists: ritanserin and setoperone. *Psychopharmacology* **88,** 434–44.

Maayani, S., Wilkinson, C. W., and Stollak, J. S. (1984). 5-Hydroxytryptamine receptor in rabbit aorta: Characterization by butyrophenone analogs. *Journal of Pharmacology and Experimental Therapeutics* **229,** 345–50.

Monod, J., Wyman, J., and Changeaux, J-P. (1965). On the nature of allosteric transitions: A plausible model. *Journal of Molecular Biology* **12,** 88–118.

Müller-Schweinitzer, E. and Engel, G. (1983). Evidence for mediation by 5-HT_2 receptors of 5-hydroxytryptamine-induced contraction of canine basilar artery. *Naunyn-Schmiedeberg's Archives of Pharmacology* **324,** 287–92.

Müller-Schweinitzer, E. and Weidmann, H. (1978). Basic pharmacological properties. In *Ergot alkaloids and related compounds* (ed. B. Berde and H. O. Schild) *Heffter-Heubner's Handbook of Experimental Pharmacology* vol. 49, pp. 87–232, Springer-Verlag, Heidelberg.

Nakaki, T., Roth, B. L., Chuang, D. M., and Costa, E. (1985). Phasic and tonic components in 5-HT_2 receptor-mediated rat aorta contraction. Participation of Ca^{++} channels and phospholipase C. *Journal of Pharmacology and Experimental Therapeutics* **234,** 442–6.

Peroutka, S. J. (1984). Vascular serotonin receptors. Correlation with 5-HT_1 and 5-HT_2 binding sites. *Biochemical Pharmacology* **33,** 2349–53.

Peroutka, S. J. and Snyder, S. H. (1979). Multiple serotonin receptors: Differential binding of [^{3}H]-5-hydroxytryptamine, [^{3}H] lysergic acid diethylamide and [^{3}H]-spiroperidol. *Molecular Pharmacology* **16,** 687–99.

Peroutka, S. J., Huang, S., and Allen, G. S. (1986). Canine basilar artery contractions mediated by 5-hydroxytryptamine$_{1A}$ receptors. *Journal of Pharmacology and Experimental Therapeutics* **237,** 901–6.

Ratz, P. H. and Flaim, S. F. (1984). Mechanism of 5-HT contraction in isolated bovine ventricular coronary arteries. Evidence for transient receptor-operated calcium influx channels. *Circulation Research* **54,** 135–43.

Sine, S. M. and Taylor, P. (1982). Local anaesthetics and histrionicotoxin are allosteric inhibitors of the acetylcholine receptor. Studies of clonal muscle cells. *Journal of Biological Chemistry* **257,** 8106–14.

Somlyo, A. V., Bond, M., and Somlyo, A. P. (1985). Inositol triphosphate induced calcium release and contraction in vascular smooth muscle. *Proceedings of the National Academy of Sciences USA* **82,** 5231–5.

Thorn, C. D. (1973). On the analysis of pharmacological experiments in terms of an allosteric receptor model. *Molecular Pharmacology* **9,** 1–9.

Warsh, J. J., Coscina, D. R., Godse, D. D., and Chan, P. W. (1979). Dependence of brain tryptamine formation on tryptophan availability. *Journal of Neurochemistry* **32,** 1191–6.

4

5-Hydroxytryptamine receptors and effector coupling mechanisms in peripheral tissues

D. HOYER

4.1 Introduction

Progress in the 5-HT research field has been hampered for several reasons: until very recently there has been a lack of agonists and antagonists selective for the various 5-HT receptor subtypes and therefore their identification has been difficult; the existence of different nomenclatures based on pharmacological, electrophysiological and biochemical studies has led to problems of communication between the different research communities; the apparent lack of correlation beween pharmacological, electrophysiological and biochemical studies rendered 5-HT research complicated and even confusing. Nevertheless, two systems of 5-HT receptor nomenclatures have emerged, the 5-HT D and 5-HT M receptor classification of Gaddum and Picarelli (1957), based on functional studies, and the 5-HT_1, 5-HT_2 classification of Peroutka and Snyder (1979), based on radioligand binding studies. Bradley *et al.* (1986), in a pragmatic approach, proposed the reconciliation of these two nomenclatures, and suggested that 5-HT receptors could be classified into three groups designated '5-HT_1-like' 5-HT_2 and 5-HT_3 (see also Chapter 5).

To provide a framework upon which a discussion of the second messengers linked to 5-HT receptors can rest, this chapter starts with a review of the different 5-HT recognition sites which have been detected in radioligand binding studies. In keeping with the overall brief, the general emphasis will, whenever possible, be on data obtained in peripheral tissues. However, it should be emphasized that relatively few biochemical studies using selective compounds have been performed in peripheral tissues and it will therefore be necessary to refer to results obtained in the brain.

4.2 Radioligand binding studies

4.2.1 5-HT_1 recognition sites

By definition [^{3}H]5-HT labels all 5-HT_1 recognition sites. At the present time, four distinct 5-HT$_1$ sites have been identified. 5-HT$_{1A}$ and 5-HT$_{1B}$ sites were defined by Pedigo *et al.* (1981) as those [^{3}H]5-HT sites showing, respectively, high and low affinity for spiperone. It was soon recognized that the sites with low affinity for spiperone were not homogeneous (Schnellman *et al.* 1984); Palacios and his collaborators (Cortes *et al.* 1984; Pazos *et al.* 1984b) subsequently identified 5-HT$_{1C}$ receptors which are found in particularly high concentrations in the choroid plexus of rat, pig, and mouse. More recently, the 5-HT$_{1D}$ recognition site, which has a clearly different pharmacology from 5-HT$_{1A}$, 5-HT$_{1B}$ or 5-HT$_{1C}$ sites, has been described in bovine, porcine, and human brain (Heuring and Peroutka 1987; Hoyer *et al.* 1988). For a recent detailed review of 5-HT recognition sites and their ligands see Peroutka (1988).

Table 4.1 lists the affinity values of various drugs for 5-HT$_{1A}$, 5-HT$_{1B}$, 5-HT$_{1C}$, and 5-HT$_{1D}$ sites. [^{3}H]8-OH-DPAT was the first radioligand which showed selectivity for 5-HT$_{1A}$ sites (Middlemiss and Fozard 1983; Gozlan *et al.* 1983; Hoyer *et al.* 1985b), followed by [^{3}H]ipsapirone (Dompert *et al.* 1985) and [^{3}H]PAPP (Ransom *et al.* 1986). Other radioligands which also label 5HT$_{1A}$ sites, but are not selective, include [^{3}H]WB 4101, [^{3}H]LSD, [^{3}H]lisuride, and [^{3}H]spiroxatrine. Selective agonists at these sites include 8-OH-DPAT, dipropyl-5-carboxamidotryptamine, ipsapirone, buspirone, PAPP, and MDL 72832, whereas spiperone is a non-selective antagonist (Table 4.1). 5-HT$_{1B}$ recognition sites may be identified using [^{125}I]iodocyanopindolol, [^{3}H]5-HT, or [^{3}H]dihydroergotamine (Hoyer *et al.* 1985a; Peroutka 1986; Hamblin *et al.* 1987). There is no selective ligand for 5-HT$_{1B}$ receptors, although they show some preferential affinity for indole β-adrenoceptor antagonists (i.e. 21-009, cyanopindolol, pindolol) and for the agonists RU 24969 or CGS 12066B (Table 4.1). 5-HT$_{1C}$ receptors were originally identified using [^{3}H]5-HT and [^{3}H]mesulergine (Pazos *et al.* 1984b; Hoyer *et al.* 1985b). [^{3}H]LSD, [^{125}I]LSD and 1-methyl-[^{125}I]LSD also label 5-HT$_{1C}$ sites in addition to 5-HT$_2$ receptors. There is at present no selective agonist at these sites; mesulergine, mianserin, and LY53857 are potent antagonists (Table 4.1). 5-HT$_{1D}$ sites have now been characterized using [^{3}H]5-HT in bovine, pig, and human brain (Heuring and Peroutka 1987; Hoyer *et al.* 1988; Waeber *et al.* 1988a and b). These sites show high affinity for 5-HT, 5-carboxamidotryptamine, 5-methoxytryptamine, rauwolscine, yohimbine, and some ergolines, but low affinity to compounds which display high affinity and/or selectivity for 5-HT$_{1A}$, 5-HT$_{1B}$, or 5-HT$_{1C}$ sites. There is no selective radioligand for 5-HT$_{1D}$ sites (Table 4.1).

Both similarities and differences in 5-HT$_1$ binding sites have been described between species (Schnellman *et al.* 1984; Heuring *et al.* 1986). 5-HT$_{1A}$ receptors show very little pharmacological difference in rat, pig, and human brain (Hoyer *et al.* 1985b and 1986a; Pazos and Palacios 1985; Pazos *et al.* 1987a). In contrast, 5-HT$_{1B}$ receptors have only been identified

Table 4.1 Affinity values of drugs for 5-HT_1 and 5-HT_2 recognition sites in mammalian brain membranes

5-HT_1 drugs	5-HT_{1A}	5-HT_{1B}	5-HT_{1C}	5-HT_{1D}	5-HT_2
1 Dipropyl-5-carboxamido-tryptamine	9.5	4.9	4.5	7.2	5.3
2 5-Carboxamidotryptamine	9.5	8.3	6.2	8.6	4.7
3 MDL 72832	9.1	6.2	6.3	6.4	6.8
4 N,N-Dimethyl-5-carboxamidotryptamine	9.0	7.3	6.0		5.5
5 8-OH-DPAT	8.7	4.2	5.2	5.9	5.0
6 5-HT	8.5	7.6	7.5	8.4	5.5
7 Iodocyanopindolol	8.3	9.5	5.0	6.8	4.9
8 Cyanopindolol	8.3	8.3	4.4	6.9	4.5
9 RU 24969	8.1	8.4	6.5	7.3	6.0
10 Spiroxatrine	8.1	3.9	5.1	5.1	6.2
11 5-Methoxytryptamine	8.0	6.4	7.6	8.4	5.5
12 (−) 21-009	8.0	9.4	5.3	6.4	5.0
13 WB 4101	7.9	5.7	6.4		5.9
14 N,N-Dimethyl-5-methoxytryptamine	7.9	5.9	7.0	7.5	6.2
15 Ipsapirone	7.7	3.9	4.5	4.9	5.1
16 (−) Pindolol	7.7	7.8	4.2	5.2	4.4
17 Bufotenine	7.6	6.0	7.2	8.1	6.4
18 Buspirone	7.6	3.9	5.1	4.5	6.1
19 Carazolol	7.5	8.0	5.2		5.4
20 5-Methyltryptamine	7.3	5.8	8.1		5.2
21 Urapidil	7.2	4.9	4.8	4.7	
22 Gepirone	7.1	3.9	4.6		5.4
23 Rauwolscine	6.9	5.3	5.8	7.7	6.1
24 Yohimbine	6.9	5.5	4.4	7.1	6.0
25 (−) Propranolol	6.8	7.3	6.8	5.5	5.7
26 Tryptamine	6.8	5.0	7.3	7.4	6.0
27 1-Methyl-5-HT	6.6	5.5	8.4		6.3
28 Oxprenolol	6.6	6.8	4.7		5.2
29 Idazoxan	6.4	4.4	4.9		4.3
30 5-Amino-tryptamine	6.3	6.3	7.1		5.5
31 Labetalol	6.3	4.4	4.9		4.9
32 (+) 21-009	6.2	7.5	5.0		4.7
33 (+) Pindolol	5.9	5.0	4.4		4.9
5-HT_2 drugs	**5-HT_{1A}**	**5-HT_{1B}**	**5-HT_{1C}**	**5-HT_{1D}**	**5-HT_2**
34 Cyproheptadine	6.5	5.3	7.9		8.5
35 SCH 23390	6.5	4.9	8.3	5.5	8.1
36 LY53857	6.4	5.5	8.1		7.3
37 Mesulergine	6.2	4.9	8.8	5.2	8.4
38 Pizotifen	6.2	5.5	8.1	5.6	7.8

39 Cinanserin	6.1	5.2	6.7	5.8	7.7
40 Mianserin	6.0	5.2	8.0	6.4	8.1
41 Pirenperone	5.9	5.3	7.3		8.8
42 Ketanserin	5.9	5.7	7.0	6.0	8.9
43 Setoperone	5.6	5.3	7.3		8.6
44 Altanserin	5.6	6.0	6.9		8.6
45 Ritanserin	5.4	4.0	8.6		9.3

Non-selective drugs	5-HT_{1A}	5-HT_{1B}	5-HT_{1C}	5-HT_{1D}	5-HT_2
46 Lisuride	9.1	6.7	7.7	7.5	8.3
47 Dihydroergotamine	8.9	9.1	7.5	7.7	8.6
48 LSD	8.6	6.8	7.9	8.2	8.6
49 Methergine	8.5	6.7	7.6		7.6
50 α-Dihydroergocryptine	8.4	6.9	7.0		8.0
51 Ergotamine	8.4	8.7	7.3	7.6	7.7
52 Ergometrine	8.3	6.5	7.5		7.9
53 β-Dihydroergocryptine	8.3	6.9	6.9		7.6
54 Metergoline	8.1	7.4	9.2	9.1	9.0
55 2-Iodo-LSD	7.7	6.3	7.4		8.9
56 Methysergide	7.6	5.8	8.6	8.4	8.6
57 Bromocriptine	7.6	7.0	6.7		7.0
58 2-Bromo-LSD	7.2	6.2	7.3		8.1
59 Spiperone	7.2	5.3	5.9	5.3	8.8
60 1-Naphthyl-piperazine	7.2	6.6	8.2	7.8	7.2
61 Methiothepin	7.1	7.3	7.6	7.3	8.8
62 Wy 26392	6.8	4.8	6.0	5.7	6.3
63 ICI 89406	6.5	5.9	5.0		5.5
64 mCPP	6.5	6.6	7.7	5.8	6.7
65 TFMPP	6.3	6.4	7.2	6.2	6.6
66 Corynanthine	6.2	4.0	5.5	5.3	5.9
67 2-Methyl-5-HT	5.6	4.4	5.9	6.4	5.0
68 Quipazine	5.5	6.5	6.7	5.9	6.2
69 Haloperidol	5.3	5.1	4.8	5.5	6.6
70 Metoclopramide	5.2	4.3	5.4		5.9
71 Domperidone	5.1	5.2	4.7	4.8	6.8
72 ICS 205-930	4.7	4.4	4.6		5.3
73 ICI 118,551	4.3	5.6	4.6		5.1
74 Harmine	3.8	3.5	4.7	3.8	4.6
75 Betaxolol	3.8	4.8	4.0		4.2
76 Atenolol	3.3	3.8	4.0		4.0

Affinity values ($pK_D = -\log$ mol/l) were determined using the following systems. 5-HT_{1A}: [^{3}H]-8-OH-DPAT in pig frontal cortex. 5-HT_{1B}: [^{125}I]cyanopindolol in rat frontal cortex in the presence of 30 μM isoprenaline in order to avoid binding to β-adrenoceptors. 5-HT_{1C}: [^{3}H]-mesulergine in pig choroid plexus. 5-HT_{1D}: [^{3}H]5-HT in calf caudate in the presence of 100 nM 8-OH-DPAT and 100 nM mesulergine. 5-HT_2: [^{3}H]-ketanserin in rat frontal cortex. Non-specific binding was determined in the presence of 10 μM 5-HT (5-HT_1 sites) or 10 μM mianserin (5-HT_2 sites) in rat, mouse, and possibly hamster brain (Hoyer *et al.* 1985a; Pazos and Palacios 1985; Heuring *et al.* 1986), and are apparently absent in pig, bovine, and human brain (Hoyer *et al.* 1985b, 1986a, and 1988; Pazos *et al.* 1987a; Heuring and Peroutka 1987). 5-HT_{1C} sites with very similar pharmacological profiles have been described in the choroid plexus and different brain regions of rat, pig, and man (Cortes *et al.* 1984; Hoyer *et al.* 1985b and 1986b; Pazos and Palacios, 1985; Pazos *et al.* 1987a). The pharmacology of 5-HT_{1D} sites is very similar in bovine, pig, and human brain (Heuring and Peroutka, 1987; Hoyer *et al.* 1988; Waeber *et al.* 1988a and b).

4.2.2 5-HT_2 recognition sites

Although 5-HT_2 recognition sites were first identified with [^{3}H]spiperone (Leysen *et al.* 1978), this ligand shows even higher affinity for dopamine D_2 receptors. [^{3}H]Ketanserin was the first selective 5-HT_2 receptor radioligand to be developed (Leysen *et al.* 1982). Since then, further 5-HT_2 site radioligands have been reported: [^{3}H]LSD, [^{3}H]mianserin, [^{125}I]LSD, [^{3}H]mesulergine, and 1-methyl-[^{125}I]LSD. These radioligands also bind to 5-HT_{1C} receptors (see above). [^{3}H]DOB (Lyon *et al.* 1987) was the first 5-HT receptor agonist to be used in 5-HT_2 binding studies. It is interesting to note that agonists show significantly higher affinity for the sites labelled with [^{3}H]DOB in comparison with antagonist radioligands. The affinities of a number of compounds for antagonist-labelled 5-HT_2 sites are listed in Table 4.1. Truly selective agonists are still lacking, whereas ketanserin, cinanserin, and pirenperone are relatively selective 5-HT_2 receptor antagonists.

5-HT_2 receptors show close pharmacological similarities in rat, pig, and human brain (Pazos *et al.* 1984a; Hoyer *et al.* 1986b; Pazos *et al.* 1987b). However, differences have also been noted. For instance, [^{3}H]mesulergine which labels 5-HT_2 receptors rat brain membranes, does not label these sites in pig brain preparations (Pazos *et al.* 1984a and b).

4.2.3 5-HT_3 recognition sites

Recently 5-HT_3 receptors were identified with [^{3}H]ICS 205-930 in neuroblastoma N1E-115 and neuroblastoma-glioma NG 108-15 cells of peripheral origin (Hoyer and Neijt 1987 and 1988; Neijt *et al.* 1988a). These sites displayed high affinity to the selective 5-HT_3 receptor antagonists, MDL 72222 (Fozard 1984b), ICS 205-930 (Richardson *et al.* 1985), and GR38032 (Brittain *et al.* 1987) (see Table 4.2). Furthermore, they showed similar affinity for 5-HT, 2-methyl-5-HT, and phenylbiguanide, the latter being selective 5-HT_3 receptor agonists. Other 5-HT receptor ligands, which are inactive at 5-HT_3 receptors, did not affect [^{3}H]ICS 205-930 binding. Func-

Table 4.2 Affinity values of drugs for $5\text{-}HT_3$ recognition sites in various tissues and cell lines

Drug	N1E-115	NG 108-15	Cat vagus	Rat brain
SDZ 206-830	9.8	9.4	8.8	
ICS 205-930	9.1	8.8	8.9	8.5
SDZ 206-792	9.1	8.8	7.9	
BRL 43694	8.9	8.7	8.5	9.2
Quipazine	8.7	8.5		8.9
BRL 24924	8.5	8.1		
GR38032	7.9			8.5
MDL 72222	8.2	7.8	7.5	7.3
mCPP	7.0	7.0		7.3
Metoclopramide	6.7	6.7		6.4
5-HT	6.4	6.0	6.7	6.9
2-Methyl-5-HT	5.9	6.0	7.0	7.1
Phenylbiguanide	6.1	5.8		
Methysergide	4.5	<4		<5
8-OH-DPAT	4.6	4.5		
5-Carboxamidotryptamine	4.1	4.7		
Ketanserin	3.6	3.9		<5

Affinity values for [^{3}H]ICS 205-930 recognition sites on membranes of N1E–115 (Hoyer and Neijt 1987) and NG 108-15 cells (Neijt *et al.* 1988a), cat vagus nerve (author's laboratory) and rat brain (data from Kilpatrick *et al.* 1987) are expressed as pK_D values ($-\log$ mol/l)

tional studies performed with N1E-115 cells show that 5-HT and 2-methyl-5-HT cause a fast depolarization which is potently antagonized by ICS 205-930 and MDL 72222 (Neijt *et al.* 1986 and 1988b), suggesting a $5\text{-}HT_3$ receptor-mediated effect. The data suggest that the $5\text{-}HT_3$ receptors from N1E-115 and NG 108–15 cells, are similar to those found on the vagus nerve and superior cervical ganglion (Round and Wallis 1986 and 1987; Ireland and Tyers 1987), and indeed similar results were obtained with [^{3}H]ICS 205-930 using membranes prepared from cat vagus nerve and superior cervical ganglion (see Table 4.2). Binding studies have also been performed using rat brain membranes, with a tritiated derivative of GR38032 (Kilpatrick *et al.* 1987), [^{3}H]quipazine (Peroutka and Hamik 1988), [^{3}H]zacopride (Barnes *et al.* 1988), and [^{3}H]quaternized ICS 205-930 (Watling *et al.* 1988). The pharmacology of these sites is in good agreement with that of [^{3}H]ICS 205-930 sites in NG 108-15 and N1E-115 cells. GppNHp or GTP did not affect the binding of agonists or antagonists, suggesting that $5\text{-}HT_3$ receptors are not coupled to G proteins.

Finally a comment on the binding of [^{3}H]5-HT to recognition sites in the myenteric plexus of the guinea-pig which has been described by Takaki *et al.* (1985) and Mawe *et al.* (1986). These sites are not $5\text{-}HT_1$ and $5\text{-}HT_2$ receptors since classical 5-HT receptor ligands show very low affinity.

Moreover, $[^3H]$5-HT binding to enteric 5-HT sites is not displaced by ICS 205-930, MDL 72222, quipazine, or metoclopramide (Mawe *et al.* 1986) which rules out that a 5-HT_3 receptor is being labelled. Further details of this site and its pharmacology are given in Chapter 11.

The Bradley *et al.* (1986) classification, whilst conceding the existence of '5-HT_1-like' receptor subtypes, did not attempt their definition. For the purpose of this chapter, the subtypes of the 5-HT_1 receptor identified by radioligand binding, will be considered under the heading '5-HT_1-like'.

4.3 5-HT receptor-mediated modulation of adenylate cyclase activity

4.3.1 Stimulation of adenylate cyclase

4.3.1.1 *'5-HT_1-like' receptors*

Peroutka *et al.* (1981) first proposed that 5-HT_1 receptors are coupled to adenylate cyclase. 5-HT-stimulated adenylate cyclase has been studied thoroughly in guinea-pig and rat hippocampus and it is clear that a '5-HT_1-like' receptor mediates this effect (Shenker *et al.* 1983 and 1985; Barbaccia *et al.* 1983). Thus, the effects of 5-HT were antagonized by metergoline and spiperone but not by mianserin or ketanserin, ruling out a 5-HT_2 receptor mechanism (Barbaccia *et al.* 1983). Furthermore, the potent 5-HT_{1A} receptor agonists, 8-OH-DPAT and 5-carboxamidotryptamine, mimicked the stimulatory effects of 5-HT on adenylate cyclase (Shenker *et al.* 1985). Markstein *et al.* (1986) used a large series of agonists and several antagonists to establish that, in the rat hippocampus, 5-HT_{1A} receptors mediate stimulation of adenylate cyclase. In addition to 5-carboxamidotryptamine, dipropyl-5-carboxamidotryptamine, 8-OH-DPAT, and lisuride were found to be potent agonists, whereas the slightly selective 5-HT_{1B} receptor ligand, RU 24969, showed a lower agonist potency. A very good correlation was found between 5-HT_{1A} binding and the potencies of the various agonists as stimulants of adenylate cyclase (see Fig. 4.1A). The rank order of potency of antagonists was also consistent with a 5-HT_{1A} receptor-mediated effect. Shenker *et al.* (1987) presented similar findings in rat and guinea-pig hippocampal preparations.

5-HT_{1A} receptors have been characterized in peripheral tissues. For instance, sound evidence suggests that 5-HT-induced inhibition of electrically stimulated $[^3H]$acetylcholine release from the guinea-pig ileum is mediated through these sites (Fozard and Kilbinger, 1985). The 5-HT-induced contraction in the canine basilar artery (Taylor *et al.* 1986; Peroutka *et al.* 1986) has also been claimed to be 5-HT_{1A} receptor mediated, although in this instance the data are controversial (see Chapter 5). At present, it is not known by which receptor-effector coupling mechanism

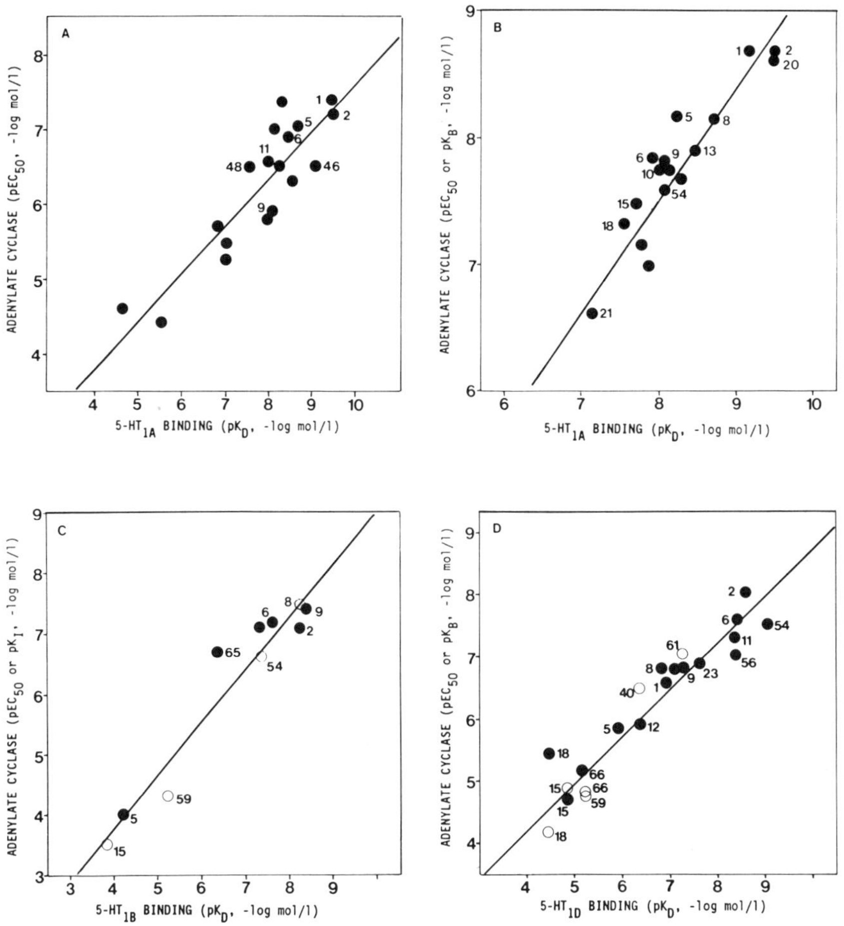

Fig. 4.1 Modulation of adenylate cyclase activity by 5-HT receptors. Fig. 1A compares pEC_{50} values of agonists on adenylate cyclase activity (rat hippocampus) and $5\text{-}HT_{1A}$ binding affinities (pK_D). The data are in part from Markstein *et al.* (1986) and from Markstein (unpublished). Fig. 1B compares pEC_{50} and pK_B values of agonists (closed symbols) and antagonists (open symbols) on forskolin-stimulated adenylate cyclase activity in calf hippocampus and $5\text{-}HT_{1A}$ binding affinities (pK_D). The data are from the author's laboratory. Fig. 1C compares pEC_{50} and pK_I values of agonists (closed symbols) and antagonists (open symbols) on forskolin-stimulated adenylate cyclase activity in rat substantia nigra with $5\text{-}HT_{1B}$ binding affinities (pK_D). The data are from the author's laboratory and Bouhelal *et al.* (1988). Fig. 1D compares pEC_{50} and pK_B values of agonists (closed symbols) and antagonists (open symbols) on forskolin-stimulated adenylate cyclase activity in calf substantia nigra with $5\text{-}HD_{1D}$ binding affiities (pK_D). The data are from Schoeffter *et al.* (1988). The numbers at the side of the individual points identify compounds which appear in Table 4.1

peripheral 5-HT_{1A} receptors mediate their effects, and there is no evidence for these sites being linked in a positive way to adenylate cyclase, as seems to be the case in certain brain areas.

5-HT_{1B} receptors have been identified in both brain and peripheral tissues of the rat. Engel *et al.* (1986) presented evidence that 5-HT_{1B} autoreceptors mediated inhibition of the release of $[^3H]$5-HT in rat cortex. Similar findings were reported in rat hippocampus and cerebellum (Maura *et al.* 1986; Raiteri *et al.* 1986). The rat hippocampus also contains presynaptic 5-HT_{1B} heteroreceptors which modulate the release of $[^3H]$acetylcholine (Maura and Raiteri 1986). In the periphery, 5-HT_{1B} receptors which modulate the release of $[^3H]$noradrenaline have been identified in the rat vena cava (Molderings *et al.* 1987). However, it has not been established that 5-HT_{1B} receptors stimulate adenylate cyclase activity in the periphery or in the brain. In particular, in the rat substantia nigra, which contains a high density of 5-HT_{1B} sites, no 5-HT_{1B} receptor-stimulated adenylate cyclase activity could be demonstrated (Schoeffter and Hoyer unpublished).

Palacios *et al.* (1986) demonstrated that 5-HT_{1C} receptors are not linked to adenylate cyclase activity in pig choroid plexus homogenates. 5-HT_{1D} receptor-mediated stimulation of adenylate cyclase activity could not be demonstrated in calf substantia nigra, where a high concentration of these sites is found (Schoeffter *et al.* 1988).

Trevethick *et al.* (1984 and 1986) have described 5-carboxamidotryptamine and 5-HT stimulated cyclic AMP production and vasorelaxation in neonatal porcine vena cava. α-Methyl-5-HT (a selective 5-HT_2 receptor agonist) was inactive. Methysergide, but not cyproheptadine or ketanserin, suppressed the effect of 5-HT. These results suggest that '5-HT_1-like' receptors, but not 5-HT_2 receptors, are involved in the activation of adenylate cyclase. From further studies it appears that this receptor is neither 5-HT_{1A} nor 5-HT_{1B} (Sumner *et al.* 1987). In cultured calf aorta smooth muscle cells, Luchins and Makman (1980) reported that 5-HT and 5-methoxytryptamine stimulated adenylate cyclase activity. Cinanserin, LSD, pizotifen, and cyproheptadine were not effective as antagonists, whereas methysergide was. This suggests that a '5-HT_1-like' receptor is involved, although which particular subtype remains to be defined.

4.3.1.2 *5-HT_2 and 5-HT_3 receptors*

There is little evidence that 5-HT_2 receptors modulate adenylate cyclase activity. 5-HT stimulates adenylate cyclase activity in rat striatal muscle and a 5-HT D (5-HT_2) receptor-mediated effect has been proposed (Maillet and Garber 1986). However, the high potency of 5-HT in this system ($pEC_{50} = 8.54$) and the rank order of potency of antagonists (LSD $\gg$ methysergide = (+)propranolol > cyproheptadine) would be more consistent with a '5-HT_1-like' receptor mechanism.

A positive link between 5-HT_3 receptors and adenylate cyclase activity has not been described. In fact, 5-HT_3 receptors are definitely not coupled to adenylate cyclase in the neuroblastoma cells, NCB 20, NG 108-15, and N1E-115 (MacDermot *et al.* 1979). Moreover, although adenylate cyclase activity is stimulated by 5-HT in NCB 20 cells, this effect is not mediated by a 5-HT_3 receptor (MacDermot *et al.* 1979). A preliminary report of Dumuis *et al.* (1988) describes 5-HT-stimulated adenylate cyclase activity in rat brain which is antagonized by ICS 205-930; however, a thorough pharmacological analysis of this system showed that other potent 5-HT_3 receptor antagonists were very weak.

4.3.2 Inhibition of adenylate cyclase

4.3.2.1 *'5-HT_1-like' receptors*

De Vivo and Maayani (1985 and 1986) described 5-HT receptor-mediated inhibition of forskolin-stimulated adenylate cyclase in rat and guinea-pig hippocampus. The rank order of potency of agonists (5-carboxamidotryptamine > 8-OH-DPAT = 5-HT), the potent antagonism displayed by spiperone, and the lack of effect of 5-HT_2 receptor antagonists suggested a role for a '5-HT_1-like' receptor, most probably the 5-HT_{1A} subtype. We have obtained similar results in calf hippocampus where potencies to inhibit adenylate cyclase correlated highly significantly with 5-HT_{1A} binding (Fig. 4.1B). Weiss *et al.* (1986) and Bockaert *et al.* (1987) described similar findings on forskolin- and vasoactive intestinal peptide-stimulated adenylate cyclase in mouse and guinea-pig hippocampal cells and membranes. Since in these preparations, 5-HT_{1A} receptors also mediate stimulation of cyclase activity, it has to be concluded that 5-HT_{1A} receptors are coupled to at least two different G or N proteins in the same tissue (though not necessarily in the same cells). Stimulation of adenylate cyclase is mediated via N_S, the stimulatory GTP binding protein, whereas inhibition is mediated most probably via N_i and possibly other G proteins, but not N_S. N_i is ADP ribosylated by pertussis toxin which blocks the effect of N_i coupled receptors (see Graziano and Gilman 1987). Clarke *et al.* (1987) described the parallel decrease of 5-HT_{1A} receptor-mediated inhibition of adenylate cyclase and inhibition of spike amplitude in CA1 hippocampus cells. Andrade *et al.* (1986) have also described the presence of a pertussis toxin sensitive G protein (G_K?) which couples 5-HT_{1A} receptors in hippocampal pyramidal cells to a K^+ channel. Makman *et al.* (1988) described pertussis toxin sensitive, 5-HT-induced inhibition of adenylate cyclase activity in mouse spinal cord cord explants. Since the effects were mimicked by 8-OH-DPAT and PAPP, the authors concluded that a 5-HT_{1A} receptor mechanism was involved.

5-HT_{1B} receptors appear also to be negatively coupled to adenylate cyclase (Bouhelal *et al.* 1988). This was demonstrated in homogenates of rat substantia nigra, which is endowed almost exclusively with 5-HT_{1B} recognition sites (Pazos and Palacios 1985). 5-HT receptor agonists inhibited forskolin-stimulated adenylate cyclase activity with the rank order of potency: RU 24969 > 5-HT > 5-carboxamidotryptamine > TFMPP > tryptamine. Cyanopindolol, propranolol, and metergoline acted as potent antagonists (Fig. 4.1C) (Schoeffter and Hoyer unpublished). In contrast, 8-OH-DPAT, ipsapirone, spiperone, mesulergine, and ketanserin, were devoid of significant agonist or antagonist activity. These results are fully consistent with a 5-HT_{1B} receptor-mediated inhibition of adenylate cyclase activity in the rat substantia nigra. In the periphery 5-HT_{1B} receptors may also be negatively coupled to adenylate cyclase. In a hamster lung cell line, Seuwen *et al.* (1988) demonstrated the mitogenic effects of 5-HT which were associated with inhibition of adenylate cyclase activity. The pharmacological profile of these effects was consistent with that of a 5-HT_{1B} receptor.

5-HT_{1C} receptors are neither negatively nor positively coupled to adenylate cyclase in the choroid plexus (Palacios *et al.* 1986).

Activation of 5-HT_{1D} receptors leads to inhibition of forskolin-stimulated adenylate cyclase activity in calf substantia nigra (Hoyer and Schoeffter 1988; Schoeffter *et al.* 1988), a tissue which contains a high proportion of 5-HT_{1D} sites (Waeber *et al.* 1988a and b). The rank order of potency of a large series of agonists and antagonists for their interaction with adenylate cyclase activity correlated very significantly with their rank order of affinity for 5-HT_{1D} sites (Fig. 4.1D). Preliminary data from a bovine smooth muscle cell line suggest the presence of inhibitory 5-HT_{1D} receptors in this tissue (Seuwen personal communication).

4.3.2.2 *5-HT_2 and 5-HT_3 receptors*

5-HT_2 or 5-HT_3 receptor-mediated inhibition of adenylate cyclase has not been documented in central or peripheral tissues.

4.4 5-HT receptor-mediated modulation of guanylate cyclase activity

Ogura *et al.* (1986) described a 5-HT-induced increase of cyclic GMP in C_6 glioma cells which was antagonized by methysergide, metergoline, and cyproheptadine. The rise in cyclic GMP induced by 5-HT was dependent on Ca^{2+} entry which then triggered the activation of guanylate cyclase. 8-OH-DPAT showed no agonist effects, whereas spiperone (0.1 μM) inhibited 5-HT-evoked membrane depolarization in these cells, indicating a possible role for 5-HT_2 receptors. In snail neurones, cyclic GMP and zaprinast, a cyclic GMP phosphodiesterase inhibitor, mimicked the action of 5-HT and caused prolongation of Ca^{2+} dependent action potentials.

Cyclic AMP and forskolin were devoid of effect (Paupardin-Tritsch *et al.* 1986). The nature of the receptor involved is not known.

4.5 5-HT receptor-mediated modulation of phospholipase C, inositol lipid metabolism and protein kinase C

4.5.1 '5-HT$_1$-like' receptors

In a recent report, Claustre *et al.* (1988) showed that 5-HT$_{1A}$ receptor agonists inhibited carbachol-stimulated accumulation of inositol phosphates in rat hippocampus. There are no published data on 5-HT$_{1B}$ or 5-HT$_{1D}$ receptor-mediated modulation of phospholipase C activity or inositol lipid metabolism. In contrast, there is good evidence that 5-HT$_{1C}$ receptor activation in rat and mouse choroid plexus leads to the activation of phospholipase C and accumulation of inositol phosphates (Conn *et al.* 1986; Conn and Sanders-Bush 1986b). We have studied the effects of a large series of agonists and antagonists on inositol phosphate accumulation in pig choroid plexus homogenates (Hoyer 1988). There was a highly significant correlation between the rank order of potency of agonists and antagonists and their affinity to 5-HT$_{1C}$ sites (Fig. 4.2A). Since 5-HT$_2$ and

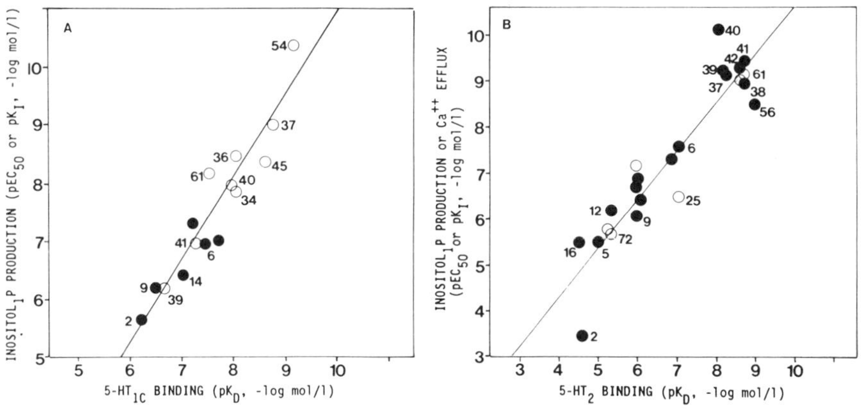

Fig. 4.2 Modulation of inositol phosphate production and Ca^{2+} efflux by 5-HT receptors. Fig. 2A compares pEC$_{50}$ and pK$_I$ values of agonists (closed symbols) and antagonists (open symbols) on inositol-1-phosphate production in pig choroid plexus with 5-HT$_{1C}$ binding affinities (pK$_D$). The data are in part from Hoyer *et al.* (1985b), Hoyer (1988) and from Dravid (unpublished). Fig. 2B compares pEC$_{50}$ and pK$_I$ values of agonists and antagonists on inositol-phosphate production (open symbols) and Ca^{2+} efflux in A$_7$r$_5$ cells (closed symbols) with 5-HT$_2$ binding affinities (pK$_D$). The data are in part from Doyle *et al.* (1986). The numbers at the side of the individual points identify compounds which appear in Table 4.1

5-HT_3 receptor antagonists failed to inhibit the effects of 5-HT, the data are consistent with a 5-HT_{1C} receptor-mediated effect.

Based on preliminary data it was suggested that 5-HT_{1C} receptors mediate 5-HT induced contraction of the rat fundus (Buchheit *et al.* 1986). However, more recent data does not support this proposal (Cohen and Wittenauer 1987) (see Chapter 9). Furthermore, it has been demonstrated that 5-HT does not stimulate phosphoinositide turnover in the rat fundus (Cohen and Fludzinski 1987). This is an interesting finding which suggests that 5-HT-induced contraction may not necessarily involve the production of inositol lipids.

4.5.2 5-HT_2 receptors

5-HT-stimulated hydrolysis of inositol lipids has been reported in various brain regions (Conn and Sanders-Bush 1985 and 1986a; Brown *et al.* 1984; Kendall and Nahorski 1985). It was established that the high affinity component of the 5-HT effect was mediated by 5-HT_2 receptors since potent and selective 5-HT_2 receptor antagonists inhibited the action of 5-HT. The low affinity component of 5-HT was not further characterized. 5-HT also increases inositol phosphate production via 5-HT_2 receptors in C_6 glioma cells (Ananth *et al.* 1987).

The various aspects of the 5-HT-induced stimulation of the phospholipase C cascade have been well characterized in peripheral tissues, in particular in platelets, smooth muscle, and smooth muscle cell cultures. In platelets, De Chaffoy de Courcelles *et al.* (1984, 1985, and 1987) demonstrated that stimulation of 5-HT_2 receptors leads to the production of phosphatidic acid, hydrolysis of polyphosphoinositides, and stimulation of protein kinase C. There was a significant correlation between the potency of various drugs in this respect and their affinity for 5-HT_2 sites. Affolter *et al.* (1984) and Erne and Pletscher (1985) showed that 5-HT, by activating 5-HT_2 receptors, caused rapid increases in intracellular calcium which could be related to the shape change in platelets.

In strips of rat aorta activation of 5-HT_2 receptors results in stimulation of phospholipase C, accumulation of inositol phosphates, and stimulation of protein kinase C (Roth *et al.* 1984 and 1986; Nakaki *et al.* 1985). In an extensive study performed in A_7r_5 smooth muscle cells (Doyle *et al.* 1986), the 5-HT receptor mediating stimulation of phospholipase C and Ca^{2+} efflux was characterized. A significant correlation was found between the potency of a large series of agonists on the two biochemical parameters and their affinity for 5-HT_2 sites (Fig. 4.2B). Moreover, '5-HT_1-like' and 5-HT_3 receptor agonists and antagonists were devoid of significant effects. Similar results have been obtained in primary cultures of aortic myocytes (Cory *et al.* 1986; Berta *et al.* 1986) and WRK_1 cells and membranes (Cory *et al.* 1987). In WRK_1 membranes, a requirement for guanine nucleotides was demonstrated, suggesting that 5-HT_2 receptors are coupled to a G-protein, as is characteristic for receptors activating phospholipase C. Bruns and Marmé (1987) also proposed a role for a G protein, since the 5-HT-induced rise in intracellular calcium in cultured rat aortic cells could be blocked with pertussis toxin. Taken together, the evidence suggests that 5-HT_2 receptors stimulate phospholipase C activity via a G protein. This leads to the production of inositol trisphosphate which mobilizes intracellular Ca^{2+} (Berridge and Irvine 1984) and diacylglycerol which activates protein kinase C (Nishizuka 1984).

4.6 5-HT receptors linked to ion channels

There is compelling evidence that 5-HT causes depolarization and hyperpolarization by acting at various ion channels via the different 5-HT receptors. The slow K^+ channel (S channel) found in aplysia neurones (Siegelbaum *et al.* 1986) is probably the model of a 5-HT receptor-modulated ion channel which has been most extensively characterized. In these neurones 5-HT causes accumulation of cyclic AMP and an increase in Ca^{2+} spike duration which is caused by a decrease in K^+ conductance. These effects are mediated by the cyclic AMP-dependent protein kinase A cascade. Pharmacologically, the 5-HT receptor subtype involved in S channel modulation is not established but, by analogy, it could be a '5-HT_1-like' receptor (see below). As mentioned in Section 4.4 above, Paupardin-Tritsch *et al.* (1986) have described 5-HT-induced increases in Ca^{2+} currents in snail neurones mediated by the production of cyclic GMP. Again, the nature of the 5-HT receptor involved in this model is not established.

4.6.1 '5-HT_1-like' receptors

5-HT and LSD, when applied to hippocampal neurones, cause hyperpolarization via a Ca^{2+} dependent increase in K^+ conductance (Aghajanian and Lakoski 1984). These effects have been thoroughly studied in rat hippocampal pyramidal cells by Andrade *et al.* (1986). The effects of 5-HT on K^+ conductance were mimicked by low concentrations of 8-OH-DPAT and ipsapirone; spiperone, in contrast to ketanserin, was a potent antagonist. The data suggest that the effect is mediated by the 5-HT_{1A} receptor, a conclusion which is strongly supported by the results of Colino and Halliwell (1987). Cyclic AMP production appears to play no role in the modulation of K^+ channels, since neither forskolin nor cyclic AMP phosphodiesterase inhibitors was able to reproduce the effects observed with 5-HT. Phorbol esters were also inactive, thus ruling out an effect via protein kinase C. However, a G-protein appears to link the 5-HT_{1A} receptor to K^+ channels (see Section 4.3.2.1 above).

5-HT_{1C} receptors also appear able to modulate the activity of Cl^- channels. Dascal *et al.* (1986) injected rat brain mRNA into Xenopus oocytes, which then expressed a 5-HT receptor characterized as a 5-HT_{1C} receptor (Lübbert *et al.* 1987a and b). 5-HT_{1C} receptors modulate the activity of phospholipase C (see Section 4.5.1 above), and it was shown that 5-HT_{1C} receptor-mediated mobilization of intracellular Ca^{2+} was initiating the activation of Ca^{2+} dependent Cl^- channels. Taken together, the data suggest that stimulation of the 5-HT_{1C} receptor induces a G protein mediated activation of phospholipase C, production of inositoltrisphosphate and modulation of intracellular Ca^{2+} which triggers an increase in Cl^- currents. Similar results have been presented by other groups using oocytes injected with mRNA from rat and human brain and rat intestine—see Hoyer (1988) for a recent review. Interestingly and surprisingly, the pharmacology of these responses suggests that only a 5-HT_{1C} receptor had been expressed, whatever the origin of the mRNA. It remains to be established whether 5-HT_{1C} receptors of the brain are actually coupled to Cl^- channels, or whether this coupling is artificially created by mRNA injection into occytes and the subsequent expression of 5-HT_{1C} receptors in these cells.

4.6.2 5-HT_2 receptors

5-HT causes a further depolarization of the partially depolarized rat facial motoneurones *in vivo*, which is antagonized by cyproheptadine, methysergide, mianserin, and metergoline (VanderMaelen and Aghajanian 1980 and 1982). This excitatory effect is due to a decrease in K^+ conductance probably mediated by 5-HT_2 receptors (for a review, see Fozard 1984a). Similarly, excitatory 5-HT receptors located on sympathetic preganglionic neurones appear to be of the 5-HT_2 type (see also Fozard 1984a).

4.6.3 5-HT_3 receptors

The presence of functional 5-HT_3 receptors has been documented in the enteric nervous system, in autonomic neurones (both sympathetic and parasympathetic), in sensory neurones, and in neuroblastoma cells. Invariably, activation of 5-HT_3 receptors leads to neuronal depolarization in these systems (see MacDermot *et al.* 1979; Fozard 1984a). The different types of ion channels modulated by 5-HT_3 receptors in enteric, autonomic and sensory neurones are dealt with in detail in Chapters 10 and 11.

In neuroblastoma or neuroblastoma-hybridoma cell lines (NCB 20, N1E-115, NG 108-15) 5-HT causes a rapid depolarization and acetylcholine release (MacDermot *et al.* 1979). In N1E-115 cells, the 5-HT-induced depolarization is due to a simultaneous increase in permeability to Na^+ and K^+, and thus has similar mechanisms to the fast depolarization induced by 5-HT_3 receptor activation in autonomic and sensory neurones (see Chapter 10). The pharmacology of this 5-HT receptor was studied in N1E-115 cells (Neijt *et al.* 1986 and 1988b). 2-Methyl-5-HT mimicked the effects of 5-HT which were antagonized by ICS 205-930 and MDL 72222 (at nanomolar concentrations). Methysergide, methiothepin, and ketanserin were virtually inactive. 8-OH-DPAT and TFMPP failed to mimic the 5-HT effects. Similar findings have been made with NG 108–15 cells (H. Neijt unpublished). Together, the data show that N1E-115 cells possess 5-HT_3 receptors which mediate rapid depolarization. Since NG 108–15 and N1E-115 cells do not possess 5-HT stimulated adenylate cyclase activity (MacDermot *et al.* 1979) the cyclic AMP protein kinase A cascade can be excluded as the mediator of this effect.

Nothing is known about the biochemical link between 5-HT_3 receptors and the ion channels that they modulate. However, since 5-HT_3 receptor activation invariably results in a 'nicotine-like' fast depolarization which is subject to rapid tachyphylaxis, one could speculate that 5-HT_3 receptors and ion channels are very tightly associated. The channel could even be part of the receptor, as seems to be the case with nicotine, $GABA_A$, or glycine receptors, which belong to a family of structurally and functionally related channels (Stevens 1987).

4.7 Concluding remarks

4.7.1 5-HT receptor classification and nomenclature

The nomenclature of Bradley *et al.* (1986) distinguishes three types of 5-HT receptor based on distinct pharmacological characteristics. It is of interest that biochemical studies would also support the evidence of three classes of 5-HT receptor: receptors linked to adenylate or guanylate cyclase, receptors linked to phospholipase C, and receptors linked to 'nicotine-like' channels. However, the receptor groups would not be entirely the same. For instance, respecting the classification of Peroutka and Snyder (1979), the site found in the choroid plexus has been named 5-HT_{1C} since it is labelled with high affinity by [^{3}H]5-HT. However, 5-HT_{1C} and 5-HT_2 receptors (for which [^{3}H]5-HT has low affinity) show similar receptor effector coupling mechanisms and their pharmacology is also rather similar. Indeed, when comparing the affinity values of a large series of drugs to 5-HT_1 and 5-HT_2 sites, it is possible to separate clearly two groups of receptors, 5-HT_{1A}, 5-HT_{1B} and 5-HT_{1D} on the one hand and 5-HT_{1C}

and 5-HT_2 on the other [see Table 4.1). Biochemically, therefore, one could make a strong case for the 5-HT_{1C} receptor and the 5-HT_2 receptor being of the same family.

If biochemical considerations are combined with a minimalistic analysis of antagonist potencies and selectivities then one might conclude that there are only two major groups of 5-HT receptors. Thus, the structural requirements for affinity at 5-HT_3 receptors are very different from those at '5-HT_1-like' and 5-HT_2 receptors. Potent 5-HT_3 receptor antagonists (ICS 205-930, MDL 72222, GR38032) have minimal effects on '5-HT_1-like' or 5-HT_2 receptors. In contrast, many antagonists (i.e. methiothepin, spiperone, mianserin, methysergide, and ergolines) interact with both '5-HT_1-like' and 5-HT_2, but not with 5-HT_3 receptors. Combine this with the fact that '5-HT_1-like' and 5-HT_2 receptors are coupled to G proteins (although this has not been demonstrated in every instance) and that 5-HT_3 receptors are (presumably) not coupled to G proteins, and one arrives essentially at two main families of 5-HT receptors as initially proposed by Gaddum and Picarelli (1957).

4.7.2 Parallels between receptors for 5-HT and acetylcholine

The fact that there may be just two major groups of 5-HT receptors is reminiscent of the situation with acetylcholine receptors, which are also subdivided into two major classes: muscarine and nicotine receptors (Nathanson 1987; Changeux *et al.* 1987). Pharmacologically, muscarine receptors are totally different from nicotine receptors, just as '5-HT_1-like' and 5-HT_2 receptors are very different from 5-HT_3 receptors. Like '5-HT_1-like' and 5-HT_2 receptors, subtypes of muscarine receptors have been described, but their pharmacological characteristics are rather similar, as are their coupling mechanisms. Muscarine receptors mediate their actions via G proteins, either directly (modulation of slow K^+ channels) or indirectly (modulation of adenylate cyclase and phospholipase C activities). Nicotine receptors are not thought to be coupled to G proteins and/or second messengers. Their modulation of ion channels is direct, since the subunits of the nicotine receptor constitute the actual ion channel.

Structurally, striking differences exist between muscarine receptors which comprise a single peptide without subunits (Kubo *et al.* 1986; Peralta *et al.* 1987) and nicotine receptors which are made up of up to five subunits (Noda *et al.* 1983; Changeux *et al.* 1987). The muscarine receptor is a member of the G protein-linked receptor family of which several members have been sequenced, cloned, and expressed, e.g. rhodopsin, α- and β-adrenoceptors, and muscarine receptors (Dixon *et al.* 1986; Yarden *et al.* 1986; Kobilka *et al.* 1987; Kubo *et al.* 1986; Peralta *et al.* 1987). The proposed structure of these receptors is very similar; they are all presumed to have seven hydrophobic transmembrane spanning regions and show a remarkable degree of homology, with similar domains of phosphorylation close to the C terminal and other similarities in their binding domains (Dixon *et al.* 1986; Kubo *et al.* 1986). From their amino acid sequence, it seems that these receptors have very similar regulatory mechanisms (Lefkowitz *et al.* 1986; see also Sibley *et al.* 1986; Yoshimasa *et al.* 1987). In contrast, nicotine receptors (Noda *et al.* 1983), like glycine (Greningloh *et al.* 1987) and $GABA_A$ receptors (Schofield *et al.* 1987), comprise several subunits which do not share structural features with the G protein-coupled receptors.

It is tempting to suggest that '$5\text{-}HT_1$-like' and $5\text{-}HT_2$ receptors will display structural similarities with muscarine receptors, whereas $5\text{-}HT_3$ receptors could be similar to nicotine receptors and form ion channels. In this context, the partial sequence of the $5\text{-}HT_{1C}$ receptor has been established (for a review see Hoyer 1988). Julius *et al.* (1988) have recently cloned a functional cDNA encoding the $5\text{-}HT_{1C}$ receptor. The single peptide chain of the functional $5\text{-}HT_{1C}$ receptor has 460 amino acids and its seven putative transmembrane spanning regions show considerable sequence homology with those of the muscarine and substance K receptors, opsines and α- and β-adrenoceptors (Julius *et al.* 1988).

However, it is by no means certain that the situation with subtypes of 5-HT receptors will become less complicated as suggested by the multiplicity of acetylcholine, catecholamine, and opsine receptor subtypes which have been cloned most recently. For instance, at least four subtypes of muscarine receptors have been sequenced, cloned, and expressed (Kubo *et al.* 1986; Peralta *et al.* 1987; Bonner *et al.* 1987). Their pharmacologies, functions, and distributions appear to be different. In addition, Bonner *et al.* (1987) have identified up to ten different DNAs for muscarine receptors in the rat genome and up to five or six DNAs in the human genome. These additional putative muscarine receptors have not yet been characterized; nevertheless, their complexity reminds one of the '$5\text{-}HT_1$-like'–$5\text{-}HT_2$ receptor family. There is now evidence for multiple functional nicotine receptor subtypes (see Stevens 1987). Similarly, the existence of subtypes has been suggested for $5\text{-}HT_3$ receptors (see Chapter 10). It is likely that other members of the 5-HT receptor family will be sequenced, cloned, and expressed within the next few years using these new techniques, and molecular biology will undoubtedly play a key role in our understanding of 5-HT receptors and their effector coupling mechanisms.

Acknowledgements

Thanks are expressed to E. Schüpbach and M. Girod for expert technical assistance, and to the many colleagues at Sandoz and elsewhere for stimulating discussions and collaborations.

References

Affolter, H., Erne, P., Bürgisser, E., and Pletscher, A. (1984). Ca^{++} as messenger of 5-HT_2 receptor stimulation in human blood platelets. *Naunyn-Schmiedeberg's Archives of Pharmacology* **325,** 337–42.

Aghajanian, G. K. and Lakoski, J. M. (1984). Hyperpolarization of serotonergic neurons by serotonin and LSD: studies in brain slices showing increased K^+ conductance. *Brain Research* **305,** 181–5.

Ananth, U. S., Leli, U., and Hauser, G. (1987). Stimulation of phosphoinositide hydrolysis by serotonin in C6 glioma cells. *Journal of Neurochemistry* **48,** 253–61.

Andrade, R., Malenka, R. C. and Nicoll, R. A. (1986). A G protein couples serotonin and $GABA_B$ receptors to the same channels in hippocampus. *Science* **234,** 1261–5.

Barbaccia, M. L., Brunello, N., Chuang, D.-M., and Costa, E. (1983). Serotonin-elicited amplification of adenylate cyclase activity in hippocampal membranes from adult rat. *Journal of Neurochemistry* **40,** 1671–9.

Barnes, N. M., Costall, B., and Naylor, R. J. (1988). ^{3}H Zacopride identifies 5-HT_3 binding sites in rat enthorinal cortex. *British Journal of Pharmacology* **94,** 319P.

Berridge, M. J. and Irvine, R. F. (1984). Inositol trisphosphate, a novel second messenger in cellular signal transduction. *Nature (London)* **312,** 315–21.

Berta, P., Sladeczek, F., Travo, P., Bockaert, J., and Haiech, J. (1986). Activation of phosphatidylinositol synthesis by different agonists in a primary culture of smooth muscle cells grown on collagen microcarriers. *FEBS Letters* **200,** 27–31.

Bockaert, J., Dumuis, A., Bouhelal, R., Sebben, M. and Cory, R. N. (1987). Piperazine derivatives including the putative anxiolytic drugs, buspirone and ipsapirone, are agonists at 5-HT_{1A} receptors negatively coupled with adenylate cyclase in hippocampal neurons. *Naunyn Schmiedeberg's Archives of Pharmacology* **335,** 588–92.

Bonner, T. I., Buckley, N. J., Young, A. C., and Brann, M. R. (1987). Identification of a family of muscarinic acetylcholine receptor genes. *Science* **237,** 527–32.

Bouhelal, R., Smounya, L., and Bockaert, J. (1988). 5-HT_{1B} receptors are negatively coupled with adenylate cyclase in rat substantia nigra. *European Journal of Pharmacology* **151,** 189–96.

Bradley, P. B., Engel, G., Feniuk, W., Fozard, J. R., Humphrey, P. P. A., Middlemiss, D. N., Mylecharane, E. J., Richardson, B. P., and Saxena, P. R. (1986). Proposals for the classification and nomenclature of functional receptors for 5-hydroxytryptamine. *Neuropharmacology* **25,** 563–76.

Brittain, R. T., Butler, A., Coates, I. H., Fortune, D. H., Hagan, R., Hill, J. M., Humber, D. C., Humphrey, P. P. A., Ireland, S. J., Jack, D., Jordan, C. C., Oxford, A., Straughan, D. W., and Tyers, M. B. (1987). GR38032F, A novel selective 5-HT_3 receptor antagonist. *British Journal of Pharmacology* **90,** 87P.

Brown, E., Kendall, D. A., and Nahorski, S. R. (1984). Inositol phospholipid hydrolysis in rat cerebral cortex slices. I. Receptor characterization. *Journal of Neurochemistry* **42,** 1379–87.

Bruns, C. and Marmé, D. (1987). Pertussis toxin inhibits the angiotensin II and serotonin-induced rise of free cytoplasmic calcium in cultured smooth muscle cells from rat aorta. *FEBS Letters* **212,** 40–4.

Buchheit, K. H., Engel, G., Hagenbach, A., Hoyer, D., Kalkman, H. O., and Seiler, M. P. (1986). The rat isolated stomach fundus strip, a model for 5-HT_{1C} receptors. *British Journal of Pharmacology* **88,** 367P.

Changeux, J. P., Giraudat, J., and Dennis, M. (1987). The nicotine acetylcholine receptor: molecular architecture of a ligand-regulated ion channel. *Trends in Pharmacological Sciences* **8,** 465–72.

Clarke, W. P., De Vivo, M., Beck, S. G., Maayani, S., and Goldfarb, J. (1987). Serotonin decreases population spike amplitude in hippocampal cells through a pertussis toxin substrate. *Brain Research* **410,** 357–61.

Claustre, Y., Benavides, J., and Scatton, B. (1988). 5-HT_{1A} receptor agonists inhibit carbachol-induced stimulation of phosphoinositideturnover in the rat hippocampus. *European Journal of Pharmacology* **149,** 149–53.

Cohen, M. L. and Fludzinski, L. A. (1987). Contractile serotonergic receptors in rat stomach fundus. *Journal of Pharmacology and Experimental Therapeutics* **243,** 264–9.

Cohen, M. L. and Wittenauer, L. A. (1987). Serotonin receptor activation of phosphoinositide turnover in uterine, fundal, vascular and tracheal smooth muscle. *Journal of Cardiovascular Pharmacology* **10,** 176–81.

Colino, A. and Halliwell, J. V. (1987). Differential modulation of three separate K-conductances in hippocampal CA1 neurons by serotonin. *Nature* **328,** 73–7.

Conn, P. J. and Sanders-Bush, E. (1985). Serotonin stimulated phosphoinositide turnover: mediation by the S2 binding site in rat cerebral cortex but not in subcortical regions. *Journal of Pharmacology and Experimental Therapeutics* **234,** 195–203.

Conn, P. J. and Sanders-Bush, E. (1986a). Regulation of serotonin-stimulated phosphoinositide hydrolysis: Relation to the serotonin 5-HT-2 binding site. *The Journal of Neuroscience* **6,** 3669–75.

Conn, P. J. and Sanders-Bush, E. (1986b). Agonist-induced phosphoinositide hydrolysis in choroid plexus. *Journal of Neurochemistry* **47,** 1754–60.

Conn, P. J., Sanders-Bush, E., Hoffman, B. J., and Hartig, P. R. (1986). A unique serotonin receptor in choroid plexus is linked to phosphatidylinositol turnover. *Proceedings of the National Academy of Science, USA* **83,** 4086–8.

Cortés, R., Palacios, J. M., and Pazos, A. (1984). Visualization of multiple serotonin receptors in the rat brain by autoradiography. *British Journal of Pharmacology* **81,** 202P.

Cory, R. N., Berta, P., Haiech, J., and Bockaert, J. (1986). 5-HT_2 receptor-stimulated inositol phosphate formation in rat aortic myocytes. *European Journal of Pharmacology* **131,** 153–7.

Cory, R. N., Rouot, B. Guillon, G., Sladeczek, F., Balestre, M-N., and Bockaert, J. (1987). The 5-hydroxytryptamine (5-HT_2) receptor stimulates inositol phosphate formation in intact and broken WRK1 cells: Determination of occupancy-response relationships for 5-HT agonists. *Journal of Pharmacology and Experimental Therapeutics* **241,** 258–67.

Dascal, N., Ifune, C., Hopkins, R., Snutch, T. P., Lübbert, H., Davidson, N., Simon, M. I., and Lester, H. A. (1986). Involvement of a GTP-binding protein in mediation of serotonin and acetycholine responses in xenopus oocytes injected with rat brain messenger RNA. *Molecular Brain Research* **1,** 201–9.

De Chaffoy de Courcelles, D., Roevens, P., and Van Belle, H. (1984). Stimulation by serotonin of 40 kDa and 20 kDa protein phosphorylation in human platelets. *FEBS Letters* **171,** 289–92.

De Chaffoy de Courcelles, D., Leysen, J. E., De Clerck, F., Van Belle, H., and Janssen, P. A. J. (1985). Evidence that phospholipid turnover is the signal transducing system coupled to serotonin-S_2 receptor sites. *Journal of Biological Chemistry* **260,** 7603–8.

De Chaffoy de Courcelles, D., Roevens, P., Wynants, J., and Van Belle, H. (1987). Serotonin-induced alterations in inositol phospholipid metabolism in human platelets. *Biochimica et Biophysica Acta* **927,** 291–302.

De Vivo, M. and Maayani, S. (1985). Inhibition of forskolin-stimulated adenylate cyclase activity by 5-HT receptor agonists. *European Journal of Pharmacology* **119,** 231–4.

De Vivo, M. and Maayani, S. (1986). Characterization of 5-hydroxytryptamine$_{1A}$-receptor-mediated inhibition of forskolin-stimulated adenylate cyclase activity in guinea-pig and rat hippocampal membranes. *Journal of Pharmacology and Experimental Therapeutics* **238,** 248–53.

Dixon, R. A. F., Kobilka, B. K., Strader, D. J., Benovic, J. L., Dohlman, H. G., Frielle, T., Bolanowski, M. A., Bennett, C. D., Rands, E., Diehl, R. E., Mumford, R. A., Slater, E. E., Sigal, I. S., Caron, M. G., Lefkowitz, R. J., and Strader, C. D. (1986). Cloning of the gene and cDNA for mammalian β-adrenergic receptor and homology with rhodopsin. *Nature* **321,** 75–9.

Dompert, W. U., Glaser, T., and Traber, J. (1985). ^{3}H-TVX Q 7821: Identification of 5-HT_1 binding sites as target for a novel putative anxiolytic. *Naunyn-Schmiedeberg's Archives of Pharmacology* **328,** 467–70.

Doyle, V. M., Creba, J. A., Rüegg, U. T., and Hoyer, D. (1986). Serotonin increases the production of inositol phosphates and mobilises calcium via the 5-HT_2 receptor in A_7r_5 smooth muscle cells. *Naunyn-Schmiedeberg's Archives of Pharmacology* **333,** 98–103.

Dumuis, A., Bouhelal, R., Sebben, M., and Bockaert, J. (1988). A 5-HT receptor in the central nervous system, positively coupled with adenylate cyclase, is antagonised by ICS 205–930 *European Journal of Pharmacology* **146,** 187–8.

Engel, G., Göthert, M., Hoyer, D., Schlicker, E., and Hillenbrand, K. (1986). Identity of inhibitory presynaptic 5-hydroxytryptamine (5-HT) autoreceptors in the rat brain cortex with 5-HT_{1B} binding sites. *Naunyn-Schmiedeberg's Archives of Pharmacology* **332,** 1–7.

Erne, P. and Pletscher, A. (1985). Rapid intracellular release of calcium in human platelets by stimulation of 5-HT_2-receptors. *British Journal of Pharmacology* **84,** 545–9.

Fozard, J. R. (1984a). Neuronal 5-HT receptors in the periphery. *Neuropharmacology* **23,** 1473–86.

Fozard, J. R. (1984b). MDL 72222, a potent and highly selective antagonist at neuronal 5-hydroxytryptamine receptors. *Naunyn-Schmiedeberg's Archives of Pharmacology* **326,** 36–44.

Fozard, J. R. and Kilbinger, H. (1985). 8-OH-DPAT inhibits transmitter release from guinea-pig enteric cholinergic neurones by activating 5-HT_{1A} receptors. *British Journal of Pharmacology* **86,** 601P.

Gaddum, J. H. and Picarelli, Z. P. (1957). Two kinds of tryptamine receptor. *British Journal of Pharmacology and Chemo therapy* **12,** 323–8.

Gozlan, H., El Mestikawy, S., Pichat, L., Glowinski, J., and Hamon, M. (1983). Identification of presynaptic serotonin autoreceptors by a new ligand: ^{3}H-PAT. *Nature (London)* **305,** 140–2.

Graziano, M. P. and Gilman, A. G. (1987). Guanine nucleotide-binding regulatory proteins: mediators of transmembrane signalling. *Trends in Pharmacological Sciences* **8,** 478–86.

Greningloh, G., Rienitz, A., Schmitt, B., Methfessel, C., Zensen, M., Beyreuther, K., Gundelfinger, E. D., and Betz, H. (1987). The strychnine-binding subunit of the glycine receptor shows homology with nicotinic acetylcholine receptors. *Nature (London)* **328,** 215–20.

Hamblin, M. W., Ariani, K., Adriaenssens, P. I., and Ciaranello, R. D. (1987). ^{3}H dihydroergotamine as a high affinity, slowly dissociating radioligand for 5-HT_{1B} binding sites in rat brain membranes: evidence for guanine nucleotide regulation of agonist affinity states. *Journal of Pharmacology and Experimental Therapeutics* **243,** 989–1001.

Heuring, R. E. and Peroutka, S. J. (1987). Characterization of a novel ^{3}H-5-hydroxytryptamine binding site subtype in bovine brain membranes. *Journal of Neuroscience* **7,** 894–903.

Heuring, R. A., Schlegel, J. R., and Peroutka, S. J. (1986). Species variations in RU 24969 interactions with non-5-HT_{1A} binding sites. *European Journal of Pharmacology* **122,** 279–82.

Hoyer, D. (1988). Molecular pharmacology and biology of 5-HT_{1C} receptors. *Trends in Pharmacological Sciences* **9,** 89–94.

Hoyer, D. and Neijt, H. C. (1987). Identification of serotonin 5-HT_3 recognition sites by radioligand binding in NG108–15 neuroblastoma-glioma cells. *European Journal of Pharmacology* **143,** 191–2.

Hoyer, D. and Neijt, H. C. (1988). Identification of serotonin 5-HT_3 recognition sites in membranes of N1E-115 neuroblastoma cells by radioligand binding. *Molecular Pharmacology* **33,** 303–9.

Hoyer, D. and Schoeffter, P. (1988). 5-HT_{1D} receptors inhibit forskolin-stimulated adenylate cyclase activity in calf substantia nigra. *European Journal of Pharmacology* **147,** 145–7.

Hoyer, D., Engel, G., and Kalkman, H. O. (1985a). Characterization of the 5-HT_{1B} recognition site in rat brain: binding studies with ^{125}I iodocyanopindolol. *European Journal of Pharmacology* **118,** 1–12.

Hoyer, D., Engel, G., and Kalkman, H. O. (1985b). Molecular pharmacology of 5-HT_1 and 5-HT_2 recognition sites in rat and pig brain membranes: radioligand binding studies with ^{3}H5-HT, ^{3}H 8-OH-DPAT, (-) ^{125}I iodocyanopindolol, ^{3}H mesulergine and ^{3}H ketanserin. *European Journal of Pharmacology* **118,** 13–23.

Hoyer, D., Pazos, A., Probst, A., and Palacios, J. M. (1986a). Serotonin receptors in the human brain. I Characterization and autoradiographic localization of 5-HT_{1A} recognition sites. Apparent absence of 5-HT_{1B} recognition sites. *Brain Research* **376,** 85–96.

Hoyer, D., Pazos, A., Probst, A., and Palacios, J. M. (1986b). Serotonin receptors in the human brain II. Characterization and autoradiographic localization of 5-HT_{1C} and 5-HT_2 recognition sites. *Brain Research* **376,** 97–107.

Hoyer, D., Waeber, C., Pazos, A., Probst, A., and Palacios, J. M. (1988). Identification of a 5-HT_1 recognition site in human brain membranes different from 5-HT_{1A}, 5-HT_{1B} and 5-HT_{1C} sites. *Neuroscience Letters* **85,** 357–62.

Ireland, S. J. and Tyers, M. B. (1987). Pharmacological characterization of 5-hydroxytryptamine-induced depolarization of the rat isolated vagus nerve. *British Journal of Pharmacology* **90,** 229–38.

Julius, D., MacDermott, A. B., Axel, R., and Jessell, T. M. (1988). Molecular characterization of a functional cDNA encoding the serotonin 1C receptor. *Science*, **241,** 558–64.

Kendall, D. A. and Nahorski, S. R. (1985). 5-hydroxytryptamine stimulated inositol phospholipid hydrolysis in rat cerebral cortex slices: pharmacological characterisation and effects of antidepressants. *Journal of Pharmacology and Experimental Therapeutics* **233,** 473–9.

Kilpatrick, G. J., Jones, B. J., and Tyers, M. B. (1987). The identification and distribution of 5-HT_3 receptors in rat brain using radioligand binding. *Nature (London)* **330,** 746–8.

Kobilka, B. K., Dixon, R. A. F., Frielle, T., Dohlman, H. G., Bolanowski, M. A., Sigal, I. S., Yang-Feng, T. L., Francke, U., Caron, M. G., and Lefkowitz, R. J. (1987). cDNA for the human β_2-adrenergic receptor: A protein with multiple membrane-spanning domains and encoded by a gene whose chromosomal location is shared with that of the receptor for platelet-derived growth factor. *Proceedings of the National Academy of Science, USA* **84,** 46–50.

Kubo, T., Fukuda, K., Mikami, A., Maeda, A., Takahashi, H., Mishina, M., Haga, T., Haga, K., Ichiyama, A., Kangawa, K., Kojima, M., Matsuo, H., Hirose, T., and Numa, S. (1986). Cloning, sequencing and expression of complementary DNA encoding the muscarinic acetycholine receptor. *Nature (London)* **323,** 411–6.

Lefkowtiz, R. J., Benovic, J. L., Kobilka, B., and Caron, M. G. (1986). β-Adrenergic receptors and rhodopsin: shedding new light on an old subject. *Trends in Pharmacological Sciences* **7,** 444–8.

Leysen, J. E., Niemegeers, C. J. E., Tollenaere, J. P., and Laduron, P. M. (1978). Serotonergic component of neuroleptic receptors. *Nature (London)* **272,** 168–71.

Leysen, J. E., Niemegeers, C. J. E., Van Nueten, J. M., and Laduron, P. M. (1982). ^{3}H ketanserin (R41 468), a selective ^{3}H lingand for serotonin 2 receptor binding sites. *Molecular Pharmacology* **21,** 301–14.

Lübbert, H., Hoffman, B. J., Snutch, T. P., VanDyke, T., Levine, A. J., Hartig, P. R., Lester, H. A., and Davidson, N. (1987a). cDNA cloning of a serotonin 5-HT_{1C} receptor by electrophysiological assays of mRNA-injected xenopus oocytes. *Proceedings of the National Academy of Science, USA* **84,** 4332–6.

Lübbert, H., Snutch, T. P., Dascal, N., Lester, H. A., and Davidson, N. (1987b). Rat brain 5-HT_{1C} receptors are encoded by a 5–6 kbase mRNA size class and are functionally expressed in injected xenopus oocytes. *The Journal of Neuroscience* **7,** 1159–65.

Luchins, A. R. and Makman, M. H. (1980). Presence of histamine and serotonin receptors associated with adenylate cyclase in cultured calf-aorta smooth muscle cells. Pharmacological characterization of the histamine response. *Biochemical Pharmacology* **29,** 3155–61.

Lyon, R. A., Davis, K. H., and Titeler, M. (1987). ^{3}H-DOB (4-bromo-2,5-dimethoxyphenylisopropylamine) labels a guanyl nucleotide-sensitive state of cortical 5-HT_2 receptors. *Molecular Pharmacology* **31,** 194–9.

MacDermot, J., Higashida, H., Wilson, S. P., Matsuzawa, H., Minna, J., and Nirenberg, M. (1979). Adenylate cyclase and acetycholine release regulated by separate serotonin receptors of somatic cell hybrids. *Proceedings of the National Academy of Science USA* **76,** 1135–9.

Makman, M. H., Dvorkin, B., and Crain, S. M. (1988). Modulation of adenylate cyclase activity of mouse spinal cord-ganglion explants by opioids, serotonin and pertussis toxin. *Brain Research* **445,** 303–13.

Maillet, C. M. and Garber, A. J. (1986). Adrenergic and serotonergic regulation of skeletal muscle metabolism in the rat: specificity of the serotonin- and isoproterenol-stimulable adenylyl cyclase in sarcolemma. *Molecular and Cellular Endocrinology* **48,** 179–90.

Markstein, R., Hoyer, D., and Engel, G. (1986). 5-HT_{1A}-receptors mediate stimulation of adenylate cyclase in rat hippocampus. *Naunyn-Schmiedeberg's Archives of Pharmacology* **333,** 335–41.

Maura, G. and Raiteri, M. (1986). Cholinergic terminals in rat hippocampus possess 5-HT_{1B} receptors mediating inhibition of acetycholine release. *European Journal of Pharmacology* **129,** 333–7.

Maura, G., Roccataglia, E., and Raiteri, M. (1986). Serotonin autoreceptor in rat hippocampus: Pharmacological characterization as a subtype of the 5-HT_1 receptor. *Naunyn-Schmiedeberg's Archives of Pharmacology* **334,** 323–6.

Mawe, G. M., Branchek, T. A., and Gershon, M. D. (1986). Peripheral neural serotonin receptors: Identification and characterization with specific antagonists and agonists. *Proceedings of the National Academy of Science USA* **83,** 9799–803.

Middlemiss, D. N. and Fozard, J. R. (1983). 8-hydroxy-2-(di-n-propylamino)-tetralin discriminates between subtypes of the 5-HT_1 recognition site. *European Journal of Pharmacology* **90,** 151–3.

Molderings, G. J., Fink, K., Schlicker, E., and Göthert, M. (1987). Inhibition of noradrenaline release in the rat vena cava via presynaptic 5-HT_{1B} receptors. *Naunyn Schmiedeberg's Archives of Pharmacology* **336,** 245–50.

Nakaki, T., Roth, B. L., Chuang, D., and Costa, E. (1985). Phasic and tonic components in 5-HT_2 receptor-mediated rat aorta contraction: participation of Ca++ channels and phospholipase C. *Journal of Pharmacology and Experimental Therapeutics* **234,** 442–6.

Nathanson, N. M. (1987). Molecular properties of the muscarinic acetylcholine receptor. *Annual Review of Neuroscience* **10,** 195–236.

Neijt, H. C., Vijverberg, H. P. M., and Van den Berken, J. (1986). The dopamine response in mouse neuroblastoma cells is mediated by serotonin 5HT receptors. *European Journal of Pharmacology* **127,** 271–4.

Neijt, H. C., Karpf, A., Schoeffter, P., Engel, G., and Hoyer, D. (1988a). Characterisation of 5-HT_3 recognition sites in membranes of NG 108–15 neuroblastoma-glioma cells by radioligand binding. *Naunyn-Schmiedeberg's Archives of Pharmacology* **337,** 493–9.

Neijt, H. C., te Duits, I. J., and Vijverberg, H. P. M. (1988b). Pharmacological characterization of serotonin 5-HT_3 receptor-mediated electrical response in cultured mouse neuroblastoma cells. *Neuropharmacology* **27,** 301–7.

Nishizuka, Y. (1984). The role of protein kinase C in cell surface signal transduction and tumor promotion. *Nature (London)* **308,** 693–8.

Noda, M., Furutani, Y., Takahashi, H., Toyosato, M., Tanabe, T., Shimizu, S.,

Kikyotani, S., Kayano, T., Hirose, T., Inayama, S., and Numa, S. (1983). Cloning and sequencing of calf cDNA and human genomic DNA encoding α-subunit precursor of muscle acetylcholine receptor. *Nature (London)* **305,** 818–23.

Ogura, A., Ozaki, K., Kudo, Y., and Amano, T. (1986). Cytosolic calcium elevation and cGMP production induced by serotonin in a clonal cell of glial origin. *Journal of Neuroscience* **6,** 2489–94.

Palacios, J. M., Markstein, R., and Pazos, A. (1986). Serotonin-1C sites in the choroid plexus are not linked in a stimulatory or inhibitory way to adenylate cyclase. *Brain Research* **380,** 151–4.

Paupardin-Tritsch, D., Hammond, C., Gerschenfeld, H. M., Nairn, A. C., and Greengard, P. (1986). cGMP-dependant protein kinase enhances Ca^{2+} current and potentiates the serotonin-induced Ca^{2+} current increase in snail neurones. *Nature (London)* **323,** 812–4.

Pazos, A. and Palacios, J. M. (1985). Quantative auto-radiographic mapping of serotonin receptors in the rat brain. I. Serotonin-1 receptors. *Brain Research* **346,** 205–30.

Pazos, A., Hoyer, D., and Palacios, J. M. (1984a). Mesulergine, a selective serotonin-2 ligand in the rat cortex, does not label these receptors, in porcine and human cortex; Evidence for species differences on brain serotonin-2 receptors. *European Journal of Pharmacology* **106,** 531–8.

Pazos, A., Hoyer, D., and Palacios, J. M. (1984b). The binding of serotonergic ligands ot the porcine choroid plexus: characterization of a new type of serotonin recognition site, *European Journal of Pharmacology* **106,** 539–46.

Pazos, A., Probst, A., and Palacios, J. M. (1987a). Serotonin receptors in the human brain-III. Autoradiographic mapping of serotonin-1 receptors. *Neuroscience* **21,** 97–122.

Pazos, A., Probst, A., and Palacios, J. M. (1987b). Serotonin receptors in the human brain-IV. Autoradiographic mapping of serotonin-2 receptors. *Neuroscience* **21,** 123–39.

Pedigo, N. W., Yamamura, H. I., and Nelson, D. L. (1981). Discrimination of multiple ^3H 5-hydroxytryptamine-binding sites by the neuroleptic spiperone in rat brain. *Journal of Neurochemistry* **36,** 220–6.

Peralta, E. G., Winslow, J. W., Peterson, G. L., Smith, D. H., Ashkenazi, A., Ramachandran, J., Schimerlik, M. I., and Capon, D. J. (1987). Primary structure and biochemical properties of an M_2 muscarinic receptor. *Science* **236,** 600–5.

Peroutka, S. J. (1986). Pharmacological differentiation and characterization of 5-HT_{1A}, 5-HT_{1B} and 5-HT_{1C} binding sites in rat frontal cortex. *Journal of Neurochemistry* **47,** 529–40.

Peroutka, S. J. (1988). 5-hydroxytryptamine receptor subtypes. *Annual Reviews in Neurosciences* **11,** 45–60.

Peroutka, S. J. and Hamik, A. (1988). ^3H Quipazine labels 5-HT_3 recognition sites in rat cortical membranes. *European Journal of Pharmacology* **148,** 297–9.

Peroutka, S. J. and Snyder, S. H. (1979). Multiple serotonin receptors: differential binding of ^3H 5-hydroxytryptamine, ^3H lysergic acid diethylamide and ^3H spiroperidol. *Molecular Pharmacology* **16,** 687–99.

Peroutka, S. J., Lebovitz, R. M., and Snyder, S. H. (1981). Two distinct central serotonin receptors with different physiological functions. *Science* **212,** 827–9.

Peroutka, S. J., Huang, S., and Allen, G. S. (1986). Canine basilar artery contractions mediated by 5-hydroxytryptamine$_{1A}$-receptors. *Journal of Pharmacology and Experimental Therapeutics* **237,** 901–6.

Raiteri, M., Maura, G., Bonanno, G., and Pittaluga, A. (1986). Differential pharmacology and function of two 5-HT$_1$ receptors modulating transmitter release in rat cerebellum. *Journal of Pharmacology and Experimental Therapeutics* **237,** 644–8.

Ransom, R. W., Asarch, K. B., and Shih, J. C. (1986). ^{3}H 1-2-(4-aminophenyl)-ethyl-4-(3/trifluoromethylphenyl)poperazine: A selective radioligand for 5-HT$_{1A}$ receptors in rat brain. *Journal of Neurochemistry* **46,** 68–75.

Richardson, B. P., Engel, G., Donatsch, P., and Stadler, P. A. (1985). Identification of serotonin M-receptor subtypes and their specific blockade by a new class of drugs. *Nature (London)* **316,** 126–31.

Roth, B. L., Nakaki, T., Chuang, D.-M., and Costa, E. (1984). Aortic recognition sites for serotonin (5-HT) are coupled to phospholipase C and modulate phosphatidylinositol turnover. *Neuropharmacology* **23,** 1223–5.

Roth, B. L., Nakaki, T., Chuang, D.-M., and Costa, E. (1986). 5-hydroxytryptamine$_2$ receptors coupled to phospholipase C in rat aorta: modulation of phosphoinositide turnover by phorbol ester. *Journal of Pharmacology and Experimental Therapeutics* **238,** 480–5.

Round, A. and Wallis, D. I. (1986). The depolarizing action of 5-hydroxytryptamine on rabbit vagal afferent and sympathetic neurones in vitro and its selective blockade by ICS 205–930. *British Journal of Pharmacology* **88,** 485–94.

Round, A. and Wallis, D. I. (1987). Further studies on the blockade of 5-HT depolarizations of rabbit vagal afferent and sympathetic ganglion cells by MDL 72222 and other antagonists. *Neuropharmacology* **26,** 39–48.

Schnellman, R. G., Waters, S. J., and Nelson, D. L. (1984). ^{3}H 5-hydroxytryptamine binding sites: Species and tissue variation. *Journal of Neurochemistry* **42,** 65–70.

Schoeffter, P., Waeber, C., Palacios, J. M., and Hoyer, D. (1988). The 5-HT$_{1D}$ receptor subtype is negatively coupled to adenylate cyclase in calf substantia nigra. *Naunyn-Schmiedeberg's Archives of Pharmacology* **337,** 602–8.

Schofield, P. R., Darlison, M. G., Fujita, N., Burt, D. R., Stephenson, F. A., Rodriguez, H., Rhee, L. M., Ramachandran, J., Reale, V., Glencorse, T. A., Seeburg, P. H., and Barnard, E. A. (1987). Sequence and functional expression of the GABA$_A$ receptor shows a ligand-gated receptor super-family. *Nature (London)* **328,** 221–7.

Seuwen, K., Magnaldo, I., and Pouyssegur, J. (1988). Serotonin stimulates DNA synthesis in fibroblasts via 5-HT$_{1B}$ receptors coupled to the inhibitory G-protein of adenylyl cyclase (Gi). Nature (London) **335,** 254–6.

Shenker, A., Maayani, S., Weinstein, H., and Green, J. P. (1983). Enhanced serotonin-stimulated adenylate cyclase activity in membranes from adult guinea-pig hippocampus. *Life Science* **32,** 2335–42.

Shenker, A., Maayani, S., Weinstein, H., and Green, J. P. (1985). Two 5-HT receptors linked to adenylate cyclase in guinea pig hippocampus are discriminated by 5-carboxamidotryptamine and spiperone. *European Journal of Pharmacology* **109,** 427–9.

Shenker, A., Maayani, S., Weinstein, H., and Green, J. P. (1987). Pharmacological characterization of two 5-hydroxytryptamine receptors coupled to adenylate cyclase in guinea pig hippocampal membranes. *Molecular Pharmacology* **31,** 357–67.

Sibley, D. R. and Lefkowitz, R. J. (1986). Molecular mechanisms of receptor desensitization using the β-adrenergic receptor-coupled adenylate cyclase as a model. *Nature (London)* **317,** 124–9.

Siegelbaum, S. A., Belardetti, F., Camardo, J. S., and Shuster, M. J. (1986). Modulation of the serotonin-sensitive potassium channel in aplysia sensory neurone cell body and growth cone. *Journal of Experimental Biology* **124,** 287–306.

Stevens, C. F. (1987). Channel families in the brain. *Nature (London)* **328,** 198–9.

Sumner, M. J., Humphrey, P. P. A. and Feniuk, W. (1987). Characterization of the 5-HT_1-like receptor mediating relaxation of porcine vena cava. *British Journal of Pharmacology* **92,** 574P.

Takaki, M., Branchek, T., Tamir, H., and Gershon, M. D. (1985). Specific antagonism of enteric neural serotonin receptors by dipetides of 5-hydroxytryptophan: Evidence that serotonin is a mediator of slow synaptic excitation in the myenteric plexus. *The Journal of Neuroscience* **5,** 1769–80.

Taylor, E. W., Duckles, S. P., and Nelson, D. L. (1986). Dissociation constants of serotonin agonists in the canine basilar artery correlate to K_i values at the 5-HT_{1A} binding site. *Journal of Pharmacology and Experimental Therapeutics* **236,** 118–25.

Trevethick, M. A., Feniuk, W., and Humphrey, P. P. A. (1984). 5-hydroxytryptamine-induced relaxation of porcine vena cava in vitro. *Life Science* **35,** 477–86.

Trevethick, M. A., Feniuk, W., and Humphrey, P. P. A. (1986). 5-Carboxamidotryptamine: a potent agonist mediating relaxation and elevation of cyclic AMP in the isolated neonatal porcine vena cava. *Life Science* 1521–8.

VanderMaelen, C. P. and Aghajanian, G. K. (1980). Intracellular studies showing modulation of facial motoneurone excitability by serotonin. *Nature* **287,** 346–7.

VanderMaelen, C. P. and Aghajanian, G. K. (1982). Serotonin-induced depolarization of rat facial motoneurons in vivo: comparison with amino acid transmitters. *Brain Research* **238,** 139–52.

Waeber, C., Dietl, M. M., Hoyer, D., Probst, A., and Palacios, J. M. (1988a). Visualization of a novel serotonin recognition site (5-HT_{1D}) in the human brain by autoradiography. *Neuroscience Letters* **88,** 11–16.

Waeber, C., Schoeffter, P., Palacios, J. M., and Hoyer, D. (1988b). Molecular pharmacology of 5-HT_{1D} recognition sites: radioligand binding studies in human pig and calf brain membranes. *Naunyn-Schmiedeberg's Archives of Pharmacology* **337,** 595–601.

Watling, K. J., Aspley, S., Swain, C. J., and Saunders, J. (1988). ^{3}H-Quaternised ICS 205–930 labels 5-HT_3 receptor binding sites in rat brain. *European Journal of Pharmacology* **149,** 397–8.

Weiss, S., Sebben, M., Kemp, D. E., and Bockaert, J. (1986). Serotonin 5-HT_1 receptors mediate inhibition of cyclic AMP production neurons. *European Journal of Pharmacology* **120,** 227–30.

Yarden, Y., Rodriguez, H., Wong, S. K. F., Brandt, D. R., May, D. C., Burnier, J., Harkins, R. N., Chen, E. Y., Ramachandran, J., Ullrich, A., and Ross, E. M. (1986). The avian β-adrenergic receptor: Primary structure and membrane topology. *Proceedings of the National Academy of Science, USA*. **83,** 6795–9.

Yoshimasa, T., Sibley, D. R., Bouvier, M., Lefkowitz, R. J., and Caron, M. G. (1987). Cross-talk beween cellular signalling pathways suggested by phorbol-ester induced adenylate cyclase phosphorylation. *Nature (London)* **327,** 67–70.

Note added in proof

The cloning of 5-HT_{1A} and 5-HT_2 receptors has now been reported (Fargin *et al.* 1988; Pritchett *et al.* 1988); they are, like the 5-HT_{1C} receptor, members of the 7 transmembrane spanning region receptor family. The similarity between 5-HT_{1C} and 5-HT_2 receptors is illustrated by significant sequence homology, especially in putative transmembrane spanning regions, of the 5-HT_{1C} and 5-HT_2 receptor clones (Pritchett *et al.* 1988; Julius *et al.* 1988). Further, 5-HT_{1C} and 5-HT_2 receptors expressed in Xenopus oocytes are both able to stimulate inositol phosphate production, Ca^{2+} mobilization, and Cl^- channels. In contrast, there is little sequence homology between the 5-HT_{1A} receptor on the one hand and 5-HT_{1C} and 5-HT_2 receptors on the other (Pritchett *et al.* 1988; Fargin *et al.* 1988), as suggested earlier from the biochemical pharmacology of these receptors.

Fargin, A., Raymond, J. R., Lohse, M. J., Kobilka, B. K., Caron, M. G., and Lefkowitz, R. J. (1988). The genomic clone G-21 which resembles a beta-adrenergic receptor sequence encodes the 5-HT_{1A} receptor. *Nature (London)* **335,** 358–60.

Pritchett, D. B., Bach, A. W. J., Wozny, M., Taleb, O., Dal Toso, R., Shih, J., and Seeburg, P. H. (1988). Structure and functional expression of cloned rat serotonin 5-HT_2 receptor. *EMBO Journal* **73,** 4235-41.

5
Mechanisms of 5-hydroxytryptamine-induced vasoconstriction

W. FENIUK and P. P. A. HUMPHREY

5.1 Introduction

Ever since the identification of 5-hydroxytryptamine (5-HT; serotonin) as a vasoconstrictor substance in blood (Rapport *et al.* 1948) countless studies have demonstrated its marked vasoconstrictor activity in a variety of different vascular beds from various species. It is now abundantly clear that many different factors can modify the vasoconstrictor action of 5-HT both *in vitro* and *in vivo* and these are dealt with in greater detail in other chapters. In the coronary arteries of pigs and dogs the vasoconstrictor action of 5-HT is highly dependent upon the integrity of the vascular endothelium (Cocks and Angus 1983), whilst the presence of 5-HT receptors which mediate vasodilatation, either directly or indirectly, might also serve to attenuate the degree of the vasoconstrictor action of 5-HT (see Feniuk *et al.* 1981a and 1983). In recent years much attention has focused on the pathological role that 5-HT may play in certain vasoconstrictor disease states. Thus, platelet-derived 5-HT has been implicated in cerebral vasospasm following subarachnoid haemorrhage, coronary vasospasm, Raynaud's phenomenon, and even hypertension (Vanhoutte and Houston 1986; see Chapter 16). Recent elegant histochemical, biochemical, and pharmacological studies (Reinhard *et al.* 1979; Moskowitz *et al.* 1982; Edvinsson *et al.* 1984) have provided good evidence for the identification of a specific 5-HT neuronal innervation of many cerebral blood vessels which has fuelled speculation that 5-HT may be involved in the maintenance of cerebral vascular tone and even involved in the pathology of migraine (Mackenzie *et al.* 1985; Young *et al.* 1986). The identification of novel 5-HT receptor agonists or antagonists as potentially useful agents for the treatment of such diseases can only be achieved with a detailed and systematic understanding of the receptors mediating the vascular actions of 5-HT both *in vitro* and *in vivo*. In recent years several extensive reviews have been published concerning the contractile actions of 5-HT on vascular smooth muscle, although these have focused largely, but not exclusively, on 5-HT_2 receptor mechanisms (e.g. Van Nueten *et al.* 1984a; Feniuk 1984). This chapter reviews the mechanisms, both direct and indirect, by which 5-HT may cause contraction of vascular smooth muscle *in vitro* and *in vivo* with a specific emphasis on the receptors involved in these responses.

5.2 Nomenclature of 5-HT receptors

The most powerful actions of 5-HT are mediated through specific receptors for this amine. We will therefore focus in the main on the 5-HT receptor types involved in its vasoconstrictor actions. Throughout this book the nomenclature of 5-HT receptors proposed by Bradley *et al.* (1986b) has been used. The basis of this receptor classification is outlined in Table 5.1.

Table 5.1 The classification of functional 5-HT receptors with selective agonists and antagonists

Receptor	Selective agonist	Antagonist
'5-HT_1-like'	5-Carboxamidotryptamine	Methiothepin Methysergide
5-HT_2	α-Methyl-5-HT has some selectivity	Ketanserin Ritanserin*
5-HT_3	2-Methyl-5-HT	MDL 72222 ICS 205-930 GR38032†

Modified from Bradley *et al.* (1986b). *Van Nueten *et al.* (1986). †Brittain *et al.* (1987a)

5.3 Vasoconstriction via stimulation of '5-HT_1-like' receptors

Bradley *et al.* (1986b) adopted the term '5-HT_1-like' to describe 5-HT receptors which mediate certain functional responses including vasoconstriction as well as vasodilatation. They were so named because they have characteristics which are similar to, but not always the same as, the 5-HT_1 binding site in rat brain (Humphrey and Feniuk 1987a). Although definitive pharmacological characterization of functional '5-HT_1-like' receptors in vascular smooth muscle is undoubtedly hampered by the lack of highly specific and selective blocking drugs, they can be characterized by the high potency of 5-carboxamidotryptamine as an agonist and by a resistance to blockade by 5-HT_2 receptor blocking drugs such as ketanserin and 5-HT_3 receptor blocking drugs such as MDL 72222. In addition, methiothepin has antagonist activity at '5-HT_1-like' receptors and methysergide may act as either an antagonist or partial agonist. As already mentioned, most recent reviews on 5-HT vasoconstrictor mechanisms have tended to focus largely on 5-HT_2 receptor mediated vasoconstriction (see Section 5.4) yet many 5-HT-induced vasoconstrictor responses both *in vitro* and *in vivo* are resistant to blockade by 5-HT_2 receptor blocking drugs (e.g. Saxena 1972; Apperley *et al.* 1980; Feniuk *et al.* 1985; Jandhyala and Kivlighn 1987).

5.3.1 Studies *in vitro*

The contractile effects of 5-HT in the dog saphenous vein, and dog, human, and rabbit basilar arteries are largely resistant to blockade by selective 5-HT_2 receptor blocking drugs (Apperley *et al.* 1980; Forster and Whalley 1982; Feniuk *et al.* 1985; Bradley *et al.* 1986a; Peroutka *et al.* 1986), demonstrating that in these preparations, 'non 5-HT_2' receptors mediate vascular smooth muscle contraction and providing evidence that 5-HT receptors mediating vasoconstriction are heterogeneous.

Feniuk *et al.* (1985) studied the actions of a series of indoleamine analogues in the dog isolated saphenous vein as well as investigating the antagonistic activity of some 5-HT_2 receptor blocking drugs and concluded, on the basis of those and earlier studies (Apperley *et al.* 1980), that the 5-HT receptor mediating contraction was similar to the 5-HT_1 site and called it '5-HT_1-like'. First, ketanserin has a low affinity for 5-HT_1 sites and was ineffective as an antagonist at 5-HT receptors in the dog saphenous vein. Secondly, methysergide was a weak partial agonist in the dog saphenous vein with an affinity similar to that at 5-HT_1 binding sites (Apperley *et al.* 1980; Leysen *et al.* 1981). Thirdly, Engel *et al.* (1983) have demonstrated that 5-carboxamidotryptamine has a high affinity for the 5-HT_1 site and have confirmed our earlier observation (Feniuk *et al.* 1981b) that 5-carboxamidotryptamine is even more potent than 5-HT in causing contraction of the dog isolated saphenous vein (see Table 5.2). Finally, and most importantly, methiothepin caused a marked and selective antagonism of the contractile effect of 5-HT in the dog isolated saphenous vein, providing the most definitive evidence for the claim that '5-HT_1-like' receptors mediate contraction in this preparation (Apperley and Humphrey 1986).

The definition of the term '5-HT_1-like' receptor has deliberately been kept as broad as possible to allow scope for subdivision and modification as our knowledge advances. Even on the basis of available data, there is some evidence that '5-HT_1-like' receptors can be subdivided. Thus, there are marked differences in the relative agonistic potency of 5-carboxamidotryptamine and α-methyl 5-HT in isolated functional preparations containing '5-HT_1-like' receptors mediating vasoconstriction compared with those mediating vasodilation (see Table 5.2). Furthermore, we have now identified a novel 5-HT receptor agonist, AH25086, which has a remarkably selective vasoconstrictor action in the dog isolated saphenous vein, but which is devoid of agonist and antagonist activity at other functional 5-HT receptors, including '5-HT_1-like' receptors in cat saphenous vein which mediate relaxation (Table 5.2; see Humphrey *et al.* 1987a). The contractile

Table 5.2 Relative agonistic potencies of some structural analogues of 5-HT in isolated tissues containing different 5-HT receptors

	Dog saphenous vein (contraction—'5-HT_1-like)'	Cat saphenous vein (relaxation—'5-HT_1-like)'	Rabbit aorta (contraction—5-HT_2)	Rat vagus nerve (depolarization—5-HT_3)
5-HT	1	1	1	1
5-Carboxamidotryptamine	0.3	0.02	26	>1000
AH25086	4.0*	>100	>192	>1000
α-Methyl-5-HT	13	571	2	>1000
5-Fluoro-α-methyl tryptamine	77	>300	7	NT
2-Methyl-5-HT	>240	>100	>46	4

Values are the geometric means of the equipotent concentration ratio (5-HT = 1).
Data derived from Humphrey (1984) and unpublished observations.
Note that the rank order of agonist potencies is very different in the two '5-HT_1-like' receptor containing preparations. The selectivity of action of AH25086 for the '5-HT_1-like' receptor in dog saphenous vein is at least as great as that for 2-methyl-5-HT at 5-HT_3 receptors.
*95 per cent confidence limits (1.5–10.5), $n = 4$ or more (Humphrey and Feniuk 1988).
NT – not tested

effect of AH25086 in the dog isolated saphenous vein is resistant to blockade by ketanserin and MDL 72222 but can be selectively antagonized by methiothepin (Fig. 5.1). This suggests that the agonistic action of AH25086 results from activation of a particular type of '$5\text{-}HT_1$-like' receptor, providing further evidence for their subdivision. The identification of the functional selectivity of AH25086 at '$5\text{-}HT_1$-like' receptors in the dog saphenous vein contrasts markedly with that of RU 24969, which is clearly $5\text{-}HT_1$-selective on the basis of ligand binding experiments (Hunt and Oberlander 1981; Middlemiss 1985) but is a potent agonist at $5\text{-}HT_2$ receptors in the rabbit isolated aorta (see Section 5.4.1). Interestingly, recent studies have shown that the archetypal '$5\text{-}HT_1$-like' receptor ligand, 5-carboxamidotryptamine, also stimulates $5\text{-}HT_2$ receptors in the renal artery of the anaesthetized rabbit (Wright and Angus, 1988). However, this effect appears to be indirect, the result of a release of endogenous 5-HT, possibly from blood platelets. This highlights the fact that selectivity is normally only relative but nevertheless our studies in a variety of $5\text{-}HT_2$ containing preparations has failed to demonstrate any agonist or antagonist activity for AH25086 at $5\text{-}HT_2$ receptors (our own unpublished observation).

Although studies from several laboratories have provided abundant evidence for the existence of 'non $5\text{-}HT_2$' receptors on cerebral blood vessels (Forster and Whalley 1982; Peroutka *et al.* 1986; Cohen and Colbert 1986; Bradley *et al.* 1986a), other workers (e.g. Van Nueten *et al.* 1981; Müller-Schweinitzer and Engel 1983) have provided conflicting

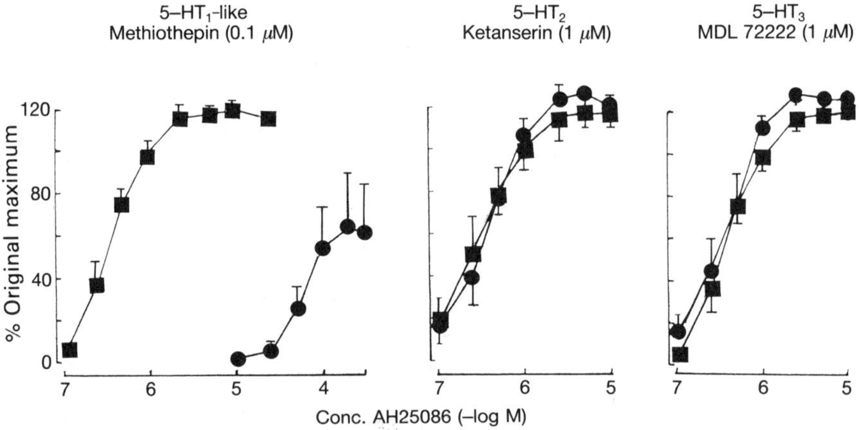

Fig. 5.1 Effect of methiothepin, ketanserin, and MDL 72222 on contractile effects of AH25086 in dog isolated saphenous vein. Concentration–effect curves are shown in the presence of antagonist (●) in comparison with time matched control responses (■). All values are mean ± SEM from at least four experiments which were conducted as described previously (Apperley *et al.* 1980; Feniuk *et al.* 1985)

evidence in support of a 5-HT_2 receptor mediated constrictor response in the canine basilar artery. Further controversy stems from a recent claim that a 5-HT_{1A} receptor mediates contraction of canine basilar artery but the interpretation of this study is complicated by the use of insufficiently selective drug tools, as well as by the problems of attempting to measure agonist affinities in a functional study (Taylor *et al.* 1986). From our own work it seems that some blood vessels including the canine basilar artery contain a mixed and variable population of 5-HT_2 and '5-HT_1-like' receptors (like those in the dog saphenous vein) which mediate contraction, and this heterogeneous receptor population undoubtedly confounds attempts to characterize the actions of 5-HT in such preparations (Connor *et al.* 1987; our own unpublished observations). Studies with more selective agonists, such as AH25086 and the more recently described GR43175 (Humphrey *et al.* 1987 and 1988), identified on the basis of sound functional studies, now provide a more definitive means of characterizing '5-HT_1-like' receptors mediating the vasoconstrictor action of 5-HT. It should be emphasized yet again that these '5-HT_1-like' receptors do not resemble the 5-HT_{1A}, 5-HT_{1B}, 5-HT_{1C}, or 5-HT_{1D} binding sites (see Heuring and Peroutka 1987). Thus, in the dog saphenous vein for example, 8-OH-DPAT, spiperone, cyanopindolol, mesulergine (all 1 μM), and metergoline (0.1 μM) are devoid of agonistic or antagonistic activity (Feniuk *et al.* 1985; Humphrey *et al.* 1988; and our own unpublished observations). A similar receptor appears to exist pre-junctionally in the dog saphenous vein and in the rat kidney where it mediates inhibition of noradrenaline release (Watts *et al.* 1981; Charlton *et al.* 1986).

5.3.2 Studies *in vivo*

Although the vasoconstrictor action of 5-HT at the arteriolar level in rats and on the major conducting arteries in a range of species is undoubtedly mediated via the activation of ketanserin-sensitive 5-HT_2 receptors (see Section 5.4.2), there are many examples where the vasoconstrictor action of 5-HT *in vivo* appears to be mediated via 'non 5-HT_2' receptors. Saxena (1974) suggested that atypical 5-HT receptors might mediate vasoconstriction in the carotid artery bed of anaesthetized dogs on the basis of the agonistic activity of methysergide and on the weak antagonistic activity of a range of 5-HT_2 receptor blocking drugs on 5-HT-induced increases in total carotid arterial vascular resistance (Saxena *et al.* 1971; Saxena 1972). Similar 5-HT receptors appear to mediate constriction in the primate cranial circulation (Mylecharane *et al.* 1978), the dog hindlimb (Feniuk *et al.* 1981a; Blackshear *et al.* 1985), and the dog forelimb (Jandhyala and Kivlighn 1987). Interestingly, in the dog forelimb, Jandhyala and Kivlighn demonstrated that methysergide, but not ketanserin, was capable of antagonising 5-HT-induced vasoconstrictor responses. Furthermore methysergide was capable of antagonizing neurogenically mediated vasoconstrictor responses, suggesting that 5-HT may be a neurotransmitter in this vascular bed. These findings would be consistent with the view that the vasoconstrictor action of 5-HT in these vascular beds is mediated by '5-HT_1-like' receptors.

Extensive studies from Saxena's group have demonstrated that a large proportion of carotid arterial blood flow bypasses the capillary circulation via arteriovenous anastomoses in anaesthetized cats and pigs (Spierings and Saxena 1979; Saxena and Verdouw 1982). The 5-HT receptors mediating vasoconstriction of carotid arteriovenous anastomoses appear to be '5-HT_1-like' since they are resistant to antagonism by ketanserin (Verdouw *et al.* 1984) and MDL 72222 (Saxena *et al.* 1986). They can be stimulated by the '5-HT_1-like' receptor agonist, 5-carboxamidotryptamine (Saxena and Verdouw 1985), and can be blocked by methiothepin in doses which themselves did not affect carotid haemodynamics (Saxena *et al.* 1986). The '5-HT_1-like' receptor mediating constriction of carotid arteriovenous anastomoses is clearly similar to the 5-HT receptor mediating contraction of the dog isolated saphenous vein. Not surprisingly, AH25086 and GR43175, which have been identified as selective '5-HT_1-like' receptor agonists in the dog isolated saphenous vein (see Section 5.3.1), also cause a selective constriction of the carotid artery bed of anaesthetized dogs and of carotid arteriovenous anastomoses in anaesthetized cats (Humphrey *et al.* 1987; Brittain *et al.* 1987b; Feniuk *et al.* 1987).

There is therefore good evidence for the presence of '5-HT_1-like' receptors mediating vasoconstriction *in vivo*; however, their involvement in the overall haemodynamic response to 5-HT itself will also be dependent upon its activation of other 5-HT receptor mechanisms, including those tending to produce vasodilatation (Feniuk *et al.* 1981a and 1983; Connor *et al.* 1986; see also Chapter 7).

5.4 Vasoconstriction via stimulation of 5-HT_2 receptors

Peroutka and Snyder (1979) introduced the term 5-HT_2 receptor to describe a ligand binding site in the rat brain with a low affinity for 5-HT but with a high affinity for spiperone and other so-called 'classical' or 5-HT D-receptor antagonists, such as 2-bromo-LSD, methysergide, cyproheptadine, mianserin, and cinanserin. Unquestionably, attempts to correlate the 5-HT_2 binding site with the functional effects of 5-HT were hampered by both the lack of specificity and/or selectivity of these compounds. The identification of ligands such as ketanserin, which has a high affinity for 5-HT_2 but negligible affinity for 5-HT_1 recognition sites, provides us with a useful tool with which to characterize 5-HT_2 receptors pharmacologically (Leysen *et al.* 1981). Although there are some potential disadvantages with the use of ketanserin as a pharmacological probe, notably its α_1-adrenoceptor blocking activity (see Bradley *et al.* 1986b), more potent and selective 5-HT_2 receptor ligands, such as ritanserin, have now been identified (Van Nueten *et al.* 1986).

5.4.1 Studies *in vitro*

There is little doubt that the activation of 5-HT_2 receptors on vascular smooth muscle mediates contraction since ketanserin appears to be a competitive antagonist of the contractile effects of 5-HT in a range of different isolated vascular preparations from a number of different species (Table 5.3). It is evident that activation of specific 5-HT receptors is involved since the pA_2 value for ketanserin clearly exceeds its affinity at α_1-adrenoceptors (see Feniuk *et al.* 1985). Furthermore, many studies have directly correlated measures of affinity of a range of 5-HT_2 receptor blocking drugs determined from pA_2 values for antagonism of the contractile effects of 5-HT with their affinities at displacing either [^{3}H]spiperone or [^{3}H]ketanserin from rat cortical membranes (e.g. Humphrey *et al.* 1982; Lemberger *et al.* 1984; Van Nueten *et al.* 1984b).

Like others (e.g. Clancy and Maayani 1985; Leff and Martin 1986), the standard preparation which we have used to investigate 5-HT_2 receptor

Table 5.3 Antagonistic activity of ketanserin in a range of different isolated vasular preparations from a range of different species

Species	Preparation	pA_2	Slope of Schild plot	Reference
Rabbit	Aorta	8.67	0.94	Feniuk *et al.* (1985)
		8.56	0.91	Leff and Martin (1986)
	Tibial artery	8.72	0.96	Van Nueten *et al.* (1984b)
	Femoral artery	8.27	0.80	
Rat	Aorta	8.40	0.90	Bradley *et al.* (1983)
	Tail artery	9.08	1.15	Bradley *et al.* (1985)
		9.10	1.10	Van Nueten *et al.* (1981)
	Jugular vein	9.05	1.12	Leff and Martin (1986)
	Portal vein	9.88	1.05	Cummings *et al.* (1986)
	Mesenteric vein	9.59	0.97	Cummings *et al.* (1986)
Dog	Gastrosplenic artery	8.70	1.22	Van Nueten *et al.* (1981)
	Gastrosplenic vein	8.88	0.86	Van Nueten *et al.* (1981)
	Coronary artery	9.30	—	Frenken and Kaumann (1985)
	Internal carotid artery	9.02	1.24	Van Nueten *et al.* (1981)
Calf	Pulmonary artery	9.38	1.05	Frenken and Kaumann (1984)
	Coronary artery	9.45	1.07	Frenken and Kaumann (1984)

function is the rabbit isolated aorta (Apperley and Humphrey 1976; Feniuk *et al.* 1985) where we have determined the antagonistic potency of a wide range of 5-HT receptor blocking drugs and compared their dissociation constants determined from these functional studies with equivalent data obtained from ligand binding studies (Table 5.4). It can be seen that there is a close agreement between the measures of affinity obtained from both the functional and the binding studies and the data indicate that 5-HT_2 receptor activation mediates contraction of vascular smooth muscle.

Table 5.4 A comparison of measures of affinity of a variety of 5-HT receptor blocking drugs for 5-HT_2 binding sites with those for 5-HT receptors mediating contraction of rabbit isolated aorta

Antagonist	5-HT_2-binding*	Contraction†
Ketanserin	9.04	8.67
Methysergide	8.48	8.49
Cyproheptadine	8.77	8.73
Pizotifen	8.87	9.42
Spiperone	9.10	8.92
BW 501C67	8.55	9.10
LY 53857	8.66	7.40
Mianserin	8.14	8.17
Phentolamine	6.73	6.21

*Mean $-\log K_i$ values calculated from [^{3}H]ketanserin and/or [^{3}H]spiperone binding studies (Leysen *et al.* 1982; Fuller *et al.* 1986).
†pA_2 values against 5-HT-induced contraction of rabbit isolated aorta (Apperley and Humphrey 1976; Humphrey *et al.* 1982; Feniuk *et al.* 1985 and unpublished observations).

At present there is little good evidence to suggest that 5-HT_2 receptors mediating contraction of vascular smooth muscle are heterogenous. Nevertheless, trazodone has been shown to be approximately 20 times more potent as a 5-HT receptor blocking drug in the rat jugular vein compared with the rabbit aorta—compare data from Black *et al.* (1983) with Cohen *et al.* (1983)—yet ketanserin has a similar affinity in both preparations (Leff and Martin 1986). It would be premature at this stage, however, to consider 5-HT_2 receptor heterogeneity without considering other potential complicating factors (Humphrey 1984; Leff and Martin 1986; see also Chapters 2 and 3).

Characterization of the 5-HT_2 receptors mediating contraction of vascular smooth muscle can also be aided by the use of agonists. Although we have investigated the effect of a wide range of tryptamine agonists in the rabbit isolated aorta, only α-methyl 5-HT and 5-fluoro-α-methyltryptamine appear to show any degree of selectivity for the 5-HT_2 receptors mediating contraction in this preparation (Table 5.2). Claims of selectivity for agonists on the basis of ligand binding affinity measures are problematical since 5-HT itself has a low affinity for the 5-HT_2 binding site. Indeed, most 5-HT receptor agonists appear 5-HT_1 selective on the basis of ligand binding experiments. For example, it has been argued that RU 24969 is a selective 5-HT_1 receptor agonist on the basis of its high affinity at the 5-HT_1 recognition site (Hunt and Oberlander 1981; Middlemiss 1985). We have found that RU 24969 has marked agonistic activity at 5-HT_2 receptors in the rabbit isolated aorta with an EC_{50} value of 2.4 (0.4–15.0) μM (geometric mean, 95 per cent confidence limits), but its maximum response amounted to only 54 ± 6 per cent of the 5-HT maximum. Furthermore the contractile effect of RU 24969 was antagonized by ketanserin (30–300 nM) in a concentration-dependent manner, although this effect was accompanied by a progressive reduction in the maximum response (see Figure 5.2). Although the mechanism responsible for this reduced maximum is not known, the results indicate that RU 24969 is not a selective 5-HT_1 receptor agonist and that its agonistic activity in the rabbit aorta involves activation of 5-HT_2 receptors. These functional studies once again highlight the potential danger of putting too much emphasis on data derived with putative selective agonists, characterized solely on the basis of ligand binding studies. The need for judicious interpretation of data derived with the use of such agonists cannot be over-emphasized.

The biochemical consequences of 5-HT_2 receptor activation of vascular smooth muscle are less clear. Although there seems little doubt that the contractile effect of 5-HT, resulting from 5-HT_2 receptor activation, can

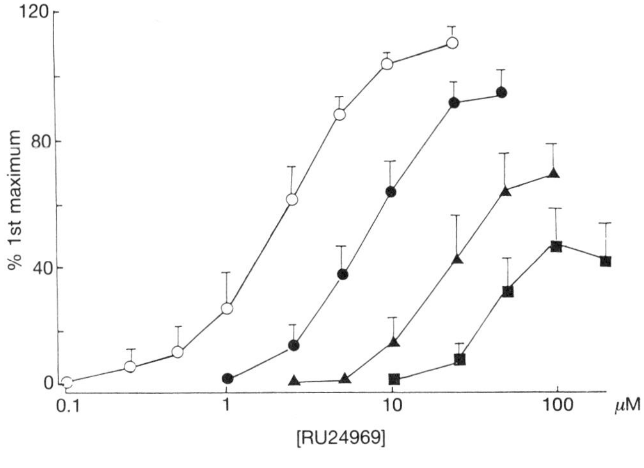

Fig. 5.2 Contractile effect of RU24969 in rabbit isolated aorta in the absence (○) and presence of increasing concentrations of ketanserin (0.03 μM, ●), (0.3 μM, ▲), and (3 μM, ■). Values are mean ± SEM from five experiments. These experiments were carried out in the presence of phentolamine (1 μM)

involve an influx of extracellular Ca^{2+} (Van Nueten and Vanhoutte 1981), the cellular mechanisms responsible for Ca^{2+} mobilization have yet to be fully elucidated. Berridge (1983) has shown that Ca^{2+} influx in the blowfly salivary gland is accompanied by the production of inositol triphosphate, and 5-HT stimulation of platelets is also accompanied by an increase in intracellular Ca^{2+} and phosphatidic acid (de Chaffoy de Courcelles *et al.* 1985). Conn and Sanders-Bush (1985) have shown that 5-HT_2 receptor activation in rat cerebral cortex caused a stimulation of phosphoinositide turnover. More germane, recent studies (Doyle *et al.* 1986) on rat aortic smooth muscle cells in culture have shown that 5-HT stimulated inositol phosphate accumulation and 5-HT stimulation of $^{45}Ca^{2+}$ efflux were potently and competitively antagonized by the 5-HT_2 receptor antagonists, ketanserin and pirenperone; furthermore, selective '5-HT_1-like' receptor agonists such as 5-carboxamidotryptamine were extremely weak. These and other data suggest that in vascular smooth muscle in general 5-HT mobilizes intracellular Ca^{2+} via phosphatidylinositol hydrolysis as a consequence of 5-HT_2 receptor activation. In addition, 5-HT_2 receptor stimulation can activate voltage-dependent calcium channels by a mechanism independent of phospholipase C mediated phosphotidylinositol hydrolysis (Roth *et al.* 1984; Nakaki *et al.* 1985). To our knowledge, no equivalent studies have been carried out in vascular preparations containing '5-HT_1-like' receptors. It would obviously be of great interest to compare the intracellular mediators involved (for a full discussion of second messengers associated with 5-HT receptors, see Chapter 4).

5.4.2 Studies *in vivo*

As has already been described, the existence of 5-HT_2 receptors mediating contraction of vascular smooth muscle *in vitro* on major conducting arteries is unequivocal. However, the involvement of vascular 5-HT_2 receptors in the overall haemodynamic response to 5-HT in different species *in vivo* is less clear.

In pithed rats, we have shown that 5-HT-induced increases in arterial blood pressure are susceptible to antagonism by a range of 5-HT_2 receptor blocking drugs (Fig. 5.3), confirming the results from many other studies (e.g. Fozard 1982). Furthermore, the 'selective' 5-HT_2 receptor agonist, α-methyl 5-HT, causes increases in diastolic blood pressure in DOCA salt hypertensive rats, an effect which can be specifically antagonized by ketanserin (Dalton *et al.* 1986). On the basis of such studies there seems little doubt about the presence of 5-HT_2 receptors on the arterioles controlling vascular resistance in rats. Indeed, studies from our laboratories have demonstrated that the vasopressor effect of 5-HT in pithed rats is associated with an increase in total peripheral resistance (Perren, unpublished observation). The importance of 5-HT_2 receptor activation in the

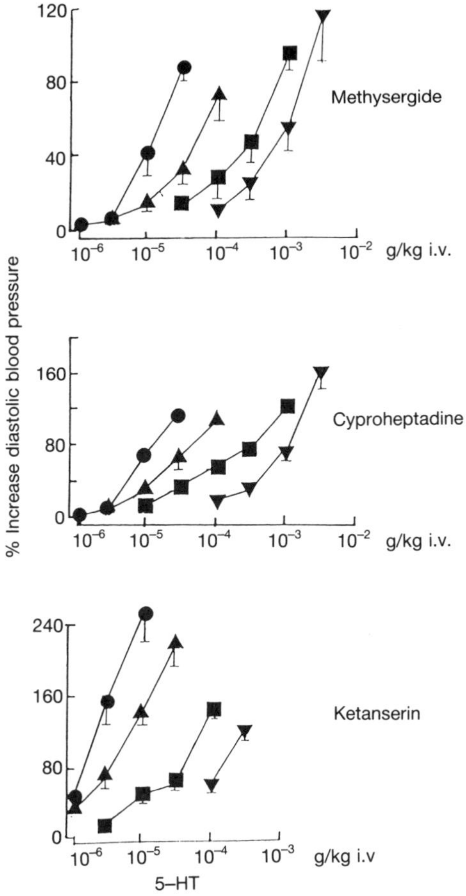

Fig. 5.3 Vasopressor effect of 5-HT in pithed rats before (●) and after treatment with 5-HT_2 receptor blocking drugs, methysergide, cyproheptadine, or ketanserin, all at 10 μg/kg (▲), 30 μg/kg (■), and 100 μg/kg (▼) i.v. Values are mean ± SEM from at least four experiments. The data with methysergide and cyproheptadine were obtained from a previously published study (see Feniuk *et al.* 1982)

mediation of 5-HT-induced arteriolar vasoconstriction in other species is less clear. Blackshear *et al.* (1985) have demonstrated, using angiography, that 5-HT induces large artery constriction in the dog hindlimb *in vivo* via activation of 5-HT_2 receptors, yet 5-HT_2 receptor blockade had little influence on the overall haemodynamic response to 5-HT, suggesting an absence of 5-HT_2 receptors at the level of the arterial resistance vessels. Indeed, it is possible, on the basis of the vasoconstrictor action of methysergide in the femoral artery bed, that '5-HT_1-like' receptors mediate vasoconstriction in this bed (Feniuk *et al.* 1981a; Bradley *et al.* 1986b).

Similarly, Saxena and Verdouw (1982 and 1984) have also suggested that 5-HT_2 receptors mediating vasoconstriction are located in the large conducting arteries of the carotid artery bed of anaesthetized cats and pigs, whilst '5-HT_1-like' receptors appear to mediate vasodilatation at the arteriolar level. The importance of 5-HT_2 receptors in the large arteries being involved in the mediation of changes in vascular resistance is questionable. O'Rourke (1982) has demonstrated that a 50 per cent reduction in calibre of major conducting arteries causes less than a 2 per cent reduction in blood flow. However, such vasoconstrictor mechanisms may have more important implications in occlusive vascular disease where arterial diameter may already be markedly reduced (Angus and Wright 1988).

Although we have focused on the direct action of 5-HT on vascular smooth muscle to produce contraction via 5-HT_2 receptors, there is some evidence that activation of 5-HT_2 receptors on adrenal medullary chromaffin cells causes the release of adrenaline and noradrenaline which then produce vasoconstriction via activation of α-adrenoceptors. Thus, in ganglion-blocked anaesthetized dogs intravenous 5-HT causes vasopressor responses which are susceptible to blockade by low doses of methysergide, cyproheptadine, and ketanserin, as well as phentolamine (Feniuk *et al.* 1981c; Humphrey and Feniuk 1987b). Although the precise mechanism involved in this catecholamine release is not known it may involve depolarization of adrenal medullary cells (see Humphrey and Feniuk 1987b).

5.5 Vasoconstriction via stimulation of 5-HT_3 receptors

To date, 5-HT_3 receptors have only been identified on neurones where they mediate depolarization (Bradley *et al.* 1986b). It is now well established that 5-HT can release noradrenaline from sympathetic nerve endings, but the mechanisms involved in such an action may differ (Humphrey *et al.* 1983; and see Table 5.5). 5-HT can release noradrenaline from the sympathetic nerve endings in the rabbit heart by a depolarizing action mediated by activation of 5-HT_3 receptors (Fozard and Mwaluko 1976; Fozard *et al.* 1979) but may also release noradrenaline in vascular smooth muscle via a tyramine-like action (Humphrey *et al.* 1983). Although both mechanisms are susceptible to blockade by cocaine, which is itself a weak 5-HT_3 receptor blocking drug (Humphrey 1978; Fozard *et al.* 1979), they can be differentiated on the basis of their susceptibility to blockade by more potent and selective 5-HT_3 receptor blocking drugs, such as MDL 72222 (Fozard 1984) and by their dependence upon the external calcium concentration (see Table 5.5). As far as we are aware, there is no good evidence for the existence of 5-HT_3 receptors, similar to those on the

Table 5.5 Characteristics of the mechanisms by which 5-HT can release noradrenaline from sympathetic nerve endings

5-HT_3-receptor activation	Tyramine-like action
Involves neuronal depolarization	Involves uptake and stoichiometric displacement
Ca^{2+} dependent (Fozard and Mwaluko 1976)	Ca^{2+} independent (Thoa *et al.* 1975)
Blockade by cocaine (Fozard *et al.* 1979)	Blockade by cocaine (Humphrey 1978)
Blockade by MDL 72222 (Fozard 1984)	Resistant to blockade by MDL 72222 (Fozard 1984)

sympathetic nerve terminals of the rabbit heart, occurring on the sympathetic nerve terminals of vascular smooth muscle either *in vitro* or *in vivo*. However, Marin *et al.* (1979) and Marin and Sanchez (1980) have demonstrated that the 5-HT-induced release of noradrenaline from cat isolated pial vessels is both sensitive to a reduction in the external Ca^{2+} concentration and is also susceptible to blockade by cocaine. These findings suggest that activation of 5-HT_3 receptors may be involved (see Table 5.5) but the definitive characterization of this mechanism requires further experimentation with the more potent and selective 5-HT_3 receptor blocking drugs now available.

5.6 Other mechanisms of 5-HT-induced vasoconstriction

5.6.1 Direct activation of α-adrenoceptors

In some isolated vascular preparations which contain few, if any, 5-HT receptors, 5-HT can be shown to produce vasoconstriction by a direct activation of α-adrenoceptors, albeit at high concentrations. Thus, in the rabbit ear artery (Apperley and Humphrey 1976) and rabbit external carotid artery (Black *et al.* 1981), phentolamine is a highly potent antagonist of the vasoconstrictor action of both 5-HT and noradrenaline. In preparations which do contain specific 5-HT receptors (e.g. rabbit aorta, dog saphenous vein) phentolamine is a much weaker antagonist of the effects of 5-HT, reflecting its much lower affinity for 5-HT receptors. Purdy *et al.* (1987) have concluded that 5-HT is an agonist at α-adrenoceptors in *all blood vessels*, but the ability of 5-HT to produce contraction via this mechanism is dependent upon the density and the coupling of the α-adrenoceptor population in any vessel.

5.6.2 Indirect activation of α-adrenoceptors

The ability of 5-HT to cause a release of noradrenaline as a consequence of 5-HT_3 receptor activation has been outlined (see Section 5.5). However 5-HT may also release noradrenaline due to a tyramine-like action, the characteristics of which are detailed in Table 5.5. This tyramine-like action occurs only with high concentrations of 5-HT, has been demonstrated in dog isolated saphenous vein (Humphrey 1978), and has recently been reviewed by Humphrey *et al.* (1983).

It is evident that many drugs demonstrate a cross reactivity between 5-HT and α-adrenoceptors, and good quantitative studies may be necessary to distinguish between the actions of both agonists and antagonists at the two amine receptors (Apperley and Humphrey 1976; Curro *et al.* 1978; Feniuk 1984).

5.6.3 Release of other mediators

Although this chapter has largely focused on the direct vasoconstrictor actions of 5-HT, these effects can be influenced by the release of other mediators. Thus, for example, the release of EDRF (Cocks and Angus 1983) serves to attenuate the coronary vasoconstrictor action of 5-HT in the dog, whilst Coughlin *et al.* (1981 and 1984) have demonstrated a 5-HT_2 receptor-mediated stimulation of prostacyclin synthesis in bovine aortic smooth muscle cells which might be expected to reduce the vasoconstrictor action of 5-HT. In contrast, recent studies on human umbilical arteries suggests that the release of vasoconstrictor prostanoids (thromboxanes) might contribute to the vasoconstrictor effect of 5-HT (McGrath and MacLennan 1986).

5.7 Concluding remarks

It is clear that 5-HT may induce vasoconstriction by a variety of different mechanisms both direct and indirect. The overall haemodynamic response to 5-HT will depend upon the summation of a variety of different effects mediated mostly via activation of 5-HT receptors of several different types, not only mediating vasoconstriction, but also vasodilatation and activation of a variety of cardio-pulmonary reflexes (e.g. Salmoiraghi *et al.*, 1956; see also Chapters 7 and 13). It is not surprising, therefore, that researchers over the last three or four decades have been perplexed by the unpredictability of its action *in vivo*.

Nevertheless, great strides have very recently been made in our understanding of 5-HT receptor classification and many new drug tools are now available (Fozard 1987). With such agents it is now clear that contraction of smooth muscle is mediated by at least two types of 5-HT receptor: the

$5\text{-}HT_2$ and the more recently identified '$5\text{-}HT_1$-like' receptor. It remains to be seen what their physiological, pathological, or pharmacological importance is, if any. Certainly the role of $5\text{-}HT_2$ receptors in hypertension is very controversial and the clinical value of specific $5\text{-}HT_2$ receptor blocking drugs questionable (see Humphrey *et al.* 1982; Reimann and Frölich 1983). Even less is known about '$5\text{-}HT_1$-like' receptors, but in animals their distribution seems more localized than $5\text{-}HT_2$ receptors, apparently being distributed predominantly in cranial vessels, notably carotid arteriovenous anastomoses (Humphrey *et al.* 1987 and 1988; Feniuk *et al.* 1987). The efficacy of the very selective $5\text{-}HT_1$-like receptor agonists, AH25086 and GR43175, in the treatment of acute migraine provides evidence that these receptors, like $5\text{-}HT_2$ receptors, exist in man as well as in animals.

Acknowledgements

We would like to acknowledge our many colleagues at Glaxo Group Research Limited who have contributed to this work.

References

Angus, J. A. and Wright, C. E. (1988). Atheromatous lesions enhance contractility responses to serotonin in large coronary arteries of the rabbit. In *Serotonin: Actions, Receptors, Pathophysiology; Proceedings of the 1987 IUPHAR Congress Satellite Meeting, Heron Island, Australia* (ed. E. J. Mylecharane, J. A. Angus, I. S. de la Lande, and P. P. A. Humphrey). Macmillan, London (in press).

Apperley, E. and Humphrey, P. P. A. (1976). Receptors for 5-hydroxytryptamine and noradrenaline in rabbit isolated ear artery and aorta. *British Journal of Pharmacology* **58,** 211–21.

Apperley, E. and Humphrey, P. P. A. (1986). The interaction of 5-hydroxytryptamine and methysergide with methiothepin at '$5\text{-}HT_1$-like' receptors in dog saphenous vein. *British Journal of Pharmacology* **87,** 131P.

Apperley, E., Feniuk, W., Humphrey, P. P. A., and Levy, G. P. (1980). Evidence for two types of excitatory receptor for 5-hydroxytryptamine in dog isolated vasculature. *British Journal of Pharmacology* **68,** 215–24.

Berridge, M. J. (1983). Rapid accumulation of inositol triphosphate reveals that agonists hydrolyse polyphosphoinositides instead of phosphotidyl inositol. *Biochemical Journal* **212,** 345–60.

Black, J. L., French, R. J., and Mylecharane, E. J. (1981). Receptor mechanisms for 5-hydroxytryptamine in rabbit arteries. *British Journal of Pharmacology* **74,** 619–26.

Black, J. W., Brazenor, R. M., Gerskowitch, V. P., and Leff, P. (1983). The problem of insurmountable antagonism in 5-hydroxytryptamine receptor classification. *British Journal of Pharmacology* **80,** 607P.

Blackshear, J. L., Orlandi, C., Garnic, J. D., and Hollenberg, N. K. (1985). Differential large and small vessel responses to serotonin in the dog hindlimb in vivo: role of the 5-HT_2 receptor. *Journal of Cardiovascular Pharmacology* **7,** 42–9.

Bradley, P. B., Humphrey, P. P. A., and Williams, R. H. (1983). Are vascular D and 5-HT_2 receptors for 5-hydroxytryptamine the same? *British Journal of Pharmacology* **79,** 295P.

Bradley, P. B., Humphrey, P. P. A., and Williams, R. H. (1985). Tryptamine-induced vasoconstrictor responses in rat caudal arteries are mediated predominantly via 5-hydroxytryptamine receptors. *British Journal of Pharmacology* **84,** 919–25.

Bradley, P. B., Humphrey, P. P. A., and Williams, R. H. (1986a). Evidence for the existence of 5-hydroxytryptamine receptors, which are not of the 5-HT_2 type, mediating contraction of isolated rabbit basilar artery. *British Journal of Pharmacology* **87,** 3–4.

Bradley, P. B., Engel, G., Feniuk, W., Fozard, J. R., Humphrey, P. P. A., Middlemiss, D. N., Mylecharane, E. J., Richardson, B. P., and Saxena, P. R. (1986b). Proposals for the classification and nomenclature of functional receptors for 5-hydroxytryptamine. *Neuropharmacology* **25,** 563–76.

Brittain, R. T., Butler, A., Coates, I. H., Fortune, D. H., Hagan, R., Hill, J. M., Humber, D. C., Humphrey, P. P. A., Ireland, S. J., Jack, D., Jordan, C. C., Oxford, A., Straughan, D. W., and Tyers, M. B. (1987a). GR38032F, a novel selective 5-HT_3 receptor antagonist. *British Journal of Pharmacology* **90,** 87P.

Brittain, R. T., Butina, D., Coates, I. H., Feniuk, W., Humphrey, P. P. A., Jack, D., Oxford, A. W., and Perren, M. J. (1987b). GR43175 selectively constricts the canine carotid arterial bed via stimulation of 5-HT_1-like receptors. *British Journal of Pharmacology* **92,** 618P.

Charlton, K. G., Bond, R. A., and Clarke, D. E. (1986). An inhibitory prejunctional '5-HT_1-like' receptor in the isolated perfused rat kidney: apparent distinction from the 5-HT_{1A}, 5-HT_{1B} and 5-HT_{1C} subtypes. *Naunyn-Schmiedeberg's Archives of Pharmacology* **332,** 8–15.

Clancy, B. M. and Maayani, S. (1985). 5-Hydroxytryptamine receptors in isolated rabbit aorta: Characterisation with tryptamine analogues. *Journal of Pharmacology and Experimental Therapeutics* **233,** 761–9.

Cocks, T. M. and Angus, J. A. (1983). Endothelium-dependent relaxation of coronary arteries by noradrenaline and serotonin. *Nature (London)* **305,** 627–30.

Cohen, M. L. and Colbert, W. E. (1986). Relationship between receptors mediating serotonin (5-HT) contractions in the canine basilar artery to 5-HT_1, 5-HT_2 and rat stomach fundus 5-HT receptors. *Journal of Pharmacology and Experimental Therapeutics* **237,** 713–8.

Cohen, M. L., Mason, N., Wiley, K. S., and Fuller, R. W. (1983). Further evidence that vascular serotonin receptors are of the 5-HT_2 type. *Biochemical Pharmacology* **32,** 567–70.

Conn, P. J. and Sanders-Bush, E. (1985). Serotonin-stimulated phosphoinositide turnover: mediation by the S_2 binding in rat cerebral but not in subcortical regions. *Journal of Pharmacology and Experimental Therapeutics* **234,** 195–203.

Connor, H. E., Feniuk, W., Humphrey, P. P. A., and Perren, M. J. (1986). 5-Carboxamidotryptamine is a selective agonist at 5-HT receptors mediating vasodilatation and tachycardia in anaesthetised cats. *British Journal of Pharmacology* **87,** 417–26.

Connor, H. E., Feniuk, W., and Humphrey, P. P. A. (1987). Characterisation of 5-HT receptors mediating contraction in dog and primate isolated basilar artery. *British Journal of Pharmacology* **92,** 757P.

Coughlin, S. R., Moskowitz, M. A., Antoniades, H. N., and Levine, L. (1981). Serotonin receptor mediated stimulation of bovine smooth muscle cell prostacyclin synthesis and its modulation by platelet-derived growth factor. *Proceedings of the National Academy of Science U.S.A.* **78,** 7134–8.

Coughlin, S. R., Moskowitz, M. A., and Levine, L. (1984). Identification of a serotonin type 2 receptor linked to prostacyclin synthesis in vascular smooth muscle cells. *Biochemical Pharmacology* **33,** 692–5.

Cummings, S. A., Groszmann, R. J., and Kaumann, A. J. (1986). Hypersensitivity of mesenteric veins to 5-hydroxytryptamine- and ketanserin-induced reduction of portal pressure in portal hypertensive rats. *British Journal of Pharmacology* **89,** 501–13.

Curro, F. A., Greenberg, S., Verbeuren, T. J., and Vanhoutte, P. M. (1978). Interaction between alpha adrenergic and serotonergic activation of canine saphenous veins. *Journal of Pharmacology and Experimental Therapeutics* **207,** 936–49.

Dalton, D. W., Feniuk, W., and Humphrey, P. P. A. (1986). An investigation into the mechanisms of the cardiovascular effects of 5-hydroxytryptamine in conscious normotensive and DOCA-salt hypertensive rats. *Journal of Autonomic Pharmacology* **6,** 219–28.

De Chaffoy de Courcelles, D., Leysen, J. E., De Clerck, F., Van Belle, H., and Janssen, P. A. (1985). Evidence that phospholipid turnover is the signal transducing system coupled to serotonin-S_2 receptor sites. *Journal of Biological Chemistry* **260,** 7603–8.

Doyle, V. M., Creba, J. A., Rüegg, U. T., and Hoyer, D. (1986). Serotonin increases the production of inositol phosphates and mobilises calcium via the 5-HT_2 receptor in A_7r_5 smooth muscle cells. *Naunyn-Schmiedeberg's Archives of Pharmacology* **333,** 98–103.

Edvinsson, L., Birath, E., Uddman, R., Lee, T. J.-F., Duverger, D., MacKenzie, E. T., and Scatton, B. (1984). Indole-aminergic mechanisms in brain vessels: Localisation, concentration, uptake and in vitro responses of 5-hydroxytryptamine. *Acta Physiologica Scandinavica* **121,** 291–9.

Engel, G., Göthert, M., Müller-Schweinitzer, E., Schliker, E., Sistonen, L., and Stadler, P. A. (1983). Evidence for common pharmacological properties of [^{3}H] 5-hydroxytryptamine binding sites, presynaptic 5-hydroxytryptamine autoreceptors in CNS and presynaptic 5-hydroxytryptamine receptors on sympathetic nerves. *Naunyn-Schmiedeberg's Archives of Pharmacology* **324,** 116–24.

Feniuk, W. (1984). An analysis of 5-hydroxytryptamine receptors mediating contraction of isolated smooth muscle. *Neuropharmacology* **23,** 1467–72.

Feniuk, W., Humphrey, P. P. A., and Watts, A. D. (1981a). Modification of the vasomotor actions of methysergide in the femoral arterial bed of the anaesthetised dog by changes in sympathetic nerve activity. *Journal of Autonomic Pharmacology* **1,** 127–32.

Feniuk, W., Humphrey, P. P. A., and Watts, A. D. (1981b). Further characterisation of pre- and post-junctional receptors for 5-hydroxytryptamine in isolated vasculature. *British Journal of Pharmacology* **73,** 191P–2P.

Feniuk, W., Hare, J., and Humphrey, P. P. A. (1981c). An analysis of the mechanism of 5-hydroxytryptamine induced vasopressor responses in ganglion blocked anaesthetised dogs. *Journal of Pharmacy and Pharmacology* **33,** 155–60.

Feniuk, W., Humphrey, P. P. A., and Perren, M. J. (1982). Tryptamine-induced vasopressor responses in pithed rats are not predominantly mediated via 5-hydroxytryptamine receptors. *British Journal of Pharmacology* **77,** 521P.

Feniuk, W., Humphrey, P. P. A., and Watts, A. D. (1983). 5-Hydroxytryptamine induced relaxation of isolated mammalian smooth muscle. *European Journal of Pharmacology* **96,** 71–8.

Feniuk, W., Humphrey, P. P. A., Perren, M. J., and Watts, A. D. (1985). A comparison of 5-hydroxytryptamine receptors mediating contraction in rabbit isolated aorta and dog saphenous vein. Evidence for different receptor types obtained by the use of selective agonists and antagonists. *British Journal of Pharmacology* **86,** 697–704.

Feniuk, W., Humphrey, P. P. A., and Perren, M. J. (1987). Selective vasoconstrictor action of GR43175 on arteriovenous anastomoses (AVAs) in the anaethetised cat. *British Journal of Pharmacology* **92,** 756P.

Forster, C. and Whalley, E. T. (1982). Analysis of the 5-hydroxytryptamine induced contraction of the human basilar arterial strip compared with the rat aortic strip in vitro. *Naunyn-Schmiedeberg's Archives of Pharmacology* **319,** 12–7.

Fozard, J. R. (1982). Mechanism of the hypotensive effect of ketanserin. *Journal of Cardiovascular Pharmacology* **4,** 829–38.

Fozard, J. R. (1984). MDL 72222, a potent and highly selective antagonist at neuronal 5-hydroxytryptamine receptors. *Naunyn-Schmiedeberg's Archives of Pharmacology* **326,** 36–44.

Fozard, J. R. (1987). 5-HT: the enigma variations. *Trends in Pharmacological Sciences* **8,** 501–6.

Fozard, J. R. and Mwaluko, G. M. P. (1976). Mechanism of the indirect sympathomimetic effect of 5-hydroxytryptamine on the isolated heart of the rabbit. *British Journal of Pharmacology* **57,** 115–25.

Fozard, J. R., Mobarok Ali, A. T. M., and Newgrosh, G. (1979). Blockade of serotonin receptors on autonomic nerves by (-)-cocaine and some related compounds. *European Journal of Pharmacology* **59,** 195–210.

Frenken, M. and Kaumann, A. J. (1984). Interaction of ketanserin and its metabolite ketanserinerol with 5-HT_2 receptors in pulmonary and coronary arteries of calf. *Naunyn-Schmiedeberg's Archives of Pharmacology* **326,** 334–9.

Frenken, M. and Kaumann, A. J. (1985). Ketanserin causes surmountable antagonism of 5-hydroxytryptamine induced contractions of large coronary arteries of the dog. *Naunyn-Schmiedeberg's Archives of Pharmacology* **328,** 301–3.

Fuller, R. W., Kurz, K. D., Mason, N. R., and Cohen, M. L. (1986). Antagonism of a peripheral vascular but not an apparently central serotonergic response by xylamidine and BW501C67. *European Journal of Pharmacology* **125,** 71–7.

Heuring, R. E. and Peroutka, S. J. (1987). Characterisation of a novel ^{3}H-5-Hydroxy tryptamine binding site subtype in bovine brain membranes. *Journal of Neuro science* **7,** 894–903.

Humphrey, P. P. A. (1978). The effects of α-adrenoceptor antagonists on contractile responses to 5-hydroxytryptamine in dog saphenous vein. *British Journal of Pharmacology* **63,** 671–5.

Humphrey, P. P. A. (1984). Peripheral 5-hydroxytryptamine receptors and their classification. *Neuropharmacology* **23,** 1503–10.

Humphrey, P. P. A. and Feniuk, W. (1987a). Classification of functional 5-hydroxytryptamine receptors. In *Proceedings of the 10th International Congress of Pharmacology, Sydney* (ed. M. J. Rand and C. Raper) pp. 277–80. Elsevier, Amsterdam.

Humphrey, P. P. A. and Feniuk, W. (1987b). The pharmacological characterisation of functional receptors for 5-hydroxytryptamine. In *Neuronal messengers in vascular function* (Fernstrom Foundation Series, Vol. 10, ed. A. Nobin, C. Owman, and B. Arneklo-Nobin) pp. 3–19. Elsevier, Amsterdam.

Humphrey, P. P. A. and Feniuk, W. (1988). The sub-classification of functional 5-HT_1-like receptors. In *Serotonin: Actions, Receptors, Pathophysiology; Proceedings of the 1987 IUPHAR Congress Satellite Meeting, Heron Island, Australia* (ed. E. J. Mylecharane, J. A. Angus, I. S. de la Lande, and P. P. A. Humphrey). Macmillan, London (in press).

Humphrey, P. P. A., Feniuk, W., and Watts, A. D. (1982). Ketanserin – a novel antihypertensive drug? *Journal of Pharmacy and Pharmacology* **34,** 541.

Humphrey, P. P. A., Feniuk, W., and Watts, A. D. (1983). Prejunctional effects of 5-hydroxytryptamine on noradrenergic nerves in the cardiovascular system. *Federation Proceedings* **42,** 218–22.

Humphrey, P. P. A., Feniuk, W., Perren, M. J., Oxford, A. W., Brittain, R. T., and Jack, D. (1987). The pharmacology of selective 5-HT_1-like receptor agonists for the acute treatment of migraine. *Cephalalgia* **7,** 400–1.

Humphrey, P. P. A., Feniuk, W., Perren, M. J., Connor, M. E., Oxford, A. W., Coates, I. H., and Butina, D., (1988). GR43175, a selective agonist for 5-HT_1-like receptor in dog isolated saphenous vein. *British Journal of Pharmacology* **94,** 1123–32.

Hunt, P. and Oberlander, C. (1981). The interaction of indole derivatives with the serotonin receptor and non-dopaminergic circling behaviour. *Advances in Experimental Medical Biology* **133,** 547–62.

Jandhyala, B. S. and Kivlighn, S. D. (1987). Antagonism by methysergide of neurogenic vasoconstriction in the dog forelimb. *Federation Proceedings* **46,** 276–80.

Leff, P. and Martin, G. R. (1986). Peripheral 5-HT_2-like receptors. Can they be classified with the available antagonists? *British Journal of Pharmacology* **88,** 585–93.

Lemberger, H. F., Mason, N., and Cohen, M. L. (1984). 5-HT_2-receptors in the rat portal vein: Desensitisation following cumulative serotonin addition. *Life Sciences* **35,** 71–7.

Leysen, J. E., Awouters, F., Kennis, L., Laduron, P. M., Vandenberk, J., and Janssen, P. A. J. (1981). Receptor binding profile of R41 468, a novel antagonist at 5-HT_2 receptors. *Life Sciences* **28,** 1015–22.

Leysen, J. E., Niemegeers, C. J. E., Van Nueten, J. M., and Laduron, P. M. (1982). [^{3}H] Ketanserin (R41468), a selective 3H-ligand for serotonin$_2$ receptor binding sites. Binding properties, brain distribution and functional role. *Molecular Pharmacology* **21,** 301–4.

McGrath, J. C. and MacLennan, S. J. (1986). Oxygen modifies the potency of 5-HT, and of 5-HT antagonists, in human umbilical artery (HUA). *British Journal of Pharmacology* **88,** 320P.

Mackenzie, E. T., Edvinsson, L., and Scatton, B. (1985). Functional bases for a central serotonergic involvement in classic migraine. *Cephalalgia* **5,** 69–78.

Marin, J. and Sanchez, C. F. (1980). Release of noradrenaline from cat cerebral arteries by different drugs and potassium. *Biochemical Pharmacology* **29,** 840–2.

Marin, J., Salaices, M., and Sanchez, C. F. (1979). Release of [^{3}H] noradrenaline induced by 5-hydroxytryptamine from cat pial arteries. *Journal of Pharmacy and Pharmacology* **31,** 818–21.

Middlemiss, D. N. (1985). The putative 5-HT$_1$ receptor agonist RU24969, inhibits the efflux of 5-hydroxytryptamine from rat frontal cortex slices by stimulation of the 5-HT autoreceptor. *Journal of Pharmacy and Pharmacology* **37,** 434–7.

Moskowitz, M. A., Liebmann, J. E., Reinhard, J. F., and Schlosberg, A. (1982). Raphe origin of serotonin containing neurones within the choroid plexus of the rat. *Brain Research* **247,** 388–92.

Müller-Schweinitzer, E. and Engel, G. (1983). Evidence for mediation by 5-HT$_2$ receptors of 5-hydroxytryptamine-induced contraction of canine basilar artery. *Naunyn-Schmiedeberg's Archives of Pharmacology* **324,** 287–92.

Mylecharane, E. J., Misbach, J., Duckworth, J. W., and Lance, J. W. (1978). Effects of methysergide, pizotifen and ergotamine in the monkey cranial circulation. *European Journal of Pharmacology* **48,** 1–9.

Nakaki, T., Roth, B. L., Chuang, D.-M., and Costa, E. (1985). Phasic and tonic components in 5-HT$_2$ receptor-mediated rat aorta contraction: participation of Ca^{++} channels and phospholipase C. *Journal of Pharmacology and Experimental Therapeutics* **234,** 442–6.

O'Rourke M. F. (1982) Function of arteries as vascular conduits. In *Arterial function in health and disease*, (ed. O'Rourke, M. F.) pp. 53–64. Churchill Livingstone, Edinburgh, London, Melbourne and New York.

Peroutka, S. J. and Snyder, S. H. (1979). Multiple serotonin receptors: differential binding of [^{3}H]5-hydroxytryptamine, [^{3}H]lysergic acid diethylamide and [^{3}H]spiroperidol. *Molecular Pharmacology* **16,** 687–99.

Peroutka, S. J., Huang, S., and Allen, G. S. (1986). Canine basilar artery contractions mediated by 5-hydroxytryptamine$_{1a}$ receptors. *Journal of Pharmacology and Experimental Therapeutics* **237,** 901–6.

Purdy, R. E., Murray, D. L., and Stupecky, G. L. (1987). Receptors for 5-hydroxytryptamine in rabbit blood vessels: Activation of alpha adrenoceptors in rabbit thoracic aorta. *Journal of Pharmacology and Experimental Therapeutics* **240,** 535–41.

Rapport, M. M., Green, A. A., and Page, I. H. (1948). Serum vasoconstrictor (serotonin) i.v. isolation and characterisation. *Journal of Biological Chemistry* **176,** 1243.

Reimann, I. W. and Frölich, J. C. (1983). Mechanism of antihypertensive action of ketanserin in man. *British Medical Journal* **287,** 381–3.

Reinhard, J. F. jr., Liebmann, J. E., Schlosberg, A. J., and Moskowitz, M. A. (1979). Serotonin neurones project to small blood vessels in the brain. *Science* **206,** 85–7.

Roth, B. L., Nakaki, T., Chuang, D.-M., and Costa, E. (1984). Aortic recognition sites for serotonin (5-HT) are coupled to phospholipase C and modulate phospha tidylinositol turnover. *Neuropharmacology* **23,** 1223–5.

Salmoiraghi, G. C., Page, I. H., and McCubbin, J. W. (1956). Cardiovascular and respiratory response to intravenous serotonin in rats. *Journal of Pharmacology and Experimental Therapeutics* **118,** 479–81.

Saxena, P. R. (1972). The effects of antimigraine drugs on the vascular responses by 5-hydroxytryptamine and related biogenic substances on the external carotid bed of dogs. Possible pharmacological implications to their antimigraine action. *Headache* **12,** 44–54.

Saxena, P. R. (1974). Selective vasoconstriction in carotid vascular bed, by methysergide: possible relevance to its antimigraine effect. *European Journal of Pharmacology* **27,** 99–105.

Saxena, P. R. and Verdouw, P. D. (1982). Redistribution by 5-hydroxytryptamine of carotid arterial blood at the expense of arteriovenous blood flow. *Journal of Physiology (London)* **332,** 501–20.

Saxena, P. R. and Verdouw, P. D. (1984). Effect of methysergide and 5-hydroxytryptamine on carotid blood flow distribution in pigs: further evidence for the presence of atypical 5-HT receptors. *British Journal of Pharmacology* **82,** 817–26.

Saxena, P. R. and Verdouw, P. D. (1985). 5-Carboxamidotryptamine, a compound with a high affinity for 5-HT_1 binding sites, dilates arterioles and constricts arteriovenous anastomoses. *British Journal of Pharmacology* **84,** 533–44.

Saxena, P. R., van Houwelingen, P., and Bonta, I. L. (1971). The effects of mianserin hydrochloride on the vascular responses evoked by 5-hydroxytryptamine and related vasoactive substances. *European Journal of Pharmacology* **13,** 295–305.

Saxena, P. R., Duncker, D. J., Bom, A. H., Heiligers, J., and Verdouw, P. D. (1986). Effects of MDL 72222 and methiothepin on carotid vascular responses to 5-hydroxytryptamine in the pig: evidence for the presence of '5-hydroxytryptamine$_1$-like' receptors. *Naunyn-Schmiedeberg's Archives of Pharmacology* **333,** 198–204.

Spierings, E. L. H. and Saxena, P. R. (1979). Effect of ergotamine on cranial arteriovenous shunting in experiments with constant flow perfusion. *European Journal of Pharmacology* **56,** 31–7.

Taylor, E. W., Duckles, S. P., and Nelson, D. L. (1986). Dissociation constants of serotonin agonists in the canine basilar artery correlate to K_i values at the 5-HT_{1A} binding site. *Journal of Pharmacology and Experimental Therapeutics* **236,** 118–25.

Thoa, N. B., Wooten, G. F., Axelrod, J., and Kopin, I. J. (1975). On the mechanism of release of norepinephrine from sympathetic nerves induced by depolarising agents and sympathomimetic drugs. *Molecular Pharmacology* **11,** 10–8.

Vanhoutte, P. M. and Houston, D. S. (1986). Serotonin, platelets and vascular function. In *Central and peripheral mechanisms of cardiovascular regulation* (ed. A. Magro, W. Osswald, D. Reis, and P. M. Vanhoutte) pp. 135–161. Plenum, New York.

Van Nueten, J. M. and Vanhoutte, P. M. (1981). Selectivity of calcium antagonism and serotonin antagonism with respect to venous and arterial tissues. *Angiology* **32,** 476–84.

Van Nueten, J. M., Janssen, P. A. J., Van Beek, J., Xhonneux, R., Verbeuren, T. J., and Vanhoutte, P. M. (1981). Vascular effects of ketanserin (R41468), a novel antagonist of 5-HT_2-serotonergic receptors. *Journal of Pharmacology and Experimental Therapeutics* **218,** 217–30.

Van Nueten, J. M., Janssen, P. A. J., De Ridder, W., and Vanhoutte, P. M. (1982). Interaction between 5-hydroxytryptamine and other vasoconstrictor substances in the isolated femoral artery of the rabbit; effect of ketanserin (R41468). *European Journal of Pharmacology* **77,** 281–7.

Van Nueten, J. M., Leysen, J. E., de Clerck, F., and Vanhoutte, P. M. (1984a). Serotonergic receptor subtypes and vascular reactivity. *Journal of Cardiovascular Pharmacology* **6,** S564–74.

Van Nueten, J. M., De Ridder, W., and Vanhoutte, P. M. (1984b). Ketanserin and vascular contractions in response to cooling. *European Journal of Pharmacology* **99,** 329–32.

Van Nueten, J. M., Schuurkes, J. A. J., De Ridder, W. J. E., Kuyps, J. J. M. D., and Janssens, W. J. (1986). Comparative pharmacological profile of ritanserin and ketanserin. *Drug Development Research* **8,** 187–95.

Verdouw, P. D., Jennewein, H. M., Heiligers, J., Duncker, D. J., and Saxena, P. R. (1984). Redistribution of carotid artery blood flow by 5-HT: effects of the 5-HT_2 receptor antagonists ketanserin and WAL 1307. *European Journal of Pharmacology* **102,** 499–509.

Watts, A. D., Feniuk, W., and Humphrey, P. P. A. (1981). A pre-junctional action of 5-hydroxytryptamine and methysergide on noradrenergic nerves in dog isolated saphenous vein. *British Journal of Pharmacology* **33,** 515–20.

Wright, C. E. and Angus, J. A. (1988). 5-Carboxamidotryptamine elicits 5-HT_2 and 5-HT_3-receptor mediated cardiovascular responses in the conscious rabbit: evidence for 5-HT release from platelets. *Journal of Cardiovascular Pharmacology* (in press).

Young, A. R., Hamel, E., and Mackenzie, E. T. (1986). Recent studies on the serotonergic innervation of the cerebral circulation: a review. In *Neural regulation of brain circulation* (ed. C. Owman and J. E. Hardebo) pp. 195–217. Elsevier, Amsterdam.

6
Amplification mechanisms in peripheral tissues

I. S. DE LA LANDE

6.1 Introduction

More than 30 years ago it was observed that 5-HT increased the contractile response of the isolated guinea-pig ileum to histamine (Bülbring and Burnstock 1955). An increased constrictor response of the perfused rabbit ear artery to noradrenaline following 5-HT was noted by de la Lande and Rand (1965) and was subjected to detailed analysis in a subsequent study (de la Lande *et al.* 1966). A similar interaction between 5-HT and contractile agents has now been observed in arteries from a variety of species. This chapter will review the present understanding of possible mechanisms and significance of the interaction in vascular smooth muscle.

6.2 Terminology and concepts

The terminology applied to agonist–agonist interactions appears to depend more on author preference than that of agonist–antagonist interactions. The following terms and concepts will be used to describe the interaction between excitatory agents A and B, where A in a low concentration influences the contractile response of smooth muscle to B.

6.2.1 Contractile response and contractile state

The distinction advocated by Mackay (1982) is followed, i.e. where the response to agonist B is elicited in the presence of A, the response refers to the change in contractile state. The 'state' refers to the absolute tension; for example, if A by virtue of its own contractile activity increases the tension from zero to 1 g, and B then increases the tension to 6 g, the response to B is 5 g and the contractile state is 6 g.

6.2.2 Types of synergism

The terms synergize, potentiate, amplify, sensitize will be used interchangeably to describe the same phenomenon, namely an increased response to B in the presence of A. In his succinct style, Gaddum (1953) described a synergistic interaction as one where 'the drugs are helping one another'. Strictly speaking, when A increases the response to B without itself producing a response, and there is no evidence that B is 'helping' A, the interaction is not synergistic. Nevertheless, in the past, synergism has been used in a very general sense to include the latter interaction, and will be used in this way in the following discussion. Types of synergism which have been either identified or predicted include the following:

6.2.2.1 *Competitive*

The two agonists produce contractile responses via the same receptor. As a result B is able to elevate the contractile state to a higher level than would be the case if A were not present. This does not mean that the response to B is increased and indeed, since the receptor population available to B is diminished by A, the response may be decreased. The term 'additive' is often applied to this type of interaction; an additive interaction was described by Gaddum (1953) as one where 'the two drugs which were thought to be different were really the same'.

6.2.2.2 *Functional*

The two drugs produce their contractile responses via different receptors, but utilize a common effector, i.e. at some stage in the sequence of reactions which result from receptor occupancies by A and B, the sequences converge and the two agonists then utilize the same effector pathway.

6.2.2.3 *Metaffinoid*

A and B act via different receptors; however, receptor occupation by A does not lead to a contractile response but instead influences receptor occupation by B.

6.2.2.4 *Metactoid*

A and B act via different receptors; however, receptor occupation by A does not lead to a contractile response but instead influences the effector system for B.

6.2.2.5 *Dispositional*

A does not act via a receptor but influences the concentration of B at its receptors by inhibiting removal of B from the extracellular compartment.

Dose-response curves derived from receptor theory for the types of synergism in Sections 6.2.2.1 to 6.2.2.4 are illustrated in the extensive review by Van Den Brink (1977). Dose-response relationships permit a crude distinction between some types of synergism, for example, the absence of a contractile response to A distinguishes metaffinoid and metactoid from competitive and functional synergism, while the potential for an increase in the maximum contractile state exists in metactoid and possibly also in metaffinoid synergism if receptor occupation by A leads to 'unmasking' of B receptors.

The phenomenon of 'threshold' synergism, where synergism rapidly attenuates with increasing agonist concentration, has received attention in two recent studies. Stupecky *et al.* (1986) showed that, when responses of the rabbit aorta to contractile agents were extremely small (of the order of 2–3 per cent of their maxima), combinations of any two agonists, or even doubling the concentration of a single agonist, generated a response which was far greater than the sum of the individual responses (see Fig. 6.1). An explanation became apparent when the relationship between agonist concentration and response was plotted on an arithmetic scale; namely that the relationship was nearly linear, but when extrapolated to zero response, the line defining the relationship did not pass through the origin as would be expected if occupation of a receptor always led to a response. Instead there was a well defined threshold concentration which varied for the different agonists. The relationship for several agonists is illustrated in Fig. 6.2, where it will be evident that the magnitude of the synergistic interaction between any two of the agonists will be a function of their threshold concentrations and the slopes of the curves relating response to concentration. It will be evident also that concentrations of each agonist, which separately do not elicit a response, may do so in combination. The presence of a threshold implies either that one of the events involved in

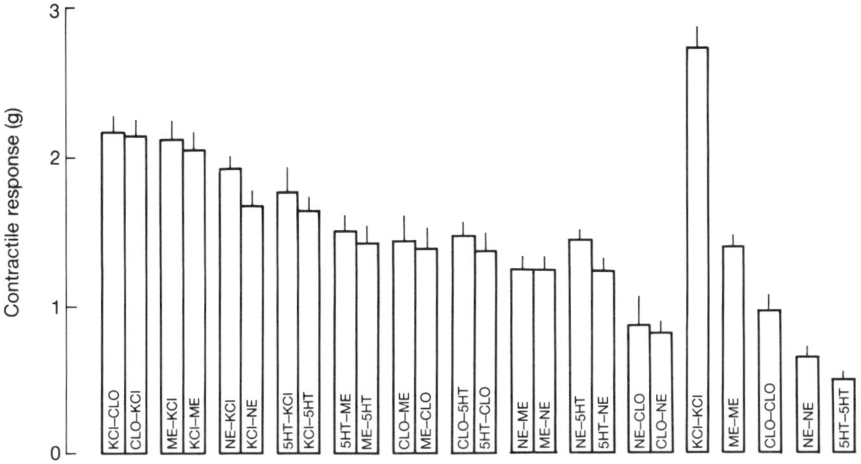

Fig. 6.1 Rabbit aorta rings—contractile responses to combinations of two threshold concentrations of agonists, each separately causing only 0.1g contraction. The maximum responses of the aorta were of the order of 4–7g. NE, noradrenaline; ME, methoxamine; CLO, clonidine. (From Stupecky *et al.* 1986 with permission)

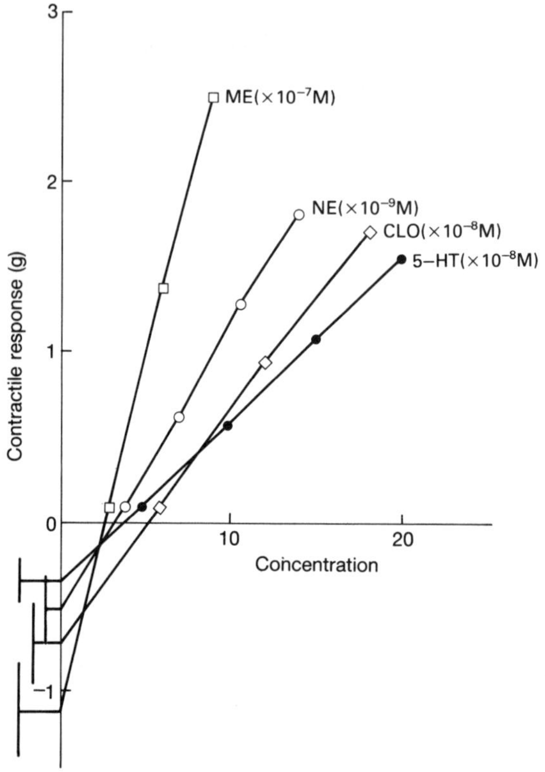

Fig. 6.2 Lower parts of DR curves to noradrenaline, methoxamine, clonidine, and 5-HT on an arithmetic dose scale. The curves in this range are almost linear and intersect the Y axis at different points below zero, indicating threshold phenomena. —○— noradrenaline (NE); —□— methoxamine (ME); —◇— clonidine (CLO); —●— 5-HT. (From Stupecky *et al.* 1986 with permission)

stimulus–response coupling must reach a critical level to initiate a following event, or that the response itself must attain a certain level in order to be detected by the recording system.

Leff (1987) has proposed a general model of the interaction between agonists acting via different receptors. The model assumes that each agonist–receptor complex gives rise to a common mediator, which represents the point at which the two effector pathways converge, and that threshold synergism occurs when there is positive co-operativity in the mediator–effect relationship; the mediator is considered to be intracellular calcium. The model successfully predicted threshold synergism in the 5-HT–histamine interaction in the rabbit aorta, and the absence of synergism in the noradrenaline–histamine interaction in the guinea-pig atria. Leff points out that interactions where an agonist exerts a synergistic effect at a sub-threshold concentration or where it produces a parallel leftward shift of the dose-response curve to a second agonist, are not encompassed by the model.

6.3 Synergism between 5-HT and contractile agents in arterial smooth muscle *in vitro*

It is important to emphasize that the following outline is confined to synergism which is rapid in onset and readily reversible. In the rabbit ear artery, increased vascular responsiveness can follow a single exposure to contractile agents, including 5-HT, but the increases are small in magnitude and persist for several hours after removing the agent (Rapoport and Bevan 1983). A long-lasting inhibitory effect of 5-HT on vascular responsiveness has also been described, and attributed to desensitization of the α-adrenoceptor (Rapoport and Bevan 1982). The prolonged nature of these excitability changes suggests that they have little bearing on the mechanisms responsible for the rapidly-reversible synergistic actions of 5-HT, and hence they will not be considered further.

6.3.1 Rat aorta

A study on the rat aorta strip by Asano and Hidaka (1980) provided evidence for functional synergism between 5-HT and a test agonist, acetylcholine. They showed that 5-HT in a concentration producing a response of the order of the ED15 of acetylcholine, potentiated the response to acetylcholine without affecting the maximum contractile state. That the effect was not additive was shown by examining the response to acetylcholine in the presence of a concentration of acetylcholine generating the same response as 5-HT (Fig. 6.3). That the 5-HT and acetylcholine were acting on different receptors was shown by the ability of methysergide to prevent the potentiating action of 5-HT without affecting responses to acetylcholine alone. Finally, Asano and Hidaka (1980) showed that noradrenaline and histamine potentiated acetylcholine to about the same extent as did 5-HT, and that their effects were also blocked by appropriate selective antagonists (phentolamine and mepyramine, respectively). The message emerging from the study is that synergism between 5-HT and other receptor-agonists in the rat aorta is by no means a property unique to 5-HT and is probably an example of a functional interaction.

6.3.2 Rat caudal artery

In this vessel, the increase in response to noradrenaline produced by 5-HT in concentrations of the order of 0.01 μM is dramatic. The ability of

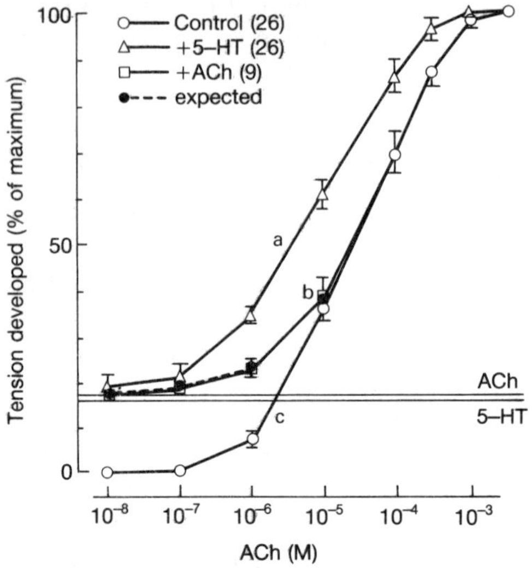

Fig. 6.3 Rat aorta strip—potentiating effect of 5-HT on cumulative responses to acetylcholine (ACh). Curve c (○), control curve for ACh in the absence of 5-HT; Curve a (△), curve for ACh in the presence of 5-HT 0.014 μM; the 5-HT itself caused an increase in tension of 16.4 per cent (lower horizontal line). Curve b (□), curve for ACh in the presence of ACh 2.3 μM which caused an increase in tension (17.8 per cent) upper horizontal line about equal to that produced by the 5-HT. The broken line (....●....) is the predicted DR curve for ACh in the presence of ACh 2.3 μM (predicted from curve c). In each DR curve, the maximum contractile tension developed by ACh 3 mM was taken as 100 per cent. (From Asano and Hidaka 1980 with permission)

ketanserin to depress both the contractile activity and synergistic activity of 5-HT without influencing the contractile activity of noradrenaline alone, indicates that both actions of 5-HT are mediated by a single receptor of the 5-HT_2 subtype (Van Nueten *et al.* 1981). Results of other studies indicate that the contractile response to noradrenaline on this vessel is mediated by α_1 and α_2 adrenoceptor subtypes (Medgett and Langer 1984). Accordingly it is reasonable to assume that a functional interaction contributes at least in part to the syngergism, but the data are insufficient to indicate whether a 'threshold' effect also contributes.

6.3.3 Rat mesenteric artery

The dose-response curve to noradrenaline in the perfused arterial bed was shifted leftwards by 5-HT at a time when complete tachyphylaxis had occurred to the vasoconstrictor response to 5-HT 1 μM (Seabrook and

Nolan 1983). Since synergism was unaffected by ketanserin in a concentration which abolished the vasoconstrictor response to 5-HT, it was concluded that the synergism was not mediated by the 5-HT_2 receptor. Nevertheless the study of Su and Uruno (1985) indicated that synergism at a lower, subconstrictor concentration of 5-HT (0.1 μM) was due to activation of the 5-HT_2 receptor.

6.3.4 Rat hind limb

Vasoconstrictor responses to bolus injections of 5-HT into the perfused hind limb increased markedly during a sustained constrictor response to noradrenaline (Wilton and McCalden 1977). The synergism was considered to be dispositional and to involve competition for monoamine oxidase since it was prevented by tranylcypromine. However, in the autoperfused rat hind limb preparation, it is reported that synergism between 5-HT and noradrenaline is mediated by 5-HT_2 receptors (Medgett 1985).

6.3.5 Rabbit aorta

When full dose-response curves to contractile agents were elicited in the study of Stupecky *et al.* (1986), referred to earlier, the interaction between noradrenaline and methoxamine proved to be additive. This accords with evidence that both agents act via the same receptor (α adrenoceptor). In contrast, the interaction between 5-HT in a threshold concentration (for contraction) and either noradrenaline or methoxamine, was characterized by a parallel leftward shift of the dose-response curves to the latter amines with no change in the maximum contractile states. Hence, the interaction between these agonists is more than additive, and resembles functional synergism. Functional synergism is also consistent with evidence that 5-HT exerts its contractile effect on the rabbit aorta primarily via 5-HT receptors, and not via α-adrenoceptors (Apperley *et al.* 1976; Purdy *et al.* 1981).

Interestingly, a threshold concentration of noradrenaline caused a parallel shift to the left of the dose-response curve to 5-HT, but now the effect was associated with an increased maximum to the 5-HT. It is not clear why the maximum response is increased in the reverse interaction, although the authors speculate that 5-HT may not be a full agonist on the rabbit aorta and an increase might occur in the event that the noradrenaline 'unmasked' 5-HT receptors.

6.3.6 Rabbit femoral artery

In artery rings, 5-HT in constrictor concentrations markedly increased responses to noradrenaline, histamine, and prostaglandin $F_{2\alpha}$ (Van Nueten *et al.* 1982). The synergistic and constrictor responses to 5-HT, but not the constrictor responses to the interacting agonists, were antagonized by ketanserin. Hence, as in the rat caudal artery, the interaction fulfills one of the requirements of functional synergism, in that 5-HT acts via a different receptor to that of the interacting agent.

6.3.7 Rabbit coronary artery

In the perfused artery, acetylcholine is the constrictor and noradrenaline the dilator. There was pronounced synergism between 5-HT and acetylcholine, but 5-HT was without effect on the dilator response to noradrenaline (de la Lande *et al.* 1974).

6.3.8 Human peripheral arteries

Synergism between 5-HT and noradrenaline in perfused segments of human digital artery and human temporal artery were both comparable in magnitude with the synergism observed in the rabbit ear artery (de la Lande *et al.* 1966; Carroll *et al.* 1974). The effects of receptor antagonists on these interactions do not appear to have been studied. However, in human digital artery strips, a synergistic interaction between 5-HT and the thromboxane A_2 receptor agonist, U466191, was insensitive to ketanserin (Young *et al.* 1986).

6.3.9 Cerebral vessels

5-HT increased the response to prostaglandin $F_{2\alpha}$ in the canine basilar-artery (Van Nueten 1985), and the response to noradrenaline in the guinea-pig pial artery (Hardebo *et al.* 1988) but, surprisingly, was without effect on the response to noradrenaline in the rabbit basilar artery (Bevan *et al.* 1975). The effect on the guinea-pig pial artery is of particular interest since the interaction is blocked by methiothepin but not by ketanserin, suggesting that it is mediated by the '5-HT_1-like' rather than the 5-HT_2 receptor; furthermore, the sympathetic nerves to this artery take up 5-HT from the surrounding cerebrospinal fluid, enabling 5-HT to act as a co-transmitter with noradrenaline (Chang *et al.* 1989).

Several studies have demonstrated synergistic interactions between anti-migraine agents and 5-HT in cerebral vessels; these are considered in Section 6.6.

6.4 Rabbit ear artery

6.4.1 Characteristics of synergism

In an early study (de la Lande *et al.* 1966) the interactions of 5-HT with other vasoconstrictor agents were examined in segments perfused at a constant rate where the 5-HT was present in a low concentration in the solutions bathing both the adventitial and the luminal surface, and the interacting agents (noradrenaline, adrenaline, histamine, angiotensin II) were administered by intraluminal bolus injections. Hence, the responses to the latter agents were transient and not steady-state. Analysis of the 5-HT–noradrenaline interaction under these conditions revealed that synergism commenced at a concentration of 5-HT (1 nM) which was about an order of magnitude less than the concentration eliciting detectable vasoconstriction. The maximum effect comprised an approximately ten-fold shift to the left of the dose-response curves to noradrenaline and to histamine, and occurred at concentrations of 5-HT approaching 0.1 μM. The effect of 5-HT was considerable greater than the effect of an equipotent vasoconstrictor concentration of noradrenaline, excluding the possibility that the interaction was additive in nature. The influence of 5-HT on the maximum responses to noradrenaline or histamine was not determined since, at high perfusion pressures, estimates of maxima are unreliable (see da la Lande 1975). However, a subsequent study on artery strips and rings has shown that synergism between 5-HT and noradrenaline can occur without a change in the maximum contractile state of the artery (de la Lande and Kennedy, unpublished observations).

In the perfused artery, it was shown qualitatively that synergism occurs irrespective of whether the amines are applied to the same or opposite surfaces (de la Lande *et al.* 1966). Subsequently Cannell (1969) noted that the increase in sensitivity to noradrenaline produced by 5-HT 0.1 μM (the latter applied to both surfaces) was about the same (of the order of ten-fold) irrespective of whether noradrenaline was applied to the adventitial surface so that the response was steady-state, or was applied as an intraluminal bolus so that the response was transient (Cannell 1969).

Recently, Meehan *et al.* (1986) examined the influence of the surface of amine entry on synergism at a lower concentration (0.01 μM) of 5-HT than was used in the early studies. Their study included analysis of the interaction between 5-HT and sympathetic nerve stimulation. Whereas intraluminal 5-HT increased the responses to intraluminal and extraluminal noradrenaline, and the response to nerve stimulation, each to about the same extent, the potentiating effect of extraluminal 5-HT was weaker and it failed to increase responses to intraluminal noradrenaline and to nerve stimulation. However, in the presence of cocaine, the potentiating effect of extraluminal 5-HT reappeared. These findings were explained in terms of two factors. One is the gradient of concentration of an agonist between its surface of entry into the vessel wall and the opposite surface; as a result, the responses to the extraluminal agonist and to sympathetic nerve stimulation tend to be localized to the smooth muscle cells in the outer region of the media, while those to the intraluminal agonist tend to be localized to the cells in the inner region (Kalsner 1972; de la Lande 1975; de la Lande *et al.* 1980). The second factor is uptake of 5-HT and noradrenaline by the sympathetic nerve terminals (Iversen 1967; Paiva *et al.* 1984), one consequence of which is that extraluminal 5-HT, but not intraluminal 5-HT, is exposed to sites of loss before it enters the smooth muscle layer. When the above factors were taken into account, it was evident that 5-HT exerted very similar effects on responses to sympathetic nerve stimulation and responses to noradrenaline. The similarity extended to inhibition by ketanserin in a low, 5-HT_2 receptor selective concentration, leading to the important conclusion that 5-HT potentiated the responses to nerve stimulation by activating post-junctional 5-HT_2 receptors (Meehan *et al.* 1986).

In a subsequent study Meehan *et al.* (1987) obtained evidence that 5-HT–noradrenaline synergism in some, although not all, smooth muscle cells required the presence of the endothelium. Key findings were that: (a) removal of the endothelium increased the sensitivity to noradrenaline and abolished synergism wth 5-HT; (b) cocaine restored synergism; and (c) endothelium removal did not affect potentiation of the response to sympathetic nerve stimulation by 5-HT. They interpret the results as evidence that there are common elements in the mechanisms by which the endothelium derived relaxing factor (EDRF) decreases, and 5-HT increases, the response to noradrenaline. (The element may be agonist-induced entry of extracellular Ca^{2+} into the smooth muscle cells; see Section 6.4.5). The restoring effect of cocaine and the failure of endothelium removal to influence the effect of 5-HT on the response to nerve stimulation, imply that the interaction between EDRF and 5-HT occurs in the inner but not in the outer smooth muscle cells. Apart from focusing attention on the role of the endothelium, the results again emphasize the importance of amine distribution in the artery wall and the modification of this distribution by neuronal uptake, as factors determining the magnitude of the synergistic interaction between 5-HT and noradrenaline.

6.4.2 Type of synergism

A dispositional basis for synergism can be excluded on a number of grounds. It is unlikely that the diversity of contractile agents reacting synergistically with 5-HT share a removal process in common. Although 5-HT and noradrenaline do share common removal processes, namely $uptake_1$ and $uptake_2$ (Paiva *et al.* 1984; Grohmann and Trendelenburg 1984), inhibitors of these processes have much less effect on sensitivity to intraluminal noradrenaline (about two-fold increases—Johnson and de la Lande 1978) than does 5-HT (increases of up to ten-fold). Furthermore, inhibitors of $uptake_1$ and $uptake_2$ (cocaine, corticosteroids) do not eliminate synergism between 5-HT and noradrenaline (Cannell 1969; Meehan *et al.* 1986; de la Lande and Kennedy, unpublished observations). As indicated earlier, $uptake_1$ inhibition can result in increased synergism.

Analysis of factors other than dispositional has focused on the question of whether synergism simply reflects a non-specific interaction between contractile agents (i.e. of the functional or threshold type) or whether it reflects a more specific sensitizing action on the part of the 5-HT. Evidence which suggests that there is both a non-specific and a 5-HT-specific component was obtained by Fozard (1973, 1975, 1976, and 1977) in the course of an analysis of the action of antimigraine drugs on responses to contractile agents which included noradrenaline, histamine, and 5-HT. He noted that two of the antimigraine drugs which were themselves vasoconstrictor, namely methysergide and clonidine, potentiated the responses to noradrenaline and to histamine to comparable degrees, and that the potentiation was prevented by appropriately selective noradrenaline and histamine receptor antagonists. He suggested that this type of synergism reflected a non-specific interaction between contractile agents. Since it commenced at subconstrictor concentrations of methysergide and clonidine, he suggested also that it included a threshold component such that subthreshold excitatory changes produced by each of the interacting agonists separately could summate in the interaction to generate a contractile response. Since the rabbit ear artery does not appear to possess post-junctional α_2 adrenoceptors (Hieble and Woodward 1984; Meehan *et al.* 1986) it is reasonable to assume that clonidine exerts its vasoconstrictor action in the rabbit ear artery via the α_1 adrenoceptor. Hence the 'non-specific' synergism defined by Fozard includes interaction between two agonists of the same receptor (clonidine and noradrenaline) and in this respect has features in common with the 'threshold synergism' described by Stupecky *et al.* (1986) in the rabbit aorta. Fozard's evidence for a 5-HT-specific component was based on the finding that clonidine potentiated responses to 5-HT far more than those to noradrenaline or histamine. In magnitude the potentiation varied with the vasoconstrictor activity of clonidine, suggesting that it stemmed from the ability of the 5-HT to increase the activity of the latter. In this respect, the interaction between clonidine and 5-HT appears to be very similar to the interaction between noradrenaline and 5-HT. De la Lande *et al.* (1966) noted that the dose-vasoconstrictor response curve to 5-HT (applied by intraluminal bolus) was extremely flat, but if the 5-HT was applied during a sustained vasoconstrictor response to noradrenaline, the dose-response curve to 5-HT increased markedly in slope and became parallel with the dose-response curve to noradrenaline alone. These changes were interpreted as evidence that 5-HT, during its transient contact with the artery, sensitized the artery to the vasoconstrictor action of noradrenaline.

A true 'sensitizing' action on the part of 5-HT is also suggested by evidence that the synergistic and contractile actions of 5-HT are mediated by different receptors. Several groups have reported that the contractile response to 5-HT in the perfused artery is mediated by the α-adrenoceptor

(Fozard 1973; Apperley *et al.* 1976; Black *et al.* 1981). The reports are open to the criticism that the 5-HT was applied to the artery as an intraluminal bolus and hence the distribution of 5-HT in the artery wall was non-uniform and transient. However, the criticism is offset by recent evidence that the pK_B values for antagonism of 5-HT and noradrenaline by prazosin in the artery ring preparation are very similar (8.6 and 8.5 respectively—Bevan and Mangiarua 1987). In this preparation, both surfaces of the artery are exposed to the same concentration of agonist, and the response recorded is steady-state; the similar pK_B values imply that under these conditions the contractile response to 5-HT is mediated only by the α_1 adrenoceptor.

In contrast the synergism between 5-HT and noradrenaline has been reported in different studies to be inhibited by 5-HT receptor antagonists, namely 2-bromo-LSD, pizotifen, and ketanserin, each in a concentration which was either without effect or produced a variable effect on the vasoconstrictor response to noradrenaline (de la Lande *et al.* 1966; Carroll *et al.* 1974; Meehan *et al.* 1986). Similarly, Fozard noted that the clonidine-5-HT synergism was inhibited by methysergide. It is noteworthy that the effects of two of the antagonists, 2-bromo-LSD and ketanserin, were measured under conditions where the 5-HT was applied to both surfaces of the artery. These conditions are equivalent to those in the artery ring preparation where only an α-adrenoceptor-mediated contractile response was detected. These findings suggest strongly that the synergistic action of 5-HT, in contrast to its contractile action, is mediated by a 5-HT receptor which is probably of the 5-HT_2 subtype.

A contrasting view has been expressed recently by Purdy and Murray (1988). They compared the potentiating effect of methoxamine on contractile responses to 5-HT in three rabbit arteries, namely aorta, femoral, and ear, and noted that the leftward shifts of the dose-response curves to 5-HT (of the order of 10, 100, and 1000 respectively) correlated well with the levels of α-adrenoceptor reserve in each vessel. Accordingly they have suggested that the action of 5-HT on the α-adrenoceptor plays an important role in synergism. However, the significance of their findings will not be clear until the receptors mediating the potentiated response to 5-HT have been identified.

On the basis of present evidence, it is reasonable to favour an hypothesis that, under conditions where synergism is measured in terms of the effects of a low sub-constrictor concentration of 5-HT on the response to noradrenaline, activation of the 5-HT_2 receptor results in excitatory events which do not lead to contraction as the final effect but which facilitate the contractile action of noradrenaline. By implication, synergism is not a functional manifestation of the interaction between the excitatory events associated with the contractile actions of 5-HT and noradrenaline. It is emphasized that the hypothesis refers only to synergism in the perfused rabbit ear artery. It would appear not to apply generally to arteries in view of the number where synergism and contraction are mediated by a common receptor (5-HT_2). However, it is pointed out that in one of these arteries (perfused rat mesenteric) there is evidence of dissociation between contraction and synergism, since the former effect of 5-HT was much more sensitive to cyproheptadine (Su and Uruno 1985).

6.4.3 Synergism in reserpinized artery rings and strips

The preceding discussion has been confined mainly to the interaction between 5-HT and contractile agents in artery segments which are intraluminally perfused and whose sympathetic terminal innervation is intact. An analysis of synergism in artery strips and rings from rabbits treated with reserpine for 24 h has revealed a number of features (summarized in Stanton and de la Lande 1982; de la Lande *et al.* 1983; de la Lande and Kennedy 1985) which are difficult to reconcile with those observed in the perfused artery. The contractile response to 5-HT was in part mediated by the 5-HT_2 receptor, an observation which accords with an earlier report that a 5-HT receptor is responsible for a small component of the response in the artery ring which had been acutely denervated by 6-hydroxydopamine (Purdy *et al.* 1981). Nevertheless, in order to inhibit synergism between 5-HT and noradrenaline in the strip, it was necessary to use ketanserin in concentrations which produced significant blockade of α-adrenoceptors. Furthermore, when the sensitivity to noradrenaline was decreased about 30-fold by prazosin, synergism between 5-HT and noradrenaline in both strips and rings was attenuated. At first sight, these results suggest that, in the strip or ring, the observed interaction is mediated more by an effect of 5-HT on α-adrenoceptors than is the case in the perfused artery. However, synergism between 5-HT and methoxamine, and between 5-HT and histamine (the latter in the presence of cimetidine to block histamine H_2 receptors) was found to be relatively insensitive to both ketanserin and prazosin. These results cannot be explained at present but are mentioned because they suggest that the mechanism of the synergistic action of 5-HT may vary with the nature of experimental procedures, and with the nature of the interacting agonist (even when the interacting agonists seemingly exert their contractile response via a common receptor, as in the case with noradrenaline and methoxamine). The probable complexity of the mechanisms underlying synergism was further highlighted by the observation that ketanserin, in a high concentration (0.5 μM) which depressed the response to histamine, increased the synergistic interaction between 5-HT and histamine. Analysis of the phenomenon indicated that it stemmed from increased stimulation of histamine H_2 receptors, occasioned by the use of higher concentrations of histamine when measuring the 5-HT–histamine interaction in the presence of H_1 receptor partial blockade (de la Lande *et al.* 1989). The possibility that an increase due to this factor may have masked inhibition of 5-HT-histamine synergism by ketanserin was excluded by studying the interaction in the presence of cimetidine, as mentioned above.

6.4.4 Role of electrolytes

An early observation, that synergism between 5-HT and noradrenaline in the perfused rabbit ear artery was abolished when Na^+ in the perfusion medium was replaced by K^+, led to the suggestion that synergism was related to an effect of 5-HT on processes involved in the maintenance of membrane potential (de la Lande *et al.* 1966). That the replacement results in virtually complete depolarization of the smooth muscle membrane has been shown by Vonderlage (1981); unfortunately, complementary information on the effects of 5-HT on membrane potential in this vessel is lacking. In other vessels, the importance of the membrane potential in synergistic interactions is indicated by the ability of K^+, in concentrations which partially depolarize the smooth muscle membrane, to increase the contractile response to noradrenaline (rabbit saphenous artery—Abel *et al.* 1981; rabbit aorta, Stupecky *et al.* 1986; rat mesenteric artery, Mulvaney *et al.* 1982).

There is considerable evidence to suggest that the post-junctional supersensitivity to contractile agents which follows chronic interruption of excitatory neurotransmission in smooth muscle, is due to a decrease in membrane potential which in turn decreases the threshold for excitation. Since the decrease appears to be confined to the component of the smooth muscle membrane potential which is maintained by Na^+,K^+ electrogenic pump activity (for discussion, see Fleming and Westfall 1988), it is of interest that a recent report (Bevan and Mangiarua 1987) suggests changes in Na^+,K^+-ATPase activity exert an important influence on the contractile response of the rabbit ear artery to 5-HT. These workers showed that chronic sympathetic denervation is associated with a marked increase in the contractile activity of 5-HT, and the increased activity is mediated by a receptor which is not the α-adrenoceptor. Since the Na^+,K^+-ATPase inhibitor, ouabain, produced a similar effect, they have proposed that, in addition to the pathway mediated by the α-adrenoceptor, a 5-HT receptor-mediated contractile pathway is present in the artery, but the activity of the latter pathway is normally suppressed by concurrent activity of the Na^+,K^+ pump. The implications of their proposal to mechanisms of synergism will be clearer when the 5-HT receptor has been identified, but intriguing possibilities are that either 5-HT itself decreases electrogenic Na^+,K^+ pump activity, thus decreasing the threshold for excitation by other agonists, or the contractile pathway is incompletely suppressed by the Na^+,K^+ pump and retains sufficient excitatory activity to facilitate responses to other agonists. Whether their results have implications to the paradoxical aspects of synergism observed in the reserpinized strips and ring preparations (Section 6.4.3) is uncertain, since short-term (24 h or less) denervation is reported to be without effect on Na^+,K^+-ATPase activity in another type of smooth muscle (guinea-pig vas deferens—Fleming and Westfall 1988).

6.4.5 Role of calcium

The role of Ca^{2+} in 5-HT–noradrenaline synergism in the perfused rabbit ear artery was examined by Manzini *et al.* (1986). The vasoconstrictor response to each of these agonists consists of a phasic component attributed to mobilization of intracellular Ca^{2+}, and a tonic component, attributed to influx of extracellular Ca^{2+} (Bevan *et al.* 1973; Steinsland *et al.* 1973; Maggi *et al.* 1983). In the experiments of Manzini *et al.* (1986) the two components were distinguished by the absence of the tonic response in high K^+,Ca^{2+}-free medium, and its transient restoration when the artery was returned to the control medium. Their results indicated that the potentiating effect of 5-HT (0.3–3 μM) was confined to the tonic response and was characterized by an increase in amplitude of the response, and by marked slowing of its relaxation phase. When added soon after relaxation commenced, nifedipine and EDTA were without effect, whereas caffeine and dibutyryl-cyclic-AMP accelerated relaxation, i.e. the latter agents opposed the effect of 5-HT on the kinetics of the tonic response. They concluded that the delayed relaxation was not due to an increased efflux of Ca^{2+} into the cell, but to an effect of 5-HT on one or more of the intracellular processes which initiate relaxation. An intracellular action was also suggested by the finding that 5-HT potentiated the transient vasoconstrictor action of caffeine. These results have focused attention on the possibility that synergism is an over-reaction to vasoconstrictor agents of smooth muscle cells with impaired relaxation processes. However, a question not directly addressed in their studies was whether calcium channel blockade influences the potentiating effect of 5-HT on the magnitude, as opposed to the duration of the tonic response to noradrenaline. It has been reported recently that diltiazem attenuates the contractile response to noradrenaline in the perfused rabbit ear artery and, at the same time, attenuates the synergistic interaction with 5-HT. The result suggests that 5-HT does in fact facilitate the noradrenaline-evoked influx of extracellular Ca^{2+} into the smooth muscle cell (Meehan and Medgett 1987). In the same report, attention is drawn to evidence that EDRF inhibits noradrenaline-evoked entry of Ca^{2+} into vascular smooth muscle, and it is speculated that this process may be involved in the effects of endothelium removal upon synergism between 5-HT and noradrenaline (see Section 6.4.1).

Manzini *et al.* (1986) and Meehan and Medgett (1987) are in agreement that 5-HT does not potentiate the phasic response to noradrenaline.

However, before it is concluded that 5-HT is without effect on noradrenaline-evoked mobilization of intracellular Ca^{2+}, it should be noted that the phasic response was elicited in a high-K^{+} and Ca^{2+}-free medium. Since the artery was depolarized it is conceivable that membrane effects of 5-HT leading to synergism were eliminated by the experimental conditions.

In vessels other than the rabbit ear artery, evidence that Ca^{2+} is involved in synergism is confined to reports that the calcium channel blocking agent, flunarizine, exerts a slowly-developing inhibitory effect on the interaction between 5-HT and prostaglandin $F_{2\alpha}$ in the canine basilar artery (Van Nueten 1985), and that calcium channel blocking agents attentuate synergism between 5-HT and noradrenaline in the rat caudal artery (Xioa and Rand 1987).

6.5 Synergism *in vivo*

In the human forearm and hand, a vasoconstrictor response to an intra-arterial infusion of noradrenaline is manifested by a decrease in blood flow. 5-HT, when infused intra-arterially, increased the response to noradrenaline in the forearm, but not in the hand (Scroop and Walsh 1968). The latter result is surprising in view of the marked synergism between the two amines seen in the isolated perfused human digital artery, and raises the possibility that the synergistic interaction between 5-HT and contractile agents in arteries is an *in vitro* phenomenon and is attentuated in some way under physiological conditions. Nevertheless, in the sheep an intravenous infusion of 5-HT increases the pressor response to noradrenaline (Myers *et al.* 1985), and an enhanced vasoconstrictor response of the ear artery to noradrenaline in the presence of 5-HT has been observed in the anaesthetized rabbit by Cannell (1969). In the latter experiments, the thickness of the ear immediately over the central artery was used as an index of vascular diameter. The thickness was increased when the blood pressure was increased by an intravenous bolus of noradrenaline or adrenaline, but when either of the amines was injected during an intravenous infusion of 5-HT the pressor response was associated with decreased ear thickness in six of the twelve rabbits. In seven of ten rabbits, the decrease in ear thickness produced by stimulation of the homolateral superior cervical nerve was enhanced by the 5-HT. The effects of the 5-HT are illustrated in Fig. 6.4. Interestingly there was no evidence of an effect of the 5-HT on either the pressor responses or the responses of the nictitating membrane to the catecholamines, or on the response of the nictitating membrane to homolateral cervical nerve stimulation. These results, while indicating that synergism does occur *in vivo,* suggest that it is confined to vessels in certain regions only, and is not a generalized phenomenon involving all vessels controlling periphal resistance.

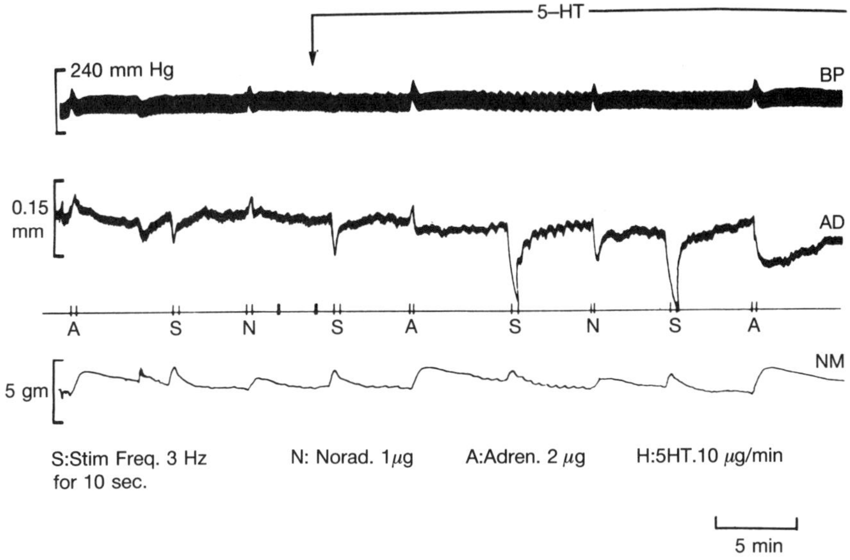

Fig. 6.4 Anaesthetized rabbit; responses of blood pressure (BP), ear thickness (AD) and of the nictitating membrane (NM) to intravenous bolus injections of adrenaline (A), noradrenaline (NA), and homolateral cervical nerve stimulation (S) At the arrow a sustained intravenous infusion of 5-HT, 10 μg/min was commenced. The small increases in ear thickness to NA and A reflected the effect of the increased blood pressure on ear artery diameter; in the presence of 5-HT the increases are converted to marked decreases, although the pressor effects are unchanged. Similarly, the decrease in ear artery thickness produced by homolateral cervical nerve stimulation is potentiated. Other details: (1) Rabbits received atropine (5mg/kg i.p.) and were then anaesthetized by a combination of paraldehyde (2ml/kg given intramuscularly) and pentobarbitone (6mg/ml infused intravenously) until surgical anaesthesia was obtained. (2) Ear thickness was recorded by means of light calipers, of which one arm was attached to the under surface of the ear. The second arm rested on the upper surface immediately over the central artery about 5cm from the base of the ear. Movement of the second arm was measured by a light beam focused on a photoelectric cell (Lippay 1956). (From Cannell 1969 with permission)

The only other reports of the occurrence of synergism *in vivo* are those of Medgett (1985 and 1987) who showed that in the autoperfused hind limb of the pithed rat, 5-HT potentiates the vasoconstrictor response to noradrenaline and produces both inhibitory and facilitatory effects on the responses to lumbar sympathetic nerve stimulation. Interestingly, the facilitory effects were more pronounced in the presence of an inhibitor of uptake$_1$, as was noted earlier in the isolated perfused rabbit ear artery (Meehan *et al.* 1986).

6.6 Pathophysiological significance

6.6.1 Migraine

Following the observation that methysergide mimicked the synergistic effects of 5-HT in the rabbit ear artery (de la Lande *et al.* 1966), a number of investigators have explored the possibility that an enhanced response to endogenous vasoconstrictors may underlie beneficial effects of drugs in migraine. However, the analyses by Fozard (1973, 1975, 1976, and 1977), which were mentioned earlier, and by Carroll *et al.* (1974) are in agreement that the action of methysergide is not shared by all antimigraine drugs, and that, in the rabbit ear artery, at least two, namely pizotifen and cyproheptadine, inhibit rather than reproduce the effects of 5-HT. Nevertheless, in both this artery and the human isolated temporal artery, potentiation of responses to noradrenaline is a feature common to various ergotamine derivatives and to methysergide (Carroll *et al.* 1974). On the basis of the interactions observed in the ear artery, Fozard has suggested the antimigraine drugs fall into three categories, according to their effects on responses to endogenous vasoconstrictor amines, namely: (a) those whose effects are inhibitory, e.g. pizotifen and cyproheptadine; (b) those which produce a non-specific increase in response, e.g. methysergide; and (c) those where an additional sensitizing component is specific for 5-HT or a related vasoconstrictor, such as tryptamine. The significance of this classification is diminished somewhat by subsequent evidence that, in extremely low concentrations (of the order of 1 pM), pizotifen and cyproheptadine are capable of sensitizing feline and human extracranial arteries to 5-HT (Hardebo *et al.* 1978). Hence, the possibility that a synergistic interaction with 5-HT is a feature common to antimigraine drugs cannot be discarded. Nevertheless, in view of evidence mentioned earlier, that 5-HT is a cotransmitter with noradrenaline in the sympathetic nerves to the guinea-pig pial artery, the interaction between these amines in intracranial rather than in extracranial arteries may be more relevant to the aetiology of migraine. The interaction between 5-HT and noradrenaline in human intracranial arteries does not appear to have been studied.

6.6.2 Vasospastic disease and hypertension

These two cardiovascular disorders can be linked for the present purpose since suggestions that the amplifying actions of 5-HT are involved are based primarily on events which occur when vascular endothelium is damaged. Experimentally, it has been shown that the vasoconstrictor response of isolated vessels caused by aggregated platelets is mediated by the released 5-HT, either directly or via its synergistic interaction with co-released vasoconstrictors such as thromboxane A_2 (de Clerck and Van

Nueten 1982). Evidence for synergism in a human vessel between 5-HT and the thromboxane A_2 receptor agonist, U466191, has been provided by Young *et al.* (1986). By augmenting vasoconstrictor tone, these interactions may contribute to localized vasospasm, or to hypertension where endothelium damage is generalized (for discussion, see Vanhoutte 1982; Vanhoutte and Houston 1986; Van Nueten and Vanhoutte 1984; Vanhoutte 1987).

The possible role of the 5-HT–noradrenaline interaction in experimental hypertension has also been explored. Of particular interest are reports that synergism is more pronounced in isolated perfused kidneys and in isolated perfused mesenteric arteries from the spontaneously hypertensive rat, than in those from the normotensive control rat (Janssens and Van Nueten 1986). However, the relevance of these findings to hypertension in man is obscured by the controversy that has surrounded the relative importance of α-adrenoceptor and 5-HT_2 receptor blockade to the antihypertensive action of ketanserin (for an appraisal of the role of 5-HT, see Vanhoutte 1985). Irrespective of the importance of 5-HT, attempts to ascribe a pathophysiological role to its amplifying action face the difficulty of distinguishing, *in vivo*, between this action and the direct constrictor effect of 5-HT on blood vessels. This difficult may be resolved when the precise relationship between the contractile and amplifying actions of 5-HT is more clearly understood than is the case at present.

6.7 Concluding remarks

A synergistic interaction between 5-HT and contractile agents appears to be a general phenomenon in arteries, but the information available permits only a glimpse of underlying mechanisms. The simplest hypothesis to account for the generality is that, in part at least, the synergism is secondary to the constrictor action of 5-HT. The evidence that the synergistic and contractile responses in a number of arteries, although not all, are mediated by a common receptor of the 5-HT_2 subtype is quite strong (e.g. in rabbit femoral, rat aorta, rat caudal), and in two of these (rat and rabbit aorta) the synergism spans sufficient of the full dose-response curve to the interacting agonist to exclude the possibility that it is only threshold in type. This is the case also in the rabbit femoral artery (Van Nueten, personal communication). In the rabbit aorta there is evidence to suggest that a functional interaction is superimposed upon a non-specific threshold effect. However, in most vessels the threshold phenomenon has not been explored, although it is a common finding that the synergistic effect of 5-HT is manifested in concentrations below these eliciting constriction, and the 'foot' of the dose-response curve to the interacting agonist is shifted to the left. In the perfused rabbit ear artery, there is evidence for a non-specific component, possibly threshold in nature, and for a second component which is more specific for 5-HT. It is to be hoped that the recent interest in smooth muscle membrane actions of 5-HT, both with respect to Na^+, K^+-ATPase activity, and the role of Ca^{2+}, foreshadows rapid advances in our understanding of the mechanisms involved in the 5-HT-specific component.

The possible pathophysiological significance of the synergistic action of 5-HT still rests heavily on the implications of *in vitro* findings. The *in vitro* interaction is now sufficiently well characterized to justify a more systematic approach to the study of its occurrence *in vivo* than has been the case in the past.

References

Abel, P. W., Urquilla, P. R., Goto, K., Westfall, D. P., Robinson, R. L., and Fleming, W. W. (1981). Chronic reserpine treatment alters sensitivity and membrane potential of the rabbit saphenous artery. *Journal of Pharmacology and Experimental Therapeutics* **217,** 430–9.

Apperley, A., Humphrey, P. P. A., and Levy, G. P. (1976). Receptors for 5-hydroxytryptamine and noradrenaline in rabbit isolated ear artery and aorta. *British Journal of Pharmacology* **58,** 211–21.

Asano, M. and Hidaka, H. (1980). Potentiation of the contractile response to acetylcholine in aorta strips by low concentrations of vascular contractile agonists. *British Journal of Pharmacology* **69,** 639–46.

Bevan, R. and Mangiarua, E. (1987). Separate receptors on smooth muscle of rabbit central ear artery for 5-hydroxytryptamine and norepinephrine revealed by inhibition of the Na^+-K^+ pump. Poster presented at the IUPHAR Satellite Meeting on Serotonin, Heron Island, Australia, September 1987.

Bevan, J. A., Gartska, C., Su, C., and Su, M. O. (1973). The bimodal basis of the contractile response of the rabbit ear artery to norepinephrine and other agonists. *European Journal of Pharmacology* **22,** 47–53.

Bevan, J. A., Duckles, S. P., and Lee, T. J-F. (1975). Histamine potentiation of nerve- and drug-induced responses of a rabbit cerebral artery. *Circulation Research* **36,** 647–53.

Black, J. L., French, R. J., and Mylecharane, E. J. (1981). Receptor mechanisms for 5-hydroxytryptamine in rabbit arteries. *British Journal of Pharmacology* **74,** 619–26.

Bülbring, E. and Burnstock, G. A. (1955). Membrane potential changes associated with tachyphylaxis and potentiation of the response to stimulating drugs in smooth muscle. *British Journal of Pharmacology and Chemotherapy* **15,** 611–24.

Cannell, V. A. (1969). The action of serotonin on blood vessels. Master of Science Thesis, University of Adelaide.

Carroll, P. R., Ebeling, P. W., and Glover, W. E. (1974). The responses of the human temporal and rabbit ear artery to 5-hydroxytryptamine and some of its antagonists. *Australian Journal of Experimental Biology and Medical Science* **52,** 813–23.

Chang, J.-Y., Hardebo, J. E., and Owman, C. (1989). Sympathetic nerves associated with brain vessels store and release serotonin which interacts with noradrenaline in cerebrovascular contraction. In *Serotonin: Actions, Receptors, Pathophysiology. Proceedings of the 1987 IUPHAR Congress Satellite Meeting, Heron Island, Australia* (ed. E. J. Mylecharane, J. A. Angus, I. S. de la Lande, and P. P. A. Humphrey). Macmillan, London. (In press).

De Clerck, F. and Van Nueten, J. M. (1982). Platelet-mediated vascular contractions: inhibition of the serotonergic component by ketanserin. *Thrombosis Research* **27,** 713–27.

De la Lande, I. S. (1975). Adrenergic mechanisms in the rabbit ear artery. *Blood Vessels* **12,** 137–60.

De la Lande, I. S. and Rand, M. J. (1965). A simple isolated nerve-blood vessel preparation. *Australian Journal of Experimental Biology and Medical Science* **43,** 639–56.

De la Lande, I. S., Cannell, Victoria A., and Waterson, J. G. (1966). The interaction of serotonin and noradrenaline on the perfused artery. *British Journal of Pharmacology and Chemotherapy* **28,** 255–72.

De la Lande, I. S., Harvey, J. A. and Holt, S. (1974). Response of the rabbit coronary arteries to autonomic agents. *Blood Vessels* **11,** 319–37.

De la Lande, I. S., Parker, D. A. S., Morris, R., Irvine, R., and Graefe K.-H. (1980). Noradrenaline diffusion across the artery. In *Vascular neuroeffector mechanisms* (ed. J. A. Bevan) pp. 148–60. Raven Press, New York.

De la Lande, I. S., Stanton, B. J., and Lungerhausen, Y. K. (1983). Receptors mediating sensitising actions of 5-hydroxytryptamine on the rabbit ear artery. *Blood Vessels* **20,** 189.

De la Lande, I. S. and Kennedy, J. A. (1985). Receptors mediating the vasoconstrictor and amplifying actions of serotonin on the rabbit ear artery. *Clinical and Experimental Physiology and Pharmacology Supplement* **9,** 6–7.

De la Lande, I. S., Kennedy, J. A., and Stanton, B. J. (1988). Amplifying action of serotonin. In *Serotonin: Actions, Receptors, Pathophysiology. Proceedings of the 1987 IUPHAR Congress Satellite Meeting, Heron Island, Australia* (eds. E. J. Mylecharane, J. A. Angus, I. S. de la Lande, and P. P. A. Humphrey). Macmillan, London. (In press.)

Fleming, W. W. and Westfall, D. P. (1988). Postjunctional supersensitivity. In *Catecholamines (Handbook of Pharmacology)*, eds. U. Trendelenburg and N. Weiner, Springer, Berlin. (In press).

Fozard, J. R. (1973). Drug interactions on an isolated artery. In *Background to Migraine* (ed. J. N. Cumings) pp. 150–69. W. Heinemann Medical, London.

Fozard, J. R. (1975). The animal pharmacology of drugs used in the treatment of migraine. *Journal of Pharmacy and Pharmacology* **27,** 297–321.

Fozard, J. R. (1976). Comparative effects of four migraine prophylactic drugs on an extracranial artery. *Europe Journal of Pharmacology* **36,** 127–39.

Fozard, J. R. (1977). The mechanism by which migraine prophylactic drugs modify vascular reactivity in vitro. In *Headache: New Vistas*, pp. 259–78. Biomedical Press, Florence.

Gaddum, J. H. (1953). In *Pharmacology* (4th edn). Oxford University Press, London.

Grohmann, M. and Trendelenburg, U. (1984). The substrate specificity of $uptake_2$

in the rat heart. *Naunyn-Schmiedeberg's Archives of Pharmacology* **328,** 164–73.

Hardebo, J. E., Edvinsson, L., Owman, C., and Svendgaard, N. (1978). Potentiation and antagonism of serotonin effects on intracranial and extracranial vessels. *Neurology* **28,** 64–70.

Hieble, J. P. and Woodward, D. F. (1984). Different characteristics of post junctional alpha adrenoreceptors on arterial and venous smooth muscle. *Naunyn-Schmiedeberg's Archives of Pharmacology* **328,** 44–50.

Iversen, L. L. (1967). *The uptake and storage of noradrenaline in sympathetic neurones.* Cambridge University Press, London.

Janssens, W. J. and Van Nueten, J. M. (1986). The direct and amplifying effects of serotonin are increased with age in the isolated perfused kidney of Wistar and spontaneously hypertensive rats. *Naunyn-Schmiedeberg's Archives of Pharmacology* **334,** 327–32.

Johnson, S. M. and De la Lande, I. S. (1978). Action of deoxycorticosterone acetate on the central artery of the rabbit ear. *Blood Vessels* **15,** 231–46.

Kalsner, S. (1972). Differential activation of the inner and outer muscle cell layers of the rabbit ear artery. *European Journal of Pharmacology* **20,** 122–4.

Leff, P. (1987). An analysis of amplifying and potentiating interactions between agonists. *Journal of Pharmacology and Experimental Therapeutics* **243,** 1035–42.

Lippay, F. (1956). Photoelectric device for the recording of diameter changes *in situ. Australian Journal of Experimental Biology and Medical Science* **34,** 301–5.

Mackay, D. (1982). Dose-response curves and mechanisms of drug action. *Trends in Pharmacological Sciences* **3,** 496–9.

Maggi, C. A., Manzini, S., and Meli, A. (1983). Contribution of cellular and extracellular Ca^{2+} during 5-hydroxytryptamine-induced contractions of rabbit ear artery. *European Journal of Pharmacology* **94,** 251–60.

Manzini, S., Maggi, C. A., and Meli, A. (1986). 5-Hydroxytryptamine delays relaxation time of norepinephrine-induced vasoconstriction. *American Journal of Physiology* **250,** H121–30.

Medgett, I. C. (1985). Amplification effect of serotonin on sympathetic vasoconstriction in rat autoperfused hind limb. *Proceedings of the Australia Physiological and Pharmacological Society* **16,** 200P.

Medgett, I. C. (1987). Effect of neuronal uptake blockade on the amplifying effect of serotonin on sympathetic vasoconstriction in rat auroperfused hind limb. *Journal of Cardiovascular Pharmacology* **10** (Supplement 3) S65–8.

Medgett, I. C. and Langer, S. Z. (1984). Hetereogeneity of smooth muscle α-adrenoceptors in rat tail artery in vitro. *Journal of Pharmacology and Experimental Therapeutics* **229,** 823–30.

Meehan, A. G. and Medgett, I. C. (1987). Effect of endothelium removal on the amplifying effect of serotonin on responses to noradrenaline in rabbit ear artery. Poster presented at the IUPHAR Satellite Meeting on Serotonin, Heron Island, Australia, September 1987.

Meehan, A. G., Rand, M. J., and Medgett, I. C. (1986). Effects of serotonin on sympathetic noradrenergic transmission in rabbit isolated ear artery. *Journal of Cardiovascular Pharmacology* **8,** 1143–53.

Meehan, A. G., Story, D. F. and Medgett, I. C. (1987). Influence of the endothelium on the amplification by serotonin of vasoconstrictor responses to noradrenaline and sympathetic nerve stimulation in rabbit isolated ear artery. *European Journal of Pharmacology* **141,** 339–46.

Mulvaney, M. J., Nilsson, H., and Flatman, J. A. (1982). Role of membrane potential in the response of rat small mesenteric arteries to exogenous noradrenaline stimulation. *Journal of Physiology* **332,** 363–73.

Myers, H., Mecca, T. E., and Clinton Webb, R. (1985). Direct and sensitizing effects of serotonin agonists and antagonists on vascular smooth muscles. *Journal of Cardiovascular Pharmacology* **7,** Supplement 7, S44-8.

Paiva, M. Q., Caramona, M., and Osswald, W. (1984). Intra- and extraneuronal metabolism of 5-hydroxytryptamine in the isolated saphenous vein of the dog. *Naunyn-Schmiedeberg's Archives of Pharmacology* **325,** 62–8.

Purdy, R. E. and Murray, D. L. (1988). Serotonin-induced vasoconstriction and contractile synergism with norapinephrine; role of adrenoceptors. In *Serotonin: Actions, Receptors, Pathophysiology. Proceedings of the 1987 IUPHAR Congress Satellite Meeting, Heron Island, Australia* (ed. E. J. Mylecharane, J. A. Angus, I. S. de la Lande, and P. P. A. Humphrey). Macmillan, London. (In press).

Purdy, R. E., Hurlbut, D. E., and Rains, L. A. (1981). Receptors for 5-hydroxytryptamine in rabbit isolated ear artery and aorta. *Blood Vessels* **18,** 16–27.

Rapoport, R. M. and Bevan, J. A. (1982). Increased contractility induced in rabbit ear artery by prior exposure to serotonin masks concomitant desensitization to norepinephrine. *Journal of Pharmacology and Experimental Therapeutics* **222,** 338–43.

Rapoport, R. M. and Bevan, J. A. (1983). Effects of contraction on the subsequent responsiveness and maximum contractility of the rabbit ear artery and saphenous vein in vitro. *Blood Vessels* **20,** 44–55.

Scroop, G. C. and Walsh, J. A. (1968). Interaction between angiotensin, noradrenaline, and serotonin on blood vessels in man. *Australian Journal of Experimental Biology and Medical Science* **48,** 573–80.

Seabrook, J. M. and Nolan, P. L. (1983). The vascular interaction of noradrenaline and 5-hydroxytryptamine. *European Journal of Pharmacology* **89,** 131–5.

Stanton, B. and de la Lande, I. S. (1982). Interaction of serotonin and ketanserin on the rabbit ear artery. *Clinical and Experimental Physiology and Pharmacology* **9,** 408.

Steinsland, O. S., Furchgott, R. F., and Kirpekar, S. M. (1973). Biphasic vasoconstriction of the rabbit ear artery. *Circulation Research* **32,** 49–58.

Stupecky, G. L., Murray, D. L., and Purdy, R. E. (1986). Vasoconstrictor threshold synergism and potentiation in the rabbit isolated thoracic aorta. *Journal of Pharmacology and Experimental Therapeutics* **238,** 802–8.

Su, C. and Uruno, T. (1985). Excitatory and inhibitory effects of 5-hydroxytryptamine in mesenteric arteries of spontaneously hypertensive rats. *European Journal of Pharmacology* **106,** 283–90.

Van den Brink, F. G. (1977). General theory of drug-receptor interactions. In *Kinetics of drug action (Handbook of Pharmacology)*, pp. 169–254. Springer, Berlin.

Vanhoutte, P. M. (1982). Does 5-hydroxytryptamine play a role in hypertension? *Trends in Pharmacological Science* **3,** 370–3.

Vanhoutte, P. M. (1985). Can peripheral serotonergic blockade explain the hypertensive effect of ketanserin? *Journal of Cardiovascular Pharmacology* **7,** (Supplement 7), S105–9.

Vanhoutte, P. M. (1987). Serotonin and the vascular wall. *International Journal of Cardiology* **14,** 189–203.

Vanhoutte, P. M. and Houston, D. S. (1986). Serotonin, platelets, and vascular function. In *Central and peripheral mechanisms of cardiovascular regulation* (eds A. Magro, W. Osswald, D. Reis, and P. Vanhoutte) pp. 135–61. Plenum, New York.

Van Nueten, J. M. (1985). Calcium entry blockers and vascular smooth muscle reactivity. In *Calcium entry blockers and tissue protection* (ed. T. Godfraind) pp. 69–79. Raven Press, New York.

Van Nueten, J. M. and Vanhoutte, P. M. (1984). Serotonin and vascular function. *Clinics in Anaesthesiology* **2,** 363–82.

Van Nueten, J. M., Janssen, P. A. J., Van Beek, J., Xhonneux, R., Verbeuren, T. J., and Vanhoutte, P. M. (1981). Vascular effects of ketanserin (R 41 468), a novel antagonist of 5-HT_2 serotonergic receptors. *Journal of Pharmacology and Experimental Therapeutics* **218,** 217–30.

Van Nueten, J. M., Janssen, P. A. J., De Ridder, W., and Vanhoutte, P. M. (1982). Interaction between 5-hydroxytryptamine and other vasoconstrictor substances in the isolated femoral artery of the rabbit; effect of ketanserin (R 41 468). *European Journal of Pharmacology* **77,** 281–7.

Vonderlage, M. (1981). Spread of contraction in rabbit ear artery preparations in response to stimulation by norepinephrine. *Circulation Research* **49,** 600–8.

Wilton, P. B. and McCalden, T. A. (1977). Interaction between noradrenaline and 5-hydroxytryptamine in the isolated perfused rat hindlimb. *European Journal of Pharmacology* **46,** 213–9.

Xiao, X. H. and Rand, M. J. (1987). Interactions between serotonin and other vasoconstrictor agents in the rat tail artery. *Clinical and Experimental Pharmacology and Physiology*. (In press).

Young, M. S., Iwanov, V. and Moulds, R. F. W. (1986). Interaction between platelet-released serotonin and thromboxane A_2 on human digital arteries. *Clinical and Experimental Pharmacology and Physiology* **13,** 143–52.

7

Mechanisms of 5-hydroxytryptamine-induced vasodilatation

EWAN J. MYLECHARANE and
CHRISTINE A. PHILLIPS

7.1 Introduction

The classical approach to gaining an understanding of the mechanisms underlying a particular pharmacological action is to define it in terms of cellular location, physiological and/or biochemical process, and receptor involvement. An emphasis is generally placed on the last of these, because it is only through a comprehensive receptor classification scheme that the unique interactions between individual pharmacological agents and their preferred receptor targets are able to be given a semblance of order. Pharmacological actions can thereby be understood in a systematic manner, and the effects of related agents may be predicted somewhat more accurately.

The vasodilator effect of 5-HT is an example of a pharmacological action that was quickly recognized, but the beginnings of a real understanding of the mechanisms involved were delayed by at least 20 years. The reasons for this may lie largely in the fact that most of the drugs which were initially available to modulate this action had little or no activity, although they had prominent effects on many of the other actions of 5-HT. Thus there were few clues as to what kind(s) of receptor might be involved, and hence little stimulus to look for subtle distinctions about the sites and nature of this action. The comparatively recent development or recognition of many drugs with actions at 5-HT receptors, and a better understanding of 5-HT receptor classification (Bradley *et al.* 1986), have helped to redress this situation.

In this chapter, the following section briefly recounts the various observations which established that 5-HT has a vasodilator action. The next section outlines the mechanisms involved in terms of sites of action and nature of the effect. The 5-HT receptors involved are discussed separately in accordance with the Bradley *et al.* (1986) classification, and some pathophysiological implications of the vasodilator effect of 5-HT are mentioned in the concluding section.

7.2 The vasodilator action of 5-HT

When 5-HT was isolated by Rapport *et al.* (1948), vasoconstriction was its best known action. Within five years, however, its vasodilating properties had been recognized. In experiments conducted *in vivo*, using techniques designed to monitor changes in vascular resistance (by measuring pressure changes in vascular beds perfused at a constant flow rate, or by measuring both blood flow and perfusion pressure), vasodilatation was able to be elicited by 5-HT in many vascular beds. Haddy (1953) described 5-HT-induced vasodilatation in the dog fore limb; similar effects were noted in the coronary, hind limb, renal, mesenteric, and hepatic circulations in dogs (Schofield and Walker 1953; McCubbin *et al.* 1962; Richardson and Withrington 1978). There have been conflicting reports of vasoconstrictor and vasodilator effects of 5-HT in the dog cranial circulation (e.g. Swank and Hissen 1964; Saxena 1972; Vidrio and Hong 1976; Mena and Vidrio 1979). A vasodilator response to 5-HT in the human forearm was reported by Roddie *et al.* (1955).

In other species, 5-HT-induced vasodilatation has been described, for example, in skin, skeletal muscle, and gastric vessels in rats (Garattini and Valzelli 1965), in the baboon external carotid vasculature (Grimson *et al.* 1969), and in the pig and cat common carotid vascular bed (Saxena and Verdouw 1982).

Vasodepressor responses to 5-HT in various species (man, dog, cat, rabbit, and rat: see Page 1958 for references) were also reported soon after its isolation. A prolonged hypotensive effect, independent of the Bezold–Jarisch-like effect, was demonstrated; this can be interpreted as vasodilatation, because 5-HT invariably produced increases in cardiac output distinct from the transient decreases associated with the Bezold–Jarisch effect (Page 1958).

In vitro studies have also demonstrated that 5-HT has vasodilator effects. Such studies are in many respects better able to define the underlying mechanisms involved, because many of the complicating factors and variables characteristic of *in vivo* studies are avoided. Surprisingly, studies of this type were not pursued during the 1950s and 1960s.

Although there was a widely held belief that 5-HT was able to induce relaxation of vascular smooth muscle (as evidenced by its frequent invoking to explain 5-HT-induced vasodilatation *in vivo*), it appears that this stemmed from a mention by Page and McCubbin (1953) of unpublished work by Twarog, suggesting that 5-HT produced relaxation in some invertebrate smooth muscle preparations. The first unequivocal demonstration of a direct vascular smooth muscle relaxant effect of 5-HT *in vitro* was that of Eyre (1975) in sheep pulmonary vein preparations.

The introduction of important new concepts in the location of the site of action for 5-HT-induced vasodilatation came with subsequent *in vitro* studies. McGrath (1977) showed that 5-HT inhibited electrically evoked noradrenaline release from adrenergic nerve terminals in dog saphenous vein and tibial artery preparations, and both Cocks and Angus (1983) and Cohen *et al.* (1983) reported that a relaxant factor derived from the endothelium was involved in 5-HT-induced relaxation in pig and dog coronary artery preparations, respectively.

7.3 Sites and functions involved in 5-HT-induced vasodilatation

7.3.1 Direct relaxant effects in vascular smooth muscle

7.3.1.1 *Involvement in the 'neurogenic tone' hypothesis* in vivo

The early *in vivo* studies on the so-called 'amphibaric' (Page 1958) or 'bidirectional' (Haddy *et al.* 1959) vascular actions of 5-HT were all interpreted in terms of contractile and relaxant effects on vascular smooth muscle, despite the absence of any firm *in vitro* evidence for the latter.

Page and McCubbin (1953) in analysing the arterial blood pressure response to 5-HT in dogs, concluded that a 'peripheral inhibition of neurogenic vasoconstriction' was responsible for the third phase of the response, a sustained vasodepression; the initial brief fall in blood pressure was due to a Bezold–Jarisch-like effect and a transient ganglion-blocking action. It needs to be emphasized that 'peripheral inhibition of neurogenic vasoconstriction', in the context used by Page and McCubbin (1953), did *not* mean a presynaptic inhibition of noradrenaline release. These authors raised the possibility that 5-HT was acting in the same manner as the adrenergic antagonists to prevent the vasoconstrictor response to noradrenaline released from adrenergic nerve terminals. The fact that 5-HT administered during noradrenaline-induced hypertension resulted in an enhanced vasopressor effect, rather than a vasodepressor response, suggested that such an action might not be responsible, so they then drew an analogy between the action of 5-HT on arterioles and its relaxant effect on invertebrate smooth muscle, and described the action of 5-HT as a 'peripheral inhibiting or relaxing effect'. In fact, Page and McCubbin (1953) specifically stated that they were unable to test the possibility that 5-HT was acting at sympathetic nerve endings to prevent the release of adrenergic neurotransmitter.

Haddy *et al.* (1959) extended these observations by studying the effects of 5-HT on vascular resistance in the dog fore limb, and developed the 'neurogenic tone' hypothesis. This was perhaps an unfortunate choice of terminology, given the later development of the concept of a presynaptic inhibitory effect of 5-HT on sympathetic nerve terminals. In essence, the hypothesis was that 5-HT causes a net vasodilatation when a vascular bed is constricted, and a net vasoconstriction when it is dilated, because of the different effects of 5-HT on vascular smooth muscle in different segments of the same vascular bed. It depended on the fact that sympathetic adrenergic activity influences vascular resistance in small arteries and arterioles to a much greater extent than in large arteries, and that 5-HT is capable of producing both contraction and relaxation of vascular smooth muscle. When neurogenic tone is lost, the smaller vessels become maximally dilated and 5-HT is unable to dilate them further but continues to constrict the larger vessels. When neurogenic tone is high, the vasodilator effect of 5-HT on the small arteries outweighs its vasoconstrictor effect on large arteries, resulting in a net vasodilatation. Thus the 'neurogenic tone' hypothesis was really a 'vascular tone' hypothesis.

This attractive explanation was accepted without question for the next 25 years. The demonstration of a direct relaxant effect of 5-HT in a vascular smooth muscle preparation (Eyre 1975) provided a somewhat belated reinforcement of the hypothesis in that it showed the actual existence of a key underlying element.

7.3.1.2 *Functional mechanisms* in vivo

The precise details of the functions involved in the vascular smooth muscle relaxant action of 5-HT *in vivo* have received relatively little attention. Haddy *et al.* (1959) considered that interference by 5-HT with the vasoconstrictor response to noradrenaline was still the most likely explanation, despite the failure of 5-HT to convert it to a vasodilator response in their experiments on the dog fore limb, and the similar findings of Page and McCubbin (1953) for the vasopressor effect of noradrenaline in dogs. Gordon *et al.* (1958) had previously noted an apparent antagonism between 5-HT and noradrenaline when monitoring small vessel tone exclusively in the dog fore limb. Haddy *et al.* (1959) suggested that the situation may be more complex in some circulations when humoral mediators are used to alter vascular tone, because a vasoconstrictor effect of noradrenaline on large vessels, in combination with that of 5-HT, might outweigh a vasodilator effect of 5-HT on small vessels brought about by interference with the vasoconstrictor response to noradrenaline.

McCubbin *et al.* (1962) explored further the way in which 5-HT opposed neurogenic vasoconstrictor activity in the dog hind limb, renal, and mesenteric circulations *in vivo*. The possibility that 5-HT blocked adrenergic contractile receptors in these circulations was discounted, because 5-HT failed to elicit a vasodilator response when vascular tone was induced by exogenous noradrenaline rather than sympathetic nerve activity. The complicating effect of noradrenaline on large arteries, as proposed by Haddy *et al.* (1959), was not considered to be relevant, because a contrasting pattern of responses had been observed; sympathetic nerve stimulation constricted the large vessels while exogenous noradrenaline had only a minor effect. McCubbin *et al.* (1962) also rejected the possibility that 5-HT was preventing the release of noradrenaline from sympathetic nerve terminals, because they found that 5-HT had no effect on tyramine-induced noradrenaline release. These authors recognized that 5-HT-induced vasodilatation occurred only during sympathetic nerve stimulation or infusion of adrenaline, but on the basis of experiments with the β-adrenoceptor antagonist, dichloroisoprenaline, they concluded that the vasodilator action of 5-HT was mediated by vascular smooth muscle adrenergic vasodilator receptors, probably via an indirect sensitization.

In the dog cranial circulation, conflicting observations of both vasoconstrictor and vasodilator responses to 5-HT had been reported (Swank and Hissen 1964; Saxena 1972; Vidrio and Hong 1976). It was noted by Mena and Vidrio (1979) that vagotomized dogs had been used in some studies but not in others; these authors were able to exclude a parasympathetic influence on the external carotid vasodilator response to 5-HT, but they established that the action of 5-HT depended upon the cervical sympathetic nerve trunks (which accompany the vagus nerves in the neck) remaining intact. Although the discrepancies in the past observations were resolved, Mena and Vidrio (1979) still explained their findings in terms of the neurogenic tone hypothesis.

Blackshear *et al.* (1985) studied responses to 5-HT and noradrenaline in large and small arterial vessels in the dog hind limb, using angiographic and electromagnetic flow probe techniques. Their findings were essentially the same as those of McCubbin *et al.* (1962), who had used a constant flow perfusion pressure technique. Additional findings, however, were that indomethacin and propranolol, at doses sufficient to inhibit the vasodilator responses to arachidonate and isoprenaline, respectively, did not affect 5-HT-induced vasodilatation; thus involvement of eicosanoid and β-adrenoceptor mechanisms was excluded. Interestingly, Blackshear *et al.* (1985) ascribed the vasodilator action of 5-HT to a presynaptic inhibition of sympathetic adrenergic nerve terminals, but cited Haddy *et al.* (1959) and McCubbin *et al.* (1962) as direct evidence for such an action.

7.3.1.3 *Assessment of relative contribution* in vivo

A series of *in vivo* experiments with 5-HT was undertaken in our laboratory, using anaesthetized dogs, to assess the relative contributions to 5-HT-induced vasodilatation made by a vascular smooth muscle relaxant component and a presynaptic inhibition of sympathetic adrenergic nerve terminals. Studies were conducted in the femoral (Phillips *et al.* 1985a) and common carotid (Markus *et al.* 1988a) arterial circulations; vascular resistances and conductances were calculated from local perfusion pressures across the circulation, and blood flow data measured using electromagnetic flow probes. It was reasoned that any direct vascular smooth muscle relaxant effect of 5-HT should be in evidence as a vasodilator response regardless of whether vascular tone was maintained by sympathetic or non-sympathetic means.

The intra-arterial infusion of 5-HT elicited reproducible vasodilator responses in both circulations, the maximal effect being approximately 75 per cent of that produced by intra-arterial infusion of acetylcholine. The response to 5-HT was abolished following removal of sympathetic vascular tone (by means of the ganglion-blocking agent mecamylamine in the femoral experiments, and bilateral cervical vagosympathectomy in the common carotid studies). In some experiments, a small vasoconstrictor response was unmasked. Removal of sympathetic vascular tone could have simply masked any vasodilator response to 5-HT, regardless of its site of action. However, the vasodilator effect of acetylcholine was still in evidence after removal of sympathetic tone, suggesting that it is unlikely that a simple masking of vasodilator effects alone is responsible. The mean responses to 5-HT and acetylcholine in the femoral arterial circulation are shown in Fig. 7.1.

When vascular tone in these circulations was restored by the intra-arterial infusion of ornipressin, a vasoconstrictor agent, the vasodilator response to 5-HT remained abolished (Fig. 7.1), but that to acetylcholine was retained (Fig. 7.1). If the vasodilator effect of 5-HT depended only on *vascular* tone, as was implied by Haddy *et al.* (1959), then 5-HT should have produced vasodilatation following restoration of vascular tone with ornipressin. Any direct relaxant effect of 5-HT would be expected to be manifested as a physiological or functional antagonism irrespective of the means by which vascular tone was maintained, as indicated by the fact that acetylcholine, which acts on the vascular endothelium to initiate vascular smooth muscle relaxation (Furchgott and Zawadzki 1980), retained its vasodilator effect in these circumstances.

The results obtained in these experiments therefore suggest that a direct relaxant effect of 5-HT makes little contribution to 5-HT-induced vasodilatation in the dog femoral and common carotid arterial circulations. The effect of 5-HT appears to be directly related to sympathetic adrenergic function. Interference with the vasoconstrictor action of noradrenaline could have accounted for our observations, and would be unable to be manifested after abolition of sympathetic function. However, in our experiments, 5-HT had no effect on the dog common carotid vasoconstrictor response to noradrenaline (before or after cervical sympathectomy), in accordance with findings that vasoconstrictor responses to noradrenaline in the dog fore limb or hind limb are either unaffected or enhanced by 5-HT (Haddy *et al.* 1959; McCubbin *et al.* 1962; Feniuk *et al.* 1981a). The results obtained therefore suggest that the vasodilator effect of 5-HT depends predominantly on a presynaptic inhibition of sympathetic vascular tone, which is discussed in more detail below in Section 7.3.3.

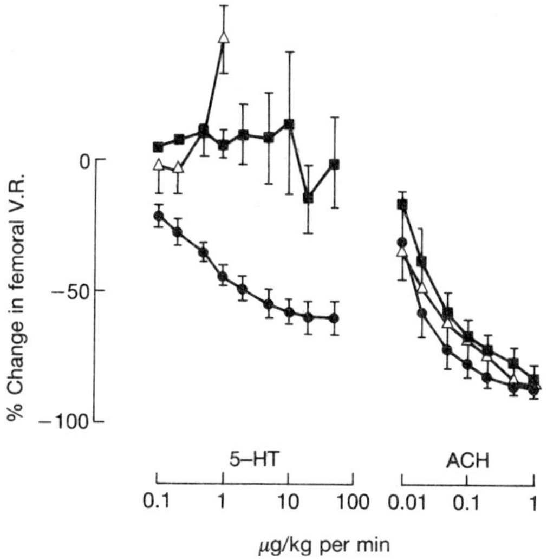

Fig. 7.1 Mean log dose–response curves for the femoral vasodilator effects of 5-HT and acetylcholine (ACH) in anaesthetized dogs, before (control; ●) and after ganglion block with mecamylamine (5 mg/kg i.v.; ■), and after restoration of femoral vascular tone by the simultaneous intra-arterial infusion of ornipressin (0.0005–0.001 units/kg per min; ▲). In each experiment, the femoral vascular resistance (V.R.) values were expressed as percentage changes from the pre-infusion femoral V.R. levels for 5-HT and acetylcholine, respectively. The points represent mean responses from all experiments (n = 5) at the cumulative intra-arterial dose levels indicated; the bars indicate SEM. Responses to all doses of 5-HT after ganglion block alone and after restoration of femoral vascular tone with ornipressin were significantly different from the corresponding control responses (Student's t test, 2-tailed, $P < 0.05$); responses to acetylcholine did not differ significantly from the corresponding control responses. (Reproduced with permission from Phillips *et al.* 1985a)

Although many investigators had recognized the involvement of sympathetic adrenergic function in the vasodilator response to 5-HT *in vivo* (e.g. Haddy *et al.* 1959; McCubbin *et al.* 1962; Mena and Vidrio 1979), they did not really entertain the possibility that the effect of 5-HT was at a site other than the vascular smooth muscle. Control experiments with direct-acting vasodilators and restoration of vascular tone by non-adrenergic means might have avoided this. With the advantage of hindsight, the complex explanation offered by Haddy *et al.* (1959), to account for the inability of 5-HT to reverse noradrenaline-induced vasoconstriction, can be seen as a classical example of question-begging. McCubbin *et al.* (1962) were perhaps unlucky in that at the time of their investigations, dichloroisoprenaline was the only β-adrenoceptor antagonist generally available. The considerable partial agonist activity of this agent (Moran 1967) may have produced sufficient vasodilatation to mask the vasodilator effect of 5-HT, irrespective of its site of action. Alternatively, dichloroisoprenaline may have been acting at specific receptors for 5-HT (again, irrespective of the site of action of 5-HT), because other β-adrenoceptor antagonists can interact with specific 5-HT receptors (see Sections 7.4.1.1 and 7.4.3.1). Likewise, subsequent investigations have shown that the absence of an effect on tyramine-induced release of noradrenaline is not necessarily relevant to presynaptic inhibition of nerve stimulation-induced release from adrenergic nerve terminals.

The fact that 5-HT-induced direct relaxation of vascular smooth muscle does not appear to be the predominant mechanism underlying its *in vivo* vasodilator action in a number of specific circulations, should not detract from the possibility that it may be important in others. Investigations to ascertain its contribution need to be conducted. Saxena and Verdouw (1982), using the radiolabelled microsphere technique, have shown that the vasodilator responses to 5-HT in the pig and cat common carotid circulations do not appear to depend on presynaptic sympathetic adrenergic inhibitory effects. 5-HT produced a substantial vasodilatation in the arteriolar–capillary (nutrient) fraction of carotid blood flow, in parallel with constriction of the arteriovenous–anastomotic component, in animals which had undergone bilateral cervical sympathectomy. In the pig, the vasodilator effect in the nutrient fraction was especially marked in the extracerebral tissues, particularly the skin and ears, which in most species are innervated predominantly by cervical sympathetic nerves (Stromberg 1964; Harper *et al.* 1972; Mylecharane *et al.* 1980). A direct relaxant effect is therefore the most likely explanation for 5-HT-induced vasodilatation in these circulations. Likewise, Kalkman *et al.* (1983) demonstrated a direct vasodepressor response to 5-HT in pithed rats following restoration of vascular tone with vasopressin.

It is likely that direct relaxation plays at least some part in 5-HT-induced vasodilatation, even in those circulations where presynaptic inhibition seems to be the dominant mechanism. In our own experiments, the vasodilator response to 5-HT in intact dogs exceeded the level to which vascular tone fell when sympathetic activity was interrupted. This could have been partly due to sympathetic tone not being completely abolished, especially in the common carotid circulation, because, as noted above, cervical sympathetic innervation controls the external carotid circulation but has little influence on the cerebral vasculature. Cerebral vessels may be modulated by an adrenergic system originating in the locus coeruleus (Goadsby *et al.* 1982); in dogs, there are extensive connections between the external carotid and the cerebral vessels, and thus 5-HT-induced changes in the latter could influence measurements made in the common carotid circulation as a whole. The fact that the vasodilator response to 5-HT was abolished or converted to a vasoconstrictor effect after interruption of sympathetic activity does not exclude the existence of a direct vascular relaxant component, since a concurrent vasoconstrictor component in the response to 5-HT could have simply outweighed it. As discussed below in Section 7.4.3.2, the different 5-HT receptors, and a variable pattern for their involvement in the effects of 5-HT which contribute to its vascular actions, could account for differences between circulations and species in the overall responses.

7.3.1.4 In vitro *studies*

The *in vitro* demonstration by Eyre (1975) of a direct vascular smooth muscle relaxant effect of 5-HT in sheep pulmonary vein preparations has been followed by similar findings in cat and human cranial arteries (Edvinsson *et al.* 1978), goat pulmonary vein (Chand 1981), cat saphenous vein (Feniuk *et al.* 1983), pig vena cava (Trevethick *et al.* 1984) and rabbit jugular vein (Martin *et al.* 1987). In most instances, the tissues were prepared as spirally cut strips in a way that would be expected to remove endothelial cells; however, Edvinsson *et al.* (1978) and Trevethick *et al.* (1984) used short cylindrical segments or rings, and thus endothelial cells could possibly have contributed to the response. All of the tissues except goat pulmonary vein required precontraction (e.g. with histamine, carbachol, methoxamine, α-methyl-5-HT, prostaglandin $F_{2\alpha}$, or K^+) before the relaxant response to 5-HT could be elicited.

Observations with a variety of pharmacological antagonists and inhibitors excluded the involvement of muscarinic cholinergic receptors, histamine H_1 and H_2 receptors, eicosanoid mediators, or neuronally derived inhibitory mediators in the relaxant response to 5-HT. The various agents were used at concentrations that would be expected to block the appropriate receptors or mechanisms, but only Eyre (1975) included data confirming that this was so for all agents used. In four of these studies, the involvement of β-adrenoceptors was ruled out by the lack of effect of propranolol, but Edvinsson *et al.* (1978) reported blockade with propranolol and INPEA. Although these authors noted that their pA_2 value for propranolol (8.29) was 'somewhat lower' than the value of 9.17 reported previously for antagonism of isoprenaline-induced relaxation in the same model (Edvinsson and Owman 1974), they concluded that the effect was mediated by β-adrenoceptors. A more plausible interpretation is that propranolol has an appreciable affinity for the specific 5-HT receptors mediating relaxation in the cat cranial vessels (see Section 7.4.1.1).

There has been very little investigation into the second messenger systems or other post-receptor events involved in the direct vascular smooth muscle relaxant effect of 5-HT. The only information available is that from the studies of Trevethick *et al.* (1984 and 1986), who have provided evidence that, in the pig vena cava, adenylate cyclase may be linked to the 5-HT relaxant receptor. Cyclic AMP levels were increased in parallel with the relaxant response, and the effects of 5-HT antagonists on both appeared to be closely related (see also Chapter 4).

7.3.2 Release of EDRF

The release of EDRF by 5-HT, first reported in pig and dog coronary arteries (Cocks and Angus 1983; Cohen *et al.* 1983), is discussed in detail in Chapter 8. Most of the investigations into this action of 5-HT have of necessity been conducted *in vitro*. Some recent *in vivo* studies, however, have shown that EDRF is involved in the overall vascular response to 5-HT, at least in large vessels such as the dog common carotid artery (Angus *et al.* 1987). Nevertheless, the contribution of 5-HT-induced EDRF release in smaller vessels and arterioles to the vasodilator response to 5-HT in the whole circulation, remains to be established. After removal of sympathetic vascular tone and restoration of tone by non-adrenergic means (as described in Section 7.3.1.3), any vascular relaxation mediated by 5-HT-induced EDRF release should be just as evident as a direct vascular smooth muscle relaxant action of 5-HT. The findings in the experiments described in Section 7.3.1.3 suggest that, as for the direct relaxant action, any EDRF component would make a minor contribution relative to that involving presynaptic inhibition of sympathetic adrenergic nerve terminals.

7.3.3 Presynaptic inhibition of sympathetic adrenergic nerve terminals

7.3.3.1 In vitro *studies in vascular preparations*

The original finding by McGrath (1977), that 5-HT inhibited the electrically evoked release of [^{3}H]noradrenaline from dog saphenous vein and tibial artery preparations, has been confirmed and extended in a range of vascular preparations. Müller-Schweinitzer (1981), Watts *et al.* (1981), Engel *et al.* (1983), and Fishburn *et al.* (1987) described similar 5-HT-induced reductions in [^{3}H]noradrenaline release from dog saphenous vein preparations stimulated at 2 Hz. Although there were minor differences in the magnitude of the inhibition and the potency of 5-HT in these experiments, in general the effect appeared to be less marked than the inhibition (approximately 80 per cent with 0.1 μM 5-HT) originally reported by McGrath (1977), where the stimulation was at 1 Hz.

Comparisons of potencies between different studies can be hazardous, because of subtle differences in experimental protocols. Göthert *et al.* (1986a and b), however, studied [^{3}H]noradrenaline release in response to electrical stimulation at both 0.66 and 2 Hz, in human saphenous vein and rat vena cava preparations. In both tissues, the presynaptic inhibitory effect of 5-HT appeared to be more marked at the lower stimulation frequency, as is commonly found with other presynaptic adrenergic modulators (Starke 1977; Langer 1980). In comparison with the dog saphenous vein studies, the magnitude of the inhibition and the potency of 5-HT appeared to be lower; at 0.66 Hz stimulation, 1 μM 5-HT produced an approximately 40 per cent inhibition in the human saphenous vein and 25 per cent (or 30 per cent; see Molderings *et al.* 1987) in the rat vena cava.

Su and Uruno (1985) reported that 5-HT (0.03 μM) produced a very small but significant inhibition of electrically evoked (2 Hz) [^{3}H]noradrenaline release in mesenteric artery preparations from normal rats, but not in preparations from spontaneously hypertensive rats. In the rat renal vasculature, Charlton *et al.* (1986) found that 5-HT inhibited the release of [^{3}H]noradrenaline in response to electrical stimulation at 2 Hz, by approximately 50 per cent at 1 μM.

Overall, the presynaptic inhibitory effect of 5-HT appears to be most potent in the dog saphenous vein, followed by rat renal vasculature, human saphenous vein, rat vena cava, and rat mesenteric artery. The effect of 5-HT cannot be an artefact brought about by enhancement of basal (or spontaneous) [^{3}H]noradrenaline release (which could give an apparent reduction in electrically evoked release). The effects of 5-HT on basal [^{3}H]noradrenaline release were monitored in most of the above studies; in all instances, presynaptic inhibitory concentrations of 5-HT had no effect on basal release, although McGrath (1977), Charlton *et al.* (1986), and Göthert *et al.* (1986b) noted moderate increases in basal release at the highest concentrations used (10, 1 and 10 μM, respectively). Electrically evoked [^{3}H]noradrenaline release was expressed as [^{3}H] efflux in all of these studies, and all except McGrath (1977), Su and Uruno (1985), and Charlton *et al.* (1986) included uptake blocking agents in the superfusate. McGrath (1977) and Watts *et al.* (1981) actually examined the basal and electrically evoked release of [^{3}H]noradrenaline relative to its [^{3}H]-labelled metabolites; [^{3}H]noradrenaline was the predominant form of [^{3}H] released in response to electrical stimulation (80 per cent according to Watts *et al.* 1981, compared with a value of only 17 per cent for basal [^{3}H] overflow). The experimental protocols used are therefore perfectly acceptable in terms of the criteria discussed by Starke (1977).

In some of the investigations in which the effect of 5-HT on [^{3}H]noradrenaline release was studied, its presynaptic inhibitory effects were also monitored indirectly (McGrath 1977; Müller-Schweinitzer 1981; Watts *et al.* 1981). Contractile responses to electrical stimulation were inhibited by 5-HT at concentrations similar to those inhibiting [^{3}H]noradrenaline release but responses to exogenous noradrenaline were unaffected or enhanced. Similar observations in dog saphenous vein preparations were made by Feniuk *et al.* (1979) and Van Nueten *et al.* (1981). Although interpretation of these experiments can be complicated by the direct contractile effects of 5-HT itself and its amplifying effect on responses to other vasoconstrictors, clear evidence is provided for a presynaptic inhibitory effect of 5-HT on vascular sympathetic adrenergic nerve terminals. In fact, Bevan *et al.* (1975) noted in the rabbit basilar artery that 5-HT (0.084 μM) reduced the contractile response to nerve stimulation but did not affect that to exogenous noradrenaline. No comment or explanation was proffered at the time, but McGrath (1977), on the basis of the direct evidence in his own experiments on other tissues, suggested that 5-HT was probably able to produce a presynaptic inhibitory effect on rabbit basilar artery adrenergic nerve terminals.

The difficulties of interpretation of indirect studies are exemplified by the observations of Kubo and Su (1983) in rat mesenteric arteries. These authors concluded that 5-HT had a presynaptic inhibitory effect on adrenergic nerve terminals, because 5-HT enhanced the contractile responses to nerve stimulation to a significantly lesser extent than it enhanced those to exogenous noradrenaline, at least in preparations from normal rats. No differences were seen in preparations from spontaneously hypertensive rats. In their study on the effect of 5-HT on electrically evoked [^{3}H]noradrenaline release in rat renal vessels, Charlton *et al.* (1986) referred to a previous study (Charlton *et al.* 1984) which implied that 5-HT potentiated the vasoconstrictor response to nerve stimulation, but no specific data were given; thus it is impossible to know if 5-HT potentiated the constrictor response to exogenous noradrenaline to a greater extent.

In some of the indirect studies in dog saphenous vein preparations, the effects of a variety of antagonists and inhibitors on the presynaptic inhibitory effect of 5-HT were tested (McGrath 1977; Feniuk *et al.* 1979; Watts *et al.* 1981). The lack of effect of these agents excluded the apparent involvement of muscarinic cholinergic receptors, histamine H_1 and H_2 receptors, dopamine receptors, and eicosanoid mediators. The effects of a number of α- and β-adrenoceptor antagonists on 5-HT-induced presynaptic inhibition have been assessed *in vitro*, in both the direct studies involving [^{3}H]noradrenaline release and the indirect studies. Because of the particular ways these antagonists might affect specific presynaptic 5-HT receptor mechanisms, these effects are discussed in Section 7.4.3.1.

Although there is considerable *in vitro* evidence for a presynaptic inhibitory effect of 5-HT on vascular sympathetic adrenergic nerve terminals, it is not universal. The notable exception is the pulmonary vasculature. Endo *et al.* (1977) found that 5-HT (0.01–10 μM) had no effect on electrically evoked [^{3}H]noradrenaline release in the rabbit pulmonary artery, and Freeman *et al.* (1981) reported a similar lack of effect of 5-HT (at concentrations of 0.1–10 μM) in human pulmonary artery preparations. As in most of the *in vitro* studies cited above, uptake blockers were present in the perfusate. Interestingly, of the seven tissues in which a direct relaxant effect of 5-HT has been demonstrated *in vitro* (see Section 7.3.1.4), two were pulmonary preparations.

One other vascular preparation in which a presynaptic inhibitory effect of 5-HT has been shown is the dog coronary artery (Cohen 1985). 5-HT, at a concentration of 0.1 μM in the absence or presence of cocaine, produced an approximately 30 per cent inhibition of [^{3}H]noradrenaline release in response to electrical stimulation at 2 Hz, without affecting basal release of [^{3}H]noradrenaline. The presynaptic inhibitory effect of 5-HT in this tissue, however, represented a component of the vasoconstrictor response to 5-HT rather than vasodilatation, because the response to nerve stimulation and exogenous noradrenaline (in preparations precontracted with prostaglandin $F_{2\alpha}$) was a β-adrenoceptor-mediated relaxation. 5-HT inhibited the relaxant response to nerve stimulation, but did not affect that to exogenous noradrenaline.

7.3.3.2 In vivo *vascular studies*

The investigations undertaken in our laboratory into the vasodilator actions of 5-HT in the femoral (Phillips *et al.* 1985a) and common carotid (Markus *et al.* 1988a) circulations in anaesthetized dogs, as described in detail in Section 7.3.1.3, led to the conclusion that 5-HT-induced vasodilatation depended predominantly on a presynaptic inhibitory effect on sympathetic vascular tone. Since 5-HT was administered intra-arterially in these experiments, the most likely site for its presynaptic inhibitory action is at the adrenergic nerve terminals, in accordance with the *in vitro* studies described in Section 7.3.3.1, although the possibility of an action at sympathetic ganglia, or even via afferent actions affecting central vascular control, cannot be completely excluded (see Section 7.3.4).

Evidence of a more direct nature for an *in vivo* presynaptic inhibitory effect of 5-HT on sympathetic vascular tone comes from studies on the effect of 5-HT on vasoconstrictor responses to sympathetic nerve stimulation. In the femoral arterial circulation of anaesthetized dogs, Feniuk *et al.* (1981a) showed that 5-HT (10 μg/kg per min intravenously) inhibited the vasoconstrictor response to lumbar sympathetic chain stimulation at 2 Hz, but did not reduce that to intra-arterial noradrenaline; it was noted that 5-HT itself tended to constrict the femoral vascular bed. Studies in our laboratory have demonstrated that 5-HT (0.2–5 μg/kg per min intra-arterially) reproducibly inhibits the vasoconstrictor responses to cervical sympathetic nerve stimulation (0.25, 0.5, 1, and 2.5 Hz) in the dog common carotid circulation (Markus *et al.* 1988b); the degree of inhibition tended to be a little more marked at the lower frequencies of stimulation. At the dose levels used, 5-HT occasionally produced a small vasoconstrictor response itself. As noted previously, the vasoconstrictor response to exogenous noradrenaline in this circulation was unaffected by 5-HT (Markus *et al.* 1988a). Some degree of vasoconstrictor response to 5-HT itself would be expected in these experiments, as discussed in Section 7.3.1.3, because the cervical sympathetic nerve trunks were sectioned at the outset. Any such direct contractile effect of 5-HT could complicate the interpretation of the data, as already pointed out for the indirect *in vitro* studies (Section 7.3.3.1); however, in our *in vivo* studies we included a series of control experiments in which a similar degree of vasoconstriction was produced by intra-arterial ornipressin. The responses to cervical sympathetic nerve stimulation were unchanged, indicating that the effect of 5-HT was due to a specific presynaptic inhibitory action.

In the external carotid circulation of anaesthetized monkeys, 5-HT normally elicits a vasoconstrictor response (Spira *et al.* 1978; Mylecharane *et al.* 1978); the sectioning of the cervical sympathetic nerve trunk does not alter base-line vascular tone in this circulation (Mylecharane *et al.* 1980), indicating that, under these experimental conditions, there is no resting sympathetic tone to the external carotid vasculature. Recent preliminary findings suggest, however, that 5-HT also has a presynaptic inhibitory effect on sympathetic adrenergic terminals in this circulation (Phillips and Mylecharane 1987). After sectioning the cervical sympathetic nerve trunk, intra-arterial 5-HT still produced vasoconstriction, but it also inhibited the vasoconstrictor responses to 0.5 and, to a lesser extent, 2.5 Hz nerve stimulation. Ornipressin, administered intra-arterially at a dose producing an equivalent degree of vasoconstriction, had no effect on the nerve stimulation-induced vasoconstrictor responses.

7.3.3.3 *Non-vascular studies*

Several investigations have demonstrated a presynaptic inhibitory effect of 5-HT at sympathetic adrenergic nerve terminals in non-vascular tissues. Although these actions are not directly relevant to the presynaptic inhibitory effect of 5-HT on vascular adrenergic nerve terminals, they nevertheless provide a wider physiological (and possibly pathological) context for the presynaptic inhibitory actions of 5-HT.

Cardio-accelerator responses to stimulation of cardiac sympathetic nerves are inhibited by 5-HT *in vivo* in dogs (Martinez and Lokhandwala 1980; Kimura and Satoh 1983). Similar experiments in rats (Göthert *et al.* 1986b) were inconclusive in that 5-HT itself induced a substantial tachycardia which was unaffected by ketanserin and which persisted to some extent after desipramine or propranolol; however, 5-methoxytryptamine, an agonist at '5-HT_1-like' and 5-HT_2 receptors, was shown to produce a presynaptic inhibitory effect following blockade of its intrinsic tachycardic effects by ketanserin. Likewise, Docherty and Warnock (1986) reported that the selective '5-HT_1-like' receptor agonist, 5-carboxamidotryptamine, produced a presynaptic inhibitory effect on cardiac sympathetic nerves in rats *in vivo*, although the effects of 5-HT were not mentioned. In guinea-pig isolated atria preparations, Adler-Graschinsky (1983) found that 5-HT inhibited the tachycardia in response to stimulation of the cardiac sympathetic nerves.

Kapur and Mottram (1979) and Docherty and Warnock (1986) found that 5-HT inhibited electrically evoked contractions of rat vas deferens preparations, and attributed this to a presynaptic inhibitory effect on noradrenergic transmission. In a subsequent investigation into this effect, Moritoki *et al.* (1986) reported that the effect was variable, but seemed to be related to age, because in older animals, augmentation rather than inhibition was observed. In another sympathetically innervated smooth muscle preparation, the bovine ovarian follicle, Kannisto *et al.* (1987) reported that 5-HT inhibited electrically evoked [^{3}H]noradrenaline release and contractile responses, without affecting the response to exogenous noradrenaline.

7.3.4 Actions at other sites

Stimulation of synthesis or release of other mediators has been suggested as a mechanism by which 5-HT might induce vasodilatation. Coughlin *et al.* (1981) found that 5-HT stimulated the synthesis of prostacyclin in cultured bovine aortic smooth muscle cells; the effect was attenuated by dexamethasone. These authors raised the possibility that this might contribute to 5-HT-induced vasodilatation, but this seems unlikely given that suppressors of eicosanoid synthesis do not influence the direct relaxant or presynaptic inhibitory effects of 5-HT in a number of *in vitro* preparations (see Sections 7.3.1.4 and 7.3.3.1).

In those *in vitro* studies, the involvement of β-adrenoceptors and histamine H_1 and H_2 receptors was ruled out with appropriate antagonists, thus a localized release of adrenaline or histamine cannot be involved. However, it is possible that adrenaline or histamine release by 5-HT from distal sites could contribute to vasodilatation *in vivo*. Saxena and Verdouw (1982) suggested a possible involvement of catecholamine release from the adrenal medulla on the basis of experimental evidence that 5-HT is capable of eliciting such a release. This mechanism has been considered before, but in the context of an indirect contribution to the vasoconstrictor and tachycardiac actions of 5-HT (see Garattini and Valzelli 1965; Feniuk *et al.* 1981b for references). The possible involvement of histamine release in 5-HT-induced vasodilatation was first suggested by Page and McCubbin (1956), but Garattini and Valzelli (1965) concluded that any such involvement was of little importance, at least in mammals, mainly on the basis of the lack of consistent blockade by the available antihistamines. This conclusion is supported by more recent studies in which cyproheptadine had no blocking effect on the vasodilator response to 5-HT in the nutrient fraction of the pig common carotid circulation (Saxena and Verdouw 1982), and pizotifen did not alter the presynaptic inhibitory effect of 5-HT in the dog common carotid circulation (Markus *et al.* 1988b); at the doses used, these 5-HT antagonists are also effective histamine H_1 receptor antagonists. In addition, the fact that intra-arterial administration of 5-HT can elicit vasodilator responses in specific circulations without concomitant systemic changes (e.g. McCubbin *et al.* 1962; Grimson *et al.* 1969; Saxena 1972; Mena and Vidrio 1979; Saxena and Verdouw 1982; Phillips *et al.* 1985a; Markus *et al.* 1988a) indicates that any contribution via systemic release of adrenaline or histamine is relatively negligible.

Page and McCubbin (1953) noted that 5-HT produced a transient inhibition of ciliary ganglion transmission, and suggested that this might contribute to the vasodilator properties of 5-HT, although they considered that it was not related to the prolonged vasodepressor response to 5-HT. Likewise, effects of 5-HT on afferent neuronal pathways can elicit cardiovascular reflexes (see Chapter 13) which might contribute to vasodilator responses *in vivo*. Again, however, any such contributions are likely to be relatively minor, given the substantial and well-maintained vasodilator effects of intra-arterial 5-HT, and the absence of concomitant systemic changes.

Although the blood–brain barrier would prevent peripherally administered (or released) 5-HT from producing vasodilatation via central nervous system mechanisms, a note of caution is necessary when the effects of 5-HT antagonists on 5-HT-induced vasodilatation are being assessed *in vivo*. Virtually all the agents available enter the central nervous system, and therefore are potentially able to modulate vascular tone themselves via central mechanisms (see Phillips *et al.* 1985b; Ramage 1985 for references).

7.4 The 5-HT receptors mediating the effects involved in 5-HT-induced vasodilatation

7.4.1 Direct relaxant effects in vascular smooth muscle

7.4.1.1 In vitro *studies*

Overall, the findings from *in vitro* preparations suggest that the direct relaxant effect of 5-HT is mediated by '5-HT_1-like' receptors. However, the results are not always consistent.

In sheep and goat pulmonary vein preparations, neither morphine (up to 100 μM) nor methysergide (up to 2 μM) antagonized the relaxant effect of 5-HT (Eyre 1975; Chand 1981); the non-specific antagonist, dibenamine, was also tested in the sheep preparations, and was ineffective at concentrations up to 2 μM. The interpretation of findings with drugs such as morphine and dibenamine ('prototype' antagonists at M and D 5-HT receptors) is now regarded as inconclusive (Bradley *et al.* 1986), but the lack of effect of relatively high concentrations of methysergide would suggest that neither '5-HT_1-like' nor 5-HT_2 receptors are involved in the pulmonary vein relaxant responses to 5-HT.

In cat cranial artery (Edvinsson *et al.* 1978) and saphenous vein (Feniuk *et al.* 1983), methysergide blocked the relaxant effect of 5-HT at concentrations consistent with an action at '5-HT_1-like' (but not 5-HT_2) receptors. Methysergide-induced blockade in the cat cranial arteries, at concentrations of 0.03–3 μM, was not competitive, but that in the saphenous vein was, yielding a pA_2 value of 6.75. The effects of antagonists on 5-HT-induced relaxation in pig vena cava preparations were not able to be tested directly, because of interference with the response to the precontracting agents, but methysergide (1 μM) inhibited the parallel 5-HT-induced elevation of cyclic AMP levels, the estimated pA_2 value being 7.19 (Trevethick *et al.* 1984). The relaxant effects of 5-HT in cat saphenous vein and pig vena cava preparations were mimicked by the selective '5-HT_1-like' receptor agonist, 5-carboxamidotryptamine, which was 30–140 times more potent than 5-HT, and was blocked by methysergide with pA_2 values of 7.75 and 7.88, respectively (Feniuk *et al.* 1984; Trevethick *et al.* 1986). 5-Carboxamidotryptamine was also approximately 100 times more potent than 5-HT in elevating pig vena cava cyclic AMP levels; methysergide blocked this effect, the pA_2 value being 7.95. In the cat cranial artery preparations, propranolol blocked the relaxant response to 5-HT (Edvinsson *et al.* 1978), but at concentrations suggesting an action at '5-HT_1-like' receptors rather than at β-adrenoceptors, given the pA_2 values and binding site affinities determined by Edvinsson and Owman (1974), Engel *et al.* (1986) and Molderings *et al.* (1987). Cyproheptadine (0.1 μM) did not affect the 5-HT- or 5-carboxamidotryptamine-induced relaxant effects or increases in cyclic AMP in cat saphenous vein or pig vena cava preparations; the relaxant response to 5-carboxamidotryptamine in the latter and in rabbit jugular vein (Martin *et al.* 1987) were unaffected by ketanserin at concentrations up to 1 μM. The findings in these tissues suggest that the relaxant effects are mediated by '5-HT_1-like', but not 5-HT_2 receptors.

7.4.1.2 In vivo *studies*

As noted previously (Section 7.3.1.3), there are very few vasculatures in which it has as yet been shown that a direct relaxant effect is primarily responsible for 5-HT-induced vasodilatation *in vivo*.

Several agents have been tested in the pig common carotid circulation, where the vasodilator response to 5-HT in the arteriolar-capillary fraction appears to be due to a direct relaxant effect. The 5-HT_2-selective antagonists, cyproheptadine (1 mg/kg), ketanserin (0.3 mg/kg), and WAL 1307 (0.3 mg/kg), resulted in a slight enhancement of 5-HT-induced vasodilatation in this fraction of the circulation, and the selective 5-HT_3 antagonist, MDL 72222 (0.5 mg/kg), was inactive (Saxena and Verdouw 1982; Verdouw *et al.* 1984; Saxena *et al.* 1986). Methysergide (0.35 mg/kg) also produced a slight enhancement of the 5-HT-induced vasodilatation (Saxena and Verdouw 1984), but methiothepin (1 mg/kg) almost abolished the response to 5-HT (Saxena *et al.* 1986). None of these antagonists themselves altered vascular tone in the arteriolar–capillary fraction. The vasodilator effect of 5-HT in this vascular bed was mimicked by 5-carboxamidotryptamine (Saxena and Verdouw 1985) and another selective '5-HT_1-like' receptor agonist, BEA 1654 (Verdouw *et al.* 1985); these agents were approximately five-fold more potent and 100-fold less potent, respectively, than 5-HT. All of these studies were conducted in bilaterally cervical vagosympathectomized pigs.

The results therefore indicate that the direct vascular relaxant effect of 5-HT in this circulation is mediated by '5-HT_1-like' receptors, and that 5-HT_2 and 5-HT_3 receptors are not involved. The dose of methysergide used was in excess of that required for 5-HT_2 receptor blockade, but may not have been quite sufficient to produce detectable effects at the vascular '5-HT_1-like' receptors mediating relaxation. The dose used, however, was sufficient to produce constriction of the arteriovenous–anastomotic fraction and a slight inhibition of the 5-HT-induced constriction. Vasoconstrictor effects in the arteriovenous–anastomotic fraction are accepted as being mediated by '5-HT_1-like' receptors (Bradley *et al.* 1986); thus there may be slight differences between the 5-HT_1 receptor subtypes present in the different fractions of the pig common carotid circulation.

In the cat common carotid vasculature, Saxena and Verdouw (1982) described a similar 5-HT-induced direct vasodilator effect in the arteriolar–capillary fraction. Connor *et al.* (1986) found that 5-carboxamidotryptamine produced vasodilatation in the common carotid circulation of cervical sympathectomized cats, although the fraction in which this was occurring was not identified. A comparison of the doses used in these investigations suggests that 5-carboxamidotryptamine is approximately 50-fold more potent than 5-HT; the effect of 5-carboxamidotryptamine was blocked by methiothepin (0.01–0.1 mg/kg) and methysergide (0.1–1 mg/kg) but was unaffected by ketanserin at doses of up to 1 mg/kg. Thus, as in the pig, '5-HT_1-like' but not 5-HT_2 receptors appear to be responsible for a direct relaxant effect in the cat common carotid circulation *in vivo*.

The only other vascular bed in which an attempt has been made to characterize the receptors mediating a direct relaxant response to 5-HT *in vivo* is the peripheral vasculature of rats (Kalkman *et al.* 1983). In pithed rats, with vascular tone restored by vasopressin and in the presence of ketanserin, 5-HT-induced vasodepressor responses were mimicked by a series of tryptamine derivatives with an order of potency which was the same as that for their 5-HT_1 binding site affinity. Quipazine, which has appreciable affinity for 5-HT_1 binding sites, was able to block the effect. Also, in anaesthetized intact rats, pretreated with ketanserin, there was a significant correlation between the vasodepressor effects of the agonists and their 5-HT_1 binding site affinities; in similar experiments in anaesthetized rats, Krstić and Katušić (1982) and Saxena and Lawang (1985) found that the vasodepressor effect of 5-HT was unaffected by ketanserin, cyproheptadine, and pirenperone (another 5-HT_2-selective antagonist). Thus '5-HT_1-like' but not 5-HT_2 receptors appear to be responsible for these effects in the rat.

7.4.2 Release of EDRF

The receptors involved in 5-HT-induced release of EDRF are discussed in detail in Chapter 8. The available evidence suggests that '5-HT_1-like' receptors are responsible.

7.4.3 Presynaptic inhibition of sympathetic adrenergic nerve terminals

7.4.3.1 In vitro *studies in vascular preparations*

Table 7.1 summarizes the effects of a variety of 5-HT receptor antagonists on the inhibition of electrically evoked [^{3}H]noradrenaline release by 5-HT or other 5-HT receptor agonists, and also the actions of 5-HT receptor agonists and partial agonists on electrically evoked [^{3}H]noradrenaline release. Ineffective agents have been listed only when they were used at concentrations which would be expected to be active at their preferred 5-HT receptor target.

In all five tissues, the involvement of 5-HT_2 receptors can be ruled out, because of the lack of effect of the selective antagonists, cyproheptadine, ketanserin, LY53857, and spiperone (Table 7.1). The lack of effect of cocaine and metoclopramide excludes an involvement of 5-HT_3 receptors in the dog coronary artery and rat renal vasculature, respectively; although the effect of cocaine was not specifically investigated in the other tissues, 5-HT elicited substantial inhibition of [^{3}H]noradrenaline release in dog and human saphenous vein preparations (Watts *et al.* 1981; Engel *et al.* 1983; Göthert *et al.* 1986a; Fishburn *et al.* 1987) in the presence of cocaine (as a neuronal uptake blocker) at concentrations sufficient to block 5-HT_3 receptors.

In the dog saphenous vein, as shown in Table 7.1, methysergide mimicked but did not block the 5-HT-induced presynaptic inhibition of electrically evoked [^{3}H]noradrenaline release. The effect of 5-HT was also mimicked by 5-carboxamidotryptamine and some other 5-HT receptor agonists with appreciable affinity for 5-HT_1 binding sites; these agents were more potent than 5-HT. These observations suggest that the presynaptic adrenergic inhibitory effect of 5-HT in this tissue is mediated by '5-HT_1-like' receptors. The results obtained indirectly in dog saphenous vein preparations, in which the effects of 5-HT or 5-carboxamidotryptamine on electrically evoked contractile responses were monitored (McGrath 1977; Feniuk *et al.* 1979 and 1981c; Müller-Schweinitzer 1981; Van Nueten *et al.* 1981; Watts *et al.* 1981), are consistent with this conclusion, the only additional points

Table 7.1 Blockade of the presynaptic inhibitory effect of 5-HT on electrically evoked [^{3}H]noradrenaline release in vascular preparations, and its mimicking by 5-HT receptor agonists and partial agonists

Tissue and references	5-HT receptor antagonists		5-HT receptor agonists and partial agonists	
	Effective (μM)[1]	Ineffective	Effective (potency relative to 5-HT)[2]	Ineffective
Dog saphenous vein				
McGrath (1977)	—	Methysergide Morphine Phentolamine	—	—
Watts *et al.* (1981)	—	Cyproheptadine[3] Phentolamine[4]	Methysergide (0.03)	—
Engel *et al.* (1983)	—	—	5-Carboxamidotryptamine (14) *N*-Methyl-5-HT (6) *N*,*N*-Dimethyl-5-HT (2)	—
Fishburn *et al.* (1987)	—	Ketanserin	—	—
Dog coronary artery				
Cohen (1985)	Methiothepin (1)[4] Metergoline (1)	Ketanserin LY53857 Cocaine Phentolamine[4]	—	—
Rat renal vasculature				
Charlton *et al.* (1986)	Methysergide (0.03–0.3) Methiothepin (0.0003–0.03)[4] Metergoline (0.03–1)[4] (−)-Propranolol (10)	Cyproheptadine Ketanserin Mesulergine Metoclopramide Phentolamine[4] Pindolol Cyanopindolol	Methysergide (0.03) 5-Carboxamidotryptamine (6) RU 24969 (0.2)	8-OH-DPAT

Tissue and references	5-HT receptor antagonists		5-HT receptor agonists and partial agonists	
	Effective (μM)[1]	Ineffective	Effective (potency relative to 5-HT)[2]	Ineffective
Rat vena cava				
Göthert *et al.* (1986b)	Methiothepin (0.1–1)[4]	Ketanserin Rauwolscine[4]	5-Methoxytryptamine (6)	8-OH-DPAT
Molderings *et al.* (1987)	Methiothepin (0.1) Metergoline (1) Quipazine (3) MK 212 (100) Propranolol (0.1) Cyanopindolol (1)[5]	Ketanserin Spiperone	5-Carboxamidotryptamine (2) RU 24969 (9) 5-Methoxytryptamine (0.5)	8-OH-DPAT Ipsapirone
Human saphenous vein				
Göthert *et al.* (1986a)	Methiothepin (0.1)[4]	Ketanserin[6] Rauwolscine[4,6] Propranolol	8-OH-DPAT (0.1)	Ipsapirone

[1] Values in parentheses are the effective concentrations tested (μM).
[2] Values in parentheses are approximate potencies relative to 5-HT.
[3] Increase in [^{3}H]noradrenaline release when used alone, attributed to calcium blockade.
[4] Increase in [^{3}H]noradrenaline release when used alone, attributed to presynaptic α-adrenoceptor blockade.
[5] Increase in [^{3}H]noradrenaline release when used alone, attributed to partial agonist effect at presynaptic β-adrenoceptors.
[6] Tested only against 8-OH-DPAT.

being that methysergide was able to block the presynaptic inhibitory effect of 5-HT to some extent, in addition to mimicking it, and propranolol was inactive.

Methysergide also both blocked and mimicked the presynaptic inhibitory effect of 5-HT on electrically evoked [^{3}H]noradrenaline release in the rat renal vasculature (Table 7.1). However, the blocking effects of methiothepin and metergoline in this tissue, and in the remaining tissues investigated (see Table 7.1), tend to confirm that '5-HT_1-like' receptors mediate the presynaptic inhibitory effect of 5-HT on sympathetic adrenergic nerve terminals. This conclusion is supported by the blocking effects of quipazine and MK 212, and by the ability of 5-HT_1-selective agonists such as 5-carboxamidotryptamine and RU 24969 to mimic 5-HT, in all tissues in which they were tested (5-carboxamidotryptamine was more potent than 5-HT).

Some attempts have been made to relate the 5-HT_1-like receptor mediating the presynaptic adrenergic inhibitory effect of 5-HT to one or other of the recognized 5-HT_1 binding site subtypes. Charlton *et al.* (1986) concluded that in the rat renal vasculature, the subtype involved did not correspond to a 5-HT_{1A}, 5-HT_{1B}, or 5-HT_{1C} binding site, because of the lack of effect of, respectively: 8-OH-DPAT; the β-adrenoceptor antagonists (−)-propranolol (which produced only a small degree of blockade at 10 μM), pindolol and cyanopindolol; and mesulergine (see Table 7.1). In the rat vena cava, Molderings *et al.* (1987) concluded that the presynaptic inhibitory receptor for 5-HT corresponded to a 5-HT_{1B} binding site, largely on the basis of the potent blocking effects of propranolol and cyanopindolol in this tissue. The data at present available for the human saphenous vein (Göthert *et al.* 1986a) are somewhat inconclusive; 8-OH-DPAT, which is inactive in the rat tissues (Table 7.1), was active in the human saphenous vein, and propranolol had no blocking effect at concentrations specific for 5-HT_{1B} binding sites, but the selective 5-HT_{1A} partial agonist ipsapirone was inactive.

Clarke *et al.* (1988) have recently suggested that the presynaptic adrenergic inhibitory receptor for 5-HT might correspond to the novel 5-HT_{1D} binding site described by Heuring and Peroutka (1987). Although there are many similarities between the functional and binding data, Clarke *et al.* (1988) have pointed out that rauwolscine, which has a high affinity for the 5-HT_{1D} binding site, has little or no functional presynaptic inhibitory activity in rat renal vasculature, or in rat vena cava (Göthert *et al.* 1986b), rat vas deferens (Docherty and Warnock 1986), and human saphenous vein (Göthert *et al.* 1986a).

Some caution is needed when interpreting functional studies of the presynaptic adrenergic inhibitory effect of 5-HT when agents with affinity for α- or β-adrenoceptors are used. Several β-adrenoceptor antagonists have appreciable affinity for 5-HT_{1B} binding sites, e.g. (−)-propranolol,

(−)-pindolol and (−)-cyanopindolol (Middlemiss 1984 and 1985; Engel *et al.* 1986). However, at the concentrations used in the presynaptic studies outlined above, these agents will have antagonist (and in some cases partial agonist) actions at the presynaptic sympathetic adrenergic β-adrenoceptors which facilitate noradrenaline release (Starke 1977; Langer 1980). The blockade of a facilitatory component activated by released noradrenaline may lead to a net decrease in noradrenaline release; hence a concomitant blockade of the presynaptic 5-HT receptor mediating inhibition of noradrenaline release might not be apparent, because the reversal of the 5-HT-induced inhibitory effect might be outweighed by the inhibitory effect of the blocking agent itself. The lack of effect of β-adrenoceptor antagonists on 5-HT-induced presynaptic inhibition in some preparations but not others might be related to the distribution of presynaptic β-adrenoceptors rather than '5-HT_1-like' receptor subtypes. Likewise, methiothepin has appreciable affinity for α_1- and α_2-adrenoceptors (see Cohen 1985; Charlton *et al.* 1986; Göthert *et al.* 1986a and b for references); blockade of presynaptic sympathetic adrenergic α-adrenoceptors which inhibit noradrenaline release may lead to a net enhancement of noradrenaline release, which could physiologically antagonize the presynaptic inhibitory effect of 5-HT; hence the apparent reversal of 5-HT-induced presynaptic inhibition might not be entirely due to an action of methiothepin at '5-HT_1-like' receptors. If the presynaptic inhibitory effect of 5-HT is being assessed indirectly, in terms of changes in the electrically evoked contractile response, further complications can obviously arise from postsynaptic α-adrenoceptor blockade, as evident for example with phentolamine, yohimbine, and ketanserin in the dog saphenous vein (Feniuk *et al.* 1979; Van Nueten *et al.* 1981).

The effects of these agents alone on noradrenaline release can be assessed; the data included in Table 7.1 were confined to experiments designed to reveal any such intrinsic actions. In most instances, methiothepin at the concentrations used enhanced noradrenaline release itself (Table 7.1). The fact that other agents with α-adrenoceptor affinity, such as phentolamine and rauwolscine, also enhanced noradrenaline release without reversing 5-HT-induced presynaptic inhibition can be used to argue that any physiological antagonism of 5-HT-induced inhibition has little or no influence. Nevertheless, subtle differences could make an important contribution; clearly, it would be preferable to use data from studies with methiothepin as corroborative rather than primary evidence, but this will not be possible until more selective '5-HT_1-like' receptor antagonists are available. Likewise, the fact that the β-adrenoceptor antagonists did not reduce noradrenaline release themselves (Table 7.1) might suggest that their β-adrenoceptor blocking actions are not interfering with the assessment of 5-HT-induced presynaptic inhibition; again, however, the potential effects may be too subtle to detect but they may still be influencing the results.

7.4.3.2 In vivo *vascular studies*

The effects of several 5-HT receptor antagonists on 5-HT-induced vasodilatation in dogs have been studied in our laboratory. Ketanserin (0.1–4 mg/kg) and pizotifen (0.1–4 mg/kg) were studied in the dog femoral (Phillips *et al.* 1985a) and common carotid (Markus *et al.* 1988a) arterial circulations; methysergide (0.02–0.4 mg/kg) and the selective 5-HT_2 receptor antagonist pirenperone (0.001–0.01 mg/kg) were also tested in the dog common carotid circulation.

The vasodilator responses to 5-HT were inhibited in a dose-dependent manner by all of the 5-HT receptor antagonists tested in each of the dog circulations. However, the degree of blockade was relatively modest, and the doses required to produce inhibition were higher than those normally used to block 5-HT_2 receptors. Pirenperone was the most effective antagonist, methysergide the least. Methysergide, however, was the only one of these antagonists which had no direct effects on resting vascular tone. Ketanserin produced appreciable vasodilatation, as did pirenperone although to a lesser extent; pizotifen markedly dilated the common carotid circulation, but did not produce consistent significant vasodilatation in the femoral vasculature. The apparent blockade of 5-HT-induced vasodilatation by these three antagonists may therefore have been due to a simple masking by their intrinsic vasodilating effects. None of these 5-HT antagonists affected the vasoconstrictor responses to noradrenaline in these experiments, except for ketanserin at the higher doses of 1–4 mg/kg. In anaesthetized dogs, ketanserin at doses of 0.1–0.4 mg/kg produces a centrally mediated inhibition of sympathetic vascular tone (Phillips *et al.* 1985b). The mechanisms responsible for the vasodilator actions of pizotifen and pirenperone are not known.

These studies in dogs indicate that it is most unlikely that a 5-HT_2 receptor mediates the presynaptic inhibitory effect of 5-HT on sympathetic vascular tone. The results with methysergide suggest that a '5-HT_1-like' receptor may be involved, but the intrinsic vasodilator effects of the other 5-HT antagonists make it difficult to assess the possibility that they may also interact with 5-HT at the presynaptic inhibitory receptor. In subsequent experiments in anaesthetized dogs, it was found that 5-carboxamidotryptamine mimicked the vasodilator effect of 5-HT in the common carotid circulation (Mylecharane and Phillips 1987). The maximal responses were similar, but 5-carboxamidotryptamine was 28 times more potent than 5-HT.

In a further series of experiments in cervical sympathectomized dogs, the effects of the four antagonists on 5-HT-induced inhibition of the vasoconstrictor responses to cervical sympathetic nerve stimulation (0.25–2.5 Hz) in the common carotid circulation were studied (Markus *et al.* 1988b). It was found that alteration of resting vascular tone with intra-arterial acetylcholine or ornipressin had no effect on the responses to nerve stimulation; thus the effects of the antagonists on the presynaptic inhibitory receptor were able to be studied directly. Ketanserin (0.4 mg/kg) and pizotifen (0.1–0.4 mg/kg) did not alter the 5-HT-induced presynaptic inhibition of the responses to nerve stimulation, but pirenperone at the higher doses used (0.005 and 0.01 mg/kg) produced a small but significant reversal of the presynaptic inhibitory effect of 5-HT. None of these agents themselves altered the vasoconstrictor responses to nerve stimulation. Methysergide acted as a partial agonist at doses of 0.2 and 0.4 mg/kg; the vasoconstrictor responses to nerve stimulation alone were inhibited, and the presynaptic inhibitory effect of 5-HT was enhanced. It was noted that methysergide increased resting common carotid vascular tone in these experiments, in contrast to its lack of effect in the intact dogs (Markus *et al.* 1988a).

Additional experiments in this experimental model (Mylecharane and Phillips 1987) have shown that the presynaptic inhibitory effect of 5-HT is mimicked by 5-carboxamidotryptamine, its potency being approximately 20-fold greater than that of 5-HT. The 5-HT_3 receptor antagonists, MDL 72222 (0.5 mg/kg) and ICS 205–930 (0.1 mg/kg), had no effect on the presynaptic inhibitory actions of these agents, and did not affect the vasoconstrictor responses to noradrenaline or nerve stimulation alone. The presynaptic inhibitory effects of 5-HT and 5-carboxamidotryptamine were partly reversed by methiothepin (0.1–0.5 mg/kg); at these dose levels, the vasoconstrictor response to noradrenaline was inhibited, but that to nerve stimulation alone was unaltered, suggesting that the postsynaptic α-adrenoceptor blocking action of methiothepin may have been balanced by an augmentation of nerve stimulation-induced noradrenaline release via a presynaptic α-adrenoceptor blocking action.

These observations in dogs indicate that neither 5-HT_2 nor 5-HT_3 receptors mediate the presynaptic sympathetic adrenergic inhibitory effect of 5-HT in the common carotid circulation. Involvement of a '5-HT_1-like' receptor is suggested by the mimicking effects of 5-carboxamidotryptamine, the partial agonist actions of methysergide, and the blocking activity of methiothepin. In the latter instance, however, the possibility of a physiological antagonism via enhanced noradrenaline release due to the presynaptic α-adrenoceptor blocking action of methiothepin (as discussed in Section 7.4.3.1) cannot be excluded. The influence of presynaptic α-adrenoceptor blockade on 5-HT-induced presynaptic inhibition in this model has not yet been investigated, although we have found that yohimbine at doses of up to 0.3 mg/kg did not affect the common carotid vasodilator response to 5-HT in intact dogs (Markus *et al.* 1988a); presynaptic α-adrenoceptors were presumably blocked, because the postsynaptic α-adrenoceptor-mediated vasoconstrictor response to noradrenaline was inhibited.

The conclusion that the presynaptic inhibitory effect of 5-HT is mediated by a '5-HT_1-like' receptor suggests in turn that pirenperone may have some affinity for '5-HT_1-like' receptors. We have also found that pirenperone acts as a competitive antagonist at the '5-HT_1-like' receptors which mediate contraction in the rat stomach fundus preparation, at concentrations approximately 2500-fold higher than those producing 5-HT_2 receptor blockade (Fong *et al.* 1987).

In the monkey external carotid circulation (after cervical sympathectomy), methysergide (0.4 mg/kg) has similar effects to those obtained in the common carotid vasculature of cervical sympathectomized dogs: inhibition of vasoconstrictor responses to nerve stimulation (0.5 and 2.5 Hz), enhancement of the presynaptic inhibitory effect of 5-HT, and an increase in resting vascular tone (Phillips and Mylecharane 1987). Feniuk *et al.* (1981a) described a similar methysergide-induced presynaptic inhibition and vasoconstriction in the dog femoral vasculature after lumbar sympathectomy, but they did not examine the interaction between methysergide and 5-HT. These observations are consistent with the involvement of a '5-HT_1-like' receptor in the presynaptic inhibition produced by 5-HT. It also seems likely that the postsynaptic vasoconstrictor response to methysergide involves a '5-HT_1-like' receptor, as suggested for example by Saxena and Verdouw (1984) and Bradley *et al.* (1986).

It is possible that the apparent absence of any effect of methysergide on resting vascular tone in the intact (i.e. non-sympathectomized) monkey external carotid (Mylecharane *et al.* 1978), dog femoral (Feniuk *et al.* 1981a), and dog common carotid (Markus *et al.* 1988a) circulations could be attributed to its postsynaptic vasoconstrictor effects being balanced by a presynaptic inhibition of sympathetic vascular tone, as suggested by Feniuk *et al.* (1981a). This may not be the entire explanation, however, at least in the monkey external carotid vasculature, because in this circulation resting vascular tone remains unchanged following cervical sympathectomy, implying that there is no resting sympathetic tone (Mylecharane *et al.* 1980). In our experiments, resting vascular tone in the dog femoral and common carotid circulations fell after mecamylamine and cervical sympathectomy, respectively (Phillips *et al.* 1985a; Markus *et al.* 1988a), but Feniuk *et al.* (1981a) found that femoral vascular resistance in the dog was unchanged after mecamylamine or lumbar sympathectomy.

The difference between the effect of 5-HT in the intact monkey external carotid vasculature (vasoconstriction) and in the intact dog femoral and common carotid circulations (vasodilatation) could be explained by differences in receptor preference and distribution. In the dog, 5-HT may activate the presynaptic sympathetic adrenergic inhibitory '5-HT_1-like' receptors in preference to the postsynaptic vasoconstricting 5-HT receptors (which may be predominantly '5-HT_1-like', and therefore activated by methysergide although without any such selectivity). In the monkey, however, 5-HT may preferentially activate the postsynaptic constrictor receptors; these may be predominantly 5-HT_2 receptors, thus the partial agonist effects of methysergide (at '5-HT_1-like' receptors) will be less apparent.

7.4.3.3 *Non-vascular studies*

The receptors mediating 5-HT-induced presynaptic inhibition in non-vascular tissues have not been extensively investigated. The findings obtained in cardiac sympathetic nerves, despite some inconsistencies, support the conclusion that '5-HT_1-like', but not 5-HT_2 receptors are involved, as in the vascular preparations. The 5-HT-induced inhibition of the tachycardic response to cardiac sympathetic nerve stimulation in dogs was blocked by cyproheptadine (0.12–2 mg/kg) but not by methysergide (2 mg/kg) according to Martinez and Lokhandwala (1980), but Kimura and Satoh (1983) reported that cyproheptadine (1 mg/kg) was ineffective. Cyproheptadine (1 mg/kg) was also ineffective in pithed rats, as was ketanserin (1.75 mg/kg), methiothepin (0.47 mg/kg), and the 5-HT_{1A}-selective agent, spiroxatrine (0.3 mg/kg); the presynaptic inhibitory effect of 5-HT on cardiac sympathetic nerves in rats was mimicked potently by 5-carboxamidotryptamine and 5-methoxytryptamine, but not by 8-OH-DPAT (Docherty and Warnock 1986; Göthert *et al.* 1986b). In guinea-pig atria preparations, the inhibition of nerve stimulation-induced tachycardia by 5-HT was only weakly antagonized by pizotifen (1 μM), and methysergide (1μM) acted as a partial agonist (Adler-Graschinsky 1983).

The presynaptic adrenergic inhibitory effect of 5-HT in rat vas deferens preparations appears to be mediated by '5-HT_1-like' receptors, because it is mimicked by 5-carboxamidotryptamine and 8-OH-DPAT, blocked by very high concentrations of methysergide, and is unaffected by cyproheptadine, ketanserin, mianserin, and rauwolscine (Kapur and Mottram 1979; Docherty and Warnock 1986; Moritoki *et al.* 1986). Docherty and Warnock (1986) concluded that the '5-HT_1-like' receptor involved was not a 5-HT_{1A} or 5-HT_{1B} subtype, because 8-OH-DPAT was not as potent as expected, and the presynaptic inhibition was not blocked by spiroxatrine, propranolol, or cyanopindolol. In the bovine ovarian follicle preparation, a '5-HT_1-like' receptor appears to be responsible for the presynaptic inhibitory effect of 5-HT, because the effect of 5-HT is mimicked by 5-carboxamidotryptamine and methysergide (which also blocks at higher concentrations), but not 8-OH-DPAT, and is blocked by methiothepin, but not cyanopindolol; ketanserin and MDL 72222 were inactive (Kannisto *et al.* 1987).

7.5 Concluding remarks

The major mechanisms by which 5-HT elicits vasodilatation are through actions at three sites: vascular smooth muscle, vascular endothelial cells, and vascular sympathetic adrenergic nerve terminals. 5-HT has a direct relaxant effect on vascular smooth muscle, and can release EDRF. At vascular sympathetic nerve terminals, 5-HT acts presynaptically to inhibit the release of noradrenaline in response to nerve stimulation. Evidence for all of these actions has been obtained both *in vitro* and *in vivo*.

Less well understood is the relative contribution made by each of these actions to 5-HT-induced vasodilatation in a particular circulation. In many vascular beds, the vasodilator response to 5-HT appears to be dependent on sympathetic vascular tone; studies on the dog femoral and common carotid circulations have shown that presynaptic adrenergic inhibition is the predominant mechanism responsible, but whether this is so for other beds remains to be proved. In some others, such as the pig and cat common carotid circulations, and in some pulmonary vascular preparations, the dominant effect of 5-HT is at the level of the vascular smooth muscle, independent of presynaptic sympathetic adrenergic inhibitory effects. In others, the question remains open; appropriate investigations have not yet been conducted. Technical limitations make it difficult to ascertain the contribution made by 5-HT-induced release of EDRF in circulations where presynaptic inhibition is not the predominant action of 5-HT.

The specific 5-HT receptors which are responsible for the vascular smooth muscle relaxant effect of 5-HT appear to be '5-HT_1-like', but not 5-HT_2 or 5-HT_3 receptors, although the data at present available are not completely conclusive. Likewise, the presynaptic inhibitory effect of 5-HT on sympathetic vascular tone appears to be mediated by a '5-HT_1-like' receptor; 5-HT_2 and 5-HT_3 receptors do not seem to be involved. Some attempts have been made to relate the presynaptic '5-HT_1-like' receptor to one or other of the 5-HT_1 binding site subtypes; that in the rat renal vasculature corresponds best to the 5-HT_{1D} subtype, and that in the rat vena cava the 5-HT_{1B} site. The limited evidence available suggests that the human saphenous vein presynaptic inhibitory receptor might resemble a 5-HT_{1A} site. 5-HT also has a presynaptic inhibitory effect on noradrenaline release from some non-vascular sympathetic nerve endings, apparently via a '5-HT_1-like' receptor. Conclusive characterization of the receptors involved in the effects of 5-HT which contribute to its vasodilator action still awaits the development of agonists and especially antagonists with more selective effects at '5-HT_1-like' receptors; better correlations between these functional effects and 5-HT_1 binding site subtypes may then be able to be made. Functional studies on noradrenaline release may also be complicated by the presynaptic α- and β-adrenoceptor effects of some of the agents currently used to characterize '5-HT_1-like' receptors.

In physiological and pathological terms, a wider question related to these vasodilator effects of 5-HT involves its origin. Clearly, platelet 5-HT may be an important source, especially in a pathological situation. There is, however, some limited evidence for the presence of 5-HT in peripheral vascular nerve terminals (see Griffith and Burnstock 1983 for references).

Hardebo *et al.* (1988) have recently reviewed investigations on 5-HT in cerebrovascular nerves in species such as rat, guinea-pigs, and monkeys. These noradrenergic nerves are able to take up, store, and release 5-HT; the vasoconstrictor effects of this 5-HT can enhance the constrictor response to noradrenaline. The possibility therefore exists that the vasodilator actions of 5-HT may also make some contribution to cerebrovascular control. The complex cranial vascular changes of migraine, for example, may depend on both vasoconstrictor and vasodilator actions of 5-HT.

The presynaptic sympathetic adrenergic inhibitory effect of 5-HT may be of some relevance to hypertension. The presence of such an effect in mesenteric vessels from normal rats, but not from spontaneously hypertensive rats, may be an indicator of a possible role. Likewise, the loss of this effect in rat vas deferens preparations in older animals may be significant. Another possibility is that the absence of this action in pulmonary vessels might be a contributing factor in pulmonary hypertension.

'Serotonin' could therefore be a particularly apt name for 5-hydroxytryptamine. Its vasoconstrictor and vasodilator actions, through a variety of mechanisms, may well be important in the overall control of vascular tone.

Acknowledgements

Investigations conducted in the authors' laboratory were supported in part by the National Health and Medical Research Council of Australia, and Janssen Pharmaceutica (Australia). Fig. 7.1 is reproduced, with permission, from European Journal of Pharmacology.

References

Adler-Graschinsky, E. (1983). Dual pre-synaptic effects of 5-hydroxytryptamine on peripheral adrenergic synapses. *Journal of Autonomic Pharmacology* **3,** 303–15.

Angus, J. A., Cocks, T. M., and Wright, C. E. (1987). Mechanisms of 5-HT induced vasospasm and vasodilatation in large and small arteries. *Clinical and Experimental Pharmacology and Physiology* **Suppl. 11,** 20.

Bevan, J. A., Duckles, S. P., and Lee, T. J.-F. (1975). Histamine potentiation of nerve- and drug-induced responses of a rabbit cerebral artery. *Circulation Research* **36,** 647–53.

Blackshear, J. L., Orlandi, C., Garnic, J. D., and Hollenberg, N. K. (1985). Differential large and small vessel responses to serotonin in the dog hindlimb in vivo: role of the 5-HT_2 receptor. *Journal of Cardiovascular Pharmacology* **7,** 42–9.

Bradley, P. B., Engel, G., Feniuk, W., Fozard, J. R., Humphrey, P. P. A., Middlemiss, D. N., Mylecharane, E. J., Richardson, B. P., and Saxena, P. R.

(1986). Proposals for the classification and nomenclature of functional receptors for 5-hydroxytryptamine. *Neuropharmacology* **25,** 563–76.

Chand, N. (1981). 5-Hydroxytryptamine induces relaxation of goat pulmonary veins: evidence for the noninvolvement of M and D-tryptamine receptors. *British Journal of Pharmacology* **72,** 233–7.

Charlton, K. G., Johnson, T. D., and Clarke, D. E. (1984). Vasoconstrictor and norepinephrine potentiating action of 5-hydroxykynuramine in the isolated perfused rat kidney: involvement of serotonin receptors and alpha$_1$-adrenoceptors. *Naunyn-Schmiedeberg's Archives of Pharmacology* **328,** 154–9.

Charlton, K. G., Bond, R. A., and Clarke, D. E. (1986). An inhibitory prejunctional 5-HT$_1$-like receptor in the isolated perfused rat kidney: apparent distinction from the 5-HT$_{1A}$, 5-HT$_{1B}$ and 5-HT$_{1C}$ subtypes. *Naunyn-Schmiedeberg's Archives of Pharmacology* **332,** 8–15.

Clarke, D. E., Bond, R. A., Charlton, K. G., and Blue, D. R. (1988). Pre-junctional 5-HT receptors mediating inhibition of transmitter release from peripheral cholinergic and noradrenergic nerves. In *Serotonin: Actions, Receptors, Pathophysiology. Proceedings of the 1987 IUPHAR Congress Satellite Meeting, Heron Island, Australia* (ed. E. J. Mylecharane, J. A. Angus, I. S. de la Lande, and P. P. A. Humphrey). Macmillan, London (in press).

Cocks, T. M. and Angus, J. A. (1983). Endothelium-dependent relaxation of coronary arteries by noradrenaline and serotonin. *Nature (London)* **305,** 627–30.

Cohen, R. A. (1985). Serotonergic prejunctional inhibition of canine coronary adrenergic nerves. *Journal of Pharmacology and Experimental Therapeutics* **235,** 76–80.

Cohen, R. A., Shepherd, J. T., and Vanhoutte, P. M. (1983). 5-Hydroxytryptamine can mediate endothelium-dependent relaxation of coronary arteries. *American Journal of Physiology* **245,** H1077–80.

Connor, H. E., Feniuk, W., Humphrey, P. P. A., and Perren, M. J. (1986). 5-Carboxamidotryptamine is a selective agonist at 5-hydroxytryptamine receptors mediating vasodilatation and tachycardia in anaesthetized cats. *British Journal of Pharmacology* **87,** 417–26.

Coughlin, S. R., Moskowitz, M. A., Antoniades, H. N., and Levine, L. (1981). Serotonin receptor-mediated stimulation of bovine smooth muscle cell prostacyclin synthesis and its modulation by platelet-derived growth factor. *Proceedings of the National Academy of Science, USA* **78,** 7134–8.

Docherty, J. R. and Warnock, P. (1986). Prejunctional inhibitory 5-hydroxytryptamine (5-HT) receptors in the rat heart and vas deferens. *British Journal of Pharmacology* **89,** 843P.

Edvinsson, L. and Owman, C. (1974). Pharmacological characterization of adrenergic alpha and beta receptors mediating the vasomotor responses of cerebral arteries in vitro. *Circulation Research* **35,** 835–49.

Edvinsson, L., Hardebo, J. E., and Owman, C. (1978). Pharmacological analysis of 5-hydroxytryptamine receptors in isolated intracranial and extracranial vessels of cat and man. *Circulation Research* **42,** 143–51.

Endo, T., Starke, K., Bangerter, A., and Taube, H. D. (1977). Presynaptic receptor mechanisms on the noradrenergic neurones of the rabbit pulmonary artery. *Naunyn-Schmiedeberg's Archives of Pharmacology* **296,** 229–47.

Engel, G., Göthert, M., Müller-Schweinitzer, E., Schlicker, E., Sistonen, L., and

Stadler, P. A. (1983). Evidence for common pharmacological properties of [^{3}H]-5-hydroxytryptamine binding sites, presynaptic 5-hydroxytryptamine autoreceptors in CNS and inhibitory presynaptic 5-hydroxytryptamine receptors on sympathetic nerves. *Naunyn-Schmiedeberg's Archives of Pharmacology* **324,** 116–24.

Engel, G., Göthert, M., Hoyer, D., Schlicker, E., and Hillenbrand, K. (1986). Identity of inhibitory presynaptic 5-hydroxytryptamine (5-HT) autoreceptors in the rat brain cortex with 5-HT_{1B} binding sites. *Naunyn-Schmiedeberg's Archives of Pharmacology* **332,** 1–7.

Eyre, P. (1975). Atypical tryptamine receptors in sheep pulmonary vein. *British Journal of Pharmacology* **55,** 329–33.

Feniuk, W., Humphrey, P. P. A., and Watts, A. D. (1979). Presynaptic inhibitory action of 5-hydroxytryptamine in dog isolated saphenous vein. *British Journal of Pharmacology* **67,** 247–54.

Feniuk, W., Humphrey, P. P. A., and Watts, A. D. (1981a). Modification of the vasomotor actions of methysergide in the femoral arterial bed of the anaesthetized dog by changes in sympathetic nerve activity. *Journal of Autonomic Pharmacology* **1,** 127–32.

Feniuk, W., Hare, J., and Humphrey, P. P. A. (1981b). An analysis of the mechanism of 5-hydroxytryptamine-induced vasopressor responses in ganglion-blocked anaesthetized dogs. *Journal of Pharmacy and Pharmacology* **33,** 155–60.

Feniuk, W., Humphrey, P. P. A., and Watts, A. D. (1981c). Further characterization of pre- and post-junctional receptors for 5-hydroxytryptamine in isolated vasculature. *British Journal of Pharmacology* **73,** 191P-2P.

Feniuk, W., Humphrey, P. P. A., and Watts, A. D. (1983). 5-Hydroxytryptamine-induced relaxation of isolated mammalian smooth muscle. *European Journal of Pharmacology* **96,** 71–8.

Feniuk, W., Humphrey, P. P. A., and Watts, A. D. (1984). 5-Carboxamidotryptamine – a potent agonist at 5-hydroxytryptamine receptors mediating relaxation. *British Journal of Pharmacology* **82,** 209P.

Fishburn, P. A., Phillips, C. A., and Mylecharane, E. J. (1987). 5-Hydroxytryptamine-induced inhibition of [^{3}H]-noradrenaline overflow responses in transmurally stimulated dog saphenous vein preparations. *Clinical and Experimental Pharmacology and Physiology* **Suppl. 10,** 72.

Fong, I., Phillips, C. A., and Mylecharane, E. J. (1987). 5-Hydroxytryptamine receptor subtypes in rabbit common carotid artery, guinea-pig trachea and rat stomach fundus preparations. *Clinical and Experimental Pharmacology and Physiology* **Suppl. 11,** 189.

Freeman, W. K., Rorie, D. K., and Tyce, G. M. (1981). Effects of 5-hydroxytryptamine on neuroeffector junction in human pulmonary artery. *Journal of Applied Physiology* **51,** 693–8.

Furchgott, R. F. and Zawadzki, J. V. (1980). The obligatory role of endothelial cells in the relaxation of arterial smooth muscle by acetylcholine. *Nature (London)* **288,**373–6.

Garattini, S. and Valzelli, L. (1965). *Serotonin,* pp. 169–98. Elsevier, Amsterdam.

Goadsby, P. J., Lambert, G. A., and Lance, J. W. (1982). Differential effects on the internal and external carotid circulation of the monkey evoked by locus coeruleus stimulation. *Brain Research* **249,** 247–54.

Gordon, P., Haddy, F. J., and Lipton, M. A. (1958). Serotonin antagonism of noradrenaline in vivo. *Science* **128,** 531–2.

Göthert, M., Kollecker, P., Rohm, N., and Zerkowski, H.-R. (1986a). Inhibitory presynaptic 5-hydroxytryptamine (5-HT) receptors on the sympathetic nerves of the human saphenous vein. *Naunyn-Schmiedeberg's Archives of Pharmacology* **332,** 317–23.

Göthert, M., Schlicker, E., and Kollecker, P. (1986b). Receptor-mediated effects of serotonin and 5-methoxytryptamine on noradrenaline release in the rat vena cava and in the heart of the pithed rat. *Naunyn-Schmiedeberg's Archives of Pharmacology* **332,** 124–30.

Griffith, S. G. and Burnstock, G. (1983). Immunohistochemical demonstration of serotonin in nerves supplying human cerebral and mesenteric blood vessels: some speculations about their involvement in vascular disorders. *Lancet* **1,** 561–2.

Grimson, B. S., Robinson, S. C., Danford, E. T., Tindall, G. T., and Greenfield, J. C. (1969). Effect of serotonin on internal and external carotid artery blood flow in the baboon. *American Journal of Physiology* **216,** 50–5.

Haddy, F. J. (1953). Physiological studies of the venous system. Unpublished Ph.D. Thesis. University of Minnesota.

Haddy, F. J., Gordon, P., and Emanuel, D. A. (1959). The influence of tone upon responses of small and large vessels to serotonin. *Circulation Research* **7,** 123–30.

Hardebo, J. E., Chang, J.-Y., and Owman, C. (1988). Sympathetic nerves associated with brain vessels store and release serotonin which interacts with noradrenaline in cerebrovascular contraction. In *Serotonin: Actions, Receptors, Pathophysiology. Proceedings of the 1987 IUPHAR Congress Satellite Meeting, Heron Island, Australia* (ed. E. J. Mylecharane, J. A. Angus, I. S. de la Lande, and P. P. A. Humphrey). Macmillan, London (in press).

Harper, A. M., Deshmukh, V. D., Rowan, J. O., and Jennett, W. B. (1972). The influence of sympathetic nervous activity on cerebral blood flow. *Archives of Neurology* **27,** 1–6.

Heuring, R. E. and Peroutka, S. J. (1987). Characterization of a novel ^{3}H-5-hydroxytryptamine binding site subtype in bovine brain membranes. *Journal of Neuroscience* **7,** 894–903.

Kalkman, H. O., Boddeke, H. W. G. M., Doods, H. N., Timmermans, P. B. M. W. M., and van Zwieten, P. A. (1983). Hypotensive activity of serotonin receptor agonists in rats is related to their affinity for 5-HT$_1$ receptors. *European Journal of Pharmacology* **91,** 155–6.

Kannisto, P., Owman, C., Schmidt, G., and Sjoberg, N.-O. (1987). Characterization of presynaptic 5-HT receptors on adrenergic nerves supplying the bovine ovarian follicle. *British Journal of Pharmacology* **92,** 487–97.

Kapur, H. and Mottram, D. R. (1979). A presynaptic inhibitory effect of 5-hydroxytryptamine on the electrically induced twitch response of the rat vas deferens. *Biochemical Pharmacology* **28,** 951–2.

Kimura, T. and Satoh, S. (1983). Presynaptic inhibition by serotonin of cardiac sympathetic transmission in dogs. *Clinical and Experimental Pharmacology and Physiology* **10,** 535–42.

Krstić, M. K. and Katušić, Z. S. (1982). Divergent effects of cyproheptadine and R 50 656, a 5-HT$_2$ antagonist, on the cardiovascular response to 5-hydroxytryptamine in rats. *European Journal of Pharmacology* **85,** 225–7.

Kubo, T. and Su, C. (1983). Effects of serotonin and some other neurohumoral agents on adrenergic neurotransmission in spontaneously hypertensive rat vasculature. *Clinical and Experimental Hypertension* **5,** 1501–10.

Langer, S. Z. (1980). Presynaptic regulation of the release of catecholamines. *Pharmacological Reviews* **32,** 337–62.

McCubbin, J. W., Kaneko, Y., and Page, I. H. (1962). Inhibition of neurogenic vasoconstriction by serotonin. *Circulation Research* **11,** 74–83.

McGrath, M. A. (1977). 5-Hydroxytryptamine and neurotransmitter release in canine blood vessels: inhibition by low and augmentation by high concentrations. *Circulation Research* **41,** 428–35.

Markus, J. K., Phillips, C. A., Mylecharane, E. J., and Shaw, J. (1988a). The mechanism of the vasodilator action of 5-hydroxytryptamine in the dog common carotid arterial circulation in vivo. *European Journal of Pharmacology* (submitted for publication).

Markus, J. K., Phillips, C. A., and Mylecharane, E. J. (1988b). Receptor mechanisms involved in the pre-synaptic sympathetic inhibitory action of 5-hydroxytryptamine in the dog common carotid arterial circulation. *European Journal of Pharmacology* (submitted for publication).

Martin, G. R., Leff, P., Cambridge, D., and Barrett, V. J. (1987). Comparative analysis of two types of 5-hydroxytryptamine receptor mediating vasorelaxation: differential classification using tryptamines. *Naunyn-Schmiedeberg's Archives of Pharmacology* **336,** 365–73.

Martinez, A. A. and Lokhandwala, M. F. (1980). Evidence for a presynaptic inhibitory action of 5-hydroxytryptamine on sympathetic neurotransmission to the myocardium. *European Journal of Pharmacology* **63,** 303–11.

Mena, M. A. and Vidrio, H. (1979). Reversal of serotonin vasodilatation in the dog external carotid bed by sympathetic denervation. *Journal of Cardiovascular Pharmacology* **1,** 149–54.

Middlemiss, D. N. (1984). Stereoselective blockade at [^{3}H]5-HT binding sites and at the 5-HT autoreceptor by propranolol. *European Journal of Pharmacology* **101,** 289–93.

Middlemiss, D. N. (1985). β-Adrenoceptor antagonists and blockade of the 5-HT autoreceptor in rat frontal cortex. *British Journal of Pharmacology* **84,** 189P.

Molderings, G. J., Fink, K., Schlicker, E., and Gothert, M. (1987). Inhibition of noradrenaline release via presynaptic 5-HT_{1B} receptors of the rat vena cava. *Naunyn-Schmiedeberg's Archives of Pharmacology* **336,** 245–50.

Moran, N. C. (1967). The development of beta adrenergic blocking drugs: a retrospective and prospective evaluation. *Annals of the New York Academy of Science* **139,** 649–60.

Moritoki, H., Iwamoto, T., Kanaya, J., Ishida, Y., and Fukuda, H. (1986). Age-related change in serotonin-mediated prejunctional inhibition of rat vas deferens. *European Journal of Pharmacology* **132,** 39–46.

Müller-Schweinitzer, E. (1981). Agonist potencies of tryptamine derivatives at pre- and postjunctional receptors in canine saphenous vein. *Postgraduate Medical Journal* **57, Suppl. 1,** 36–44.

Mylecharane, E. J. and Phillips, C. A. (1987). 5-HT_1-like receptor-mediated presynaptic sympathetic inhibition in the dog common carotid arterial circulation. *Clinical and Experimental Pharmacology and Physiology* **Suppl. 11,** 30.

Mylecharane, E. J., Spira, P. J., Misbach, J., Duckworth, J. W., and Lance, J. W. (1978). Effects of methysergide, pizotifen and ergotamine in the monkey cranial circulation. *European Journal of Pharmacology* **48,** 1–9.

Mylecharane, E. J., Duckworth, J. W., Lord, G. D. A., and Lance, J. W. (1980). Effects of low doses of clonidine in the monkey cranial circulation. *European Journal of Pharmacology* **68,** 163–73.

Page, I. H. (1958). Serotonin (5-hydroxytryptamine); the last four years. *Physiological Reviews* **38,** 277–335.

Page, I. H. and McCubbin, J. W. (1953). The variable arterial pressure response to serotonin in laboratory animals and man. *Circulation Research* **1,** 354–62.

Page, I. H and McCubbin, J. W. (1956). Arterial pressure response to infused serotonin in normotensive dogs, cats, hypertensive dogs and man. *American Journal of Physiology* **184,** 265–70.

Phillips, C. A. and Mylecharane, E. J. (1987). The actions of 5-hydroxytryptamine in the external carotid circulation of the monkey. *Clinical and Experimental Pharmacology and Physiology* **Suppl. 11,** 188.

Phillips, C. A., Mylecharane, E. J., and Shaw, J. (1985a). Mechanisms involved in the vasodilator action of 5-hydroxytryptamine in the dog femoral arterial circulation in vivo. *European Journal of Pharmacology* **113,** 325–34.

Phillips, C. A., Mylecharane, E. J., Markus, J. K., and Shaw, J. (1985b). Hypotensive actions of ketanserin in dogs: involvement of a centrally mediated inhibition of sympathetic vascular tone. *European Journal of Pharmacology* **111,** 319–27.

Ramage, A. G. (1985). The effects of ketanserin, methysergide and LY 53857 on sympathetic nerve activity. *European Journal of Pharmacology* **113,** 295–303.

Rapport, M. M., Green, A. A., and Page, I. H. (1948). Serum vasoconstrictor (serotonin). IV. Isolation and characterization. *Journal of Biological Chemistry* **176,** 1243–51.

Richardson, P. D. I. and Withrington, P. G. (1978). Responses of the simultaneously-perfused hepatic arterial and portal venous vascular beds of the dog to histamine and 5-hydroxytryptamine. *British Journal of Pharmacology* **64,** 581–8.

Roddie, I. C., Shepherd, J. T., and Whelan, R. F. (1955). The action of 5-hydroxytryptamine on the blood vessels of the human hand and forearm. *British Journal of Pharmacology* **10,** 445–50.

Saxena, P. R. (1972). The effects of antimigraine drugs on the vascular responses by 5-hydroxytryptamine and related biogenic substances on the external carotid bed of dogs: possible pharmacological implications to their antimigraine action. *Headache* **12,** 44–54.

Saxena, P. R. and Lawang, A. (1985). A comparison of cardiovascular and smooth muscle effects of 5-hydroxytryptamine and 5-carboxamidotryptamine, a selective agonist of 5-HT_1 receptors. *Archives Internationales de Pharmacodynamie et de Therapie* **277,** 235–52.

Saxena, P. R. and Verdouw, P. D. (1982). Redistribution by 5-hydroxytryptamine of carotid arterial blood at the expense of arteriovenous anastomotic blood flow. *Journal of Physiology* **332,** 501–20.

Saxena, P. R. and Verdouw, P. D. (1984). Effects of methysergide and 5-hydroxytryptamine on carotid blood flow distribution in pigs: further evidence for the presence of atypical 5-HT receptors. *British Journal of Pharmacology* **82,** 817–26.

Saxena, P. R. and Verdouw, P. D. (1985). 5-Carboxamide tryptamine, a compound with high affinity for 5-hydroxytryptamine$_1$ binding sites, dilates arterioles and constricts arteriovenous anastomoses. *British Journal of Pharmacology* **84,** 533–44.

Saxena, P. R., Duncker, D. J., Bom, A. H., Heiligers, J., and Verdouw, P. D. (1986). Effects of MDL 72222 and methiothepin on carotid vascular responses to 5-hydroxytryptamine in the pig: evidence for the presence of '5-hydroxytryptamine$_1$-like' receptors. *Naunyn-Schmiedeberg's Archives of Pharmacology* **333,** 198–204.

Schofield, B. M. and Walker, J. M. (1953). Perfusion of the coronary arteries of the dog. *Journal of Physiology* **122,** 489–97.

Spira, P. J., Mylecharane, E. J., Misbach, J., Duckworth, J. W., and Lance, J. W. (1978). Internal and external carotid vascular responses to vasoactive agents in the monkey. *Neurology* **28,** 162–73.

Starke, R. (1977). Regulation of noradrenaline release by presynaptic receptor systems. *Reviews of Physiology, Biochemistry and Pharmacology* **77,** 1–124.

Stromberg, M. W. (1964). The autonomic nervous system. In *Anatomy of the dog* (ed. M. E. Miller, G. C. Christensen, and H. E. Evans), pp. 626–44. Saunders, Philadelphia.

Su, C. and Uruno, T. (1985). Excitatory and inhibitory effects of 5-hydroxytryptamine in mesenteric arteries of spontaneously hypertensive rats. *European Journal of Pharmacology* **106,** 283–90.

Swank, R. L. and Hissen, W. (1964). Influence of serotonin on cerebral circulation. *Archives of Neurology* **10,** 468–72.

Trevethick, M. A., Feniuk, W., and Humphrey, P. P. A. (1984). 5-Hydroxytryptamine-induced relaxation of neonatal porcine vena cava in vitro. *Life Sciences* **35,** 477–86.

Trevethick, M. A., Feniuk, W., and Humphrey, P. P. A. (1986). 5-Carboxamidotryptamine: a potent agonist mediating relaxation and elevation of cyclic AMP in the isolated neonatal porcine vena cava. *Life Sciences* **38,** 1521–8.

Van Nueten, J. M., Janssen, P. A. J., Van Beek, J., Xhonneux, R., Verbeuren, T. J., and Vanhoutte, P. M. (1981). Vascular effects of ketanserin (R 41 468), a novel antagonist of 5-HT$_2$ serotonergic receptors. *Journal of Pharmacology and Experimental Therapeutics* **218,** 217–30.

Verdouw, P. D., Jennewein, H. M., Heiligers, J., Duncker, D. J., and Saxena, P. R. (1984). Redistribution of carotid artery blood flow by 5-HT: effects of the 5-HT$_2$ receptor antagonists ketanserin and WAL 1307. *European Journal of Pharmacology* **102,** 499–509.

Verdouw, P. D., Jennewein, H. M., Mierau, J., and Saxena, P. R. (1985). N-(3-Acetylaminophenyl)piperazine hydrochloride (BEA 1654), a putative 5-HT$_1$ agonist, causes constriction of arteriovenous anastomoses and dilatation of arterioles. *European Journal of Pharmacology* **107,** 337–46.

Vidrio, H. and Hong, E. (1976). Vascular tone and reactivity to serotonin in the internal and external carotid vascular beds of the dog. *Journal of Pharmacology and Experimental Therapeutics* **197,** 49–56.

Watts, A. D., Feniuk, W., and Humphrey, P. P. A. (1981). A pre-junctional action of 5-hydroxytryptamine and methysergide on noradrenergic nerves in dog isolated saphenous vein. *Journal of Pharmacy and Pharmacology* **33,** 515–20.

8
Interactions of 5-hydroxytryptamine with endothelial cells

RICHARD A. COHEN

8.1 Introduction

The single cell layer of endothelium serves as a complete lining of the cardiovascular system. In large elastic arteries the endothelium is only the first layer of a many layered vascular wall; in the capillaries the endothelium may be the only barrier separating the blood from the tissues. The endothelium excludes blood cells and proteins, and filters nutrients to the tissues. The role of the endothelium is not simply passive; it is now recognized that it acts as an important modulator of the function of the circulating blood cells, of the other cells within the blood vessel wall, and of the tissues perfused by the vessel. In order to accomplish its roles the endothelium has multiple membrane receptors, transport proteins, ion exchange pumps, and synthetic and degradative enzymes (Nossel and Vogel 1982; Cohen and Vanhoutte 1986).

8.2 The platelet as source of 5-HT

Any discussion of 5-HT and endothelial cells must consider the interaction of the endothelium with platelets. Although some endothelial cells have been demonstrated to synthesize 5-HT in culture (Maruki *et al.* 1984), it is likely that the primary source of 5-HT reaching endothelial cells *in vivo* is the platelet. The 5-HT content of the platelet (1–5 nmol/10^9 platelets; D'Amore and Shepro 1982) is localized in dense granules where it exists in a semi-crystalline complex with ADP (Holmsen 1986) (see Chapter 1). As platelets avidly take up any free circulating 5-HT (Lingjaerde 1977), it is not certain the extent to which under normal circumstances endothelial cells are exposed to 5-HT. For instance, taking great care to eliminate any platelet activation during blood sampling, free 5-HT levels of only approximately 5–10 pM have been measured in dog plasma (Benedict *et al.* 1986). When the platelet is activated 5-HT is secreted; if all the platelets in the blood were to release their 5-HT, the concentration may reach 0.6 μM (Cohen 1986a). This concentration is sufficient to activate cell surface receptors on platelets as well as endothelial and subendothelial vascular smooth muscle and nerve cells in the vicinity (Cohen, 1987). The amount of 5-HT adjacent to the luminal surface of a blood vessel localized beneath a mural thrombus may be greater than that diluted in the flowing blood.

8.3 Anti-aggregatory role of the endothelium

The endothelial cell has the important role of preventing platelet activation and aggregation. The endothelial cell accomplishes this function by virtue of its surface membrane, which lacks the proteins that on other cell types trigger platelet activation, as well as by the production of potent anti-aggregatory substances, including prostacyclin (Moncada 1982). Whether by disease or trauma, when endothelial cells dysfunction platelets release their contents, including 5-HT. The interaction between platelets and endothelial cells is complex. Activation and aggregation of platelets is influenced by positive feedback control by pro-aggregatory substances released by other platelets including calcium, ADP, thromboxane A_2, catecholamines, and 5-HT (Holmsen 1986). The positive feedback in the case of catecholamines and 5-HT may be tempered by catabolism of the amines by monoamine oxidase (see Chapter 1) and in the case of ADP by phosphatases on the endothelial cell (Goldman *et al.* 1986). The negative feedback influence of prostacyclin produced by the endothelial cell may itself be influenced by the platelets. Thus, cyclic endoperoxides produced by the platelets may serve as a synthetic precursor for prostacyclin produced by endothelial cells (Marcus *et al.* 1980). Platelet proteins (Poggi *et al.* 1983), catecholamines, and 5-HT may increase prostacyclin production by stimulating calcium influx into endothelial cells (Cohen and Vanhoutte 1986).

8.4 Influence of the endothelial cell on smooth muscle contractility

Although it was previously recognized that the endothelium synthesized and/or degraded several vasoactive substances, the first direct evidence that the endothelium plays an obligatory role in the vascular response to several vasoactive agonists was provided by Furchgott and his colleagues who demonstrated that arteries whose endothelium was removed failed to relax when acetylcholine was administered *in vitro* (Furchgott and Zawadski 1980; Furchgott *et al.* 1981). The list of substances for which the endothelium may mediate or modulate vascular relaxation continues to grow but includes ADP and ATP, bradykinin, histamine, substance P, thrombin, and vasopressin (Chand and Altura 1981; Cherry *et al.* 1982; DeMey and

Vanhoutte 1981; Furchgott *et al.* 1981; Katusic *et al.* 1984; Van de Voorde and Leusen 1982). Each of these agents probably interacts with endothelial cells via specific receptors (Fig. 8.1) and may exert its effect by a common intracellular mechanism which involves an increase in free calcium (Singer and Peach 1982). The subsequent relaxation of vascular smooth muscle is mediated by a short-lived chemical agent(s) generated by the endothelial cell known as EDRF (Furchgott 1983). EDRF is a soluble agent since it can be passed in a cascade system from 'donor' endothelial cells on one blood vessel to the smooth muscle of another used for assay (Rubanyi *et al.* 1985). The effect of EDRF on the smooth muscle bears many resemblances to that of nitrate vasodilators, including the capacity to induce an increase in the level of cyclic GMP (Furchgott and Jothianandan 1983; Rapaport and Murad 1983). In fact, endothelial cells have been shown to produce nitric oxide which has chemical characteristics similar to EDRF and is a powerful vasodilator (Palmer *et al.* 1987). In addition to relaxing the blood vessel, EDRF has been shown to assist prostacyclin in inhibiting platelet aggregation (Radomski *et al.* 1987; Alheid *et al.* 1987). Several vasoactive

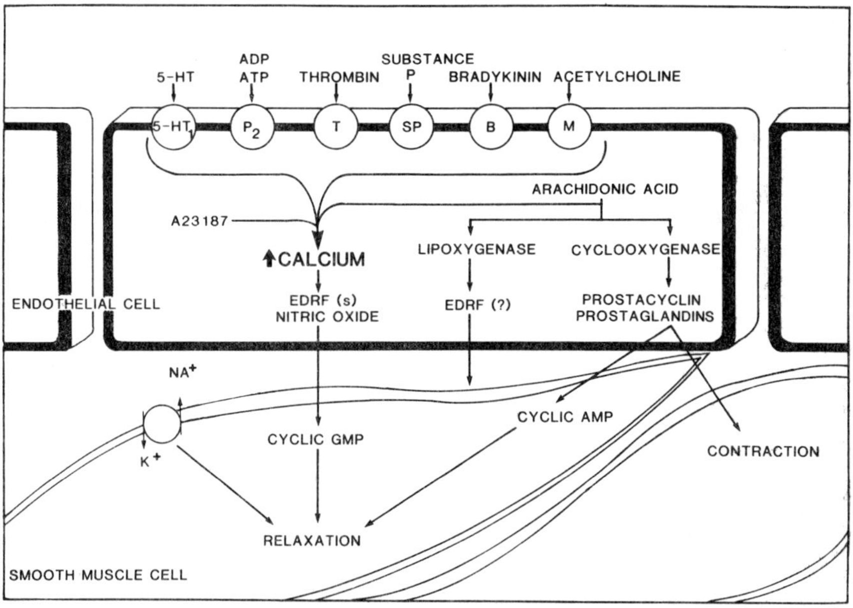

Fig. 8.1 Summary of known vasoactive mechanisms mediated by endothelial cells. Relaxation of smooth muscle may be initiated by endothelial cells in response to activation of a number of cell surface receptors, to calcium influx caused by the ionophore, A23187, or to endogenous or exogenous arachidonic acid. Contraction may be initiated by endothelial cell-derived vasoconstrictor metabolites of arachidonic acid

substances increase the release of prostacyclin from the endothelium; in the case of exogenous arachidonic acid the released prostacyclin is known to contribute significantly to the response of the smooth muscle (Miller and Vanhoutte 1985).

8.5 Biochemical interaction of 5-HT with endothelial cells

8.5.1 Uptake and degradation of 5-HT

5-HT is taken up by endothelial cells by an active uptake mechanism with an apparent K_m of 2–5 μM (see Chapter 1). The uptake mechanism is inhibited by ouabain, cocaine, amitriptyline, and imipramine (Gillis 1985). Following its uptake, 5-HT is rapidly metabolized primarily to 5-hydroxyindoleacetic acid by the A isozyme of monoamine oxidase. The removal of 5-HT by the endothelial cell, by reducing the effective local concentration, may influence the response to 5-HT of endothelial cell, smooth muscle cell, and platelet receptors. 5-HT uptake by the endothelial cell may be influenced by local conditions. Thus, hyperoxia inhibits (Block *et al.* 1985) and hypoxia stimulates (Lee and Fanburg 1986) the uptake of 5-HT, and prostacyclin generation can attenuate the accumulation of the amine (Shepro and Hechtman 1985). Disease states such as adult respiratory distress syndrome result in a decrease in the ability of the lung endothelium to clear 5-HT from the blood (Gillis *et al.* 1986).

8.5.2 Receptors for 5-HT

In addition to the uptake mechanism for 5-HT, there are receptors for 5-HT on the endothelial cell membrane of at least two subtypes (Fig. 8.2). '5-HT_1-like' receptors on endothelial cells mediate release of EDRF which may be responsible for vasodilator responses to 5-HT (see Sections 8.6.2 and 8.6.3). 5-HT_2 receptors have been recognized to mediate changes in endothelial cell membrane structural characteristics and permeability (see Sections 8.5.3.2 and 8.5.3.3). As no 5-HT ligand binding studies have yet been done on endothelial cells, the receptors have been identified either by their activation by selective 5-HT receptor agonists or by virtue of blockade of physiological responses to 5-HT by selective antagonists.

5-HT_2 receptor-mediated responses have been identified in smooth muscle (Van Nueten *et al.* 1984) and endothelial cells by selective blockade with drugs such as ketanserin or ritanserin. No selective antagonists are yet available for '5-HT_1-like' receptors, but antagonists which bind to both '5-HT_1-like' and 5-HT_2 receptors, such as methysergide and methiothepin, have been shown to block vascular responses to 5-HT against which selec-

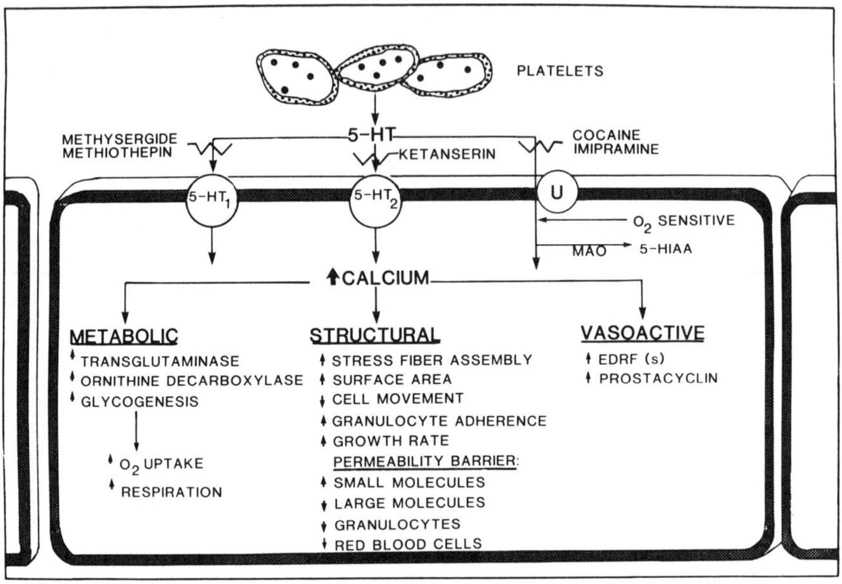

Fig. 8.2 Summary of known metabolic, structural, and vasoactive actions of 5-HT on endothelial cells. U = amine uptake pump, MAO = monoamine oxidase, 5-HIAA = 5-hydroxyindoleacetic acid

tive 5-HT_2 receptor antagonists have no effect. In this way, functional correlates of '5-HT_1-like' receptors have been suggested on endothelial cells (see Section 8.6.2), certain smooth muscle cells (Peroutka 1984; Feniuk 1984; Cohen 1986b), and adrenergic nerves (Engel *et al.* 1983; Cohen 1985b) (see also Chapter 7).

It is not yet known if 5-HT receptors which are responsible for physiological responses of endothelial cells are distinct from the binding sites responsible for amine uptake. For instance, in other cell types several drugs which act as 5-HT receptor antagonists may also inhibit amine uptake into adrenergic nerves (e.g. methiothepin, amitriptyline). However, 5-HT receptor-mediated contraction of smooth muscle, as well as inhibition of noradrenaline release from adrenergic nerves, is not prevented by uptake blockade (Cohen 1985b and 1986b).

8.5.3 Endothelial cellular changes induced by 5-HT

8.5.3.1 *Biochemical changes*

A number of important biochemical alterations occur when endothelial cells are exposed to 5-HT (Fig. 8.2). The initiating event in many or all of these changes may be the activation of a cell surface receptor and subsequent rise in intracellular calcium (D'Amore and Shepro 1978). On exposure of endothelium to 5-HT, an increase in the activity of many enzymes occurs, including transglutaminase (Carson *et al.* 1983) and ornithine decarboxylase (D'Amore *et al.* 1978), as well as those involved in glycogenesis (Spatz *et al.* 1986). As a result there is also likely to be stimulation of oxygen uptake and an increase in respiratory rate of the cell (Kjellstrom *et al.* 1985).

8.5.3.2 *Structural changes*

Changes in the structural characteristics of endothelial cells in culture have been noted to occur following exposure to 5-HT. These include an increase in structural integrity due to an increase in actin stress fibre assembly (Shepro *et al.* 1980) and an increase in cell surface area (Welles *et al.* 1985). A decrease in cell movement in the culture plate occurs which is mediated by 5-HT_2 receptors as it is attenuated by ketanserin (Bottaro *et al.* 1985). These changes are accompanied by an increase in endothelial cell growth rate. The changes in cell structure are likely to be mediated by an influx of calcium since they resemble the changes induced by the calcium ionophore, A23187, and may be prevented by chelating calcium (D'Amore and Shepro 1977). The structural changes may participate in pathological tissue responses such as the increased granulocyte adherence which develops during exposure to 5-HT released by platelets (Boogaerts *et al.* 1982). The later response may involve 5-HT uptake, as well as receptor activation, as it is attenuated by imipramine and by the non-selective 5-HT receptor antagonist, methysergide.

8.5.3.3 *Permeability changes*

Platelets, when injected intradermally, cause an increase in vascular permeability, oedema, and inflammation. There are many mediators of this reaction but, at least in the rat, a role for 5-HT_2 receptor activation is indicated by the attenuation of the response following ketanserin or depletion of platelet 5-HT by reserpine (DeClerck and Reneman 1985). The endothelium plays the key role in determining microvascular permeability. As demonstrated in frog brain microvessels (Olesen 1985), intravenous 5-HT causes decreases in the electrical resistance of endothelial cell membranes. This increase in ionic permeability is mediated by 5-HT_2 receptors and dependent on extracellular calcium. Local oedema and stasis may be favoured by this increased capillary endothelial cell ionic permeability, particularly when combined with post-capillary venular constriction which may also be mediated by 5-HT_2 receptors (Vanhoutte and Cohen 1983).

The effect of 5-HT on permeability of the endothelial cell barrier may depend on molecular size of the diffusing substance, as the amine decreases the permeability of endothelial cells to albumin (Bottaro *et al.* 1986). The endothelium is also a barrier to blood cells which would otherwise enter the tissues; 5-HT may inhibit the migration of polymorphonuclear leucocytes through the endothelium (Doukas *et al.* 1987). 5-HT also reduces capillary endothelial cell leakiness to red blood cells in thrombocytopenic hamsters. The effect is prevented by ketanserin, suggesting involvement of $5\text{-}HT_2$ receptors. Furthermore, ketanserin increases capillary fragility in normal hamsters, suggesting that 5-HT released by platelets may have a homeostatic role in maintaining endothelial cell integrity (Shepro *et al.* 1984).

8.6 Endothelial cells in the vascular response to 5-HT

8.6.1 Inhibition of contraction

Contractions of isolated rings of canine coronary artery caused by platelets are larger after mechanically removing the endothelial cells (Cohen *et al.* 1983b). These contractions are mediated primarily by 5-HT released by the platelets, suggesting an inhibitory role of the endothelium in the vascular response to the amine (Cohen *et al.* 1983b; Cohen 1986b). Indeed, the presence of endothelial cells is associated with the attenuation of contractions of the canine coronary artery caused by 5-HT itself (see Fig. 8.3) (Cohen *et al.* 1983a; Cocks and Angus 1983; Gutterman *et al.* 1986). This attenuation is selective in that contractions caused by phenylephrine, prostaglandin $F_{2\alpha}$, or K^+ are not different in rings of the same artery, with or without

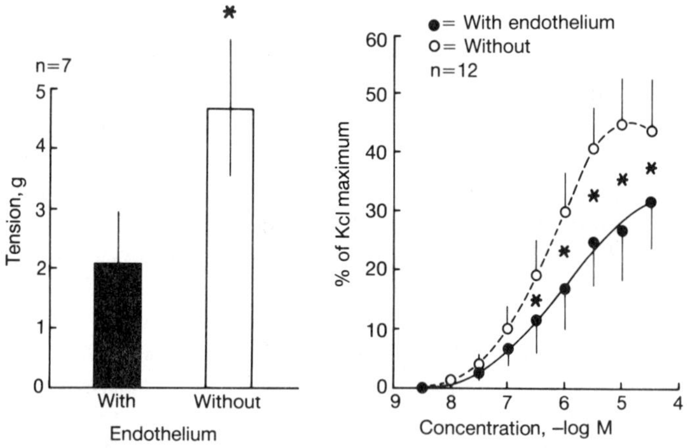

Fig. 8.3 Effect of mechanical removal of endothelium on contractions of canine left circumflex artery to aggregating platelets (left) and to cumulative increases in 5-HT concentration (right). Responses to platelets and 5-HT are of less magnitude in rings with intact endothelium. (From Cohen and Vanhoutte 1986, with permission)

endothelial cells (Cohen *et al.* 1983a). Although metabolism by endothelial cell monoamine oxidase might be expected to attenuate contractions caused by 5-HT, and thereby explain this observation, blockade of 5-HT metabolism does not effect the inhibition (Cohen *et al.* 1983a). Furthermore, prostacyclin generation by the endothelial cells cannot explain the inhibition, as it is unaffected by cyclo-oxygenase blockade (Cohen *et al.* 1983a). In perfused segments of coronary arteries, the endothelial cells also inhibit contractions caused by intraluminal 5-HT and platelets, but contractions caused by addition of the amine extraluminally are not influenced by the presence or absence of endothelium. This indicates that access of the amine to the endothelial cells is required for the inhibitory action (Cohen *et al.* 1984a). Since access of 5-HT to the smooth muscle on the luminal aspect of the vessel provides the most potent route for its contraction (Cohen *et al.* 1984a), endothelial cell function is most influential on smooth muscle contractility when 5-HT is present in the lumen of perfused vessels.

In the rabbit isolated coronary artery, contractions caused by 5-HT are also inhibited by the endothelium. The contractions are associated with depolarization of the smooth muscle cells, which is the same regardless of the presence or absence of endothelial cells (Garland 1985). This suggests that the inhibitory influence of the endothelium is not exerted on membrane potential, but rather on intracellular mechanisms which control contraction.

8.6.2 5-HT-induced endothelium-dependent relaxation

Direct evidence of an inhibitory influence by endothelial cells has been obtained by demonstrating that isolated rings of blood vessels with intact endothelium can relax when exposed to 5-HT. This occurs in canine coronary artery (Fig. 8.4) (Cohen *et al.* 1983a; Cocks and Angus 1983) and chick jugular vein (Imaizumi *et al.* 1984). In the porcine coronary artery (Cocks and Angus 1983) and Wistar–Kyoto rat aorta (Lüscher and Vanhoutte 1986), if the contractile effect of 5-HT on the smooth muscle is blocked with the 5-HT_2 receptor antagonist, ketanserin, an endothelium-dependent relaxation to 5-HT is unmasked. The relaxation caused by 5-HT may be blocked by the non-selective 5-HT receptor antagonists, methysergide (Cohen *et al.* 1983a) or methiothepin (Houston *et al.* 1985), suggesting that a '5-HT_1-like' receptor mediates the response. Platelets also cause endothelium-dependent relaxations (Cohen *et al.* 1983b) which are mediated in part by ADP released by the platelets (Houston *et al.* 1985). Released 5-HT may be implicated as well in contributing to the endothelium-dependent relaxation caused by platelets (Cohen *et al.* 1983a and b) since the concentration of 5-HT released by the platelets when given alone, also relaxes the blood vessel (Fig. 8.4). When the smooth muscle of perfused coronary

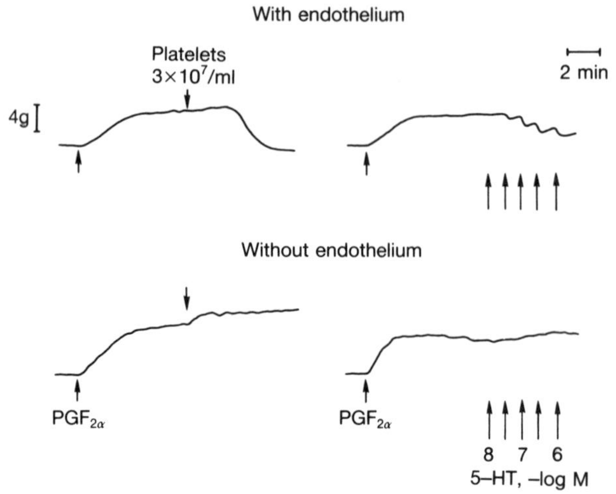

Fig. 8.4 Response to aggregating platelets and 5-HT of two canine coronary artery rings contracted with prostaglandin $F_{2\alpha}$ ($PGF_{2\alpha}$, 2 μM). In the ring with intact endothelium (upper) platelets (left) cause a brief increase in tension, followed by relaxation, whereas only contraction occurs in the ring denuded of endothelium (lower). The relaxation may be attributed in part to 5-HT released by the platelets, since 5-HT causes concentration- and endothelium-dependent relaxation (upper, right) in the same concentration range as that released by the platelets (i.e. approximately 0.4 μM). (From Cohen and Vanhoutte 1986, with permission)

arteries has pre-existing tone, 5-HT may cause relaxation provided the amine has luminal access to the endothelium (Fig. 8.5) (Cohen *et al.* 1984a).

Although it has been determined that the relaxation caused by acetylcholine, bradykinin, and A23187 may be explained by the release of a diffusible vasodilator from endothelial cells (Rubanyi *et al.* 1985; Palmer *et al.* 1987), direct evidence that the endothelium releases a vasodilator when exposed to 5-HT has been lacking. In the experiment shown in Fig. 8.6, a canine coronary artery with endothelium served as a donor artery segment, while force was recorded from a ring of coronary artery without endothelium contracted with prostaglandin $F_{2\alpha}$. When the ring was directly superfused with acetylcholine or 5-HT no response was observed since the ring had no endothelium and had been treated with atropine and methiothepin. However, when the ring was superfused with perfusate from the donor segment, the ring relaxed, indicating that under basal conditions the endothelium releases vasodilator substance(s). When the segment was perfused with acetylcholine or 5-HT a further relaxation was observed, indicating that each agonist may induce a further release of vasodilator from the endothelial cells of the donor segment. This experiment is consistent with the release by 5-HT of a diffusible vasodilator from endothelial

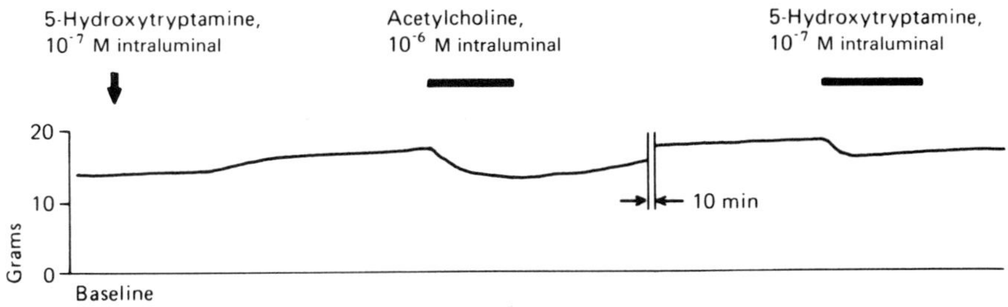

Fig. 8.5 Response of a perfused segment of canine left circumflex coronary artery to 5-HT and acetylcholine. Recording of isometric tension measured in the blood vessel whose branches had been ligated to prevent leaks while being perfused with physiological salt solution. Addition of 5-HT (0.1 μM) to the salt solution surrounding the artery (extraluminal) caused contraction of the smooth muscle. Addition of acetylcholine (1 μM) to the perfusate of the blood vessel caused relaxation indicating the presence of a functional endothelium; the contraction due to the 5-HT returned following cessation of the acetylcholine. Perfusion with 5-HT (0.1 μM, intraluminal) then caused relaxation. The results indicate separate routes of access, extraluminal to the smooth muscle and intraluminal to the endothelium, lead to opposite responses. (From Cohen and Vanhoutte 1984b with permission)

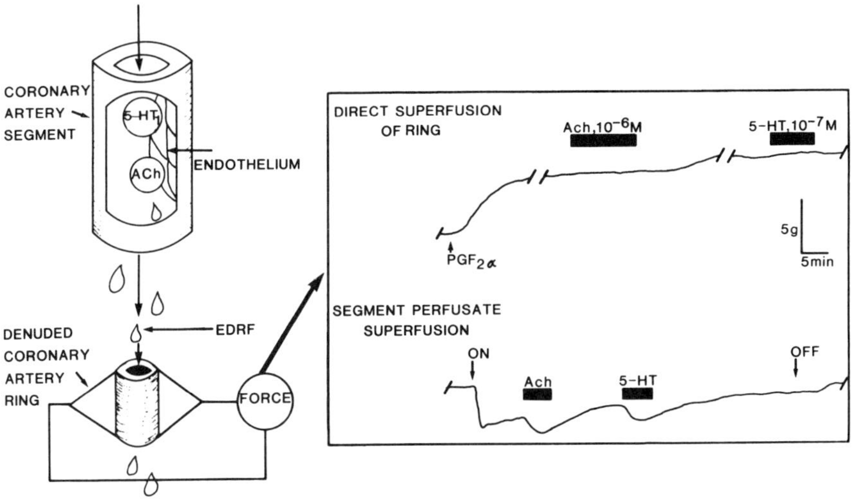

Fig. 8.6 Cascade-superfusion experiment to demonstrate that 5-HT, like acetylcholine, releases a diffusible vasodilator from endothelial cells. A segment of coronary artery with endothelium served as donor, whereas a ring of artery without endothelium served as assay tissue in which isometric tension was recorded. Vasodilator was released into the perfusate of the segment with endothelium under basal conditions and the relaxation was increased by the addition of acetylcholine (Ach) or 5-HT to the perfusate (see text)

cells which may, like that released by acetylcholine, account for the relaxation caused by 5-HT.

8.6.3 Role of the endothelium in the vascular response to 5-HT in vivo

The observation that the endothelium inhibits the contraction of isolated canine coronary arteries caused by 5-HT has been confirmed *in vivo*. Thus, 5-HT causes greater decreases in canine epicardial coronary artery diameter after removing the endothelium with a balloon catheter (Lamping *et al.* 1985). It is possible that the endothelium mediates vasodilator responses to 5-HT *in vivo* as well. When 5-HT is injected intravenously a prolonged hypotensive phase follows an initial depressor phase mediated by the Bezold–Jarish reflex, and a pressor phase mediated by activation of smooth muscle 5-HT_2 receptors (Page and McCubbin 1953; Van Nueten *et al.* 1981). Like the endothelium-dependent arterial relaxation caused by 5-HT, the hypotensive response to 5-HT may be mediated by endothelial cells as it is not prevented by ketanserin (Van Nueten *et al.* 1981), but is mediated rather by '5-HT_1-like' receptors (Kalkman *et al.* 1983) (see also Chapter 7). This hypothesis would require that the endothelium of resistance vessels responds to 5-HT as does that of larger arteries.

8.6.4 Influence of disease on the endothelial cell response to 5-HT

8.6.4.1 *Atherosclerosis*

Studies in isolated rabbit aorta indicate that, with the development of hypercholesterolaemia-induced atherosclerosis, endothelium-dependent relaxations caused by acetylcholine are attenuated (Verbeuren *et al.* 1986). Since the release of EDRF was preserved in these studies, the reduction in relaxation was attributed to the intimal thickening associated with the disease, rather than to endothelial cell dysfunction. An increase in the sensitivity of isolated blood vessels to contractions caused by 5-HT has been noted in several studies of hypercholesterolaemic animals, but the role of altered endothelial cell or smooth muscle dysfunction has not been resolved (Henry and Yokoyama 1980; Yokoyama *et al.* 1983; Heric and Tackett 1985). Contractile responses to low concentrations of 5-HT were augmented in the aortas from the hypercholesterolaemic animals, and the sensitivity to 5-HT resembled that of normal rings without endothelium (Verbeuren *et al.* 1986).

The recordings in Fig. 8.7 show relaxations caused by 5-HT of rings of coronary artery with morphologically intact endothelium from a normal and a hypercholesterolaemic pig. Since the contractile response of the smooth muscle was blocked by ketanserin and the relaxations to 5-HT are endothelium-dependent, the attenuated relaxation caused by 5-HT in the

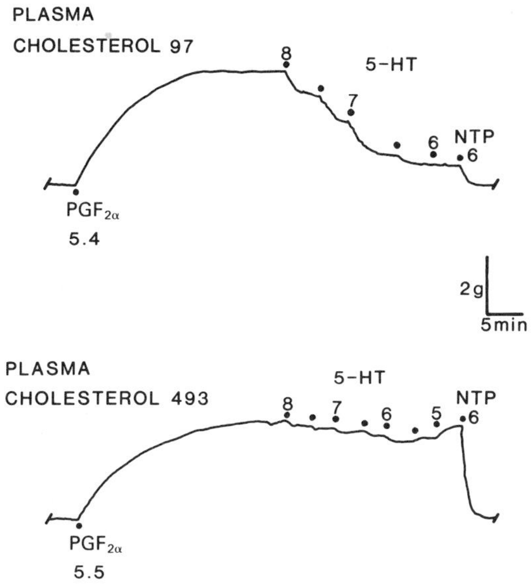

Fig. 8.7 Effect of hypercholesterolaemia on the response of porcine coronary artery to 5-HT. Recordings of isometric tension are those of rings of coronary artery from a normal (upper) and hypercholesterolaemic pig which had been fed a cholesterol-rich diet for eight weeks (lower). In each case, the rings were contracted with prostaglandin $F_{2\alpha}$ ($PGF_{2\alpha}$) and treated with ketanserin (1 μM) to block contractions caused by 5-HT to unmask the endothelium-dependent response to cumulative increases in concentration of 5-HT. The relaxations caused by 5-HT are attenuated in the artery from the hypercholesterolaemic pig despite the presence of intact endothelium demonstrated by light microscopy and a normal maximal relaxation to nitroprusside (NTP, 1 μM)

artery from the hypercholesterolaemic animal suggests endothelial cell dysfunction. Since the same vessels responded similarly to A23187 (not shown), the endothelial cell dysfunction may be selective for 5-HT. Endothelial cells of the pig coronary artery which regenerate following catheter-induced denudation do not respond normally to 5-HT, although other endothelium-dependent responses are restored, suggesting that the response to 5-HT is particularly susceptible to cell injury (Shimokawa *et al.* 1987). Reversible endothelial cell dysfunction could explain the augmented large artery vasoconstrictor responses to 5-HT observed in the hind limbs of atherosclerotic monkeys which resolve after regression of the atherosclerosis (Heistad *et al.* 1984 and 1987). Although it is as yet uncertain whether 5-HT causes vasodilation of smaller vessels which regulate vascular resistance by an endothelium-dependent mechanism, rabbit hind limb vasodilator responses to 5-HT are unaffected by hypercholesterolaemia (Wright and Angus 1986). Thus, it may be that hypercholesterolaemia affects endothelial cell function, particularly in large arteries which are also those prone to the development of atherosclerosis.

8.6.4.2 *Hypertension*

An increase in contractions caused by 5-HT has been noted in several animal models of hypertension, but it is unclear if endothelial cell dysfunction accounts for the hyper-responsiveness (Webb and Vanhoutte 1986). In the hypertensive Dahl rat, endothelium-dependent relaxations of the aorta caused by acetylcholine, ADP, and thrombin are attenuated. The endothelium markedly inhibited contractions caused by 5-HT in the normotensive animals, but the inhibitory effect was less in the hypertensive animals. Since the response of aortas without endothelium was similar in normotensive and hypertensive animals, endothelial cell dysfunction in response to 5-HT accounts for the greater contractions of aortic rings from the hypertensive rat (Lüscher *et al.* 1987). Rings of aorta with endothelium from the spontaneously hypertensive rat exhibit enhanced contractions, as compared to rings without endothelium. This may represent a 5-HT-induced endothelium-dependent contractile influence seen in the hypertensive, but not the normotensive, control animals (Lüscher and Vanhoutte 1986). As with hypercholesterolaemia, hypertension may not markedly affect rabbit hind limb vasodilator responses to 5-HT (Wright and Angus 1986).

8.7 Concluding remarks

The endothelial cell is poised to play a critical rôle in responding to 5-HT released from platelets into the blood. Although it can be released in high concentrations, metabolism of the amine by the endothelium may eliminate any further response. Once 5-HT receptors on endothelial cells are activated, the cells may, in response to a rise in cell calcium, respond in a variety of ways. The stimulated cells may release prostaglandins and EDRF into the lumen, inhibiting platelet activation and the further release by platelets of 5-HT. The same endothelial cell products can powerfully influence smooth muscle cell contractility. Beyond the immediate biochemical response to 5-HT, the amine may initiate structural, growth, and permeability changes altering the homeostatic function of the endothelial cell. In vascular diseases, platelet activation and 5-HT release may result. If the ability of the endothelial cell to metabolize or respond to 5-HT is also affected by the disease, the amine may potentiate platelet activation, and 5-HT may exert untempered vasoconstrictor effects on vascular smooth muscle and deleteriously decrease tissue perfusion. Disruption of normal endothelial cell responses to 5-HT could play a role in altered haemostasis or oedema formation, and may contribute to the development of vascular diseases such as hypertension and atherosclerosis.

Acknowledgements

The author acknowledges the great help of Katherine Boris in typing the manuscript and Marie Ryan in preparing the figures. Dr. Cohen is supported by an Established Investigator Award from the American Heart Association and NIH grants HL 31607, HL 26320, HL 38731 and DK 39080.

References

Alheid, U., Frolich, J. C., and Forstermann, U. (1987). Endothelium-derived relaxing factor from cultured human endothelial cells inhibits aggregation of human platelets. *Thrombosis Research* **47,** 561–71.

Benedict, C. R., Mathew, B., Rex, K. A., Cartwright, J., jun., and Sordahl, L. A. (1986). Correlation of plasma serotonin changes with platelet aggregation in an in vivo dog model of spontaneous occlusive coronary thrombus formation. *Circulation Research* **58,** 58–67.

Block, E. R., Patel, J. M., and Sheridan, N. P. (1985). Effect of oxygen and endotoxin on lactate dehydrogenase release, 5-hydroxytryptamine uptake, and antioxidant enzyme activities in endothelial cells. *Journal of Cellular Physiology* **122,** 240–8.

Block, E. R., Patel, J. M., Angelides, K. J., Sheridan, N. P., and Garg, L. C. (1986). Hyperoxia reduces plasma membrane fluidity: a mechanism for endothelial cell dysfunction. *Journal of Applied Physiology* **60,** 826–35.

Boogaerts, M. A., Yamada, O., Jacob, H. S., and Moldow, C. F. (1982). Enhancement of granulocyte-endothelial cell adherence and granulocyte-induced cytotoxicity by platelet release products. *Proceedings of the National Academy of Science USA* **79,** 7019–23.

Bottaro, D., Shepro, D., Peterson, S., and Hechtman, H. B. (1985). Serotonin, histamine, and norepinephrine mediation of endothelial and vascular smooth muscle cell movement. *American Journal of Physiology* **248,** C252–7.

Bottaro, D., Shepro, D., Peterson, S., and Hechtman, H. B. (1986). Serotonin, norepinephrine, and histamine mediation of endothelial cell barrier function in vitro. *Journal of Cellular Physiology* **128,** 189–94.

Carson, M. P., Peterson, S. W., Moynahan, M. E., and Shepro, D. (1983). Binding, internalization, and degradation of [^{125}I]insulin by cultured bovine aortic endothelial cells: effects of serotonin. *In Vitro* **19,** 833–40.

Chand, N. and Altura, B. M. (1981). Acetylcholine and bradykinin relax intrapulmonary arteries by acting on endothelial cells: Role in lung vascular diseases. *Science* **213,** 1376–9.

Cherry, P. D., Furchgott, R. F., and Zawadski, J. V. (1982). Role of endothelial cells in relaxation of isolated arteries by bradykinin. *Proceedings of the National Academy of Science USA* **79,** 2106–10.

Cocks, T. M. and Angus, J. A. (1983). Endothelium-dependent relaxation of coronary arteries by noradrenaline and serotonin. *Nature (London)* **305,** 627–30.

Cohen, R. A. (1985a). Platelet-induced neurogenic coronary contractions due to accumulation of the false neurotransmitter, 5-hydroxytryptamine. *Journal of Clinical Investigation* **75,** 286–92.

Cohen, R. A. (1985b). Serotonergic prejunctional inhibition of canine coronary adrenergic nerves. *Journal of Pharmacology and Experimental Therapeutics* **235,** 76–80.

Cohen, R. A. (1986a). Adenine nucleotides and 5-hydroxytryptamine released by aggregating platelets inhibit adrenergic neurotransmission in canine coronary artery. *Journal of Clinical Investigation* **77,** 369–75.

Cohen, R. A. (1986b). Contractions of isolated canine coronary arteries resistant to S_2-serotonergic blockade. *Journal of Pharmacology and Experimental Therapeutics* **237,** 548–52.

Cohen, R. A. (1987). Serotonin and the blood-vessel wall. *Journal of Hypertension* **4,** S116–8.

Cohen, R. A. and Vanhoutte, P. M. (1986). Platelets, serotonin and endothelial cells. In *Serotonin and the cardiovascular system* (ed. P. M. Vanhoutte) pp. 105–12, Raven Press, New York.

Cohen, R. A., Shepherd, J. T., and Vanhoutte, P. M. (1983a). 5-Hydroxytryptamine can mediate endothelium-dependent relaxation of coronary arteries. *American Journal of Physiology* **245** (Heart and Circulatory Physiology 14), H1077–80.

Cohen, R. A., Shepherd, J. T., and Vanhoutte, P. M. (1983b). Inhibitory role of the endothelium in the response of isolated coronary arteries to platelets. *Science* **221,** 273–4.

Cohen, R. A., Shepherd, J. T., and Vanhoutte, P. M. (1984a). Endothelium and asymmetrical responses of the coronary arterial wall. *American Journal of Physiology* **247** (Heart and Circulatory Physiology 16), H403–8.

Cohen, R. A., Shepherd, J. T., and Vanhoutte, P. M. (1984b). Vasodilation mediated by the coronary endothelium in response to aggregating platelets. *Proc. III International Mechanisms of Vasodilatation, Bibliotheca Cardiologica*, No. 38, pp. 35–42. Karger, Basle.

Cook, D. R., Howell, R. E., and Gillis, C. N. (1982). Xanthine oxidase-induced lung injury inhibits removal of 5-hydroxytryptamine from the pulmonary circulation. *Anesthesia and Analgesia* **61,** 666–70.

D'Amore, P. and Shepro, D. (1977). Stimulation of growth and calcium influx in cultured, bovine, aortic endothelial cells by platelets and vasoactive substances. *Journal of Cellular Physiology* **92,** 177–84.

D'Amore, P. and Shepro, D. (1978). Calcium flux and ornithine decarboxylase activity in cultured endothelial cells. *Life Sciences* **22,** 571–6.

D'Amore, P. and Shepro, D. (1982). Captation of 5-hydroxytryptamine and effects on endothelial cell membrane. In *5-Hydroxytryptamine in peripheral reactions* (ed. F. DeClerck and P. M. Vanhoutte) pp. 37–47. Raven Press, New York.

D'Amore, P. A., Hechtman, H. B., and Shepro, D. (1978). Ornithine decarboxylase activity in cultured endothelial cells stimulated by serum, thrombin and serotonin. *Thrombosis and Haemostasis* **39,** 496–503.

DeClerck, F. and Reneman, R. S. (1985). Platelet-derived serotonin and abnormal tissue perfusion. In *Serotonin and the cardiovascular system* (ed. P. M. Vanhoutte) pp. 155–64. Raven Press, New York.

DeMey, J. G. and Vanhoutte, P. M. (1981). Role of the intima in cholinergic and purinergic relaxation of isolated canine femoral arteries. *Journal of Physiology* (London) **316,** 347–55.

Doukas, J., Shepro, D., and Hechtman, H. B. (1987). Vasoactive amines directly modify endothelial cells to affect polymorphonuclear leukocyte diapedesis in vitro. *Blood* **69,** 1563–9.

Engel, G., Göthert, M., Müller-Schweinitzer, E., Schlicker, E., Sistonen, L., and Stadler, P. A. (1983). Evidence for common pharmacological properties of [^{3}H]5-hydroxytryptamine binding sites, presynaptic 5-hydroxytryptamine autoreceptors in CNS an inhibitory presynaptic 5-hydroxytryptamine receptors on sympathetic nerves. *Naunyn-Schmiedeberg's Archives of Pharmacology* **325,** 116–24.

Feniuk, W. (1984). An analysis of 5-hydroxytryptamine receptors mediating contraction of isolated smooth muscle. *Neuropharmacology* **23,** 1467–72.

Furchgott, R. F. (1983). Role of the endothelium in responses of vascular smooth muscle. *Circulation Research* **53,** 557–72.

Furchgott, R. F. and Jothianandan, D. (1983). Relation of cyclic GMP levels to endothelium-dependent relaxation in rabbit aorta. *Federation Proceedings* **42,** 619.

Furchgott, R. F. and Zawadski, J. V. (1980). The obligatory role of endothelial cells in the relaxation of arterial smooth muscle by acetylcholine. *Nature (London)* **288,** 373–6.

Furchgott, R. F., Zawadski, J. V., and Cherry, P. D. (1981). Role of endothelium in vasodilator response to acetylcholine. In *Vasodilatation* (ed. P. M. Vanhoutte and I. Leusen) pp. 49–66. Raven Press, New York.

Garland, C. J. (1985). Endothelial cells and the electrical and mechanical responses of the rabbit coronary artery to 5-hydroxytryptamine. *Journal of Pharmacology and Experimental Therapeutics* **233,** 158–62.

Gillis, C. N. (1985). Uptake and storage of serotonin by platelets. In *Serotonin and the cardiovascular system* (ed. P. M. Vanhoutte) pp. 27–36. Raven Press, New York.

Gillis, C. N., Pitt, B. R., Wiedemann, H. P., and Hammond, G. L. (1986). Depressed prostaglandin E_1 and 5-hydroxytryptamine removal in patients with adult respiratory distress syndrome. *American Review of Respiratory Disease* **134,** 739–44.

Goldman, S. J., Gordon, E. L., and Slakey, L. L. (1986). Hydrolysis of diadenosine 5′, 5′–P′, P′-triphosphate (Ap_3A) by porcine aortic endothelial cells. *Circulation Research* **59,** 362–6.

Gutterman, D. D., Rusch, N. J., Hermsmeyer, K., and Dole, W. P. (1986). Differential reactivity to 5-hydroxytryptamine in canine coronary arteries. *Blood Vessels* **23,** 165–72.

Heistad, D. D., Armstrong, M. L., Marcus, M. L., Piegors, D. J., and Mark, A. L. (1984). Augmented responses to vasoconstrictor stimuli in hypercholesterolemic and atherosclerotic monkeys. *Circulation Research* **54,** 711–8.

Heistad, D. D., Mark, A. L., Marcus, M. L., Piegors, D. J., and Armstrong, M. L. (1987). Dietary treatment of atherosclerosis abolishes hyperresponsiveness to serotonin: implications for vasospasm. *Circulation Research* **61,** 346–51.

Henry, P. D. and Yokoyama, M. (1980). Supersensitivity of atherosclerotic rabbit aorta to ergonovine. Mediation by a serotonergic mechanism. *Journal of Clinical Investigation* **66,** 306–13.

Heric, E. and Tackett, R. L. (1985). Altered vascular reactivity in the rabbit during hypercholesterolemia. *Pharmacology* **31,** 72–81.

Holmsen, H. (1986). Platelet activation and serotonin. In *Serotonin and the cardiovascular system* (ed. P. M. Vanhoutte) pp. 75–86. Raven Press, New York.

Houston, D. S., Shepherd, J. T., and Vanhoutte, P. M. (1985). Adenine nucleotides, serotonin, and endothelium-dependent relaxations to platelets. *American Journal of Physiology* **248,** H389–95.

Houston, D. S., Shepherd, J. T., and Vanhoutte, P. M. (1986). Aggregating human platelets cause direct contraction and endothelium-dependent relaxation of isolated canine coronary arteries. Role of serotonin, thromboxane A_2, and adenine nucleotides. *Journal of Clinical Investigation* **72,** 539–44.

Imaizumi, Y., Baba, M., Imaizumi, Y., and Watanebe, M. (1984). Involvement of endothelium in the relaxation of isolated chick jugular vein by 5-hydroxytryptamine. *European Journal of Pharmacology* **97,** 335–6.

Kalkman, H. O., Boddeke, H. W. G. M., Doods, H. N., Timmermans, P. B. M. W. M., Van Zwieten, P. A. (1983). Hypotensive activity of serotonin receptor agonists in rats is related to their affinity for 5-HT_1 receptors. *European Journal of Pharmacology* **91,** 155–6.

Katusic, Z. S., Shepherd, J. T., and Vanhoutte, P. M. (1984). Vasopressin causes endothelium-dependent relaxations of the canine basilar artery. *Circulation Research* **55,** 575–9.

Kjellstrom, T., Ahlman, H., Dahlstrom, A., Hansson, G. K., and Risberg, B. (1984). The uptake of 5-hydroxytryptamine in endothelial cells cultured from the pulmonary artery in rats. A cytochemical study. *Acta Physiologica Scandinavica* **120,** 243–50.

Kjellstrom, T., Ortenwall, P., and Risberg, B. (1985). Effects of 5-hydroxytryptamine and related agents on endothelial cell oxidative metabolism in vitro. *Acta Physiologica Scandinavica* **125,** 501–4.

Lamping, K. G., Marcus, M. L., and Dole, W. P. (1985). Removal of the endothelium potentiates canine large coronary artery constrictor responses to 5-hydroxytryptamine in vivo. *Circulation Research* **57,** 46–54.

Lee, S.-L. and Fanburg, B. L. (1986). Serotonin uptake by bovine pulmonary artery endothelial cells in culture. II. Stimulation by hypoxia. *American Journal of Physiology* **250** (Cellular Physiology 19), C766–70.

Lingjaerde, O. (1977). Platelet uptake and storage of serotonin. In *Serotonin in health and disease* (ed. W. B. Essman) pp. 139–99. Spectrum, New York.

Lüscher, T. F. and Vanhoutte, P. M. (1986). Endothelium-dependent responses to platelets and serotonin in spontaneously hypertensive rats. *Hypertension* **88,** 55–60.

Lüscher, T. F., Raij, L., and Vanhoutte, P. M. (1987). Endothelium-dependent vascular responses in normotensive and hypertensive Dahl rats. *Hypertension* **9,** 157–63.

Marcus, A. J., Weksler, B. B., Jaffe, E. A., and Broekman, M. J. (1980). Synthesis of prostacyclin from platelet-derived endoperoxide by cultured endothelial cells. *Journal of Clinical Investigation* **66,** 979–86.

Maruki, C., Spatz, M., Ueki, Y., Nagatsu, I., and Bembry, J. (1984). Cerebrovascular endothelial cell culture: metabolism and synthesis of 5-hydroxytryptamine. *Journal of Neurochemistry* **43,** 316–9.

Miller, V. M. and Vanhoutte, P. M. (1985). Endothelium-dependent contractions to arachidonic acid are mediated by products of cyclooxygenase. *American Journal of Physiology* **248** (Heart and Circulatory Physiology 17), H432–7.

Moncada, S. (1982). Biological importance of prostacyclin. *British Journal of Pharmacology* **76,** 3–31.

Nossel, H. L. and Vogel, H. J. (1982). *Pathobiology of the endothelial cell.* Academic Press, New York.

Olesen, S.-P. (1985). A calcium-dependent reversible permeability increase in microvessels in frog brain, induced by serotonin. *Journal of Physiology* **361,** 103–13.

Page, I. H. and McCubbin, J. W. (1953). Modification of vascular responses to serotonin. *American Journal of Physiology* **174,** 436–40.

Palmer, R. M. J., Ferrige, A. G., and Moncada, S. (1987). Nitric oxide release accounts for the biological activity of endothelium-derived relaxing factor. *Nature (London)* **237,** 524–6.

Peroutka, S. J. (1984). 5-HT_1 receptor sites and functional correlates. *Neuropharmacology* **23,** 1487–92.

Poggi, A., Niewiarowski, S., Steward, G. J., Sobel, E., and Smith, J. B. (1983). Human platelet secreted proteins and prostacyclin production by bovine aortic endothelial cells. *Proceedings of Social and Experimental Biological Medicine* **172,** 543–50.

Radomski, M. W., Palmer, R. M. J., and Moncada, S. (1987). Comparative pharmacology of endothelium-derived relaxing factor, nitric oxide and prostacyclin in platelets. *British Journal of Pharmacology* **92,** 181–7.

Rapaport, R. M. and Murad, F. (1983). Agonist-induced endothelium dependent relaxation in rat thoracic aorta may be mediated through cGMP. *Circulation Research* **52,** 352–7.

Rubanyi, G. M., Lorenz, R. R., and Vanhoutte, P. M. (1985). Bioassay of endothelium-derived relaxing factor(s): inactivation by catecholamines. *American Journal of Physiology* **249,** H95–101.

Shepro, D. and Dunham, B. (1986). Endothelial cell metabolism of biogenic amines. *Annual Review of Physiology* **48,** 335–45.

Shepro, D. and Hechtman, H. B. (1985). Endothelial serotonin uptake and mediation of prostanoid secretion and stress fiber formation. *Federation Proceedings* **44,** 2616–7.

Shepro, D., Li, S., Hechtman, H. B. (1980). Plasminogen activator activity of isolated cardiac muscle microvessel endothelial cells. *Thrombosis Research* **18,** 609–16.

Shepro, D., Welles, S. L., and Hechtman, H. B. (1984). Vasoactive agonists prevent erythrocyte extravasation in thrombocytopenic hamsters. *Thrombosis Research* **35,** 421–30.

Shimokawa, H., Aarhus, L. L., and Vanhoutte, P. M. (1987). Porcine coronary arteries with regenerated endothelium have a reduced endothelium-dependent responsiveness to aggregating platelets and serotonin. *Circulation Research* **61,** 256–70.

Singer, H. A. and Peach, M. J. (1982). Calcium- and endothelial-mediated vascular smooth muscle relaxation in rabbit aorta. *Hypertension* **4,** 19–25.

Spatz, M., Mrsulja, B. B., Wroblewska, B., Merkel, N., and Bembry, J. (1986). Modulation of glycogen metabolism in cerebromicrovascular smooth muscle and endothelial cultures. *Biochemical and Biophysical Research Communications* **134,** 484–91.

Van de Voorde, J. and Leusen, I. (1982). Vascular endothelium and the relaxation effect of histamine on aorta of the rat. *Archives internationales de Pharmacodynamie et Therapie* **256,** 329–30.

Vanhoutte, P. M. and Cohen, R. A. (1983). The elusory role of serotonin in vascular function and disease. *Biochemical Pharmacology* **32,** 3671–4.

Van Nueten, J. M., Janssen, P. A. J., VanBeek, J., Xhonneux, R., Verbeuren, T. J., and Vanhoutte, P. M. (1981). Vascular effects of ketanserin (R 41 468), a novel antagonist of 5-HT_2 serotonergic receptors. *Journal of Pharmacology and Experimental Therapeutics* **218,** 217–30.

Van Nueten, J. M., Leysen, J. E., De Clerck, F., and Vanhoutte, P. M. (1984). Serotonergic receptor subtypes and vascular reactivity. *Journal of Cardiovascular Pharmacology* **6,** S564–74.

Verbeuren, T. J., Jordaens, F. H., Zonnekeyn, L. L., VanHove, C. E., Coene, M.-C., and Herman, A. G. (1986). Effect of hypercholesterolemia on vascular reactivity in the rabbit. I. Endothelium-dependent and endothelium-independent contractions and relaxation in isolated arteries of control and hypercholesterolemic rabbits. *Circulation Research* **58,** 552–64.

Webb, R. C. and Vanhoutte, P. M. (1986). Responsiveness of the hypertensive blood vessel wall to serotonergic stimuli. In *Serotonin and the cardiovascular system* (ed. P. M. Vanhoutte) pp. 113–21. Raven Press, New York.

Welles, S. L., Shepro, D., and Hechtman, H. B. (1985). Vasoactive amines modulate actin cables (stress fibers) and surface area in cultured bovine endothelium. *Journal of Cellular Physiology* **123,** 337–42.

Wright, C. E. and Angus, J. A. (1986). Effects of hypertension and hypercholesterolemia on vasodilatation in the rabbit. *Hypertension* **8,** 361–71.

Yokoyama, M., Akita, H., Mizutani, T., Fukuzaki, H., and Watanabe, Y. (1983). Hyperreactivity of coronary arterial smooth muscle in response to ergonovine from rabbits with hereditary hyperlipidemia. *Circulation Research* **53,** 63–71.

9

5-hydroxytryptamine and non-vascular smooth muscle contraction and relaxation

MARLENE L. COHEN

9.1 Introduction

5-HT can be found in virtually all smooth muscles of the body and induces a prominent contractile response in smooth muscle preparations from the gastro-intestinal, pulmonary, and genito-urinary systems. Fig. 9.1 details the relative *in vitro* potency of 5-HT in contracting several smooth muscle preparations from the rat, comparing the effectiveness of 5-HT in non-vascular smooth muscle to its contractile efficacy in a vascular tissue, the rat jugular vein. In addition to contraction, 5-HT can also evoke a relaxant response in some smooth muscle preparations.

Because 5-HT affects multiple types of smooth muscle, and can produce both contraction and relaxation in certain preparations, efforts are currently directed towards understanding the receptor interactions and the second messenger signal transducing system for these responses. Evaluation of receptors responsible for 5-HT receptor mediated responses in smooth muscle preparations is hampered by the difficulty of using radiolabelled ligand binding procedures in such tissues. Although radiolabelled ligand binding to multiple binding sites is a well-established technique in brain tissue, homogenization of smooth muscle preparations generally results in a dramatic loss in specific binding to various ligands and high non-specific ligand binding. Thus, the use of radiolabelled ligand binding procedures for identifying receptors in smooth muscle has not kept pace with the plethora of data emerging with these techniques using brain tissue.

Autoradiographic procedures adapted to smooth muscle may permit more rapid progress in receptor localization and identification based on the use of radiolabelled ligands. Such procedures are only beginning to be utilized with smooth muscle. So far, the bulk of our understanding of 5-HT receptors in non-vascular smooth muscle has been derived from measurement of contractile and relaxant responses to 5-HT and the block of such responses by antagonists in isolated smooth muscle preparations.

This chapter will review and identify the receptors thought to be involved in contractile and relaxant responses to 5-HT in certain non-vascular

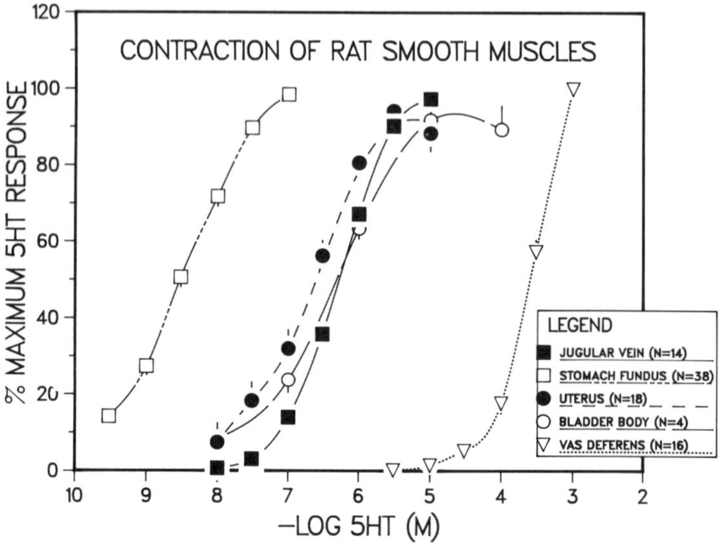

Fig. 9.1 Contractile concentration response curves to 5-HT in several non-vascular smooth muscle preparations as compared with responses in the rat jugular vein. Points are mean values and vertical bars represent SEM for the number of tissues indicated in parentheses

smooth muscles. Receptors in smooth muscle will be considered in the light of the information available on known binding sites for 5-HT in the brain. More than one type of reeptor may be responsible for a contractile or relaxant response to 5-HT. 5-HT exerts both direct effects on smooth muscle and indirect effects by modulating neurotransmission in non-vascular smooth muscle. Furthermore, 5-HT receptors mediating responses in smooth muscle may not be identical to binding sites identified in brain tissue. Finally, this chapter describes the available information with regard to second messenger systems and sources of calcium utilized by 5-HT in non-vascular smooth muscle.

9.2 5-HT receptors in gastro-intestinal smooth muscle

9.2.1 Oesophageal smooth muscle sphincter

5-HT receptors controlling oesophageal smooth muscle function have been less extensively studied than 5-HT receptors in other areas of the gastro-intestinal tract. However, 5-HT can contract the lower oesophageal sphincter of the dog, and contraction is blocked by selective 5-HT_2 receptor antagonists such as ketanserin and LY53857, supporting the presence of 5-HT_2

receptors in the control of oesophageal sphincter function (Lombardi *et al.* 1987). Agonists at such receptor sites might be useful in the treatment of gastro-oesophageal reflux disease (Lombardi *et al.* 1987). In the opossum oesophageal sphincter, 5-HT may also stimulate cholinergic excitatory neurones and/or inhibit sphincter function by stimulation of non-adrenergic inhibitory neurones (Rattan and Goyal 1977). The receptor(s) mediating these latter effects of 5-HT have not been characterized.

9.2.2 Stomach fundus

Of all smooth muscle preparations from the rat, 5-HT shows the highest potency to contract the rat stomach fundus (Fig. 9.1). Considerable effort has been directed towards characterizing the receptor responsible for 5-HT-induced contractions in the fundus. This receptor is not a 5-HT_2 receptor because selective 5-HT_2 receptor antagonists are not potent competitive antagonists of 5-HT-induced fundal contractions (Van Nueten *et al.* 1982; Cohen *et al.* 1985; Clineschmidt *et al.* 1985). Furthermore, no correlation occurred between binding affinity at 5-HT_2 receptors in the brain and ED_{50} values in the rat fundus for a series of 5-HT agonists (Fig. 9.2). The possibility that the receptor is similar to a tryptamine binding site

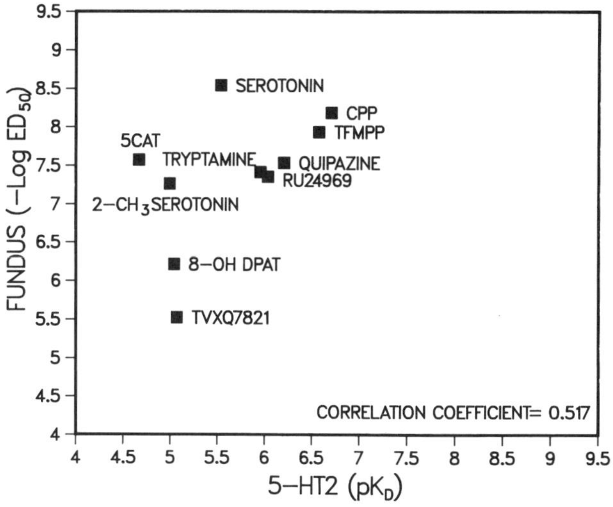

Fig. 9.2 Lack of correlation for ten 5-HT agonists between activity in the rat stomach fundus (-log ED_{50}) and binding affinity (pK_D) at 5-HT_2 receptors in brain membranes. The binding data are taken from Table 4.1 (page 74) with permission. Abbreviations: 5CAT = 5-carboxamidotryptamine; CPP = 1-(m-chlorophenyl) piperazine; serotonin = 5-HT; 2-CH_3-serotonin = 2-methyl-5-HT; TVXQ7821 = ipsapirone

has also been examined and ruled out (Cohen and Wittenauer 1985). The selective 5-HT_3 receptor antagonist, ICS 205–930, also did not antagonize 5-HT-induced contractions in the rat fundus, indicating that the 5-HT receptor responsible for such contractions is not a 5-HT_3 receptor (Cohen and Fludzinski 1987).

Because the receptor is not 5-HT_2 or 5-HT_3, several investigators have pointed out certain similarities of 5-HT-mediated responses in the rat stomach fundus to the interaction of 5-HT with subtypes of the 5-HT_1 recognition sites (Clineschmidt *et al.* 1985; Cohen and Colbert 1986). For example, 5-HT shows high affinity at 5-HT_1 binding sites and high potency in contracting the rat fundus. Thus, the fundal receptor has been classified as a '5-HT_1-like' receptor (Bradley *et al.* 1986).

However, recent evidence suggests that the 5-HT receptor which induces contraction in the rat fundus is not identical to the 5-HT_{1A} or 5-HT_{1B} site. Studies using phenylpiperazine derivatives somewhat selective for the 5-HT_{1A} and 5-HT_{1B} binding sites demonstrated that the fundal 5-HT receptor was not identical to 5-HT_{1A} or 5-HT_{1B} binding sites (Cohen and Wittenauer 1986). These results were corroborated using a larger series of agonists with known affinity and selectivity for 5-HT_{1A} and 5-HT_{1B} binding subtypes (Cohen and Fludzinski 1987). Furthermore, the potent 5-HT_{1A} receptor antagonists, spiroxatrine (Nelson and Taylor 1986) and WB4101 (Norman *et al.* 1985), and the non-selective but highly potent 5-HT_{1A} and 5-HT_{1B} receptor antagonist, cyanopindolol (Schlicker *et al.* 1985; Engel *et al.* 1986; Hamon *et al.* 1986), did not potently antagonize 5-HT-induced contractions in the rat stomach fundus. In fact, a more detailed analysis using ten structurally diverse 5-HT agonists confirmed a lack of correlation between their activities (-log ED_{50}) in the fundus and their binding affinities to 5-HT_{1A} sites in the brain (Fig. 9.3). A similar analysis suggested a modest but significant correlation between agonist activity in the fundus and binding affinity to 5-HT_{1B} sites in the brain (Fig. 9.4). Although a weak correlation occurred, the estimates of agonist affinities at 5-HT_{1B} sites and in the fundus were markedly disparate. Thus, the marked disparity of the slope of the correlation from 1, the weak correlation of agonist activity to affinity at 5-HT_{1B} sites, coupled with the ineffectiveness of cyanopindolol as an antagonist of 5-HT-induced contractions in the fundus argue against contractile 5-HT_{1B} sites in the fundus. Thus, the 5-HT contractile receptor in the rat stomach fundus appears to be distinct from the 5-HT_{1A} and 5-HT_{1B} binding sites in brain tissue.

Is the fundal 5-HT receptor similar to 5-HT_{1C} or 5-HT_{1D} binding sites in brain tissue, proposals advanced by Buchheit *et al.* (1986) and Heuring and Peroutka (1987), respectively? Again, we have analysed a series of ten structurally diverse 5-HT agonists and found a good to excellent correlation between activities in the fundus and binding affinities to 5-HT_{1C} (Fig. 9.5) and 5-HT_{1D} (Fig. 9.6) sites. These data are consistent with the pro-

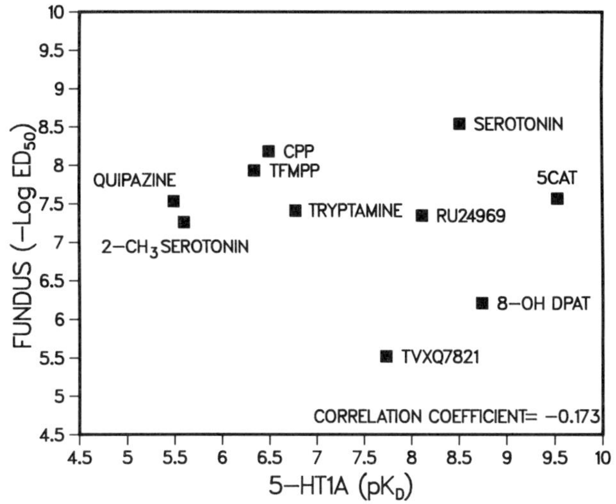

Fig. 9.3 Lack of correlation for ten 5-HT agonists between activity in the rat stomach fundus (-log ED_{50}) and binding affinity (pK_D) at $5\text{-}HT_{1A}$ receptors in brain membranes. The binding data are taken from Table 4.1 (page 74) with permission. Abbreviations: 5CAT = 5-carboxamidotryptamine; CPP = 1-(m-chlorophenyl) piperazine; serotonin = 5-HT; 2-CH_3-serotonin = 2-methyl-5-HT; TVXQ7821 = ipsapirone

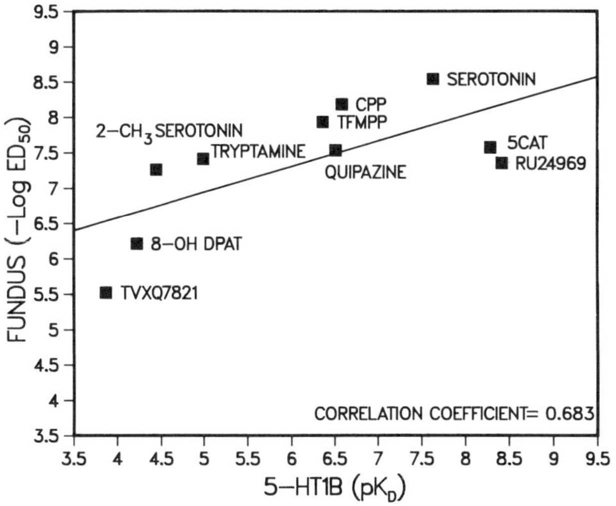

Fig. 9.4 Weak but statistically significant correlation (corr. coefficient = 0.683) for ten 5-HT agonists between activity in the rat stomach fundus (-log ED_{50}) and binding affinity (pK_D) at $5\text{-}HT_{1B}$ receptors in brain membranes. The binding data are taken from Table 4.1 (page 74) with permission. Abbreviations: 5CAT = 5-carboxamidotryptamine; CPP = 1-(m-chlorophenyl)piperazine; serotonin = 5-HT; 2-CH_3-serotonin = 2-methyl-5-HT; TVXQ7821 = ipsapirone

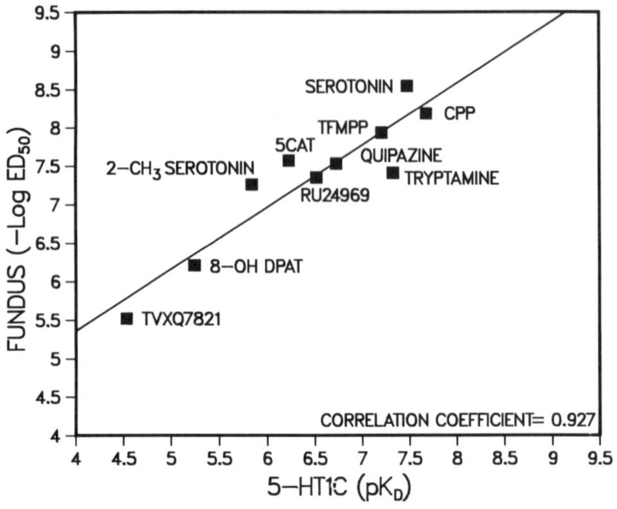

Fig. 9.5 Statistically significant correlation (corr. coefficient = 0.927) for ten 5-HT agonists between activity in the rat stomach fundus (-log ED_{50}) and binding affinity (pK_D) at 5-HT_{1C} receptors in brain membranes. The binding data are taken from Table 4.1 (page 74) with permission. Abbreviations: 5CAT = 5-carboxamidotryptamine; CPP = 1-(m-chlorophenyl)piperazine; serotonin = 5-HT; 2-CH_3-serotonin = 2-methyl-5-HT; TVXQ7821 = ipsapirone

posed similarity of the contractile 5-HT receptor in the fundus to either the 5-HT_{1C} or 5-HT_{1D} sites in brain tissue.

However, mianserin, a relatively potent and selective 5-HT_{1C} receptor ligand (which also possesses some affinity for 5-HT_2 receptors) only non-competitively blocked 5-HT contractile responses in the fundus, and this occurred in concentrations higher than anticipated based on its affinity at 5-HT_{1C} binding sites (Frankhuyzen and Bonta 1974a and b; Cohen and Fludzinski 1987). These antagonist data suggest that the 5-HT contractile receptor in the fundus may not be identical to the 5-HT_{1C} binding site. Furthermore, activation of 5-HT_{1C} binding sites in choroid plexus has been associated with increases in phosphoinositide turnover (Conn *et al.* 1986). However, 5-HT did not increase phosphoinositol hydrolysis in concentrations up to 10^{-4}M in the rat stomach fundus (Cohen and Wittenauer 1987) under conditions in which the muscarine receptor agonist, carbachol, produced marked concentration-dependent increases in phosphoinositide turnover. Thus, taken in concert, the available data suggest that the 5-HT-induced contractile receptor in the rat stomach fundus is not identical to the 5-HT_{1A} or 5-HT_{1B} sites, and although the receptor shows some similarities to both the 5-HT_{1C} and 5-HT_{1D} sites, it may be a distinct and novel 5-HT receptor.

Similarities between contractile responses to 5-HT in the rat stomach

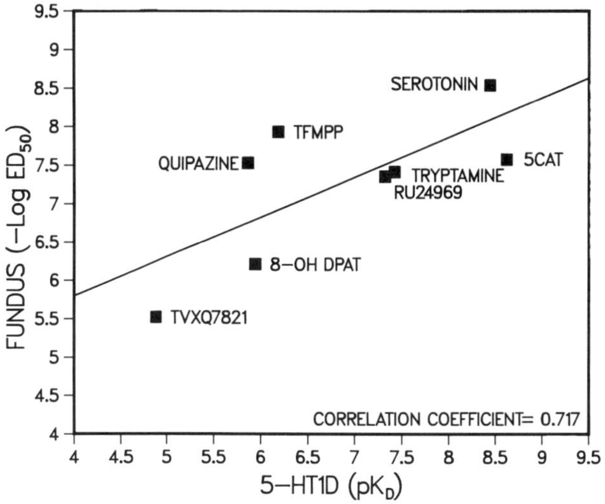

Fig. 9.6 Statistically significant correlation (corr. coefficient = 0.717) for eight 5-HT agonists between activity in the rat stomach fundus (-log ED_{50}) and binding affinity (pK_D) at $5\text{-}HT_{1D}$ receptors in brain membranes. The binding data are taken from Table 4.1 (page 74) with permission. Abbreviations: 5CAT = 5-carboxamidotryptamine; serotonin = 5-HT; $2\text{-}CH_3$-serotonin = 2-methyl-5-HT; TVXQ7821 = ipsapirone

fundus and in the canine basilar artery have been described (Cohen and Colbert 1986). Although the 5-HT receptor in the canine basilar artery resembled the $5\text{-}HT_{1A}$ site based on activity of several agonists, not all results are consistent with this interpretation (Taylor *et al.* 1986; Peroutka *et al.* 1986). Thus, the receptors of the canine basilar artery and rat stomach fundus, although possessing similarities to the $5\text{-}HT_{1A}$ binding site, may not be identical to this binding site but may be similar to each other (Cohen and Colbert 1986). Clearly, additional studies will be required to characterize the receptor mediating contraction to 5-HT in the fundus and to determine if this receptor is similar to known brain binding sites or to other 5-HT receptors. The development of selective agonists and/or antagonists will be immensely useful in this regard.

The transducing mechanism for 5-HT-mediated contraction in the rat stomach fundus has not been extensively studied. Recent evidence demonstrated that 5-HT did not increase phosphoinositide turnover in this tissue (Cohen and Wittenauer 1987) which renders it unlikely that this mechanism can be the signal transducing system. The 5-HT contractile response in the rat stomach fundus, however, is highly sensitive to calcium channel blockade, as indicated by the pronounced inhibitory effects of calcium channel antagonists (Mukai and Kubota 1980) as illustrated with nitrendipine and diltiazem (Fig. 9.7). Nevertheless, 5-HT is able to elicit a small but significant residual contractile response (~20–30 per cent of the maximum response) even in the presence of high concentrations of calcium channel antagonists (Fig. 9.7), suggesting that a small portion of the contractile response to 5-HT is maintained in the absence of extracellular calcium influx. Clearly, more research will be required to understand the signal transducing mechanisms used by 5-HT in the rat fundus.

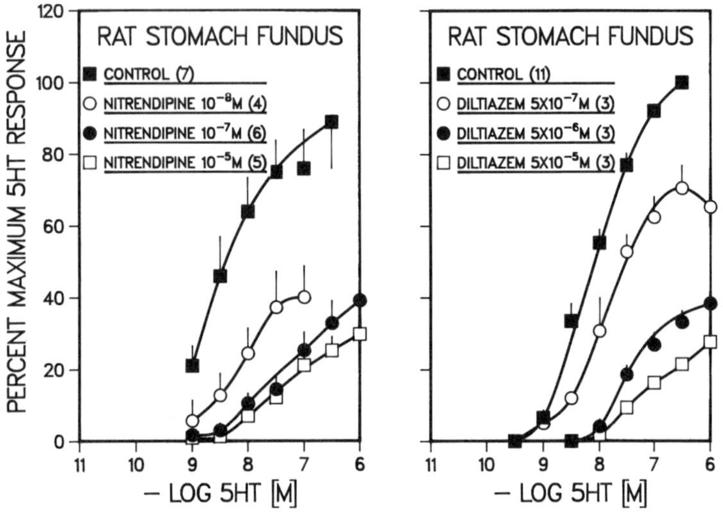

Fig. 9.71 Effect of the calcium channel antagonists, nitrendipine (left) and diltiazem (right), on the contractile response to 5-HT in the rat stomach fundus. Points are mean values and vertical bars represent SEM for the number of tissues indicated in parentheses

9.2.3 Ileal muscle

Two of the more prominent synthetic sources of 5-HT are enterochromaffin cells and enteric nerves in the intestinal tract (see Chapters 1 and 11). Thus, it is not surprising that 5-HT exerts prominent and multiple effects in intestinal smooth muscle, including actions on absorptive and secretory processes (see Chapter 12). With regard to the effects of 5-HT on motility, a minor component of the 5-HT-induced contractile response in the ileum may result from activation of 5-HT_2 receptors (Engel *et al.* 1984; Buchheit *et al.* 1985). However, the major component of the contractile response to 5-HT in the ileum results from an excitatory effect of 5-HT to enhance acetylcholine and possibly substance P release (Buchheit *et al.* 1985), an effect attributed to activation of 5-HT_3 receptors (Richardson *et al.* 1985). The neuronal excitatory responses of 5-HT in gastro-intestinal smooth muscle are discussed in detail elsewhere in this volume (see Chapters 10 and 11).

In addition to excitatory contractile responses produced by 5-HT, either through activation of smooth muscle 5-HT_2 receptors or indirectly via neurotransmitter release, 5-HT has also been reported to produce an inhibitory effect in ileal smooth muscle. Although low concentrations of 5-HT evoked a contractile response in guinea-pig ileum, higher concentrations of 5-HT decreased contractile responses to electrical field stimulation but not to exogenous acetylcholine (Sanger 1985). Such functional effects of 5-HT are consistent with the demonstration that 5-HT can inhibit electrically evoked outflow of radiolabelled material from longitudinal ileal strips pre-incubated with [^{3}H]-choline (Kilbinger and Pfeuffer-Friederich 1985). Both methysergide and methiothepin antagonized the inhibitory effect of 5-HT on [^{3}H]-acetylcholine release, whereas the 5-HT_2 receptor antagonist, ketanserin, did not. Thus, the receptor responsible for these inhibitory neuronal effects of 5-HT in the ileum is not a 5-HT_2 receptor. Inhibition of neurotransmitter release was also demonstrated with 8-OH-DPAT, an effect that was blocked by ligands with high affinity at 5-HT_{1A} sites in brain membranes (Fozard and Kilbinger 1985; Mir *et al.* 1988). Thus, 5-HT-induced inhibition of neurotransmitter release in the guinea-pig ileum may result from activation of 5-HT_{1A} receptors.

Recently, the selective 5-HT_3 receptor agonist, 2-methyl-5-HT, was shown to produce both a contractile and relaxant response in the ileum, and both effects were blocked by the selective 5-HT_3 receptor antagonist, GR38032, suggesting the presence of 5-HT_3 receptor-mediated inhibitory responses in the ileum (Gunning and Humphrey 1987).

9.2.4 Colon

As with other smooth muscles from the gastro-intestinal tract, 5-HT can exhibit both excitatory and inhibitory effects on colonic smooth muscle which may be mediated by effects on: (a) nerves; (b) circular smooth muscle; and (c) longitudinal smooth muscle. In the presence of atropine and guanethidine, 5-HT produced an initial relaxation followed by a contractile response in guinea-pig colon (Kojima and Shimo 1986). Contractile effects of 5-HT in the colon were due to activation of 5-HT_2 receptors as indicated by blockade of this response with ketanserin and spiperone and also to an indirect effect of 5-HT that could be blocked by tetrodotoxin (Kojima and Shimo 1986). The relaxant response to 5-HT in the colon was not blocked by 5-HT_2 receptor antagonists, and the receptor mediating this effect remains to be characterized. Multiple effects of 5-HT related to activation of several receptors have also been described in human colon (Burleigh 1977).

9.3 5-HT receptors in genito-urinary smooth muscle

9.3.1 Bladder

In the rabbit (Longhurst *et al.* 1984), cat (Saxena *et al.* 1985), pig (Klarskov and Horby-Petersen 1986), and human (Klarskov and Horby-Petersen 1986) detrusor muscle from the bladder, 5-HT produces an initial rapid, followed by a prolonged tonic, contractile response. In the cat bladder, the early rapid contraction produced by 5-HT is clearly mediated by activation of 5-HT_3 receptors since it is blocked by the selective 5-HT_3 receptor antagonist, MDL 72222, whereas the tonic response is sensitive to 5-HT_2 antagonists such as ketanserin, and therefore is mediated by 5-HT_2 receptors (Saxena *et al.* 1985). Although 5-HT_3 receptor antagonists have not been studied in human or pig bladder preparations, 5-HT-induced contractions in these tissues were partially antagonized by 5-HT_2 antagonists, suggesting that 5-HT_2 receptors mediate a portion of the contractile response to 5-HT in these species (Klarskov and Horby-Petersen 1986).

Mouse bladder also responds to 5-HT, although direct contractile responses are observed only at high concentrations. In concentrations of 1 μM and lower, 5-HT potentiates stimulation-evoked release of neurotransmitter without affecting the direct response to acetylcholine, suggesting excitatory neuronal receptor activation by 5-HT (Holt *et al.* 1986). This response to 5-HT in the mouse bladder is not blocked by MDL 72222 nor by ketanserin (Holt *et al.* 1986). No effect of the 5-HT_{1A} selective agonist, 8-OH-DPAT, was observed, although 5-carboxamidotryptamine was a potent agonist in the mouse bladder and pindolol was a relatively potent competitive antagonist, suggesting that this receptor may be similar to the 5-HT_{1B} subtype (Holt *et al.* 1986).

Relaxant effects of 5-HT may occur in certain urinary tract preparations such as the trigone, bladder neck, and urethra (Klarskov and Horby-Petersen 1986). These relaxant responses to 5-HT were not potently antagonized by 5-HT_2 receptor antagonists, such as ketanserin, suggesting that they are not mediated by activation of 5-HT_2 receptors. The precise characterization of these receptors has not been accomplished.

9.3.2 Uterus

5-HT also contracts the uterus (Fig. 9.1) and the response can be inhibited by 5-HT_2 receptor antagonists such as ketanserin and LY53857 (Millar *et al.* 1982; Ichida *et al.* 1983; Cohen *et al.* 1985). Wrigglesworth (1983) has documented the dissimilarity between the contractile 5-HT receptor in the fundus (see Section 9.2.2) and the uterine receptor. Contractile responses to 5-HT in the rat uterus are heavily dependent upon the utilization of extracellular sources of calcium, since 5-HT-induced contraction is potently antagonized by calcium channel antagonists (Cohen *et al.* 1986). As with other vascular 5-HT_2 receptor activated responses, 5-HT activates phosphoinositide turnover in the uterus, an effect that can also be antagonized by the 5-HT_2 receptor antagonist, LY53857 (Cohen and Wittenauer 1987). Preliminary data suggest that hormones such as oestradiol may regulate 5-HT_2 receptors in rat uterus (Hayashi *et al.* 1984). Future research must be directed towards elucidating the functional role of 5-HT_2 receptor activation in uterine smooth muscle.

9.3.3 Vas deferens

As with other smooth muscle preparations, 5-HT exerts several effects in the vas deferens that may be mediated by multiple receptor mechanisms. 5-HT can contract the rat vas deferens by a mechanism that may involve 5-HT receptors at low concentrations and α-adrenoceptors at high concentrations (Hay and Wadsworth 1982; Lucchelli *et al.* 1984; Cohen *et al.* 1985; Moritoki *et al.* 1986a).

In addition to contracting the rat vas deferens, 5-HT can affect transmitter release from nerves in this preparation. In this regard, 5-HT can inhibit stimulation-evoked responses in the rat (Karasawa *et al.* 1985; Moritoki *et al.* 1986b) and guinea-pig (Yoshida and Kuga 1986) vas deferens, effects which are potentiated by 5-HT_2 receptor blockade by ketanserin (Karasawa *et al.* 1985; Moritoki *et al.* 1986b). Potentiation of such presynaptic inhibitory effects of 5-HT with ketanserin and the observation that the contractions produced by 5-HT are less in tissues from reserpine-pretreated animals in which stores of noradrenaline are depleted (Lucchelli *et al.* 1984) suggests that 5-HT may also act indirectly in this preparation to enhance noradrenaline release.

5-HT-induced responses in the vas deferens are complicated further by apparent tachyphylaxis that may occur to certain of its effects (Lucchelli *et al.* 1984); the extent of this appears to depend on the portion of the vas deferens studied (Moritoki *et al.* 1986b) and the age of the animal (Moritoki *et al.* 1986a). For instance, the inhibitory effect of 5-HT on stimulation-evoked responses in the vas deferens appears to diminish with increasing age (Moritoki *et al.* 1986a). Clearly then, the effects of 5-HT in the vas deferens are complex and only beginning to be dissected with regard to receptor mechanisms.

9.4 5-HT receptors in respiratory smooth muscle

9.4.1 Trachea

The lung is known to play an important role in the removal of 5-HT and other biogenic amines via uptake from the circulation and subsequent inactivation by monoamine oxidase (Gillis and Roth 1976). In addition, small amounts of 5-HT may be synthesized from L-tryptophan by lung tissue (Gonmori *et al.* 1986).

Effects of 5-HT have been most extensively studied in the guinea-pig trachea where it produces a marked contractile response that is mediated by interaction with 5-HT_2 receptors (Van Nueten, *et al.* 1982; Sutherland and Mylecharane 1983; Cohen *et al.* 1985; Lemoine and Kaumann 1986). In Ringer-perfused guinea-pig lung, 5-HT increased both pulmonary vascular and airway constriction; both responses were due to 5-HT_2 receptor activation (Selig *et al.* 1988). 5-HT does not produce a prominent contractile response in rat trachea unlike the guinea-pig trachea (Cohen, unpublished observation).

Although the contractile response to 5-HT in guinea-pig trachea is mediated by activation of 5-HT_2 receptors, 5-HT did not increase phosphoinositide turnover in this tissue (Cohen and Wittenauer 1987). This observation suggests that inositol phospholipid hydrolysis is not the second messenger system for activation of 5-HT_2 receptors in tracheal smooth muscle. Recent data indicate that 5-HT-induced contractions in guinea-pig and cat trachea are allosterically regulated (Lemoine and Kaumann 1986) (see also Chapter 3), and that 5-HT-induced tracheal contraction may be altered with age (Bayol *et al.* 1985). As with other smooth muscle preparations, 5-HT may also exert presynaptic neuronal effects in tracheal smooth muscle (McCraig 1986), and may exert a relaxant effect in certain respiratory preparations (Bayol *et al.* 1985). These actions of 5-HT have not been extensively investigated.

Some information is available regarding the physiological and/or clinical significance of 5-HT receptors in respiratory smooth muscle. Although 5-HT is a potent bronchoconstrictor in guinea-pigs, it does not produce a pronounced bronchoconstriction in clinical asthmatics (Tonnesen 1985; Cushley *et al.* 1986). Furthermore, intravenous administration of ketanserin was ineffective in limiting exercise-induced asthma in young extrinsic asthmatics (So *et al.* 1985). Therefore, 5-HT does not appear to be an important mediator in human asthma, but the possibility remains that 5-HT may be important in acute respiratory failure (Huval *et al.* 1984; Olson 1987). Intravenous administration of ketanserin in patients with acute respiratory failure resulted in improvement in some parameters in patients with early respiratory failure. However, ketanserin also blocks α_1-adrenoceptors and in these patients produced a significant decline in mean arterial pressure which may have improved circulatory flow. Thus, the value of 5-HT_2 receptor antagonism in acute respiratory failure awaits the evaluation of more selective 5-HT_2 receptor antagonists in this pathological state.

9.5 5-HT receptors in other smooth muscles

9.5.1 Anococcygeus muscle

5-HT contracts the anococcygeus muscle from rat (Gillespie 1972; Doggrell and Woodruff 1978) and rabbit (Creed *et al.* 1977). Contractile responses to 5-HT were blocked by α-adrenoceptor antagonists (Doggrell and Woodruff 1978; Goyal *et al.* 1981) and by depletion of noradrenaline with 6-hydroxydopamine (Doggrell and Woodruff 1978), suggesting an indirect mechanism resulting from release of noradrenaline. In addition to an indirect effect, 5-HT-induced contractions of the anococcygeus muscle may possess a direct component that is blocked by methysergide, although not in a competitive fashion (Oriowo 1981; Tayo 1982). The receptor responsible for the direct contractile response to 5-HT has not yet been characterized.

9.5.2 Nictitating membrane

5-HT contracts the cat and dog nictitating membrane (Trendelenburg 1963; Wakade *et al.* 1970; Weinstock and Schechter 1975; Adler-Graschinsky *et al.* 1986) and has a modest effect to facilitate contraction to nerve stimulation (Adler-Graschinsky *et al.* 1986). The responses to 5-HT are both directly (not blocked by phentolamine) and indirectly mediated by noradrenaline release (Pluchino 1972). Receptors responsible for the direct contractile effect have not been characterized.

9.5.3 Spleen

5-HT concentrations in spleen are roughly 10- to 50-fold higher than those in cardiac tissue (Sturkie *et al.* 1972). The high 5-HT concentration in spleen may be related to removal of platelets from the circulation (Gershon and Ross 1966). Cat splenic capsule contracts to 5-HT with a response that is inhibited by reserpinization and by phentolamine (Pluchino 1972) and hence, due to an indirect mechanism, to release noradrenaline. 5-HT may also profoundly reduce splenic blood flow (Davies and Withrington 1973). The receptors mediating the vasoconstrictor effects and splenic smooth muscle contraction have not yet been characterized.

9.6 Concluding remarks

5-HT exerts multiple effects in non-vascular smooth muscle which result from both direct and indirect actions of 5-HT. The net effect of 5-HT in a given smooth muscle preparation will involve a summation of its excitatory and inhibitory actions in most tissues. In several non-vascular tissues bladder, trachea, uterus), receptors that mediate contractile response to 5-HT resemble 5-HT_2 receptors. 5-HT_3 receptors have been documented in guinea-pig ileum and cat bladder preparations and are invariably associated with neuronal elements. Although the receptors mediating responses to 5-HT in other smooth muscle preparations show similarities to 5-HT_1 sites identified in brain binding studies, precise characterization of those receptors awaits identification of selective agonists and antagonists for each of the remaining responses to 5-HT. As such tools become available (Fozard 1987), our knowledge of the receptors responsible for the actions of 5-HT in non-vascular smooth muscle will move rapidly forward.

Acknowledgement

The author is grateful to Ms. Gretchen Beckhelm for her expert editorial/secretarial assistance in the preparation of this chapter.

References

Adler-Graschinsky, E., Butta, N. V., and Elgoyhen, A. B. (1986). Serotonin uptake inhibitors and the prejunctional effects of serotonin on peripheral sympathetic nerves. *Life Sciences* **39,** 61–8.

Bayol, A., Benveniste, J., Brink, C., Cerrina, J., Gateau, O., Labat, C., and Raffestin, B. (1985). Response and sensitivity of guinea-pig airway muscle preparations to 5-hydroxytryptamine during ontogenesis. *British Journal of Pharmacology* **85,** 569–74.

Bradley, P. B., Engel, G., Feniuk, W., Fozard, J. R., Humphrey, P. P. A., Middlemiss, D. N., Mylecharane, E. J., Richardson, B. P., and Saxena, P. R. (1986). Proposals for the classification and nomenclature of functional receptors for 5-hydroxytryptamine. *Neuropharmacology* **25,** 563–76.

Buchheit, K.-H., Engel, G., Mutschler, E., and Richardson, B. (1985). Study of the contractile effect of 5-hydroxytryptamine (5-HT) in the isolated longitudinal muscle strip from guinea-pig ileum. *Naunyn-Schmiedeberg's Archives of Pharmacology* **329,** 36–41.

Buchheit, K. H., Engel, G., Hagenbach, A., Hoyer, D., Kalkman, H. O., and Seiler, M. P. (1986). The rat isolated stomach fundus strip, a model for 5-HT_{1C} receptors. *British Journal of Pharmacology* **88,** 367P.

Burleigh, D. W. (1977). Evidence for more than one type of 5-hydroxytryptamine receptor in the human colon. *Journal of Pharmacy and Pharmacology* **29,** 538–41.

Clineschmidt, B. V., Reiss, D. R., Pettibone, D. J., and Robinson, J. L. (1985). Characterization of 5-hydroxytryptamine receptors in rat stomach fundus. *Journal of Pharmacology and Experimental Therapeutics* **235,** 696–708.

Cohen, M. L. and Colbert, W. E. (1986). Relationship between receptors mediating serotonin (5-HT) contractions in the canine basilar artery to 5-HT_1, 5-HT_2 and rat stomach fundus 5-HT receptors. *Journal of Pharmacology and Experimental Therapeutics* **243,** 264–9.

Cohen, M. L. and Fludzinski, L. A. (1987). Contractile serotonergic receptor in rat stomach fundus. *Journal of Pharmacology and Experimental Therapeutics* **243,** 264–9.

Cohen, M. L. and Wittenauer, L. A. (1985). Relationship between serotonin and tryptamine receptors in the rat stomach fundus. *Journal of Pharmacology and Experimental Therapeutics* **233,** 75–9.

Cohen, M. L. and Wittenauer, L. A. (1986). Further evidence that the serotonin receptor in the rat stomach fundus is not $5HT_{1A}$ or $5HT_{1B}$. *Life Sciences* **38,** 1–5.

Cohen, M. L. and Wittenauer, L. A. (1987). Serotonin receptor activation of phosphoinositide turnover in uterine, fundal, vascular, and tracheal smooth muscle. *Journal of Cardiovascular Pharmacology* **10,** 176–81.

Cohen, M. L., Schenck, K. W., Colbert, W., and Wittenauer, L. (1985). Role of 5-HT_2 receptors in serotonin-induced contractions of nonvascular smooth muscle. *Journal of Pharmacology and Experimental Therapeutics* **232,** 770–4.

Cohen, M. L., Carpenter, R., Schenck, K., Wittenauer, L., and Mason, N. (1986). Effect of nitrendipine, diltiazem, trifluoperazine and pimozide on $serotonin_2$ (5-HT_2) receptor activation in the rat uterus and jugular vein. *Journal of Pharmacology and Experimental Therapeutics* **238,** 860–7.

Conn, P. J., Sanders-Bush, E., Hoffman, B. J., and Hartig, P. R. (1986). A unique serotonin receptor in choroid plexus is linked to phosphatidylinositol turnover. *Proceedings of the National Academy of Science* **83,** 4086–8.

Creed, K. E., Gillespie, J. S., and McCaffery, H. (1977). The rabbit anococcygeus muscle and its response to field stimulation and to some drugs. *Journal of Physiology* **273,** 121–35.

Cushley, M. J., Wee, L. H., and Holgate, S. T. (1986). The effect of inhaled 5-hydroxytryptamine (5-HT, serotonin) on airway calibre in man. *British Journal of Clinical Pharmacology* **22,** 487–90.

Davies, B. N. and Withrington, P. G. (1973). The actions of drugs on the smooth muscle of the capsule and blood vessels of the spleen. *Pharmacological Reviews* **25,** 373–413.

Doggrell, S. A. and Woodruff, G. N. (1978). Effect of 6-hydroxydopamine incubation on noradrenaline accumulation and on contractile responses in the rat anococcygeus muscle. *Research Communications in Chemical Pathology and Pharmacology* **20,** 253–64.

Engel, G., Hoyer, D., Kalkman, H. O., and Wick, M. B. (1984). Identification of $5HT_2$-receptors on longitudinal muscle of the guinea pig ileum. *Journal of Receptor Research* **4,** 113–26.

Engel, G., Göthert, M., Hoyer, D., Schlicker, E., and Hillenbrand, K. (1986). Identity of inhibitory presynaptic 5-hydroxytryptamine (5-HT) autoreceptors in the rat brain cortex with 5-HT_{1B} binding sites. *Naunyn-Schmiedeberg's Archives of Pharmacology* **332,** 1–7.

Fozard, J. R. (1987). 5-HT: The enigma variations. *Trends in Pharmacological Sciences* **8,** 501–6.

Fozard, J. R. and Kilbinger, H. (1985). 8-OH-DPAT inhibits transmitter release from guinea-pig enteric cholinergic neurones by activating 5-HT_{1A} receptors. *British Journal of Pharmacology* **86,** 601P.

Frankhuyzen, A. L. and Bonta, I. L. (1974a). Effect of mianserin, a potent anti-serotonin agent, on the isolated rat stomach fundus preparation. *European Journal of Pharmcology* **25,** 40–50.

Frankhuyzen, A. L. and Bonta, I. L. (1974b). Receptors involved in the action of 5-HT and tryptamine on the isolated rat stomach fundus preparation. *European Journal of Pharmacology* **26,** 220–30.

Gershon, M. D. and Ross, L. L. (1966). Location of sites of 5-hydroxytryptamine storage and metabolism by radioautography. *Journal of Physiology* **186,** 477–92.

Gillespie, J. S. (1972). The rat anococcygeus muscle and its response to nerve stimulation and to some drugs. *British Journal of Pharmacology* **45,** 404–16.

Gillis, C. N. and Roth, J. A. (1976). Pulmonary disposition of circulating vaso-active hormones. *Biochemical Pharmacology* **25,** 2547–53.

Gonmori, K., Prasada, Rao, K. S., and Mehendale, H. M. (1986). Pulmonary synthesis of 5-hydroxytryptamine in isolated perfused rabbit and rat lung preparations. *Experimental Lung Research* **11,** 295–305.

Goyal, R. K., Patel, N. M., and Verma, S. C. (1981). Comparative studies of the effects of 5-hydroxytryptamine and noradrenaline on the rat anococcygeus muscle. *Agents and Actions* **11,** 667–72.

Gunning, S. J. and Humphrey, P. P. A. (1987). Evidence for 5-HT_3 receptor mediated release of an inhibitory transmitter in guinea pig isolated ileum. *British Journal of Pharmacology* **91,** 359P.

Hamon, M., Cossery, J.-M., Spampinato, U., and Gozlan, H. (1986). Are there selective ligands for $5HT_{1A}$ and $5HT_{1B}$ receptor binding sites in brain? *Trends in Pharmacological Sciences* **7,** 336–8.

Hay, D. W. P. and Wadsworth, R. M. (1982). The contractile effects of 5-hydroxytryptamine on the rat isolated vas deferens. *British Journal of Pharmacology* **77,** 605–13.

Hayashi, T., Iwana, A., Nagai, K., and Ichida, S. (1984). Increase of serotonin receptors (specific [^{3}H]ketanserin binding) in rat uterine membrane fractions induced by estradiol. *Japanese Journal of Pharmacology* **36,** P180.

Heuring, R. E. and Peroutka, S. J. (1987). Characterization of a novel ^{3}H 5-hydroxytryptamine binding site subtype in bovine brain membranes. *Journal of Neuroscience* **7,** 894–903.

Holt, S. E., Cooper, M., and Wyllie, J. H. (1986). On the nature of the receptor mediating the action of 5-hydroxytryptamine in potentiating responses of the mouse urinary bladder strip to electrical stimulation. *Naunyn-Schmiedeberg's Archives of Pharmacology* **334,** 333–40.

Huval, W. V., Lelcuk, S., Shepro, D., and Hechtman, H. B. (1984). Role of serotonin in patients with acute respiratory failure. *Annals of Surgery* **200,** 166–72.

Ichida, S., Hayashi, T., and Terao, M. (1983). Selective inhibition by ketanserin and spiroperidol of 5-HT-induced myometrial contraction. *European Journal of Pharmacology* **96,** 155–8.

Karasawa, A., Kubo, K., Shuto, K., and Nakamizo, N. (1985). Interaction of 5-hydroxytryptamine and ketanserin in rat vas deferens subjected to low frequency field stimulation. *Japanese Journal of Pharmacology* **37,** 285–91.

Kilbinger, H. and Pfeuffer-Friederich, I. (1985). Two types of receptors for 5-hydroxytryptamine on the cholinergic nerves of the guinea-pig myenteric plexus. *British Journal of Pharmacology* **85,** 529–39.

Klarskov, P. and Horby-Petersen, J. (1986). Influence of serotonin on lower urinary tract smooth muscle *in vitro*. *British Journal of Urology* **58,** 507–13.

Kojima, S. and Shimo, Y. (1986). The sites of action of 5-hydroxytryptamine in the longitudinal muscle of the guinea pig proximal colon. *Asia Pacific Journal of Pharmacology* **1,** 111–16.

Lemoine, H. and Kaumann, A. J. (1986). Allosteric properties of 5-HT_2 receptors in tracheal smooth muscle. *Naunyn-Schmiedeberg's Archives of Pharmacology* **333,** 91–7.

Lombardi, D. M., Barone, F. C., and Ormsbee, H. S. (1987). Excitatory serotonin receptors mediate contraction of isolated canine lower esophageal sphincter (LES). *Gastroenterology* **92,** 1509.

Longhurst, P. A., Belis, J. A., O'Donnell, J. P., Galie, J. R., and Westfall, D. P. (1984). A study of the atropine-resistant component of the neurogenic response of the rabbit urinary bladder. *European Journal of Pharmacology* **99,** 295–302.

Lucchelli, A., Santagostino-Barbone, M. G., Modesto, F., and Grana, E. (1984). Direct and indirect actions of 5-hydroxytryptamine on the rat isolated vas deferens. *Archives of International Pharmacodynamics* **269,** 236–51.

McCraig, D. J. (1986). Autonomic responses of the isolated, innervated trachea of the guinea-pig; interaction with autonomic drugs, histamine and 5-hydroxytryptamine. *British Journal of Pharmacology* **88,** 239–48.

Millar, J. A., Facoory, B. D., and Laverty, R. (1982). Mechanism of action of ketanserin. *Lancet* **2,** 1154.

Mir, A. K., Hibert, M., Tricklebank, M. D., Middlemiss, D. N., Kidd, E. J., and Fozard, J. R. (1988). MDL 72832: a potent and stereoselective ligand at central and peripheral 5-HT_{1A} receptors. *European Journal of Pharmacology* **149,** 107–20.

Moritoki, H., Iwamoto, T., Kanaya, J., Ishida, Y., and Fukuda, H. (1986a). Age-related change in serotonin-mediated prejunctional inhibition of rat vas deferens. *European Journal of Pharmacology* **132,** 39–46.

Moritoki, H., Fukuda, H., Kanaya, J., and Ishida, Y. (1986b). Ketanserin potentiates the prejunctional inhibitory effect of 5-hydroxytryptamine on rat vas deferens. *Journal of Pharmacy and Pharmacology* **38,** 737–41.

Mukai, S. and Kubota, K. (1980). Differences in contractile responses to acetylcholine and serotonin in rat stomach fundus strips. *European Journal of Pharmacology* **65,** 157–63.

Nelson, D. L. and Taylor, E. W. (1986). Spiroxatrine: a selective serotonin 1_A receptor antagonist. *European Journal of Pharmacology* **124,** 207–8.

Norman, A. B., Battaglia, G., and Creese, I. (1985). [^{3}H]WB4101 labels the 5-HT_{1A} serotonin receptor subtype in rat brain. Guaninine nucleotide and divalent cation sensitivity. *Molecular Pharmacology* **28,** 487–94.

Olson, N. C. (1987). Role of 5-hydroxytryptamine in endotoxin-induced respiratory failure of pigs. *American Review of Respiratory Disorders* **135,** 93–9.

Oriowo, M. A. (1981). Direct and indirect actions of 5-hydroxytryptamine in the rat anococcygeus muscle. *Archives of International Pharmacodynamics* **254,** 13–9.

Peroutka, S. J., Huang, S., and Allen G. S. (1986). Canine basilar artery contractions mediated by 5-hydroxytryptamine$_{1A}$ receptors. *Journal of Pharmacology and Experimental Therapeutics* **237,** 901–6.

Pluchino, S. (1972). Direct and indirect effects of 5-hydroxytryptamine and tyramine on cat smooth muscle. *Naunyn-Schmiedeberg's Archives of Pharmacology* **272,** 189–224.

Rattan, S. and Goyal, R. K. (1977). Effects of 5-hydroxytryptamine on the lower esophageal sphincter in vivo. *Journal of Clinical Investigation* **59,** 125–33.

Richardson, B. P., Engel, G., Donatsch, P., and Stadler, P. A. (1985). Identification of serotonin M-receptor subtypes and their specific blockade by a new class of drugs. *Nature (London)* **316,** 126–31.

Sanger, G. J. (1985). Three different ways in which 5-hydroxytryptamine can affect cholinergic activity in guinea-pig isolated ileum. *Journal of Pharmacy and Pharmacology* **37,** 584–6.

Saxena, P. R., Heiligers, J., Mylecharane, E. J., and Tio, R. (1985). Excitatory 5-hydroxytryptamine receptors in the cat urinary bladder are of the M- and 5-HT$_2$-type. *Journal of Autonomic Pharmacology* **5,** 101–7.

Schlicker, E., Göthert, M., and Hillenbrand, K. (1985). Cyanopindolol is a highly potent and selective antagonist at the presynaptic serotonin autoreceptor in the rat brain cortex. *Naunyn-Schmiedeberg's Archives of Pharmacology* **331,** 398–401.

Selig, W. M., Bloomquist, M. A., Cohen, M. L., and Fleisch, J. H. (1988). Serotonin (5HT)$_2$ receptor mediated pulmonary vascular and airway constriction in isolated guinea pig lung. *The Faseb Journal* **2,** A1560.

So, S. Y., Lam, W. K., and Kwan, S. (1985). Selective 5-HT$_2$ receptor blockade in exercise-induced asthma. *Clinical Allergy* **15,** 371–6.

Sturkie, P. D., Woods, J. J., and Meyer, D. (1972). Serotonin levels in blood, heart, and spleen of chickens, ducks, and pigeons. *Proceedings of the Society for Experimental Biology and Medicine* **139,** 364–6.

Sutherland, R. McD. and Mylecharane, E. J. (1983). Antagonist activity of ketanserin in smooth muscle preparations. *Clinical and Experimental Pharmacology and Physiology* **10,** 701.

Taylor, E. W., Duckles, S. P., and Nelson, D. L. (1986). Dissociation constants of serotonin agonists in the canine basilar artery correlate to K_i values at the 5-HT$_{1A}$ binding sites. *Journal of Pharmacology and Experimental Therapeutics* **236,** 118–25.

Tayo, F. M. (1982). Effect of atropine on the rat anococcygeus muscle. *Journal of Pharmacy and Pharmacology* **34,** 202–3.

Tonnesen, P. (1985). Bronchial challenge with serotonin in asthmatics. *Allergy* **40,** 136–40.

Trendelenburg, U. (1963). Supersensitivity and subsensitivity to sympathomimetic amines. *Pharmacological Reviews* **15,** 225–76.

Van Nueten, J. M., Leysen, J. E., Vanhoutte, P. M., and Janssen, P. A. J. (1982). Serotonergic responses in vascular and non-vascular tissues. *Archives internationales de Pharmacodynamie et de Therapie* **256,** 331–4.

Wakade, A. R., Kanwar, R. S., and Gulati, O. D. (1970). Supersensitivity of the nictitating membrane to 5-hydroxytryptamine and norepinephrine after various procedures. *Journal of Pharmacology and Experimental Therapeutics* **175,** 189–96.

Weinstock, M. and Schechter, Y. (1975). Antagonism by propranolol of the ganglion stimulant action of 5-hydroxytryptamine. *European Journal of Pharmacology* **32,** 293–301.

Wrigglesworth, S. J. (1983). Heterogeneity of 5-hydroxytryptamine receptors in the rat uterus and stomach strip. *British Journal of Pharmacology* **80,** 691–7.

Yoshida, S. and Kuga, T. (1986). Probably pre- and postsynaptic modifications by 5-hydroxytryptamine of contractile responses to electrical stimulation of isolated guinea-pig vas deferens. *Japanese Journal of Pharmacology* **41,** 315–23.

10
Interaction of 5-hydroxytryptamine with autonomic and sensory neurones

D. I. WALLIS

10.1 Introduction

This chapter will attempt to consider the range of functional effects produced by 5-HT in the autonomic nervous system, especially in its sympathetic division and on sensory neurones, and the kind of 5-HT receptor through which these effects are mediated.

The wide distribution of 5-HT receptors and 5-HT-mediated effects might suggest that the amine has an important physiological role in peripheral nervous system function. Although this is far from established at present, recent discoveries about the localization and release of 5-HT within the peripheral nervous system (see Chapters 1 and 11) suggest that a functional role for 5-HT in regulating discharge of autonomic neurones is now considerably more likely. The principal mechanism of action of 5-HT on postganglionic sympathetic neurones, and the receptors involved, seem very similar to those on some sensory neurones, especially vagal afferent C fibres, and this chapter also deals with the latter. For a consideration of 5-HT action in modulating reflex sensory discharge, the reader is referred to Chapter 13.

It is not my intention to provide an exhaustive survey of the literature but to review certain aspects of the biology of 5-HT in relation to peripheral neurones. The localization of 5-HT, its range of actions and their possible functional significance, and the association of a particular functional mechanism with a particular 5-HT receptor will be considered. Some recent results from my laboratory and from collaboration with Dr. N. J. Dun will be described. Much of the work discussed focuses on effects mediated via 5-HT M receptors, recently reclassified as 5-HT_3 receptors (Bradley *et al.* 1986). 5-HT_3 receptors are defined as receptors which are blocked by ICS 205-930 and/or MDL 72222 and at which 2-methyl 5-HT is an effective mimic of 5-HT. Reviews of earlier studies on the effects of 5-HT in the autonomic nervous system are to be found in Wallis (1979 and 1981), Fozard (1984a and 1986), Richardson and Engel (1986), and Wallis and Dun (1989).

Many recent studies of 5-HT interactions with neurones have involved electrophysiological methods, although modulation of neurotransmitter release may be measured by biochemical or bioassay techniques. In the older literature, there are many papers in which indirect measures of neuronal excitation or inhibition were used. The electrophysiological approach has the merit of measuring a direct effect of interaction of the amine with its receptor and this approach is the one emphasized here. Although the most direct measure would be of the current which flows as a consequence of the opening or closing of ion channels, the voltage change generated by current flow may often be a more convenient measure of agonist action. Electrophysiological techniques, however, impose their own limitations on obtaining quantitative pharmacological data, especially if long equilibration periods of tissue with antagonist are required, and this is discussed in the appropriate sections below.

10.2 Localization of 5-HT in or nearby peripheral neurones

Given that the neurones of the autonomic nervous system are well-endowed with 5-HT receptors, the sources of 5-HT able to activate these should be considered, albeit briefly. These sources include 5-HT carried in the blood and arising from platelets, 5-HT in paracrine cells, and 5-HT stored in nerve fibres.

Although circulating levels of free 5-HT in plasma are low (Franzen and Eysell 1969; Garattini and Valzelli 1965), platelet 5-HT may be released on damage of blood vessel walls and in other circumstances. For instance, the platelets of migraineurs lose some 30–40 per cent of their 5-HT content at the onset of a migraine attack (see Fozard 1985). 5-HT may be released from mast cells by agents such as substance P (Johnson and Erdos 1973) to cause effects locally. Other paracrine cells which contain stores of 5-HT are the small intensely fluorescent cells whose physiological function is at present unclear. Although some may act as paracrine cells, others may be more properly considered interneurones (Williams 1967). Using immunohistochemical techniques, 5-HT has been shown to be present in the small intensely fluorescent cells of rat superior cervical ganglia (Verhofsted *et al.* 1981), and 5-HT and its synthetic enzyme tryptophan hydroxylase have been demonstrated in homogenates of rat superior cervical ganglia (Liuzzi *et al.* 1977). The presence of 5-HT in rabbit superior cervical ganglia has been demonstrated using high-performance liquid chromatography (Dun *et al.* 1980). Intrinsic neurones containing 5-HT have now been found in parasympathetic ganglia (Neel and Parsons 1986) and 5-HT-containing satellite cells in trigeminal ganglia (Moskowitz *et al.* 1979). 5-HT-containing neurones have also been found in the nodose ganglion of the cat (Gaudin-Chazal *et al.* 1978) and in dorsal root ganglia (Di Carlo 1983).

The presence of 5-HT-containing nerve fibres in close association with autonomic neurones, especially sympathetic neurones, is now well documented. For instance, the lateral horn of the spinal cord where preganglionic sympathetic somata are located, contains a dense network of nerve fibres exhibiting 5-HT fluorescence (Dahlstrom and Fuxe 1964 and 1965). A substantial proportion of these fibres have their cell bodies in the caudal raphe nuclei (Dahlstrom and Fuxe 1965; Oliveras *et al.* 1977), but some are located in the central grey and midbrain reticular formation (Bowker *et al.* 1981). It is unclear whether these 5-HT-containing nerve fibres make direct synaptic contact with preganglionic sympathetic neurones, since the descending 5-HT neuronal pathways appear to inhibit firing of the latter (Neumayr *et al.* 1974; Gilbey *et al.* 1981) whereas iontophoretic application of 5-HT on to these cells generally causes excitation (De Groat and Ryall 1967; Coote *et al.* 1981). *In vitro*, the direct action of 5-HT has been shown to be a depolarization, often accompanied by cell discharge (Ma and Dun 1986).

Recently, 5-HT immunoreactivity has been identified in both small intensely fluorescent cells and nerve fibres within coeliac superior and inferior mesenteric sympathetic ganglia of the guinea-pig (Dun *et al.* 1984; Ma *et al.* 1985). The number, as well as the intensity, of the 5-HT immunoreactive fibres was increased by preincubating ganglia with L-tryptophan and a monoamine oxidase inhibitor. The origin of these 5-HT-containing neurones is unknown, but the cell bodies might be located in the intramural plexuses of the intestine.

The terminals of postganglionic sympathetic fibres innervating cranial blood vessels have been shown to be in close association with 5-HT-containing nerve fibres (Chan-Palay 1976; Reinhard *et al.* 1979; Edvinsson *et al.* 1983), although any functional relation between the nerve fibres has not been established. There is also the possibility that sympathetic nerve terminals themselves, or perhaps a subpopulation of them, may store and release 5-HT (Jaim-Etcheverry and Zieher 1982) (see also Chapter 1). This may be the case for sympathetic nerve endings in the pineal gland (Pellegrino de Iraldi *et al.* 1965; Owman 1964; Neff *et al.* 1969) and result from the relatively unspecific character of the noradrenaline uptake system in the nerve terminals. The uptake system can transport 5-HT and leads to co-localization of 5-HT with noradrenaline in storage vesicles (Jaim-Etcheverry and Zieher 1982). 5-HT-like immunoreactivity has also been demonstrated after incubation with a monoamine oxidase inhibitor in postganglionic sympathetic neurones arising from the gerbil superior cervical ganglion and innervating cranial blood vessels (Cowen *et al.* 1986). Whether the amine is co-localized with noradrenaline in these fibres is uncertain.

10.3 Distribution of 5-HT receptors in the autonomic nervous system

Figs. 10.1 and 10.2 show the sites on autonomic and sensory neurones where 5-HT is reported to exert actions mediated through specific receptors. Other

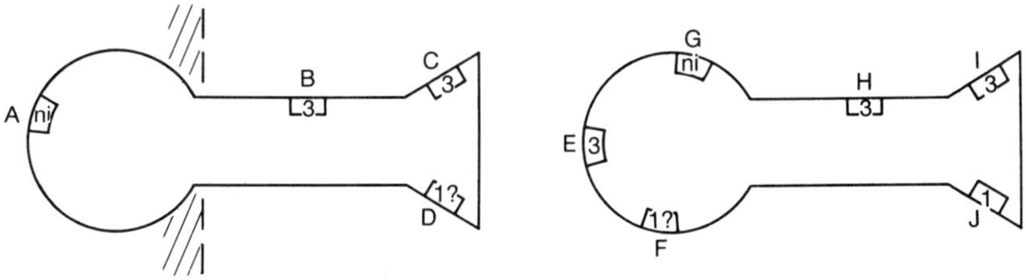

A Excitation/Depolarization	De Groat and Ryall (1967) Coote *et al.* (1981) Yoshimura and Nishi (1982) Ma and Dun (1986)	G Slow Depolarization	Kiraly *et al.* (1983) Dun *et al.* (1984) Wallis and Dun (1989)
B Depolarization	Wallis and Dun (1989)	F Hyperpolarization	Watanabe and Koketsu (1973) Haefely (1974) Wallis and North (1978) Dun and Karczmar (1981) Fortune *et al.* (1985) Ireland (1987)
C Depolarization	Wallis and Woodward (1975)		
D Reduction in ACh Release (? also mediated via C)	Hirai and Koketsu (1980) Dun and Karczmar (1981) Wallis and Dun (1989)	H Depolarization	Nash and Wallis (1981)
E Phasic Depolarization	Machova and Boska (1969) Wallis and Woodward (1973) Haefely (1974) Wallis and North (1978) Skok and Selyanko (1979)	I Noradrenaline release	Fozard and Mwaluko (1976) Fozard and Mobarok Ali (1978) Fozard (1979)
		I* Parasympathetic	Aas (1983) Fozard (1984b)
E* Parasympathetic	Saum and De Groat (1973 Akasu *et al* (1987)	J Reduction in Nor-adrenalin Release	McGrath (1977) Feniuk *et al.* (1979) Martinez and Lokhandwala (1980) Watts et al. (1981) (see Chapter 7)

Fig. 10.1 Distribution of 5-HT receptors and effects of 5-HT on autonomic neurones. The letters indicate various receptor mediated effects of 5-HT, as described by the authors indicated. 3:5-HT_3 receptor; 1?:'5-HT_1-like' receptor, ni:receptor not identified

effects of 5-HT, such as interaction with nicotine receptors (Akasu *et al.* 1981), will not be considered for reasons of space.

The wide distribution suggests that 5-HT receptors and the associated receptor-gated ion channels may be, for whatever reason, ubiquitous components of the membrane in these nerve cells. However, detailed consideration (see below) reveals that not all individual neurones have a full complement, or the same complement, of receptors and channels. Amongst sensory neurones, it is possible that cells with particular functions possess 5-HT receptors, but association of function with the possession by a neurone of a particular type of pharmacological receptor has in general been hindered by the technical difficulty of determining this.

5-HT actions at some of these sites are considered in more detail in later sections of this chapter or in other chapters of this book. This section sum-

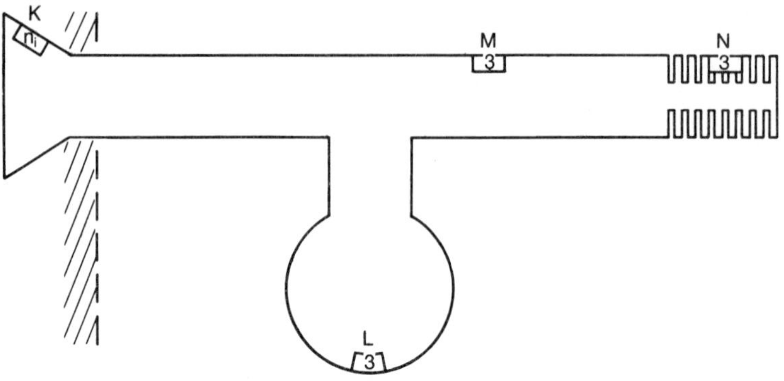

K Depolarization and/or Change in Electrical Threshold	Shirasawa and Koketsu (1977) Carstens *et al*. (1981)
L Depolarization	Higashi (1977 Higashi and Nishi (1982) Stansfeld and Wallis (1984, 1985)
M Depolarization	Neto (1978) Ireland and Tyers (1987)
N Excitation	Paintal (1954) Douglas and Ritchie (1957) Fjallbrant and Iggo (1961) Andrews (1968) Nishi (1975) (see Chapter 13)

Fig. 10.2 Distribution of 5-HT receptors and effects of 5-HT on sensory neurones. Letters and numbers as in Fig. 10.1

marizes, in so far as they are known, the key features of the receptors and the response mechanism.

10.3.1 Sympathetic neurones

Reference has already been made to the excitatory action of 5-HT on the cell bodies of preganglionic sympathetic neurones (Fig. 10.1, site A). 5-HT induces a depolarization of these cells (Ma and Dun 1986), which is accompanied by an increase in membrane input resistance; the likely mechanism was thought to be closure of 5-HT receptor-gated K^+ channels. Ma and Dun showed that methysergide (0.1–1 μM) and cyproheptadine (1–10 μM) blocked the responses reversibly. In studies *in vivo*, iontophoretic and/or systemic administration of methysergide, metergoline, or cinanserin selectively attenuated responses to 5-HT (Kadzielawa 1983; McCall 1983), which led to the suggestion that the receptors on these cells and on ventral horn motoneurones are similar and likely to be 5-HT_2 receptors (Fozard 1984a). However, the suggestion should be viewed with caution. Thus, recent studies on the isolated hemisected spinal cord of the neonate rat have revealed motoneuronal excitatory responses to 5-HT which are insensitive to ketanserin and mimicked by the '5-HT_1-like' receptor agonist, 5-carboxamidotryptamine (Connell and Wallis 1987).

5-HT depolarizes preganglionic axons of rabbit cervical sympathetic trunk via 5-HT_3 receptors (Fig. 10.1, site B) (Wallis and Dun 1989). Extracellular records from the proximal pole of the ganglion suggest that 5-HT may also depolarize the terminals of these neurones within the ganglion (Fig. 10.1, site C) (Wallis and Woodward 1975). Elliott and Wallis (unpublished) have shown that the 5-HT receptor mediating depolarization (Fig. 10.1, site B) is blocked by ICS 205-930 and MDL 72222; both this site and site C within the ganglion may involve 5-HT_3 receptors.

There are a number of electrophysiological studies where extracellular techniques have been used to identify modulating effects on ganglionic transmission. As discussed by Wallis and Dun (1989), these methods are, in the main, insufficiently incisive to distinguish a presynaptic from a postsynaptic site of modulation of transmission. Studies employing intracellular recording in conjunction with quantal analysis of the excitatory postsynaptic potentials are necessary to confirm a reduction or increase in transmitter release mediated via presynaptic receptors. By this means, a reduction in acetylcholine release brought about by 5-HT acting presynaptically has been demonstrated (Dun and Karczmar 1981; Elliott and Wallis unpublished). The receptor involved has not been characterized, although methysergide acted as an agonist and, thus, it is unlikely to be a 5-HT_3 receptor; it is shown as site D in Fig. 10.1. Preliminary observations by Elliott and Wallis (unpublished) indicate that 5-carboxamidotryptamine is a potent agonist at this receptor. Modulation of transmission might be mediated via both site C and D receptors. It is important to note that when the inputs to individual ganglion cells are investigated, both in the rabbit superior cervical ganglion (Elliott and Wallis unpublished) and in the guinea-pig coeliac ganglion (Wallis and Dun 1989), excitatory postsynaptic potentials generated by an input may be depressed, facilitated, or unaffected, depending upon the cell being studied. Thus, the presynaptic receptors are apparently distributed selectively, so that some, but not all inputs may be inhibited via 5-HT receptors.

Surface recordings in earlier studies of autonomic ganglia *in situ*, particularly in the cat, revealed a triphasic potential change following intra-arterial injection of 5-HT (Machova and Boska 1969; Haefely 1974). The potential change consisted of a brief depolarization, followed by a hyperpolarization which in turn was succeeded by a late depolarization. It is only recently that intracellular recording has clarified the nature of the potential changes in single sympathetic neurones in response to bath application or pressure ejection of 5-HT. Again, it is important to note that not every sympathetic neurone displays all three types of membrane potential change. Further, a proportion of superior cervical ganglion cells seem insensitive to 5-HT (Wallis and North 1978; Skok and Selyanko 1979). Nevertheless, three distinct 5-HT receptors seem likely to be involved (Fig. 10.1, sites E, F, and G). The nature of ganglion cell responses to 5-HT is dealt with more fully in Section 10.4.

A rapid depolarizing action of 5-HT on postganglionic axons (superior cervical ganglia Fig. 10.1, site H) has been reported for the rabbit (Nash and Wallis 1981). Since this response shows similarities to the phasic depolarization arising in ganglion somata (Fig. 10.1, site E) it may involve a similar 5-HT receptor, but this remains to be substantiated. The action of 5-HT in causing release of noradrenaline from cardiac sympathetic nerves may also involve membrane depolarization (Fozard and Mwaluko 1976) and the receptor involved (Fig. 10.1, site I) is a 5-HT_3 receptor (Fozard 1984a). The question as to whether this constitutes a sub-type of 5-HT_3 receptor is addressed in Section 10.4. This terminal excitatory receptor has not been identified on all sympathetic neurones, although it is present on sympathetic neurones other than those of the stellate ganglion (e.g. superior cervical ganglion, Adler-Graschinsky 1983). Other receptors (Fig. 10.1, site J) bring about a reduction in noradrenaline release and are widely distributed. This action of 5-HT and the receptor involved is discussed in Chapter 7.

10.3.2 Parasympathetic neurones

5-HT receptors on parasympathetic neurones have been less studied, except for sites within the enteric plexuses of the gut wall. The latter can be regarded as belonging to the enteric nervous system and are considered in detail in Chapter 11. An excitant effect of 5-HT has been demonstrated on the parasympathetic innervation of the bladder in dog, rat, and cat (Gyermek 1962; Vanov 1965; Saum and de Groat 1973). Recently 5-HT has been shown to release a non-cholinergic, non-adrenergic, probably purinergic neurotransmitter from postganglionic parasympathetic neurones in the urinary bladder (Burnstock *et al.* 1978) and an excitatory action on neurones (Fig. 10.1, site E*) has been attributed to activation of 5-HT_3 receptors (Saxena *et al.* 1985; Akasu *et al.* 1987). Release of acetylcholine by 5-HT may result from excitation of 5-HT_3 receptors on nerve terminals (Fig. 10.1, site I*); this has been reported for rat bronchi (Aas 1983) and rabbit heart (Fozard 1984b). However, 5-HT receptors possibly located on the terminals of parasympathetic neurones of the mouse bladder, which enhance responses to electrical stimulation, are not 5-HT_3 but appear to be '5-HT_1-like' (Holt *et al.* 1986).

10.3.3 Sensory neurones

Sensory neurones may possess 5-HT receptors over their total membrane surface if their axons are non-myelinated and if they are of particular functional types. There are reports of 5-HT inducing depolarization of primary afferent terminals (Fig. 10.1, site K) in the spinal cord (Shirasawa and Koketsu 1977) and of increasing electrical threshold (Carstens *et al.* 1981), but the receptor involved does not seem to have been characterized. The receptor on the soma of sensory neurones (Fig. 10.1, site L) has received more attention. It mediates a depolarization and the 5-HT-gated conductance mechanism has been analysed (Higashi and Nishi 1982). This response is considered in more detail in Sections 10.4 and 10.5. The axons appear to possess identical receptors (Fig. 10.1, site M), which mediate a depolarization with a similar ionic basis. Many sensory endings in the periphery (Fig. 10.1, site N) are potently excited by 5-HT and elicit prominent reflex effects on respiration and the cardiovascular system (Douglas and Toh 1953; Comroe *et al.* 1953; Ginzel and Kottegoda 1954) and generate pain in man (see Chapter 15). Endings associated with small myelinated and non-myelinated axons (Groups III and IV) have been shown to be sensitive to 5-HT in cutaneous (Beck and Handwerker 1974) and muscle nerves (Mense 1977). Non-myelinated visceral afferents are particularly sensitive to 5-HT (Higashi 1977; Stansfeld and Wallis 1984 and 1985).

In vivo, the Bezold–Jarisch effect of 5-HT provides a convenient measure of vagal afferent stimulation (Fozard 1984a and b). This reflex is highly sensitive to the 5-HT_3 receptor antagonist, MDL 72222, as are phasic responses of cat carotid body chemoreceptors (Kirby and McQueen 1984; see also Chapter 13). However, in many cases the pharmacological identity of the 5-HT receptor located at sensory endings has not been explored.

10.4 5-HT_3 receptor-gated mechanisms

There has been a number of studies in both autonomic and sensory ganglia in which the action of 5-HT has been investigated with intracellular electrodes, thus permitting analysis of the ionic basis of the responses.

10.4.1 Fast and slow responses

Information about 5-HT_3 receptor-gated ion channels has been obtained *in vitro* from studies of rabbit superior cervical ganglia, guinea-pig coeliac ganglion, and rabbit nodose ganglion cells. In rabbit superior cervical ganglion (Wallis and North 1978) and rabbit nodose ganglion (Higashi and

Nishi 1982), 5-HT applied by iontophoresis evokes a rapid depolarization associated with increased Na^+ and K^+ conductances (see Table 10.1). Guinea-pig coeliac ganglion neurones respond to 5-HT applied by superfusion with a slow depolarization which is associated with either an increase or little change in membrane resistance (Dun *et al.* 1984). This discrepancy between the behaviour of superior cervical ganglion and guinea-pig coeliac ganglion neurones was resolved when 5-HT was pressure ejected from a micropipette positioned close to the impaled cell (Wallis and Dun unpublished). Using this technique it was demonstrated that many guinea-pig coeliac ganglion cells displayed both fast and slow depolarizing responses; cells from guinea-pig superior cervical ganglion appeared to respond in a similar manner. In a sample of 120 guinea-pig coeliac ganglion cells, 43 per cent responded with a fast depolarization alone, 30 per cent responded with a fast and slow depolarization to give a biphasic response, and 25 per cent responded with a slow depolarization alone. In some cells a hyperpolarization was seen following the fast component of depolarization. In earlier studies, it had been noted that fast depolarizations of rabbit superior cervical ganglion cells were commonly followed by an after-hyperpolarization, probably generated by the electrogenic extrusion of Na^+ from the cell (Nash and Wallis 1981). The triphasic responses recorded from the surface of cat superior cervical ganglion *in vivo* probably correspond to these fast and slow depolarizations with a phase of hyperpolarization intervening.

The fast depolarization of guinea-pig coeliac ganglion cells appears to be generated by a conductance increase to Na^+ and K^+ ions, with a reversal potential of -22.8 ± 2.4 mV ($n = 14$) (Wallis and Dun unpublished). It seems analogous to the nicotinic acetylcholine depolarization evoked in these same cells with respect to its time course and ionic basis.

The pharmacological properties of the fast depolarization recorded from guinea-pig superior cervical ganglion and guinea-pig coeliac ganglion neurones are entirely consistent with the involvement of 5-HT_3 receptors. Characterization of 5-HT_3 receptors in terms of antagonist potency is more readily achieved using the sucrose-gap technique or some other form of extracellular recording, as described below. The effect of MDL 72222 on the fast depolarization evoked in superior cervical ganglion cells by pressure ejected 5-HT is shown in Fig. 10.3. In this cell the response was reduced by exposure to the antagonist at a concentration of 1 μM for 10 and 15 min and greatly reduced by superfusion of 5 μM (Wallis and Dun unpublished). The fast depolarization of guinea-pig coeliac ganglion cells showed marked tachyphylaxis, was completely blocked by superfusion of 5-HT (100 μM), and was reduced in amplitude by MDL 72222 (1 μM), quipazine (1 μM), metoclopramide (10 μM), and by d-tubocurarine (50 μM). The apparent order of potency of the antagonists was quipazine > MDL 72222 >

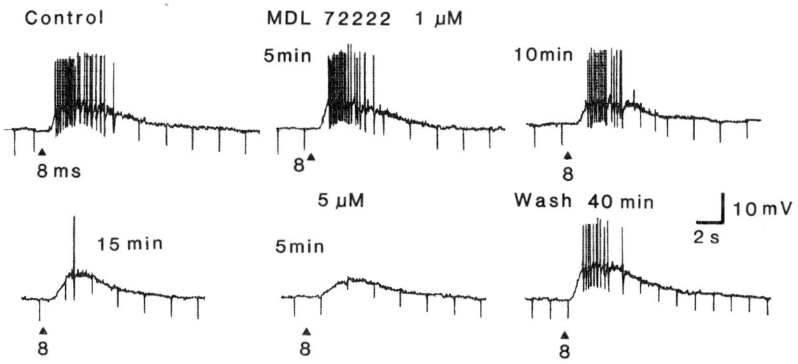

Fig. 10.3 Effects of MDL 72222 on phasic responses to 5-HT recorded from guinea-pig superior cervical ganglion neurones. Intracellular records showing depolarizing responses, with action potential discharge, to application of 5-HT by pressure ejection from a micropipette in the vicinity of the cell. Resting membrane potential was 56 mV. A solution of 10 mM 5-HT was ejected using 8 ms pulses. MDL 72222 was applied by superfusion for 15 min at a concentration of 1 μM and for 5 min at five times this concentration. Note recovery on washing

metoclopramide >> d-tubocurarine. Fast depolarizations were unaffected by methysergide or cyproheptadine.

The slow depolarization induced by 5-HT in guinea-pig coeliac ganglion neurones is in some respects similar to the muscarinic acetylcholine response of the cells, i.e. the response develops slowly, lasts for seconds or minutes, and is accompanied in many cells by a decrease in membrane conductance (Kiraly *et al.* 1983; Dun *et al.* 1984). Potent and selective antagonists for this response have not been identified, although methysergide (1 μM) was an effective blocking agent whereas d-tubocurarine was not (Wallis and Dun unpublished).

As can be seen from Table 10.1, several studies have shown that 5-HT, acting at two pharmacologically distinct receptors on autonomic neurones, generates two types of depolarizing response which possess different mechanisms of electrogenesis, e.g. in postganglionic sympathetic neurones, and enteric ganglion cells of submucous and myenteric plexuses.

10.4.2 Co-operativity of action, single channel studies

The most detailed electrophysiological investigations of 5-HT_3-gated ion channels have been carried out on rabbit primary afferent neurones and on murine neuroblastoma cells. The evidence for the latter possessing 5-HT_3 receptors is presented in Chapter 4. Higashi and Nishi (1982) held rabbit nodose C neurones under voltage clamp and obtained Hill plots of log 5-HT-induced current against log concentration of 5-HT, whose slope was

Table 10.1 Fast and slow depolarizations evoked by 5-HT and their associated ion channels

Tissue	Deplorization	Receptor	Reversal potential (mV)	Conductance change	Selective agonists	Selective antagonists	Ineffective agents	Species	Reference
Sympathetic ganglia	Fast	$5HT_3$	−24[1] −15[2] −23[3]	↑ Na,K	—	MDL 72222* Quipazine* ICS 205-930[4]	Methysergide Cypropheptadine	Rabbit scg Guinea-pig cmg[3] Guinea-pig scg	1. Wallis and North (1978) 2. Skok and Selyanko (1979) 3. Wallis and Dun unpublished 4. Round and Wallis (1986)
	Slow	?	—	↓ ?	—	Methysergide Cyproheptadine?	MDL 72222	Guinea-pig cmg	5. Wallis and Dun unpublished
Parasymp. ganglia	Fast	$5HT_3$	—	↑ Na, K	—	ICS 205-930	Methysergide	Rabbit Vesical pelvic g.	Akasu *et al.* (1987)
Sensory ganglia	Fast	$5HT_3$ Co-operativity	+7	↑ Na,K	2-methyl-5-HT	MDL 72222* ICS 205-930*	LSD Methysergide	Rabbit Nodose	Higashi and Nishi (1982) Ireland and Tyers (1987) Round and Wallis (1987)
Submucous plexus	Fast	$5HT_3$	0 to −30	↑	2-methyl-5-HT	ICS 205-930	Ketanserin Spiperone Cypropheptadine Methysergide	Guinea-pig	Surprenant and Crist (1986)
	Slow	?	—	↓ K	—	—			
Myenteric plexus	Fast	$5HT_3$	—	↑	2 Me 5HT	ICS 205-930	—		Mawe *et al.* (1986)
	Slow	?	—	↓ K?	5 or 6-hydroxy-indalpines 2-methyl-5-HT	5-HTP-DP	ICS 205-930	Guinea-pig	
Neuroblastoma cell NIE-115	Fast	$5HT_3$?	0 to −9	↑ Na,K	—	Quipazine	—	Mouse	Guharay *et al.* (1985) Peters and Usherwood (1983)

cmg: Coeliac mesenteric ganglion.
scg: Superior cervical ganglion.
* Based on sucrose-gap experiments by author.

around 2. They concluded that at least two molecules of 5-HT are required to interact at the receptor in order to produce a unit conductance change, i.e. there is co-operativity by each 5-HT molecule at the recognition site. In murine neuroblastoma cells, single channels currents have been recorded from intact cells and from isolated membrane patches (Guharay *et al.* 1985). The channels displayed a large conductance (about 140 pS) and were not sensitive to membrane potential. The channels recorded from intact cells revealed more than one kinetically distinguishable open state.

10.4.3 Hyperpolarizing responses

Hyperpolarizing responses evoked by 5-HT in ganglion cells are of complex origin. As mentioned above, an after-hyperpolarization may result from the electrogenic extrusion of Na^+ which has entered the cell during the depolarization. Alternatively, Ca^{2+} entering during the depolarizing phase of the response to the amine may induce a Ca^{2+}-activated K^+ current, at least in primary afferent neurones (Higashi and Nishi 1982). In a proportion of autonomic neurones, however, 5-HT may hyperpolarize the membrane via an additional mechanism which is independent of prior depolarization. In intracellular studies of rabbit superior cervical ganglion cells, a few cells showed a hyperpolarizing response alone (Wallis and North 1978; Dun and Karczmar 1981). This phenomenon has been attributed in bullfrog sympathetic neurones to activation of an electrogenic pump by 5-HT (Watanabe and Koketsu 1973). Recently, Ireland (1987) demonstrated that in rat superior cervical ganglion cells the hyperpolarization is not mediated via a 5-HT_3 receptor but by a '5-HT_1-like' receptor, and may be due to a K^+ conductance increase.

10.4.4 Quantitative pharmacology

When recording intracellularly, the difficulty in holding cells for the requisite period of time constrains study of antagonists and makes especially difficult the study of antagonist potency at equilibrium. Further, cell sampling is biased by the need to study neurones at or near the ganglion surface. In the case of these cells, a micropipette containing 5-HT can be manoeuvred under visual control into the vicinity of the cell and the amine ejected by iontophoresis or by pressure. This avoids desensitization of other cells in the ganglion. A consequence is biasing of the cell sample. Superfusion of a known concentration of 5-HT also yields more consistent results when surface cells are investigated, presumably because the problem of diffusion through the tissue is minimized. The technique of superfusion, however, produces a different bias, because responses which manifest rapid desensitization may fail to be detected (see above).

In order to obtain quantitative pharmacological data, we have employed an extracellular recording technique on both sympathetic neurones of the rabbit superior cervical ganglion and visceral primary afferents of the rabbit nodose ganglion. This is the sucrose-gap (Wallis *et al.* 1975 and 1982) (see also Fig. 10.4) in which potential changes are recorded from a population of neurones; a ganglion is suspended across a series of three Perspex chambers through which various superfusion media are passed. The method depends on an aqueous solution of sucrose reducing the extracellular shunt of potential between two regions on the surface of the tissue. It has been shown experimentally that the potential recorded arises principally from neurones lying at the interface between the test chamber and the chamber containing sucrose solution (Wallis *et al.* 1975). 5-HT is introduced to the tissue via the medium entering the test chamber. Quantitative measures of antagonist potency are possible using this technique and the selectivity of antagonists can be examined by using control agonists, e.g. γ-amino butyric acid, dimethylphenylpiperazinium and noradrenaline.

In two recent studies (Round and Wallis 1986 and 1987), we examined the effects of MDL 72222 and ICS 205-930 on 5-HT responses of rabbit superior cervical ganglion and nodose ganglion using the sucrose-gap technique. Typical responses obtained from the two ganglia to bolus injections of 40–320 nmol 5-HT, dissolved in a small volume of Krebs solution, are shown in Fig. 10.4. The use of bolus injections minimizes tachyphylaxis and produces dose-related depolarizations. 80 nmol 5-HT induces maximal responses in nodose ganglion neurones and 320 nmol 5-HT maximal responses of superior cervical ganglion neurones. In both tissues, the response is a brisk depolarization, followed by a repolarization and an after-hyperpolarization. The depolarizations of nodose ganglion neurones have a longer time course than those of superior cervical ganglion neurones.

With appropriate intervals between injections, it is possible to elicit reproducible dose-related depolarizations and construct repeated full or partial dose-response curves. In nodose ganglia, repeated control dose-response curves show a small rightward parallel shift with one hour wash periods interposed between curves. In superior cervical ganglia repeated dose-response curves are superimposable. In assessing the effect of an antagonist, any rightward shift under control conditions can be taken into account.

MDL 72222 proved a potent and selective antagonist of 5-HT responses in vagal primary afferents and in sympathetic neurones (Fig. 10.4) (Azami *et al.* 1985; Round and Wallis 1987). The responses to control agonists were unaffected by MDL 72222. A parallel shift in dose-response curves was found to occur with concentrations of 0.1 μM or lower, after equilibration with the antagonist for one hour (Fig. 10.4). However, more prolonged exposure to the agent caused a further rightward shift of dose-response curves and, in nodose ganglia, depressed the maximum. Concentrations above 0.1 μM also depressed the maximum of the dose-response curve.

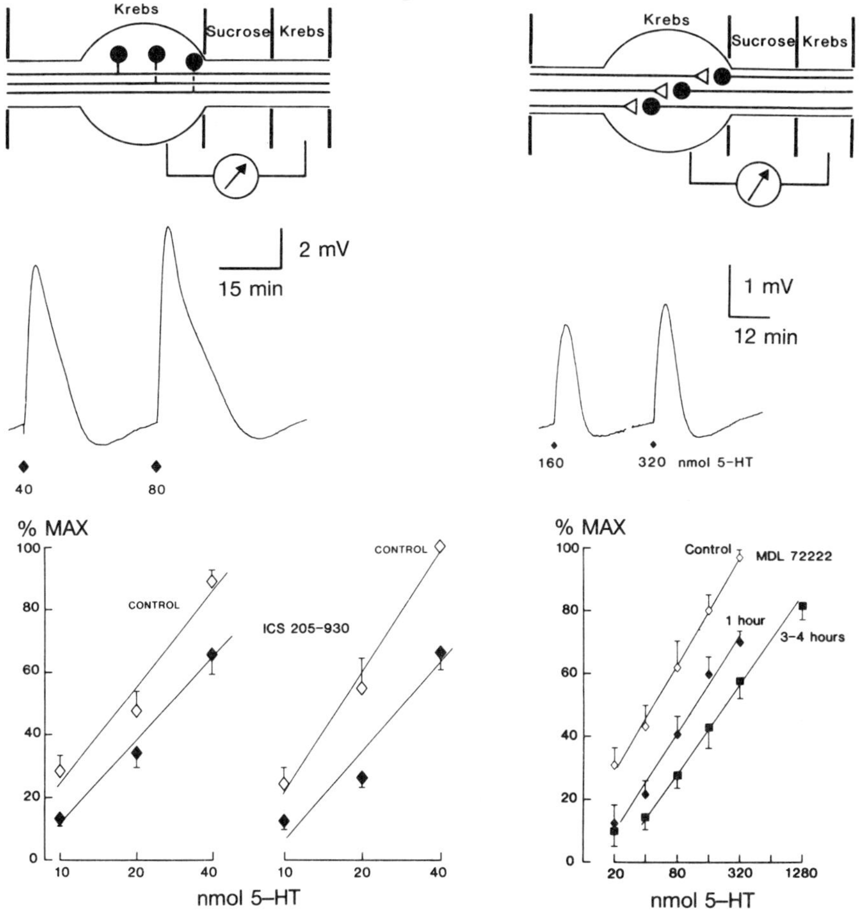

Fig. 10.4 Effects of ICS 205-930 and MDL 72222 on depolarizations evoked by 5-HT on rabbit vagal and superior cervical ganglion neurones recorded by the sucrose-gap method. The schematic diagrams show the arrangement of the ganglia in the apparatus, nodose ganglion of the vagus to the left. The traces on the left show the responses of vagal neurones to bolus injections into the superfusate of 40 and 80 nmol 5-HT, those on the right of superior cervical neurones to injections of 160 and 320 nmol 5-HT. The larger responses evoked from the two ganglia were maximal. The graph (bottom left) shows pooled dose–response curves from vagal neurones illustrating the effects of 0.01 nM (on left, six ganglia) and 0.1 nM (on right, nine ganglia) ICS 205-930 on 5-HT responses. Ganglia were equilibrated with the antagonist for 1 hour. Solid symbols indicate values obtained in the presence of the antagonist; vertical lines indicate SEM. Lines were fitted by the method of least squares. There was no statistically significant deviation from parallelism. The graph (bottom right) shows pooled dose–response curves from superior cervical neurones illustrating the effects of MDL 72222 (0.01 μM), after equilibration for 1 hour and 3–4 hours (nine ganglia), on 5-HT responses. Ordinates: response amplitude expressed as percentage of maximal or near maximal control response to 5-HT

Although our results are consistent with an irreversible mode of action in a tissue with substantial receptor reserve, it may be that MDL 72222 displays two modes of blockade: an initial, competitive blocking action and a second slower action, which is responsible for the further blockade in the superior cervical ganglion and the insurmountable blockade in the nodose ganglion (Round and Wallis 1987).

For lower concentrations of MDL 72222 (0.01–0.1 μM), dose-ratios were measured on the linear portion of individual dose-response curves and apparent pA_2 values calculated in an attempt to quantify the antagonism. Values are shown for the two ganglia in Table 10.2, which also gives our estimate of apparent pA_2 values for other antagonists which produced parallel shifts in dose-response curves.

ICS 205-930 is the most potent selective antagonist that we have tested against 5-HT responses of nodose and sympathetic neurones. It produced parallel shifts of dose-response curves from both ganglia when concentrations of 0.1 nM or lower were tested (Fig. 10.4); its potency on the two ganglia was very similar (Table 10.2). On the other hand, responses in these ganglia were virtually unaffected by the 5-HT receptor antagonists, methysergide, methiothepin, and ketanserin (Wallis 1981; Round and Wallis 1987; Ireland and Tyers 1987).

10.4.5 Sub-types of 5-HT_3 receptor

Part of the evidence for the heterogeneity of 5-HT_3 receptors rests on substantial differences between the 5-HT receptor(s) in the guinea-pig ileum and the 5-HT_3 receptors at other sites. This point is dealt with in detail in Chapters 11 and 15 and will not, therefore, be discussed further here. That 5-HT_3 receptors on different regions of sympathetic neurones may be heterogeneous is a possibility. The evidence for this has been reviewed by Richardson *et al.* (1985), Fozard (1986), and Richardson and Engel (1986). Table 10.2 summarizes apparent pA_2 values from our own unpublished work and from the literature which suggest: (a) homogeneity between receptors on the somatic region of sympathetic neurones and those on sensory neurones; and (b) a heterogeneity between 5-HT receptors on the somatic as opposed to the terminal (and possibly axonal) region of sympathetic neurones. Thus, somatic 5-HT_3 receptors on sympathetic neurones and 5-HT_3 receptors on vagal sensory neurones are not distinguished by MDL 72222, ICS 205-930, quipazine, or metoclopramide; there seems to be no substantial evidence for a difference between these receptors. It can also be seen that ICS 205-930 and metoclopramide showed similar antagonist potencies at the various receptor sites on sympathetic and vagal afferent neurones, although possible species differences need exploring. Quipazine was a potent antagonist, but reversible antagonism was seen only at low concentrations and may not be displayed in all

Table 10.2 Potencies of putative 5-HT receptor antagonists on sympathetic and vagal neurones. Apparent pA_2 values are shown

	Species	MDL 72222	ICS 205-930	SDZ 206-830[6]	Quipazine	Metoclopramide	Reference
Superior cervical ganglion somata	Rabbit	7.8 ± 0.1[1]	10.4 ± 0.1[5]	—	7.6 ± 0.1[1]	7.2 ± 0.2[1]	1. Round and Wallis (1987)
	Rat					6.3 ± 0.1[7]	
Cervical sympathetic axons	Rabbit	9.2 ± 0.1[2]	10.5 ± 0.1[2]	11.2 ± 0.1[2]	—	—	2. Elliott and Wallis unpublished
Sympathetic terminals (heart)	Rabbit	8.9[3] – 9.3[4]	10.6 ± 0.2[6]	10.1 ± 0.1[6]	—	7.1 – 7.2[4]	3. Donatsch *et al.* (1984)
Nodose somata	Rabbit	7.7 ± 0.1[1]	10.2 ± 0.2[5]	—	7.5 ± 0.1[1]	7.1 ± 0.1[1]	4. Fozard (1984c)
Vagal axons	Rabbit	7.9[3]	10.2[6]	13.1 ± 0.8[6]	8.5[8]	7.3[3]	5. Round and Wallis (1986)
	Rat		11.0[8]			6.6[7]	6. Richardson *et al.* (1985) 7. Ireland *et al.* (1982) 8. Ireland and Tyers (1987)

tissues (Peters and Usherwood 1983; Round and Wallis 1987; Ireland and Tyers 1987).

On the other hand, MDL 72222 and SDZ 206-830 (Compound I of Richardson *et al.* 1985) appear to show some selectivity. Thus, MDL 72222 was some 20 times more potent at 5-HT_3 receptors on the terminals and axons of preganglionic sympathetic neurones than at those on the somata of cells or at 5-HT_3 receptors on sensory neurones. Further, SDZ 206-830 was 1000 times more potent on 5-HT_3 receptors on vagal axons than on 5-HT_3 receptors on postganglionic sympathetic nerve terminals. However, it is probably premature to conclude that sub-types of 5-HT_3 receptor occur on peripheral neurones, since information from a larger range of antagonists and further studies with selective agonists are required to substantiate this view (see also next section).

10.5 Functional classification of 5-HT receptors

It is generally held that receptors can best be differentiated on the basis of equilibrium dissociation constants of competitive antagonists, although the rank order of agonist potencies may help in qualitative discrimination. However, this approach may not be entirely straightforward (see Chapter 2). Even in the case of 5-HT_2 receptors where potent and selective competitive antagonists are supposedly available, the antagonist may express affinities which differ according to the kind of tissue used. The variable affinities expressed by the same antagonist could either be interpreted as evidence of receptor heterogeneity or as an indication that the antagonists in question are inadequate receptor probes, e.g. they may interact with binding sites at the receptor accessory to those with which 5-HT interacts (see Chapter 2).

Although further work may clarify this issue, the difficulties of the approach using antagonist affinities may remind us that in practice there is a tendency to equate different mechanisms of action with different receptors. Electrophysiologists in particular are prone to the view that a different mode of action suggests involvement of a different receptor. This approach might be justified if the membrane component bearing the recognition site, i.e. the receptor, is usually linked with only one particular kind of membrane component which comprises the ionophore or effector mechanism.

5-HT_3 receptor-gated mechanisms seem to have in common that they involve membrane depolarization and, thus, excitation. As far as is known, they are only located on nerve cells. Further, the 5-HT_3-mediated fast responses on superior cervical ganglia somata, parasympathetic ganglia, nodose ganglia and probably, in addition, enteric submucous and myenteric plexus cells and murine neuroblastoma cells (Table 10.1) appear to be generated by very similar ionophores. This ionophore permits passage of both Na^+ and K^+, although the reversal potential may be different in different tissues (Table 10.1). The receptors mediating fast depolarizing responses at certain of these sites (superior cervical ganglion; nodose ganglion) were considered to comprise a subgroup of 5-HT_3 receptors by Richardson and Engel (1986). In the absence of any substantial evidence to the contrary it is parsimonious to conclude that the same receptor is present in all these tissues. Note that slow depolarizations evoked in several of these tissues involve a different mechanism (channel closure) and do not appear to be mediated by 5-HT_3 receptors.

The mechanism by which 5-HT releases transmitter following activation of sympathetic neuronal receptor sites has not been analysed, although it is thought to involve membrane depolarization (Fozard and Mwaluko 1976). The ionophore may gate calcium entry directly or might do so indirectly via voltage-sensitive calcium channels. Thus, it remains to be established whether it is different from the ionophore linked to 5-HT_3 receptors on the soma of sympathetic neurones.

Thus, it is feasible that all 5-HT_3 receptors are linked to the same ionophore. It is not possible to consider adequately other types of receptor since, in many cases, the mode of generation of the response is unknown. Receptors mediating a reduction in transmitter release are presumably associated with some mechanism which reduces Ca^{2+} entry directly or indirectly. These receptors are, thus, linked to a further kind of ionophore and may be '5-HT_1-like' (see Chapter 7).

Demonstration of a different mode of action should perhaps be regarded as an indicator that a different kind of 5-HT receptor may be involved; it is not a substitute for a direct demonstration based upon antagonist dissociation constants or affinities of agonists. It may be a useful indicator simply because the membrane components responsive to 5-HT may normally involve association of particular sorts of recognition site sub-unit with particular sorts of operant sub-unit, assuming the receptor-ionophore is a complex of sub-units. Alternatively, the membrane protein constituting a particular kind of ion channel may be likely to express a receptor site in only one of its sub-type manifestations.

10.6 Concluding remarks

10.6.1 Significance of different kinds of 5-HT action on neurones

One puzzling aspect of the effects of 5-HT on sympathetic and sensory neurones is their multiplicity. At present, we have little idea whether part or all (or none) of these effects have physiological significance, although a pathophysiological role is indicated by the effectiveness of 5-HT_3 receptor antagonists in certain conditions, e.g. cytotoxic drug-induced vomiting, carcinoid syndrome, and migraine (see Chapter 15). Since one novel aspect of 5-HT action is the ability to provoke both rapid and slow responses from autonomic neurones, the possible functional significance of these two kinds of response is of interest. These responses are reminiscent of the nicotinic and muscarinic actions of acetylcholine on sympathetic neurones (see also Chapter 4). They seem to be mediated by different kinds of 5-HT receptor, the fast response being mediated via a 5-HT_3 receptor. If the 5-HT-containing neurones mentioned earlier release 5-HT in the vicinity of ganglion cells, it is possible that 5-HT_3 receptors may mediate a phasic transmission of information. The nature of the response would be such as to lead to a brief discharge with the transiently high membrane conductance shunting other excitatory or inhibitory inputs. On the other hand, the slow response would produce a sustained excitation of the cell and the reduced membrane conductance would enhance the effect of other inputs; this action is well suited to a modulatory role in the transmission process. When considering functional significance, it is important to remember that certain neurones showed one but not both responses; specific functional pathways might be affected by 5-HT in one way while others may be affected in another.

10.6.2 Neuronal 5-HT receptors in specific locations

The wide distribution of 5-HT receptors on autonomic and visceral sensory neurones has already been noted. The sensory neurones may be part of reflex pathways where an efferent limb involves autonomic neurones. Thus, 5-HT may initiate or modulate activity in a reflex path via either set of neurones. For instance, 5-HT is believed to be released locally in the intestine and to stimulate vagal afferents to elicit vomiting as the key step in the induction of cytotoxic drug-induced vomiting (see Chapter 15). The ability of 5-HT to affect sensory and autonomic neurones confers a considerable potential for modulatory action. Nowhere is this more marked, perhaps, than in neural pathways to and from the heart. The potential modulatory roles available to 5-HT on reflexes involving the heart and on cardiac performance are shown in Fig. 10.5.

The heart is a source of 5-HT and increased levels in the extracellular fluid of the heart have been reported following myocardial infarction and ischaemia (Kordenat and Kedzi 1979; Sole *et al.* 1979). Mast cells around coronary arteries may be one source of 5-HT, while platelets at the site of a thrombus may be another (Kordenat and Kedzi 1979). There appears to be a third source of 5-HT within the heart, independent of mast cell and platelet 5-HT (Sole *et al.* 1979).

Released 5-HT would be likely to stimulate cardiac sensory endings involved in the Bezold–Jarisch-like effect, which mediate bradycardia and

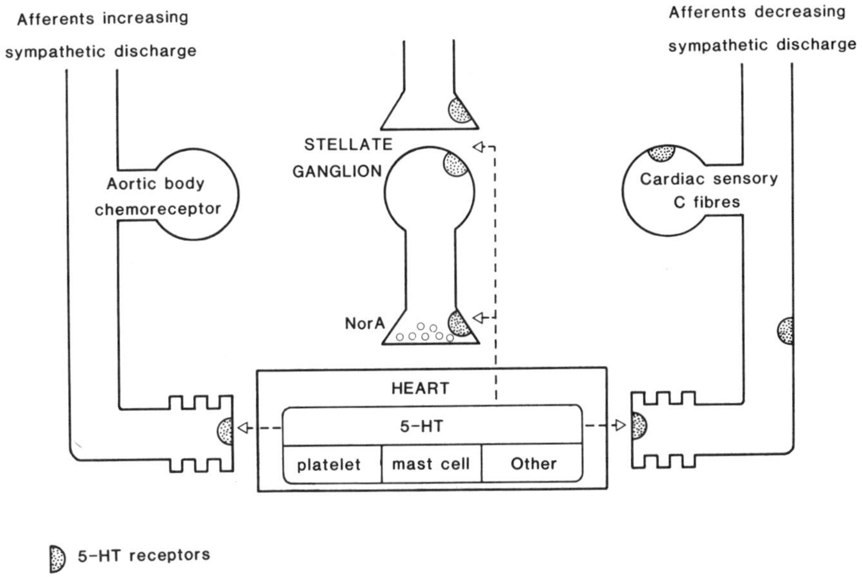

Fig. 10.5 Possible involvement of 5-HT in modulating activity in neural pathways to the heart and in cardiac function. Dashed lines show the possible targets of 5-HT released from the ischaemic heart on sympathetic or sensory neurones; see text for further details. Nor A = noradrenergic neurone

a fall in blood pressure. The reflex is evoked principally from endings in the right ventricle (Paintal 1973). The chemoreceptor afferents of the aortic bodies are also extremely sensitive to 5-HT (Eckstein *et al.* 1971). The aortic bodies may act as the chemoreceptors of the coronary artery. Activation of these afferents by 5-HT evokes the 'cardiogenic hypertensive reflex', which doubles the blood pressure in a few seconds and produces tachycardia. The afferent pathway is the vagus nerve and the efferent pathway mainly the sympathetic nerves (Hageman *et al.* 1976). Thus, reflex responses of opposite character may be elicited by 5-HT acting on different afferent pathways from the heart. 5-HT would also have modulatory effects on the sympathetic and parasympathetic nerve supply as already described. It may be concluded that although the roles for 5-HT in the pathophysiology of the failing heart remain to be elucidated, its potential role as an important modulator is very clear.

10.6.3 Conclusion

This chapter has attempted to describe the range of functional effects produced by 5-HT on autonomic and sensory neurones. The nature of both the 5-HT receptor-gated mechanisms and the 5-HT receptors has been considered. Some attention has been drawn to the sources of 5-HT within the body and to the likelihood of the involvement of 5-HT in physiological or pathophysiological processes.

References

Aas, P. (1983). Serotonin-induced release of acetylcholine from neurons in the bronchial smooth muscle of the rat. *Acta Physiologica Scandinavica* **117**, 477–80.

Adler-Graschinsky, E. (1983). Dual presynaptic effects of 5-hydroxytryptamine on peripheral noradrenergic synapses. *Journal of Autonomic Pharmacology* **3,** 303–15.

Akasu, T., Hirai, K., and Koketsu, K. (1981). 5-hydroxytryptamine controls ACh-receptor sensitivity of bullfrog sympathetic ganglion cells. *Brain Research* **211,** 217–20.

Akasu, T., Hasuo, H., and Tokimasa, T. (1987). Activation of 5-HT_3 receptor subtypes causes rapid excitation of rabbit parasympathetic neurones. *British Journal of Pharmacology* **91,** 453–5.

Andrews, W. W. (1968). Afferent impulses in the hepatic nerve of perfused livers elicited by 5-hydroxytryptamine and other substances. *British Journal of Pharmacology* **32,** 421–2P.

Azami, J., Fozard, J. R., Round, A. A., and Wallis, D. I. (1985). The depolarizing action of 5-hydroxytryptamine on rabbit vagal primary afferent and sympathetic neurones and its selective blockade by MDL 72222. *Naunyn-Schmiedeberg's Archives of Pharmacology* **328,** 423–9.

Beck, P. W. and Handwerker, H. O. (1974). Bradykinin and serotonin effects on various types of cutaneous nerve fibres. *Pflugers Archives Physiologie* **347,** 209–22.

Bowker, R. M., Westlund, K. N., and Coulter, J. D. (1981). Origin of serotonergic projections to the spinal cord in rat: an immunocytochemical, retrograde transport study. *Brain Research* **226,** 187–99.

Bradley, P. B., Engel, G., Feniuk, W., Fozard, J. R., Humphrey, P. P. A., Middlemiss, D. N., Mylecharane, E. J., Richardson, B. P., and Saxena, P. R. (1986). Proposals for the classification and nomenclature of functional receptors for 5-hydroxytryptamine. *Neuropharmacology* **25,** 563–76.

Burnstock, G., Cocks, T., Crowe, R., and Kasakov, L. (1978). Purinergic innervation of the guinea-pig urinary bladder. *British Journal of Pharmacology* **63,** 125–38.

Carstens, E., Klumpp, D., Randic, M., and Zimmermann, A. (1981). Effect of ionophoretically applied 5-hydroxytryptamine on the excitability of single primary afferent C- and A-fibres in the cat spinal cord. *Brain Research* **220,** 151–8.

Chan-Palay, V. (1976). Serotonin axons in the supra- and subependymal plexuses in the leptomeninges. Their roles in local alterations of cerebrospinal fluid and vasomotor activity. *Brain Research* **102,** 103–30.

Comroe, J. H., Van Lingen, B., Stroud, R. C., and Roncoroni, H. (1953). Reflex and direct cardiopulmonary effects of 5-OH-tryptamine (serotonin). *American Journal of Physiology* **173,** 379–89.

Connell, L. A. and Wallis, D. I. (1987). Effects of 5-hydroxytryptamine (5-HT) in the isolated, hemisected spinal cord of the neonate rat. *Journal of Physiology* **390,** 44P.

Coote, J. H., MacLeod, V. H., Fleetwood-Walker, S., and Gilbey, M. P. (1981). The response of individual sympathetic preganglionic neurones to micro-electrophoretically applied endogenous monoamines. *Brain Research* **215,** 135–45.

Cowen, T., Alafaci, C., Crockard, H. A., and Burnstock, G. (1986). 5-HT-containing nerves to major cerebral arteries of the gerbil originate in the superior cervical ganglia. *Brain Research* **384,** 51–9.

Dahlstrom, A. and Fuxe, K. (1964). Evidence for the existence of monoamine-containing neurons in the central nervous system. I. Demonstration of monoamines in the cell bodies of brainstem neurons. *Acta Physiologica Scandinavica* **62,** Suppl. 232, 1–55.

Dahlstrom, A. and Fuxe, K. (1965). Evidence for the existence of monoamine neurons in the central nervous system. II. Experimentally-induced changes in the intraneuronal amine levels of bulbospinal neuron systems. *Acta Physiologica Scandinavica* **64,** Suppl. 247, 1–36.

De Groat, W. C. and Ryall, R. W. (1967). An excitatory action of 5-hydroxy-tryptamine on sympathetic preganglionic neurones. *Experimental Brain Research* **3,** 299–305.

Di Carlo, V. (1983). Serotonergic fibres in dorsal roots of the spinal cord. *Neuro-science Letters* **43,** 233–44.

Donatsch, P., Engel, G., Richardson, B. P., and Stadler, P. (1984). ICS 205–930: A highly selective and potent antagonist at peripheral neuronal 5-hydroxy-tryptamine (5-HT) receptors. *British Journal of Pharmacology* **81,** 34P.

Douglas, W. W. and Ritchie, J. M. (1957). Discharges of non-myelinated afferent fibres in the cat's saphenous nerve in response to touch and to drugs. *Journal of Physiology* **139,** 9–10P.

Douglas, W. W. and Toh, C. C. (1953). The respiratory stimulant action of 5-hydroxytryptamine (serotonin) in the dog. *Journal of Physiology* **120,** 311–8.

Dun, N. J. and Karczmar, A. G. (1981). Evidence for a presynaptic inhibitory action of 5-hydroxytryptamine in a mammalian sympathetic ganglion. *Journal of Pharmacology and Experimental Therapeutics* **217,** 714–8.

Dun, N. J., Karczmar, A. G., and Ingerson, A. (1980). A neurochemical and neurophysiological study of serotonin in the superior cervical ganglia. *Society for Neuroscience Abstracts* **6,** 216.

Dun, N. J., Kiraly, M., and Ma, R. C. (1984). Evidence for a serotonin-mediated slow excitatory potential in the guinea-pig coeliac ganglia. *Journal of Physiology* **351,** 61–76.

Eckstein, R. W., Shintani, F., Rown, H. E. Jr., Shimomura, K., and Ohya, N. (1971). Identification of left coronary blood supply of aortic bodies in anaesthet-ized dogs. *Journal of Applied Physiology* **30,** 488–92.

Edvinsson, L., Degueurce, A., Duverger, D., MacKenzie, E. T., and Scatton, B. (1983). Central serotonergic nerves project to the pial vessels of the brain. *Nature (London)* **306,** 55–7.

Feniuk, W., Humphrey, P. P. A., and Watts, A. D. (1979). Presynaptic inhibitory action of 5-hydroxytryptamine in dog isolated saphenous vein. *British Journal of Pharmacology* **67,** 247–54.

Fjallbrant, N. And Iggo, A. (1961). The effect of histamine, 5-hydroxytryptamine and acetylcholine on cutaneous afferent fibres. *Journal of Physiology* **156,** 578–90.

Fortune, D. H., Ireland, S. J., and Tyers, M. B. (1985). 5-hydroxytryptamine-induced hyperpolarizations of the rat isolated superior cervical ganglion and vagus nerve. *British Journal of Pharmacology* **84,** 49P.

Fozard, J. R. (1979). Blockade of receptors for 5-hydroxytryptamine on the sympathetic nerves of the rabbit heart. In *Advances in the biosciences,* Vol. 8 (ed. S. Z. Langer, K. Starke, and M. L. Dubocovich) pp. 327–32. Pergamon, Oxford.

Fozard, J. R. (1984a). Neuronal 5-HT receptors in the periphery. *Neuropharmacology* **23,** 1473–86.

Fozard, J. R. (1984b). Characteristics of the excitatory 5-HT receptor on the cholinergic nerves of the rabbit heart. *Proceedings of IUPHAR Ninth International Congress of Pharmacology, London,* July 1984.

Fozard, J. R. (1984c). MDL 72222: A potent and highly selective antagonist at neuronal 5-hydroxytryptamine receptors. *Naunyn-Schmiedeberg's Archives of Pharmacology* **326,** 36–44.

Fozard, J. R. (1985). Serotonin, migraine and platelets. In *Drugs and platelets* (ed. P. A. van Zwieten and E. Schönbaum), *Progress in pharmacology*, Vol. 44, pp. 135–46. Fischer Verlag, Stuttgart.

Fozard, J. R. (1986). Peripheral neuronal 5-hydroxytryptamine receptors and their biological significance. *Proceedings of the Third SCI-RSC Medical Chemistry Symposium* (ed. R. W. Lambert) pp. 37–56. Royal Society of Chemistry, London.

Fozard, J. R. and Mobarok Ali, A. T. M. (1978). Receptors for 5-hydroxytryptamine on the sympathetic nerves of the rabbit heart. *Naunyn-Schmiedeberg's Archives of Pharmacology* **301,** 223–35.

Fozard, J. R. and Mwaluko, G. M. P. (1976). Mechanism of the indirect sympathomimetic effect of 5-hydroxytryptamine on the isolated heart of the rabbit. *British Journal of Pharmacology* **57,** 115–25.

Franzen, F. and Eysell, K. (1969). *Biologically active amines found in man.* Pergamon, Oxford.

Garattini, S. and Valzelli, L. (1965). *Serotonin.* Elsevier, Amsterdam.

Gaudin-Chazal, G., Dszuta, A., Segu, L., Ternaux, J. P., and Puizillout, J. J. (1978). Serotonin-containing neurones in the nodose ganglia of the cat. *Waking and Sleeping* **2,** 149–51.

Gilbey, M. P., Coote, J. H., MacLeod, V. H., and Peterson, D. F. (1981). Inhibition of sympathetic activity by stimulating in the raphe nucleus and the role of 5-hydroxytryptamine in this effect. *Brain Research* **226,** 131–42.

Ginzel, K. H. and Kottegoda, S. R. (1954). The action of 5-hydroxytryptamine and tryptamine on aortic and carotid sinus receptors in the cat. *Journal of Physiology* **123,** 277–88.

Guharay, F., Ramsay, R. L., and Usherwood, P. N. R. (1985). 5-Hydroxytryptamine-activated single channel currents recorded from murine neuroblastoma cells. *Brain Research* **340,** 325–32.

Gyermek, L. (1962). Action of 5-hydroxytryptamine on the urinary bladder of the dog. *Archives internationales de Pharmacodynamie et de Therapie* **87,** 137–44.

Haefely, W. (1974). The effects of 5-hydroxytryptamine and some related compounds on the cat superior cervical ganglion in situ. *Naunyn-Schmiedeberg's Archives of Pharmacology* **281,** 145–65.

Hageman, G. R., Urthaler, F., and James, T. N. (1976). Afferent and efferent neural pathways and biochemical blockade of a cardiogenic hypertensive chemoreflex. *American Journal of Cardiology* **37,** 140.

Higashi, S. (1977). 5-hydroxytryptamine receptors on visceral primary afferent neurones in the nodose ganglion of the rabbit. *Nature (London)* **267,** 448–9.

Higashi, S. and Nishi, S. (1982). 5-hydroxytryptamine receptors of visceral primary afferent neurones in rabbit nodose ganglia. *Journal of Physiology* **323,** 543–67.

Hirai, K. and Koketsu, K. (1980). Presynaptic regulation of the release of acetylcholine by 5-hydroxytryptamine. *British Journal of Pharmacology,* **70,** 499–500.

Holt, S. E., Cooper, M., and Wyllie, J. H. (1986). On the nature of the receptor mediating the action of 5-hydroxytryptamine in potentiating responses of the mouse urinary bladder strip to electrical stimulation. *Naunyn-Schmiedeberg's Archives of Pharmacology* **334,** 333–40.

Ireland, S. J. (1987). Origin of 5-hydroxytryptamine-induced hyperpolarization of the rat superior cervical ganglion and vagus nerve. *British Journal of Pharmacology* **92,** 407–16.

Ireland, S. J. and Tyers, M. B. (1987). Pharmacological characterization of 5-hydroxytryptamine-induced depolarization of the rat isolated vagus nerve. *British Journal of Pharmacology* **90,** 229–38.

Ireland, S. J., Straughan, D. W., and Tyers, M. B. (1982). Antagonism by metoclopramide and quipazine of 5-hydroxytryptamine-induced depolarizations of the rat isolated vagus nerve. *British Journal of Pharmacology* **75,** 16P.

Jaim-Etcheverry, G. and Zieher, L. M. (1982). Coexistence of monoamines in peripheral adrenergic neurones. In *Co-transmission* (ed. A. C. Cuello), Macmillan, London.

Johnson, A. R. and Erdos, E. G. (1973). Release of histamine from mast cells by vasoactive peptides. *Proceedings of the Society for Experimental Biology and Medicine* **142,** 1252–6.

Kadzielawa, K. (1983). Antagonism of the excitatory effects of 5-hydroxytryptamine on sympathetic preganglionic neurones and neurones activated by visceral afferents. *Neuropharmacology* **22,** 19–27.

Kiraly, M., Ma, R. C., and Dun, N. J. (1983). Serotonin mediates a slow excitatory potential in mammalian coeliac ganglion. *Brain Research* **275,** 378–83.

Kirby, G. C. and McQueen, D. S. (1984). Effects of the antagonists MDL 72222 and ketanserin on responses of cat carotid body chemoreceptors to 5-hydroxytryptamine. *British Journal of Pharmacology* **83,** 259–69.

Kordenat, R. K. and Kedzi, P. (1979). Serotonin blockade during experimental coronary thrombosis. *American Heart Journal* **97,** 329–38.

Liuzzi, A., Foppan, F. H., Saavedra, J. M., Levi-Montalcini, R., and Kopin, I. J. (1977). Gas chromatographic-mass-spectrometric assay of serotonin in rat superior cervical ganglia. Effects of nerve growth factor and 6-hydroxydopamine. *Brain Research* **133,** 354–7.

Ma, R. C. and Dun, N. J. (1986). Excitation of lateral horn neurons of the neonatal rat spinal cord by 5-hydroxytryptamine. *Developmental Brain Research* **24,** 89–98.

Ma, R. C., Horwitz, J., Kiraly, M., Perlman, R. L., and Dun, N. J. (1985).

Immunohistochemical and biochemical detection of serotonin in the guinea pig celiac-superior mesenteric plexus. *Neuroscience Letters* **56,** 107–12.

McCall, R. B. (1983). Serotonergic excitation of sympathetic preganglionic neurones: a microiontophoretic study. *Brain Research* **289,** 121–7.

McGrath, M. A. (1977). 5-hydroxytryptamine and neurotransmitter release in canine blood vessels. *Circulation Research* **41,** 428–35.

Machova, J. and Boska, D. (1969). The effects of 5-hydroxytryptamine, dimethylphenylpiperazinium and acetylcholine on transmission and surface potential in the cat sympathetic ganglion. *European Journal of Pharmacology* **7,** 152–8.

Martinez, A. A. and Lokhandwala, M. F. (1980). Evidence for a presynaptic inhibitory action of 5-hydroxytryptamine on sympathetic neuro-transmission to the myocardium. *European Journal of Pharmacology* **63,** 303–11.

Mawe, G. M., Branchek, T. A., and Gershon, M. D. (1986). Identification of two types of peripheral neural receptor for serotonin. *Society for Neuroscience Abstracts* **12,** 337.7.

Mense, S. (1977). Nervous outflow from skeletal muscle following chemical noxious stimulation. *Journal of Physiology* **267,** 75–88.

Moskowitz, M. A., Reinhard, J. F., Romero, J., Melamed, E., and Pettibone, D. J. (1979). Neurotransmitters and the fifth cranial nerve. Is there a relation to the headache phase of migraine? *Lancet* **ii,** 883–5.

Nash, H. L. and Wallis, D. I. (1981). Effects of divalent cations on responses of a sympathetic ganglion to 5-hydroxytryptamine and 1, 1-dimethyl-4-phenyl piperazinium. *British Journal of Pharmacology* **73,** 759–72.

Neel, D. S. and Parsons, R. L. (1986). Catecholamine, serotonin and substance P-like peptide containing intrinsic neurones in the mud puppy parasympathetic cardiac ganglion. *Journal of Neuroscience* **6,** 1970–5.

Neff, N. H., Barrett, R. E., and Costa, E. (1969). Kinetic and fluorescent histochemical analysis of the serotonin compartments in rat pineal gland. *European Journal of Pharmacology* **5,** 348–56.

Neto, F. R. (1978). The depolarizing action of 5-HT on mammalian non-myelinated nerve fibres. *European Journal of Pharmacology* **49,** 351–6.

Neumayr, R. J., Hare, B. D., and Franz, D. N. (1974). Evidence for bulbospinal control of sympathetic preganglionic neurons by monoaminergic pathways. *Life Sciences* **14,** 793–806.

Nishi, K. (1975). The action of 5-hydroxytryptamine on chemoreceptor discharges of the cat's carotid body. *British Journal of Pharmacology* **55,** 27–40.

Oliveras, J. L., Bourgoin, S., Héry, F., Besson, J. M., and Hamon, M. (1977). The topographical distribution of serotonergic terminals in the spinal cord of the cat: biochemical mapping by the combined use of microdissection and microassay procedures. *Brain Research* **138,** 393–406.

Owman, C. (1964). Sympathetic nerves probably storing two types of monoamines in the rat pineal gland. *International Journal of Neuropharmacology* **2,** 105–12.

Paintal, A. S. (1954). The response of gastric stretch receptors and certain other abdominal and thoracic vagal receptors to some drugs. *Journal of Physiology* **126,** 271–85.

Paintal, A. S. (1973). Vagal sensory receptors and their reflex effects. *Physiological Reviews* **53,** 159–227.

Pellegrino de Iraldi, A., Zieher, L. M., and De Robertis, E. (1965). Ultrastructural and pharmacological studies of nerve endings of the pineal gland. *Progress in Brain Research* **10,** 389–421.

Peters, J. and Usherwood, P. N. R. (1983). 5-hydroxytryptamine responses of murine neuroblastoma cells. Ions and putative antagonists. *British Journal of Pharmacology* **80,** 523P.

Reinhard, J. F. Jr., Liebmann, J. E., Schlosberg, A. J., and Moskowitz, M. A. (1979). Serotonin neurons project to small blood vessels in the brain. *Science* **206,** 85–7.

Richardson, B. P. and Engel, G. (1986). The pharmacology and function of 5-HT_3 receptors. *Trends in Neurosciences* **9,** 424–8.

Richardson, B. P., Engel, G., Donatsch, P., and Stadler, P. A. (1985). Identification of serotonin M-receptor subtypes and their specific blockade by a new class of drugs. *Nature (London)* **316,** 126–31.

Round, A. A. and Wallis, D. I. (1986). The depolarizing action of 5-hydroxytryptamine on rabbit vagal afferent and sympathetic neurones in vitro and its selective blockade by ICS 205–930. *British Journal of Pharmacology* **88,** 485–94.

Round, A. A. and Wallis, D. I. (1987). Further studies on the blockade of 5-HT depolarizations of rabbit vagal afferent and sympathetic ganglion cells by MDL 72222 and other antagonists. *Neuropharmacology* **26,** 39–48.

Saum, W. R. and de Groat, W. C. (1973). The actions of 5-hydroxytryptamine on the urinary bladder and on vesical autonomic ganglia in the cat. *Journal of Pharmacology and Experimental Therapeutics* **185,** 70–83.

Saxena, P. R., Heiligers, J., Mylecharane, E. J., and Tio, R. (1985). Excitatory 5-hydroxytryptamine receptors in the cat urinary bladder are of the M- and 5-HT_2 type. *Journal of Autonomic Pharmacology* **5,** 101–7.

Shirasawa, Y. and Koketsu, K. (1977). Action of 5-hydroxytryptamine on isolated spinal cord of bullfrogs. *Japanese Journal of Pharmacology* **27,** 23–9.

Skok, V. and Selyanko, A. A. (1979). Acetylcholine and serotonin receptors in mammalian sympathetic ganglion neurons. In *Integrative functions of the autonomic nervous system* (ed. C. Mc. C. Brooks, K. Koizumi, and A. Sato) pp. 248–53. University of Tokyo Press, Tokyo.

Sole, M. J., Shum, A., and Van Loon, G. R. (1979). Serotonin metabolism in the normal and failing hamster heart. *Circulation Research* **45,** 629–34.

Stansfeld, C. E. and Wallis, D. I. (1984). Generation of an unusual depolarizing response in rabbit primary afferent neurones in the absence of divalent cations. *Journal of Physiology* **352,** 49–72.

Stansfeld, C. E. and Wallis, D. I. (1985). Properties of visceral primary afferent neurones in the nodose ganglion of the rabbit. *Journal of Neurophysiology* **54,** 245–60.

Surprenant, A. and Crist, J. (1986). Serotonergic receptors on individual neurones in the guinea-pig submucous plexus. *Society for Neuroscience Abstracts* **12,** 337.1.

Vanov, S. (1965). Response of the rat urinary bladder in situ to drugs and to nerve stimulation. *British Journal of Pharmacology* **24,** 591–600.

Verhofstad, A. A. J., Steinbusch, H. W. M., Penke, B., Varga, J., and Joosten, H. W. J. (1981). Serotonin-immunoreactive cells in the superior cervical ganglion of the rat. Evidence for the existence of separate serotonin- and catecholamine-containing small ganglionic cells. *Brain Research* **212,** 39–49.

Wallis, D. I. (1979). The apparent multiplicity of ganglionic receptors. In *Trends in autonomic pharmacology* (ed. S. Kalsner) pp. 21–57. Urban and Schwarzenberg, Baltimore.

Wallis, D. I. (1981). Neuronal 5-hydroxytryptamine receptors outside the central nervous system. *Life Sciences* **29,** 2345–55.

Wallis, D. I. and Dun, N. J. (1989). Presynaptic action of 5-hydroxytryptamine on autonomic ganglia. In *Presynaptic regulation of neurotransmitter release* (ed. J. J. Feigenbaum, and M. Hanani). Freund Publishing Co., London and Tel Aviv, in press.

Wallis, D. I. and North, R. A. (1978). Intracellular recording of responses of rabbit superior cervical ganglion cells to 5-hydroxytryptamine applied by iontophoresis. *Neuropharmacology* **17,** 1023–8.

Wallis, D. I. and Woodward, B. (1973). The depolarizing action of 5-hydroxytryptamine on sympathetic ganglion cells. *British Journal of Pharmacology* **49,** 168P.

Wallis, D. I. and Woodward, B. (1975). Membrane potential changes induced by 5-hydroxytryptamine in the rabbit superior cervical ganglion. *British Journal of Pharmacology* **55,** 199–212.

Wallis, D. I., Lees, G. M., and Kosterlitz, H. W. (1975). Recording resting and action potentials by the sucrose-gap method. *Comparative Biochemistry and Physiology* **50C,** 199–216.

Wallis, D. I., Stansfeld, C. E., and Nash, H. L. (1982). Depolarizing responses recorded from nodose ganglion cells of the rabbit evoked by 5-hydroxytryptamine and other substances. *Neuropharmacology* **21,** 31–40.

Watanabe, S. and Koketsu, K. (1973). 5-HT hyperpolarization of bullfrog sympathetic ganglion cell membrane. *Experientia* **29,** 1370–2.

Watts, A. D., Feniuk, W., and Humphrey, P. P. A. (1981). A prejunctional action of 5-hydroxytryptamine and methysergide on noradrenergic nerves in dog isolated saphenous vein. *Journal of Pharmacy and Pharmacology* **33,** 515–20.

Williams, T. H. (1967). Electron microscopic evidence for an autonomic interneuron. *Nature (London)* **214,** 309–10.

Yoshimura, M. and Nishi, S. (1982). Intracellular recordings from lateral horn cells of the spinal cord in vitro. *Journal of the Autonomic Nervous System* **6,** 5–11p.

11

5-Hydroxytryptamine and enteric neurones

MICHAEL D. GERSHON, GARY M. MAWE, and THERESA A. BRANCHEK

11.1 Introduction

The enteric nervous system was classified by Langley (1921) as a third autonomic division. This classification was based on his belief that most of the intrinsic neurones of the enteric nervous system receive no direct preganglionic innervation from the central nervous system; therefore, the enteric nervous system could not be accommodated within the definitions of the sympathetic or parasympathetic autonomic divisions, which were classified according to their sources of preganglionic fibres from the central nervous system. The physiological correlate of the separate and autonomous nature of the enteric nervous system is that the enteric nervous system can mediate reflex activity in the absence of input from the central nervous system (for reviews see Gershon 1981; Costa *et al.* 1987; Furness and Costa 1987). This ability, in turn, implies that the enteric nervous system contains intrinsic primary afferent neurones and a variety of interneurones. The structure and organization of the enteric nervous system are unique in the peripheral nervous system and reflect the independent nature of the system. In fact, the central nervous system has many characteristics in common with the enteric nervous system, such as a lack of internal collagen, a blood–nerve barrier, support from glial cells that resemble astrocytes, and an abundance of different types of neurone (defined by established or putative neurotransmitter). One of the distinguishing types of enteric neurone is one that utilizes 5-HT as a neurotransmitter (Gershon 1982). This neurone has been as intensively studied as any in the enteric nervous system and a review of what has been learned about enteric 5-HT-containing neurones illustrates how much progress has been made in understanding the enteric nervous system and how much is still to be learned.

11.2 5-HT-containing neurones in the enteric nervous system

11.2.1 Storage of 5-HT in enteric neurones

The idea that there might be 5-HT-containing neurones in the bowel was not easily accepted (see, for example, Costa and Furness 1979a). Although it was widely known that the largest store of 5-HT in the body is that of the gastro-intestinal tract (Erspamer 1966), all of this 5-HT was thought to be present in the entero-endocrine cells of the gastro-intestinal mucosa. As a result, the first suggestion that 5-HT might also be located in enteric neurones and act as a neurotransmitter within the enteric nervous system (Gershon *et al.* 1965) was greeted with considerable scepticism. The evidence upon which this suggestion was based included the radioautographic demonstration of labelling of axons in the myenteric plexus, but not sympathetic nerve fibres, following the intravenous injection of [^{3}H]5-hydroxytryptophan. Moreover, stimuli applied to the mucosa that activated the peristaltic reflex led to the release of [^{3}H]5-HT from labelled segments of gut *in vitro* and this release could be blocked by addition of local anaesthetics (Gershon *et al.* 1965). Thus, it appeared that the myenteric plexus contained nerves that might utilize 5-HT as a transmitter and that they might function in the mediation of the peristaltic reflex. Although it is now apparent that this early suggestion that 5-HT is an enteric neurotransmitter was correct, the actual evidence confirming the role of 5-HT as a neurotransmitter in the bowel took a great deal of time to accumulate, and the precise role of 5-HT in the physiology of the gut is still unknown.

11.2.1.1 *Measurement of endogenous 5-HT*

A critical step in the establishment of a candidate substance as a neurotransmitter is to show that the compound is actually present in the nerve terminals in question. Since the mucosal epithelial concentration of 5-HT is so high, it was difficult to measure the extramucosal concentration of 5-HT in the wall of the gut. Nevertheless, when this was done by dissecting the mucosa from the bowel, smaller but significant amounts of 5-HT were found in the remaining tissue (Feldberg and Toh 1953). On the other hand, the discovery of the ability of enteric neurones to take up [^{3}H]5-HT (Gershon and Altman 1971; Robinson and Gershon 1971) raised the possibility that the 5-HT measured in the wall of mucosa-free preparations of intestine represented material released artifactually from the mucosa during the dissection and taken up by neurones of the enteric nervous system. This potential artifact can be avoided by dissecting the bowel in the cold because the uptake of 5-HT by enteric neurones is highly temperature dependent (Gershon and Altman 1971). It is now apparent that there is

5-HT (57–110 ng/g) in strips of adequately dissected guinea-pig small intestinal longitudinal muscle with adherent myenteric plexus (Juorio and Gabella 1974; Robinson and Gershon 1971; Gershon and Tamir 1981a; Holzer and Skofitsch 1984; Takaki *et al.* 1985b). A much higher level (~ 500 ng/g) has been reported in the muscular wall of the caecum (Legay *et al.* 1983 a and b). Since the neuronal tissue represents only a small fraction of these preparations, which are largely composed of muscle and connective tissue, the 5-HT content of the myenteric plexus is greatly underestimated in these preparations and the actual concentration of 5-HT in the myenteric plexus is probably similar to or greater than (Legay *et al.* 1983 a and b) that of 5-HT in the brain.

11.2.1.2 *Localization of enteric 5-HT-containing neurones*

Measurements of 5-HT in enteric tissue do not establish the identity of the cell(s) in which 5-HT is stored. That demonstration requires a cytochemical or immunocytochemical method. Although 5-HT can be demonstrated cytochemically by aldehyde-induced fluorescence, the method is not very sensitive (Fuxe and Jonsson 1967). In order to visualize 5-HT by this means, the neuronal content of 5-HT has to be raised (Aghajanian and Asher 1971; Aghajanian *et al.* 1973). In the absence of such pharmacological amplification 5-HT cannot be detected (Robinson and Gershon 1971; Baumgarten *et al.* 1973; Ahlman and Enerbäck 1974; Dubois and Jacobowitz 1974; Costa and Furness 1979a). On the other hand, when the pharmacological procedures used for the cytochemical detection of central 5-HT are applied to the gut, 5-HT can be visualized in enteric neurones (Robinson and Gershon 1971; Feher 1974 and 1975; Dreyfus *et al.* 1977a). Moreover, 5-HT is readily seen in the enteric nervous system of lower vertebrates (Baumgarten *et al.* 1973; Goodrich *et al.* 1980; Watson 1979; Anderson 1983) and cephalochordates (Salimova 1978). Moreover, immunocytochemistry has also revealed 5-HT-containing neurones in the myenteric plexus of adult guinea-pigs (Costa *et al.* 1982; Furness and Costa 1982; Legay *et al.* 1984; Saffrey *et al.* 1984), adult cats (Grønstad *et al.* 1985), adult and developing mice (Rothman and Gershon 1982; Gershon and Sherman 1987), adult rats (Nada and Toyohara 1987), and humans (Kurian *et al.* 1983).

11.2.1.3 *Detection of 'serotonin binding protein' in enteric neurones*

Serotonin binding protein was first identified in 5-HT-containing neurones of the brain (Tamir and Huang 1974; Tamir and Kuhar 1975). Serotonin binding protein is translocated proximo-distally by fast axonal transport (Tamir and Gershon 1979), which in turn implies that serotonin binding protein is contained within vesicles (Grafstein and Forman 1980). This implication has been confirmed (Barasch *et al.* 1987a; Gershon and Tamir 1981b; Gershon *et al.* 1983; Liu *et al.* 1985; Tamir and Gershon 1979).

More recent analyses have indicated that there are two forms of serotonin binding protein, one of 45 kDa, and other 56 kDa (Gershon *et al.* 1983; Liu *et al.* 1985). It has been suggested that the 56 kDa serotonin binding protein may be a precursor of the 45 kDa material.

Both the 56 kDa and the 45 kDa forms of serotonin binding protein have been found outside the brain. Serotonin binding protein has been located in the thyroid gland (Jonakait *et al.* 1977; Bernd *et al.* 1979 and 1981) and in the gut (Jonakait *et al.* 1977 and 1979). Within the thyroid, serotonin binding protein is located in the secretory granules of parafollicular ('C') cells, where it is co-localized with 5-HT and calcitonin (Barasch *et al.* 1987b). Within the gut, serotonin binding protein appears to be contained in enteric 5-HT-containing neurones because it makes its appearance during development coincidentally with the expression of other 5-HT-related neuronal markers (Jonakait *et al.* 1977), it is depleted by administration of 5,7-dihydroxytryptamine (Jonakait *et al.* 1977), which lesions enteric 5-HT-containing neurones (Gershon *et al.* 1980), and it is released, along with 5-HT, in a Ca^{2+}-dependent manner by electrical stimulation (Jonakait *et al.* 1979). Both of the peripheral serotonin binding protein-containing cell types develop from the neural crest (Le Douarin 1982; Rothman *et al.* 1986). One is a neurone and the other, the parafollicular cell, retains the capacity to respond like neurones even in adult life (Barasch *et al.* 1987a). Serotonin binding protein is not found in peripheral sites of 5-HT storage, such as mast cells, platelets, and entero-endocrine cells (Gershon and Tamir 1985), which do not develop from the neuro-ectoderm. The function of serotonin binding protein in 5-HT-containing neurones and parafollicular cells is unknown; however, its properties suggest that it may serve as a 5-HT storage protein. Serotonin binding protein binds 5-HT with extremely high affinity under intracellular conditions (low Na^+, Ca^{2+}) but loses its affinity for 5-HT when exposed to the salt concentrations found in extracellular fluid (Gershon and Tamir 1985). Thus, serotonin binding protein can be expected to bind 5-HT very tightly within synaptic vesicles, but to release 5-HT and not to re-bind it when exposed to the ambient medium during the process of exocytosis.

11.2.1.4 *Culture of enteric 5-HT-containing neurones: proof that the neurones are intrinsic components of the enteric nervous system*

The first demonstration that enteric 5-HT-containing neurones are intrinsic to the enteric nervous system was accomplished by showing that enteric 5-HT-containing neurones persisted in long-term organotypic tissue cultures of bowel, even after the degeneration of extrinsic nerves and the mucosa (Dreyfus *et al.* 1977a and b). These neurones contain 5-HT, synthesize it from L-tryptophan, and take up [^{3}H]5-HT as do enteric 5-HT-containing neurones in the intact gut. Cultured enteric 5-HT-containing neurones have also been found to be strikingly similar in their morphology to those found in the myenteric plexus *in situ* (Saffrey *et al.* 1984). The concentration of 5-HT in whole cultured enteric neurones has been estimated to be about 7 mM (~60 fmol/cell) (Gershon 1982); the intravesicular concentration must be much greater. Such a high concentration necessitates the existence of a mechanism to prevent the osmotic swelling of vesicles. It is possible that the 45 kDa serotonin binding protein functions as a component of this mechanism.

11.2.2 High affinity uptake of 5-HT into enteric neurones

Enteric neurones take up [^{3}H]5-HT (Robinson and Gershon 1971; Gershon and Altman 1971). This uptake is saturable, has a K_M of about 0.7 μM, a Q_{10} (27–37°C) of 3.6, and is energy dependent. [^{3}H]5-HT uptake is also Na^+-dependent, inhibited as an exponential function of the $[K^+]_e$, and sensitive to $[Ca^{2+}]_e$. Affinity of compounds for the [^{3}H]5-HT uptake site is much reduced when the alkyl amino group of 5-HT is substituted and when the 5-hydroxyl group on the indole ring is missing or masked (Gershon *et al.* 1976). Ring hydroxylated analogues of 5-HT compete with [^{3}H]5-HT for entry into enteric neurones and are themselves taken up by them. The uptake of some of these analogues can be demonstrated histologically. For example, 6-hydroxytryptamine can be visualized by aldehyde-induced histofluorescence (Jonsson *et al.* 1969), and 5,7-dihydroxytryptamine can be identified immunocytochemically using anti-5-HT sera (Costa *et al.* 1982; Furness and Costa 1982), or by its characteristic reaction with sodium permanganate (Gershon and Sherman 1982a and b). Although cell bodies of enteric 5-HT-containing neurones also take up [^{3}H]5-HT, the uptake is mainly into nerve terminals (Gershon and Sherman 1982a and 1987). Like central nervous system neuronal uptake of [^{3}H]5-HT, the enteric 5-HT uptake mechanism is inhibited by chlorimipramine more potently than by imipramine or desmethylimipramine (Gershon *et al.* 1976), and is specifically inhibited by fluoxetine (Gershon and Jonakait 1979) and zimelidine (Takaki *et al.* 1985b). The specificity of the uptake mechanism for 5-HT and its distinction from that of noradrenaline makes it possible to use indolic neurotoxins to ablate enteric 5-HT-containing axons selectively (Gershon *et al.* 1980).

11.2.3 Biosynthesis of 5-HT from L-tryptophan and the metabolism of 5-HT in enteric neurones

The synthesis of [^{3}H]5-HT from [^{3}H]L-tryptophan by mucosa-free strips of intestinal wall was first demonstrated by Dreyfus *et al.* (1977a) and, although not detected by Lees *et al.* (1979; quoted in Furness and Costa

1987), it has since been confirmed by Legay *et al.* (1983b). This synthesis is apparently dependent upon a tryptophan hydroxylase similar to that of the brain, because, like the biosynthesis of 5-HT in the central nervous system, it is inhibited by p-chlorophenylalanine. Since [^{3}H]L-tryptophan is also metabolised by myenteric plexus-containing intestinal strips to [^{3}H]5-hydroxyindoleacetic acid (Dreyfus *et al.* 1977a; Legay *et al.* 1983b), at least some of the 5-HT in the enteric nervous system is probably metabolised by monoamine oxidase. [^{3}H]L-tryptophan is similarly metabolised to [^{3}H]5-HT and [^{3}H]5-hydroxyindoleacetic acid by explanted enteric 5-HT-containing neurones in organotypic tissue cultures (Dreyfus *et al.* 1977a). On the other hand, monoamine oxidase does not fully account for the metabolism of 5-HT by the bowel; conversion of 5-HT to 5-HT-*O*-glucuronide occurs in the gut and is a significant route of enteric 5-HT metabolism (Gershon and Ross 1966). Enteric 5-HT-containing neurones also take up [^{3}H]L-tryptophan by a mechanism that is saturable ($K_M \sim$ 50μM), energy dependent, and stimulated by 5-HT itself (Gershon and Dreyfus 1980). The 5-HT stimulation of [^{3}H]L-tryptophan uptake is blocked by tetrodotoxin and is mimicked by elevating $[K^+]_e$, suggesting that it is due to neuronal activation.

11.2.4 Release of 5-HT and 'serotonin binding protein' from enteric neurones

5-HT is released from the stomach by stimulation of the vagus nerves (Paton and Vane 1963). Much of this 5-HT, as well as that released by vagal stimulation into the intestinal lumen, is probably derived from entero-endocrine cells (Tansy *et al.* 1971; Ahlman *et al.* 1976; Grønstad *et al.* 1985); however, stimulation of the vagi continues to release 5-HT from the stomach even after the mucosa, and thus the entero-endocrine cells, have been destroyed by asphyxiation (Bülbring and Gershon 1967). [^{3}H]5-HT taken up by nerve terminals of the dissected myenteric plexus *in vitro* is released by electrical stimulation (Schulz and Cartwright 1974; Jonakait *et al.* 1979); moreover, this release is Ca^{2+}-dependent and inhibited by tetrodotoxin (Jonakait *et al.* 1979). More recently, the release of endogenous 5-HT from stimulated myenteric neurones has been demonstrated, and this release, like that of [^{3}H]5-HT, is Ca^{2+}-dependent (Gershon and Tamir 1981a; Holzer and Skofitsch 1984). The neuronal release of 5-HT appears to be increased by peristalsis (Gwee and Yeoh 1968), an observation that is consistent with the idea that neuronal 5-HT functions in the mediation of the peristaltic reflex (Julé 1980). Physiologically active enteric neuronal 5-HT is also released from enteric neurones by the application of tryptamine (Takaki *et al.* 1985b); thus, tryptamine can be used to evoke 5-HT-mediated neuronal responses, to desensitize enteric neuronal 5-HT receptors, and to deplete enteric neuronal stores of 5-HT.

11.3 Actions of 5-HT on the gut

11.3.1 Actions of 5-HT on intact preparations of bowel

The release of 5-HT in large amounts from entero-endocrine cells by vagal stimulation (Ahlman *et al.* 1976; Grønstad *et al.* 1985) and by application of pressure to the gastro-intestinal mucosa (Bülbring and Lin 1958; Bülbring and Crema 1959) raises the question of how 5-HT can act as an enteric neurotransmitter if the enteric nervous system is continually bathed in 5-HT. In fact, application of 5-HT to the serosal surface of the bowel (Bülbring and Crema 1958) does block the peristaltic reflex, whereas introduction of 5-HT into the lumen of the gut activates the peristaltic reflex and can restore it when the reflex has been partially blocked by intraluminal administration of procaine. These observations indicate that the actions of 5-HT on the gut differ depending on the structures of the bowel to which the amine has access. The difference in effects, between mucosal and serosal application of 5-HT, implies that there is an intra-enteric barrier that protects the myenteric plexus from the 5-HT released from the mucosal epithelium. Such a barrier has been demonstrated using [^{3}H]5-HT (Gershon and Tamir 1981a). The barrier is located between the submucosa and the muscularis externa and does not prevent the trans-mucosal movement of 5-HT (Cooke *et al.* 1983) or access to mucosal afferent nerve fibres of 5-HT released from enterochromaffin cells (Bülbring and Crema 1959).

It is now clear that 5-HT has many sites of action within the bowel, affecting epithelium and muscle, as well as neurones. For example, 5-HT stimulates chloride secretion by crypt cells of the mucosal epithelium (Sheerin 1979; Donowitz *et al.* 1980; Hardcastle *et al.* 1981; Cooke and Carey 1985; Baird and Cuthbert 1987). Although tetrodotoxin does not fully block the effect of 5-HT on chloride secretion, a portion of the response is indirect and due to stimulation by 5-HT of submucosal neurones (Cassuto *et al.* 1982; Cooke and Carey 1985; Cooke 1987; Baird and Cuthbert 1987; see Chapter 12). 5-HT also potently activates the enteric musculature (Vane 1957; Gaddum and Picarelli 1957; Gershon 1967; see Chapter 9); however, the ability of the smooth muscle to respond directly to 5-HT varies widely in different regions of the gastro-intestinal tract and in different species (Bülbring and Gershon 1967; Gershon 1967; Costa and Furness 1979b). Neuronal actions of 5-HT are much more constant and include: excitation of afferent nerve fibres from the bowel from which recordings can be made in the abdomen (Paintal 1954 and 1964; Lew and Longhurst 1986); vagus nerves (Ireland and Tyers 1987) and nodose ganglion (Wallis *et al.* 1982); as well as primary afferent neurones within the wall of the gut itself (Bülbring and Lin 1958; Bülbring and Crema 1958 and

1959). The action of 5-HT on enteric afferent nerves appears to be an extension to the bowel of a general ability of 5-HT to activate sensory nerve fibres (see Fozard 1984; and Chapter 10). Additional neuronally mediated effects of 5-HT on the bowel include the stimulation of cholinergic ganglion cells innervating the muscularis externa (Gaddum and Picarelli 1957; Brownlee and Johnson 1963; Harry 1963; Bülbring and Gershon 1967; Gershon 1967; Vizi and Vizi 1978; Drakontides and Gershon 1968; Costa and Furness 1979b) and muscularis mucosa (Kamikawa and Shimo 1983), the activation of intrinsic inhibitory non-adrenergic, non-cholinergic ganglion cells (Bülbring and Gershon 1967; Gershon 1967; Drakontides and Gershon 1968; Furness and Costa 1973; Costa and Furness 1979b), and a presynaptic inhibition of the release of acetylcholine (North *et al.* 1980; Sanger 1985). A variety of neuronal responses of the gastro-intestinal tract have been attributed to 5-HT. These include vagal relaxation of the stomach in mice and guinea-pigs (Bülbring and Gershon 1967), vagal relaxation of the lower oesophageal sphincter of the opposum (Rattan and Goyal 1978), post-train synaptic excitation in the myenteric plexus (Dingledine and Goldstein 1976), ascending peristaltic excitation in the colon (Furness and Costa 1973; Costa and Furness 1976), descending inhibition of vagal excitation of the colon (Julé 1980), cholera toxin-induced intestinal secretion (Cassuto *et al.* 1982), and regulation of the migrating myo-electric complex (Ormsbee *et al.* 1984; Davidson and Pilot 1986).

11.3.2 Actions of 5-HT on single enteric neurones

Neurones of the myenteric plexus can be classified physiologically into three, possibly four, groups (Holman *et al.* 1972; Nishi and North 1973; Hirst *et al.* 1974; Wood 1987). The criteria used for this classification in the present review are those of Takaki *et al.* (1985a), are similar to those of Wood (1987), and are as follows: Type I/S neurones have a high input resistance, spike repeatedly during the injection of depolarizing current, display anodal break excitation, have tetrodotoxin-sensitive somal spikes, and do not display prolonged hyperpolarizing after-potentials. Type II/AH cells have a low input resistance, fail to spike repeatedly during the injection of depolarizing current, do not exhibit anodal break excitation, have tetrodotoxin-resistant action potentials, and display prolonged hyperpolarizing after-potentials. Type III/NS neurones have a low input resistance, do not spike at all in response to injection of depolarizing current pulses, but show prominent cholinergic, nicotinic excitatory postsynaptic potentials that never trigger action potentials. A fourth type of cell is initially inexcitable when first impaled with a recording micropipette, but eventually behaves like a type II/AH neuron.

Application of 5-HT to single myenteric neurones by pressure micro-ejection or by superfusion induces a long lasting depolarization in type II/

AH neurones (Wood and Mayer 1979; Johnson *et al.* 1980b; Bornstein *et al.* 1984; Takaki *et al.* 1985a). This response is associated with an increased input resistance. During the response the after-hyperpolarization that characterizes type II/AH neurones is attenuated. As a consequence, the cells become hyperexcitable. This effect of 5-HT is concentration-dependent and is not inhibited by hexamethonium. Other types of effect have been seen in response to micro-ejection of 5-HT. The most common is a rapid depolarization of type I/S or type II/AH cells associated with a fall in input resistance. This response is significantly shorter than the slow depolarization evoked by 5-HT, enabling a single cell to display both types of response. The fast response to 5-HT is not antagonized by hexamethonium, tetrodotoxin, low Ca^{2+}/high Mg^{2+}-containing medium, or naloxone; moreover, both the fast and slow responses to 5-HT are blocked by receptor desensitization using prolonged exposures to high concentrations of 5-HT itself. Like the slow response to 5-HT, therefore, the fast response is mediated directly on the surfaces of responding neurones through a subtype of 5-HT receptor. More rarely, a hyperpolarization, associated with a decrease in input resistance, is evoked by application of 5-HT. Responses, if any, of type III/NS neurones to 5-HT have not been reported.

11.4 Enteric neuronal 5-HT receptors

11.4.1 M and D receptors for 5-HT

Enteric receptors for 5-HT were first classified by Gaddum and Picarelli (1957). Mechanical responses of the guinea-pig ileum to 5-HT that were antagonized by morphine were said to be mediated by M receptors, while those that were blocked by dibenzyline were attributed to D receptors. Neither morphine nor dibenzyline are specific 5-HT receptor antagonists; however, the distinction between these two types of receptor has been found to be real and to correspond to a difference in receptor location. M receptors are neuronal and lead to a contractile response of the gut because, when activated by 5-HT, they directly or indirectly lead to the release of acetylcholine from the common excitatory motor neurones of the gut (Gershon 1967; Drakontides and Gershon 1968; Vizi and Vizi 1978; Buchheit *et al.* 1985). M receptor-mediated responses to 5-HT have thus come to mean those effects of 5-HT that can be blocked by neuronal paralysis with an agent such as tetrodotoxin (Gershon 1967; Drakontides and Gershon 1968; Fozard and Mobarok Ali 1978; Richardson *et al.* 1985). The tetrodotoxin-resistant D receptor-mediated responses to 5-HT are due to a direct activation of receptors located on the intestinal smooth muscle itself (Gershon 1967). Since the drugs that block D receptor-mediated responses to 5-HT also are antagonists at the 5-HT_2 class of receptor that has been found in the central nervous system (Cohen *et al.* 1983; Peroutka

1984), it is probable that D receptors and 5-HT_2 receptors correspond (Engel *et al.* 1984; Richardson and Engel 1986; Bradley *et al.* 1986). The M receptor has only recently been studied in detail and, as described in the sections which follow, the situation is complex.

11.4.1.1 *Recent development of selective agonists and antagonists at enteric neuronal 5-HT receptors*

The electrophysiological observations, discussed above, that 5-HT has multiple actions on enteric neurones suggest that it is an oversimplification to assume that there is only a single type of enteric neuronal M receptor for 5-HT. This is particularly apparent in those myenteric neurones that display both fast and slow responses to 5-HT, which are associated with different ion conductances. Moreover, the enteric nervous system is extremely complex, containing many different neurotransmitter-defined neuronal phenotypes (Gershon 1981; Costa *et al.* 1987). Many of these are interneurones and are acted upon by 5-HT. As a result, it is difficult to interpret the results of pharmacological experiments in which a neuronally active drug is applied simultaneously to many elements of the enteric nervous system in an organ bath where muscle contraction is used to assess drug action. Muscle contraction or relaxation reflects the output of final common excitatory or inhibitory myenteric neurones and may be the net result of a complicated series of actions on the interneurones that comprise enteric microcircuits. These experiments can be likened to administration of drugs into the carotid arteries and trying to analyse central nervous system receptors from the resulting effects on animal behaviour. To characterize effectively enteric neuronal receptors for 5-HT, therefore, it is advantageous to study each of the actions of the amine on individual enteric neurones. Receptor ligand binding assays that could be correlated with electrophysiological results would also be most helpful.

Recently, dipeptides of 5-hydroxytryptophan, *N*-hexanoyl- and *N*-acetyl-5-hydroxytryptophyl-5-hydroxytryptophan amide (5-HTP-DP) have been shown to be antagonists of the slow response to 5-HT of myenteric type II/AH neurones (Takaki *et al.* 1985a). 5-HTP-DP is a highly specific antagonist of 5-HT and does not antagonize those responses of myenteric type II/AH neurones to substance P or to acetylcholine (mediated via muscarine receptors), which are similar to the slow responses of these neurones to 5-HT; nor does 5-HTP-DP affect fast excitatory postsynaptic potentials, nicotinic responses to acetylcholine, or fast responses to 5-HT (Takaki *et al.* 1985a; Mawe *et al.* 1986).

In addition to 5-HTP-DP, other compounds have been reported to be antagonists of 5-HT at neuronal receptors. These include indoletropanyl esters e.g. ICS 205-930 (Richardson *et al.* 1985). ICS 205-930 blocks a part of the tetrodotoxin-sensitive contractile responses of the bowel to 5-HT (Buchheit *et al.* 1985). While it is thus clear that the site of action of ICS

205-930 is within the enteric nervous system, the particular 5-HT receptor at which this compound acts has to be distinguished from that acted upon by 5-HTP-DP; which of the several electrophysiologically demonstrable actions of 5-HT are affected by ICS 205-930 is considered in Section 11.4.3.

11.4.2 Assay of enteric neuronal 5-HT receptors using [^{3}H]5-HT as a radioligand

11.4.2.1 *General properties of [^{3}H]5-HT binding to isolated enteric membranes*

Radioligand binding techniques for the assay of enteric neuronal 5-HT receptors have been developed. These methods employ [^{3}H]5-HT as a ligand and have been applied to the enteric nervous system of guinea-pigs, rabbits (Branchek *et al.* 1984a; Gershon *et al.* 1985), and mice (Branchek and Gershon 1987). The binding of [^{3}H]5-HT to membranes derived from neurones of the myenteric plexus is saturable and dissociable with an equilibrium dissociation constant (K_D) that ranges in different species between 1.4 and 3.7 nM. The dissociation rate constant of the [^{3}H]5-HT-receptor complex (K_{-1}) is about 0.11/min. The association rate constant (K_{+1}) is about 7.5×10^7 /mol/min. Kinetic estimates of the K_D for the binding of [^{3}H]5-HT, from the ratio K_{-1}/K_{+1}, 1.5 nM, is close to that obtained from analysis of saturation isotherms. Moreover, the Hill coefficient approximates unity. These data suggest that there is only a single high affinity binding site in enteric neuronal membranes for [^{3}H]5-HT with no positive or negative co-operativity. The specificity of the binding of [^{3}H]5-HT to enteric neuronal membranes is indicated by its lack of inhibition by compounds known to combine with receptors for other neurotransmitters, '5-HT_1-like' or 5-HT_2 receptors, or [^{3}H]5-HT uptake sites. Classes of compounds that have been tried and that have failed to reduce significantly the binding of [^{3}H]5-HT to enteric neuronal membranes include agonists and/or antagonists at: muscarine (atropine), nicotine (hexamethonium), D_1 and D_2 dopamine (metoclopramide, spiroperidol, domperidone, butaclamol), H_1 and H_2 histamine (cyproheptadine, diphenhydramine, cimetidine, mianserin), opiate (naloxone, codeine), and GABA (muscimol) receptors and α- and β-adrenoceptors (phentolamine, propranolol). Compounds that have activity at other types of 5-HT receptor, but which have been found to be unable to antagonize the binding of [^{3}H]5-HT to enteric neuronal membranes include: cinanserin, LSD, metergoline, 5-methoxytryptamine, methysergide, 6-nitroquipazine, trazodone, cyproheptadine, ketanserin, mianserin, spiroperidol, pirenpirone, 8-OH-DPAT, MDL 72222, and ICS 205-930. On the other hand, the binding of [^{3}H]5-HT to enteric neuronal membranes is antagonized by 5-HTP-DP and by hydroxylated indalpines (Takaki *et al.* 1985a; Mawe *et al.* 1986; Branchek and Gershon 1987). The high affinity receptor for [^{3}H]5-HT in enteric neuronal membranes, therefore, is different from the 5-HT receptors that have previously been described in the central nervous system and is not the receptor responsible for the actions of 5-HT antagonized by benzoyltropine analogues (Fozard 1984) and indoletropanyl esters (Richardson *et al.* 1985; Buchheit *et al.* 1985). On the other hand, it is likely to be the receptor responsible for those actions of 5-HT that are antagonized by 5-HTP-DP.

11.4.2.2 *Structure–activity relationships*

The structure–activity relationship of requirements for ^{3}H-5-HT binding matches that in the older literature for agonists at neuronal (M) 5-HT receptors. Hydroxylated tryptamines are known to evoke neuronally mediated enteric contractions, while tryptamine itself neither mimics nor antagonizes neuronally mediated responses to 5-HT (Gyermek 1966; Drakontides and Gershon 1968; Fozard and Mobarok Ali 1978; Costa and Furness 1979b). Similarly, the binding of [^{3}H]5-HT to enteric neuronal membranes from rabbit, guinea-pig, or mouse bowel is antagonized only by hydroxylated tryptamines and not by analogues bearing other substituents, or by tryptamine itself (Branchek *et al.* 1984a; Gershon *et al.* 1985; Mawe *et al.* 1986; Branchek and Gershon 1987). In like manner, compounds that are substituted on the aliphatic amino group retain both the ability to act on neuronal 5-HT receptors and their ability to interfere with the binding of [^{3}H]5-HT. The inability of classical 5-HT receptor antagonists, such as LSD, methysergide, cyproheptadine, metergoline, and cinanserin to block [^{3}H]5-HT binding to enteric neuronal membranes also corresponds with the failure of these same compounds to antagonize neuronally mediated responses of the bowel (or other peripheral organs) to 5-HT (Gyermek 1966; Drakontides and Gershon 1968; Fozard and Mobarok Ali 1978; Costa and Furness 1979b; Humphrey 1983; Humphrey *et al.* 1983). A possible exception is 5-methoxytryptamine, which has been reported to be a 5-HT M receptor agonist (Fozard and Mobarok Ali 1978; Fozard 1985), but fails to block the binding of [^{3}H]5-HT to enteric neuronal membranes (Branchek *et al.* 1984a). Although Gyermek (1966) classified 5-methoxy tryptamine as active at D (or muscle) receptors, the compound does, in fact, act on enteric neurones; however, its neuronal action, assessed electrophysiologically, is not similar to that of 5-HT (Takaki *et al.* 1985a). 5-Methoxytryptamine produces a long-lasting depolarization of enteric neurones with a conductance change (increase) opposite to that seen in the slow response to 5-HT and longer-lasting than that seen in the fast response to 5-HT. The receptor acted upon by 5-methoxytryptamine, therefore, is probably not the receptor responsible for the slow or the fast response to 5-HT. The observation that tryptamine, which is an agonist at muscle receptors for 5-HT and methysergide, which is an antagonist at muscle 5-HT receptors (Drakontides and Gershon 1968), fail to block binding of [^{3}H]5-HT to enteric neuronal membranes (Branchek *et al.* 1984a), further argues that the binding sites are enteric neuronal receptors for 5-HT. Tryptamine, in fact, has a highly unusual effect on enteric 5-HT-containing neurones, which can lead to both activation and transient desensitization of enteric neuronal 5-HT receptors. Tryptamine releases endogenous 5-HT from enteric 5-HT-containing neurones in such a way that the released 5-HT reaches and acts on these receptors in postganglionic neurones (Takaki *et al.* 1985b). This release of 5-HT is Ca^{2+}-independent. Neuronal actions of tryptamine, therefore, are not seen in tissue from which 5-HT has been depleted and tryptamine can be used to deplete enteric neuronal stores of 5-HT.

11.4.2.3 *Autoradiography*

Binding sites for [^{3}H]5-HT can be located in peripheral tissues by dry mount autoradiography (Branchek *et al.* 1984a). The binding of [^{3}H]5-HT assessed by autoradiography shows identical properties to those of [^{3}H]5-HT binding to enteric neuronal membranes in rapid filtration assays (Branchek *et al.* 1984a; Gershon *et al.* 1985; Mawe *et al.* 1986; Branchek and Gershon 1987). It can therefore be assumed that the same receptor is demonstrated by each of these methods. Autoradiography has made it possible to study the development of enteric neuronal 5-HT receptors (Branchek and Gershon 1987). [^{3}H]5-HT binding sites appear in the fetal murine bowel, from stomach to mid-jejunum at day 14 of development and then slowly spread distally down the bowel. [^{3}H]5-HT binding sites, however, do not reach the distal colon until postnatal day 2 and the terminal rectum until the animals are weaned. An investigation to discover what abilities are acquired by the colon in the period around weaning may provide considerable insight into the function of neuronal 5-HT receptors in the physiology of the lower bowel.

[^{3}H]5-HT binding sites have been found by autoradiography not only in the myenteric and submucosal plexuses, but also in the lamina propria of the enteric mucosa (Branchek and Gershon 1987). In the lamina propria the [^{3}H]5-HT binding sites lie just beneath the gastro-intestinal epithelium in regions that contain nerve fibres. Autoradiography resolution is not sufficient to permit these nerve fibres to be positively identified as the location of [^{3}H]5-HT binding sites; however, the mucosal [^{3}H]5-HT binding sites are virtually absent from the terminal colon of the lethal spotted (*ls*/*ls*) mutant mouse, which is aganglionic (Branchek *et al.* 1984b). The aganglionic zone of the *ls*/*ls* gut is deficient mainly in intrinsic neurites; the tissue is heavily innervated by extrinsic axons, which are the enteric projections of neurones, the cell bodies of which lie outside the bowel (Payette *et al.* 1987). Moreover [^{3}H]5-HT does not bind to mucosal epithelial cells themselves (Gaginella *et al.* 1983). These observations suggest that high affinity 5-HT receptors are found on mucosal intrinsic nerve fibres as well as on enteric ganglion cells. A possible role for 5-HT receptors on mucosal axons is to function in the initiation of the peristaltic reflex (Bülbring and Lin 1958). When pressure is applied to the mucosal surface of the gut, 5-HT is released from entero-endocrine cells and reaches the lamina propria, stimulating intrinsic mucosal afferent nerve processes that activate the peristaltic reflex. The location of mucosal [^{3}H]5-HT binding sites is thus appropriate for them to be the 5-HT receptors upon which musocal 5-HT acts. The observation that these sites are virtually absent from the aganglionic bowel of *ls*/*ls* mice (Branchek *et al.* 1984b) is consistent with this hypothesis.

11.4.3 5-HT$_{1P}$ and 5-HT$_3$ (5-HT$_{2P}$) receptors

11.4.3.1 *Correlation between the binding of ^{3}H-5-HT and the actions of 5-HT on single enteric neurones*

There is a very good correlation between the structure of compounds that antagonize the binding of [^{3}H]5-HT and the structure of compounds able either to induce responses of myenteric type II/AH neurones that mimic the slow response of these neurones to 5-HT, or to block slow responses to 5-HT (Takaki *et al.* 1985a; Mawe *et al.* 1986). These observations suggest that compounds that are able to displace [^{3}H]5-HT from its binding sites on enteric neuronal membranes will be agonists or antagonists at the receptor responsible for the induction of slow responses to 5-HT. Furthermore, the selective 5-HT antagonist, 5-HTP-DP, effectively blocks the binding of [^{3}H]5-HT, as do hydroxylated indalpines, which mimic the slow response to 5-HT (Mawe *et al.* 1986). On the other hand, ICS 205-930, which blocks the fast response to 5-HT, but not the slow (see Fig. 11.1), fails to interfere with the binding of [^{3}H]5-HT. 2-Methyl-5-HT is an agonist that, at low concentrations, mimics the fast response to 5-HT, but at higher concentrations 2-methyl-5-HT is an agonist of slow responses. Correspondingly, 2-methyl-5-HT is a weak antagonist of the binding of [^{3}H]5-HT. Two depolarizing responses to 5-HT, essentially identical to those found in neurones of the myenteric plexus, have recently been observed in neurones of the submucosal plexus (Surprenant and Crist 1988). Like their counterparts in the myenteric plexus, these responses can also be differentiated on the basis of their differing ionic mechanisms and sensitivity to ICS 205-930.

These observations have led to the proposal that enteric neuronal receptors for 5-HT be classified as 5-HT$_{1P}$ and 5-HT$_{2P}$ (Mawe *et al.* 1986). This nomenclature continues the convention that has been developed for classifying 5-HT receptors in the central nervous system. Those 5-HT receptors that can be labelled by [^{3}H]5-HT are generally classified as one of the subsets of the 5-HT$_1$ receptor (Pedigo and Nelson 1981; Pazos *et al.* 1985a and b; Yagaloff and Hartig 1985; Heuring and Peroutka 1987). On the other hand, the 5-HT$_2$ class of receptor has a relatively low affinity for 5-HT itself and cannot be labelled by ^{3}H-5-HT (Pazos *et al.* 1985a and b).

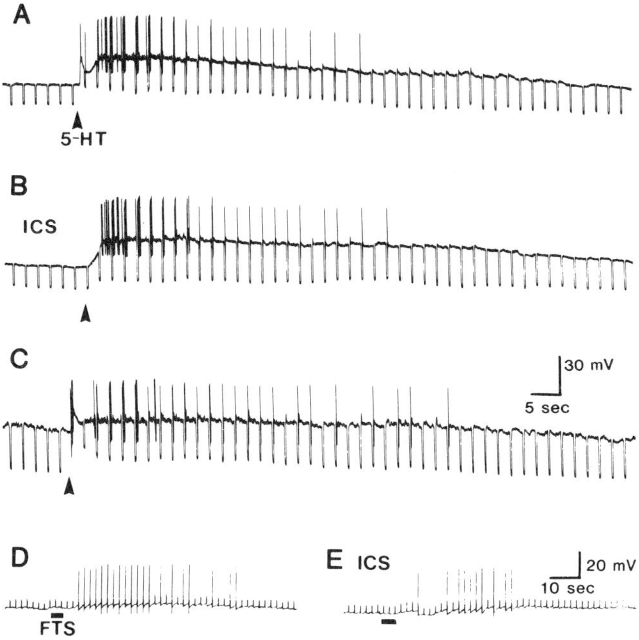

Fig. 11.1 Responses of two type II/AH neurones of the guinea-pig myenteric plexus to the micro-ejection of 5-HT and to stimulation of an interganglionic fibre tract (FTS). The effect on these responses of the addition of ICS 205-930 is also illustrated. Constant hyperpolarizing (A-C) or depolarizing (D and E) current pulses were injected into the cells through the recording micro-electrode. **A.** Control response to micro-ejection of 5-HT (arrowhead): there is a transient depolarization (the fast response) and a partial recovery of the membrane potential, followed by a prolonged depolarization of the membrane (the slow response). During the slow response the neurone is hyperexcitable and spikes frequently. Note the increase in input resistance during the slow response to 5-HT. **B.** Superfusion with ICS 205-930 (1 μM) blocks the initial fast response to 5-HT, but does not affect the slow response. **C.** One hour following the washout of ICS 205-930 the fast response to micro-ejection of 5-HT (arrowhead) has recovered. Note that input resistance is diminished during the fast response and the partial recovery of the membrane potential that follows it. **D.** A slow excitatory postsynaptic potential is evoked by FTS. Note that although there is only a modest depolarization of the membrane (5 mV) the cell is hyperexcitable throughout the response, which long outlasts the application of stimuli. **E.** Superfusion of ICS 205-930 (1 μM) does not affect the slow excitatory postsynaptic potential (6 mV). These observations indicate that the 5-HT receptor blocked by ICS 205-930 (5-HT_3; 5-HT_{2P}) mediates fast responses to 5-HT, but that these receptors are not responsible for the physiological effect of 5-HT released by FTS, to generate a slow excitatory postsynaptic potential in follower neurones. This effect is mediated by the high affinity 5-HT receptor at which hydroxylated indalpines are agonists and 5-HTP-DP is an antagonist (5-HT_{1P}). (From Mawe *et al.* 1986 with permission)

The 5-HT_2 receptor can instead be labelled by antagonists, such as [^{3}H] ketanserin (Leysen *et al.* 1982). The enteric neuronal 5-HT receptor that is responsible for the slow response of type II/AH neurones to 5-HT appears to be the high affinity receptor that can be labelled by [^{3}H]5-HT; logically, therefore, it is a subtype of 5-HT_1 receptor. In view of the fact that the receptor has, thus far, been found only in the periphery, it has been named the 5-HT_{1P} receptor (1 indicates its ability to be labelled by [^{3}H]5-HT and P indicates peripheral). In contrast, the receptor responsible for the mediation of fast responses to 5-HT cannot be labelled by [^{3}H]5-HT and so may be considered a subset of the 5-HT_2 receptor (5-HT_{2P}). This receptor can be defined pharmacologically (ICS 205-930 is a selective antagonist and 2-methyl-5-HT a selective agonist at this receptor); therefore, the 5-HT_{2P} receptor is probably identical to that classified as 5-HT_3 (Bradley *et al.* 1986).

11.4.3.2 *Contributions of knowledge of enteric neuronal 5-HT receptors to defining a physiological role for 5-HT in the enteric nervous system: 5-HT mediates slow excitatory postsynaptic potentials*

For a long time research on the physiological role played by 5-HT in the enteric nervous system was complicated by the lack of a specific antagonist at neuronal 5-HT receptors. This led some to doubt whether 5-HT or a similar compound is an enteric neurotransmitter (Costa and Furness 1979a). The observation that slow excitatory postsynaptic potentials on myenteric type II/AH neurones resemble the slow response to 5-HT and could be blocked by methysergide suggested that these responses might be mediated by 5-HT (Wood and Mayer 1979). On the other hand, methysergide is not a specific antagonist at enteric neuronal 5-HT receptors and has the additional action of reducing the height of cholinergic fast excitatory postsynaptic potentials in both type I/S and type II/AH neurones. Since methysergide does not block nicotinic responses to acetylcholine itself, its effect on fast excitatory postsynaptic potentials raises the possibility that it antagonizes slow excitatory postsynaptic potentials, not at postjunctional 5-HT receptors, but presynaptically, reducing the release of the authentic transmitter. Moreover, 5-HT is not the only neuroactive substance found in the enteric nervous system that mimics slow excitatory postsynaptic potentials. The response is also mimicked by acetylcholine acting at muscarinic receptors (North and Tokimasa 1982), by neuropeptides, such as substance P (Katayama and North 1978; Katayama *et al.* 1979; Johnson *et al.* 1981), bombesin, cholecystokinin, gastrin releasing peptide, vasoactive intestinal polypeptide, and somatostatin, as well as by histamine (Wood 1987). Since these other substances evoke responses that resemble slow excitatory postsynaptic potentials and are found in the myenteric plexus, each must be considered a possible mediator of the response.

More recent evidence confirms that the original suggestion that 5-HT is a mediator of slow excitatory postsynaptic potentials was correct. Every type II/AH neuron in which a slow excitatory postsynaptic potential can be elicited by stimulating interganglionic fibre tracts in the myenteric plexus in fact receives 5-HT-containing synapses (Erde *et al.* 1985). These have been demonstrated by combining the electron microscopic autoradiographic demonstration of specific [^{3}H]5-HT uptake into axon terminals with the intracellular injection of horseradish peroxidase into physiologically identified neurones. Because relatively few myenteric neurones are contacted by 5-HT-containing synapses, the distribution of these synapses specifically on those cells that display slow excitatory postsynaptic potentials is highly significant and non-random. As yet, no comparable study has been made of cells showing slow excitatory postsynaptic potentials to determine if they receive peptidergic or cholinergic synapses. In addition, exposure of the gut to tryptamine, in concentrations and for a duration of time sufficient to deplete endogenous 5-HT, leads to a loss of the ability to evoke slow, but not fast, excitatory postsynaptic potentials on myenteric type II/AH cells (Takaki *et al.* 1985b). More compelling evidence that slow excitatory postsynaptic potentials are mediated by 5-HT has been provided by analyses of the effects of 5-HTP-DP on these responses (Takaki *et al.* 1985a). 5-HTP-DP blocks slow excitatory postsynaptic potentials as well as slow responses to 5-HT, but as noted above, fails to antagonize responses of these neurones to acetylcholine or substance P. Furthermore, unlike methysergide, 5-HTP-DP has no presynaptic effect to reduce the size of fast excitatory postsynaptic potentials. Thus, its action in blocking slow excitatory postsynaptic potentials cannot be attributed to interference with the release of transmitter or to a non-specific action on other possible mediators. These studies demonstrate, therefore, that 5-HT is an enteric neurotransmitter and that one of its roles is to serve as a mediator of slow excitatory postsynaptic potentials in myenteric type II/AH neurones. It is unlikely that this will turn out to be the only role of 5-HT in the enteric nervous system because the amine is found in synapses on other types of neurone (Erde *et al.* 1985) and 5-HT has many different actions in the enteric nervous system (Johnson *et al.* 1980a and b). It is also probable that slow excitatory postsynaptic potentials have other mediators besides 5-HT (Bornstein *et al.* 1984). Wood (1987) has proposed that 5-HT-containing neurones, which are all very long, relatively few in number, and which all project proximo-distally (Costa *et al.* 1982; Gershon and Sherman 1987), are responsible for a 'feed-forward' activation of synaptic networks that mediate the descending wave of oral contraction that comprises the peristaltic reflex. His idea is that 5-HT-mediated slow excitatory postsynaptic potentials function in conducting the activity down the bowel, while peptide-mediated slow excitatory postsynaptic potentials propagate the activity circumferentially in locally contracted regions. This postulate requires that 5-HT-containing neurones make synapses on one another, which they do

(Erde *et al.* 1985; Gershon and Sherman 1987) and that peptide-containing neurones that might be responsible for evoking slow excitatory postsynaptic potentials be both more numerous and shorter than 5-HT-containing neurones; substance P-containing neurones fulfill these requirements (Costa *et al.* 1987).

11.5 Concluding remarks

Since the first suggestion in 1965 that 5-HT might be an enteric neurotransmitter a great deal has been learned about enteric 5-HT-containing neurones and the role played by 5-HT in the physiology of the gut. The cell bodies of these neurones are located in ganglia in the myenteric plexus; the 5-HT-containing neurones are interneurons and they send extremely long axonal projections down the gut in a predominantly oral–anal direction. 5-HT is synthesized within the bowel from L-tryptophan, is released along with a specific binding protein, 'serotonin binding protein', and is evidently inactivated by a high affinity, Na^+-dependent uptake mechanism, that is specifically inhibited by drugs such as fluoxetine or zimelidine. Roles for 5-HT have been proposed in many of the functions of the gut; however, deciding which of these are of physiological significance is difficult because 5-HT has many sites of action in the bowel. The analysis of the neuronal actions of 5-HT has been aided by electrophysiological studies of its effects on single enteric neurones. Two of these, a prolonged depolarization associated with a rise in input resistance (the slow response) and a short-lived depolarization associated with a fall in input resistance (the fast response) have been linked to specific 5-HT receptors. The slow response has been associated with a receptor called '5-HT_{1P}', which displays a high affinity for $[^3H]$5-HT and thus can be studied by radioligand binding assays. This receptor is present in both myenteric and submucosal ganglia and in the mucosa, where there is evidence that the receptor may be present on intrinsic enteric axons. Hydroxylated indalpines appear to be agonists, and dipeptides of 5-hydroxytryptophan, antagonists at 5-HT_{1P} receptors. It is likely that 5-HT is one of the mediators of slow excitatory postsynaptic potentials in type II/AH neurones of the myenteric plexus and that these responses to endogenously released 5-HT are mediated by 5-HT_{1P} receptors. The fast response to 5-HT has been associated with 5-HT_3 (5-HT_{2P}) receptors, since it can be mimicked by 2-methyl-5-HT and blocked by ICS 205-930. Studies with compounds such as 5-methoxytryptamine, detailed analyses of the concentration-effect relationship of responses of the bowel to 5-HT, and investigations of the development of 5-HT_{1P} receptors, all suggest that there are additional 5-HT receptors in the enteric nervous system that have yet to be linked to an electrophysiologically characterized response. Considerable excitement and some controversy have fuelled research on 5-HT in the past. Despite the impressive progress made in a relatively short period of time, the number of unresolved questions that still remain provide reason to believe that neither the excitement nor the controversy will be less in the future.

Acknowledgements

Work presented in this review was supported in part by grants NS 12969, NS 15547, NS 22637, and NS 07062 from the National Institutes of Health. Additional support was provided by the Council for Tobacco Research and the Pharmaceutical Manufacturer's Association Foundation.

References

Aghajanian, G. K. and Asher, I. M. (1971). Histochemical fluoresence of raphe neurones: selective enhancement by tryptophan. *Science* **172,** 1159–61.

Aghajanian, G. K., Kuhar, M. J., and Roth, R. H. (1973). Serotonin-containing neuronal perikarya and terminals: differential effects of p-chlorophenylalanine. *Brain Research* **54,** 85–101.

Ahlman, H. and Enerbäck, L. (1974). A cytofluorometric study of the myenteric plexus in the guinea-pig. *Cell Tissue Research* **153,** 419–34.

Ahlman, H., Lundberg, J., Dahlström, A., and Kewenter, J. (1976). A possible vagal adrenergic release of serotonin from enterochromaffin cells in the cat. *Acta Physiologica Scandinavica* **98,** 366–75.

Anderson, C. (1983). Evidence for 5-HT containing intrinsic neurons in the teleost intestine. *Cell Tissue Research* **230,** 377–86.

Baird, A. W. and Cuthbert, A. W. (1987). Neuronal involvement in type 1 hypersensitivity reactions in gut epithelia. *British Journal of Pharmacology* **93,** 647–55.

Barasch, J. M., Mackey, H., Tamir, H., Nunez, E. A., and Gershon, M. D. (1987a). Induction of neural phenotype in a serotonergic endocrine cell derived from the neural crest. *Journal of Neuroscience* **7,** 2874–83.

Barasch, J. M., Tamir, H., Nunez, E. A., and Gershon, M. D. (1987b). Serotonin-storing secretory granules from thyroid parafolicular cells. *Journal of Neuroscience* **7,** 4017–4033.

Baumgarten, H. G., Bjorklund, A., Lachenmayer, L., Nobin, A., and Rosengren, E. (1973). Evidence for existence of serotonin-, dopamine- and noradrenaline-containing neurons in the gut of Lampetra fluviatilis. *Zeitschrift für Zellforschung* **141,** 33–54.

Bernd, P., Nunez, E. A., Gershon, M. D., and Tamir, H. (1979). Serotonin binding protein; characterization and localization in parafollicular cells of the sheep thyroid. In *Molecular endocrinology* (eds. I. MacIntyre and M. Szelke) pp. 123–32. Elsevier/North Holland, Amsterdam, New York, Oxford.

Bernd, P., Gershon, M. D., Nunez, E. A., and Tamir, H. (1981). Separation of dissociated thyroid follicular and parafollicular cells: association of serotonin binding protein with parafollicular cells. *Journal of Cell Biology* **88,** 499–508.

Bornstein, J., North, R. A., Costa, M., and Furness, J. B. (1984). Excitatory synaptic potentials due to activation of neurons with short projection in the myenteric plexus. *Neuroscience* **11,** 723–31.

Bradley, P. B., Engel, G., Feniuk, W., Fozard, J. R., Humphrey, P. P. A., Middlemiss, D. N., Mylecharane, E. J., Richardson, B. P., and Saxena, P. R. (1986). Proposals for the classification and nomenclature of functional receptors for 5-hydroxytryptamine. *Neuropharmacology* **25,** 563–76.

Branchek, T. A. and Gershon, M. D. (1987). Development of neural receptors for serotonin in the murine bowel. *Journal of Comparative Neurology* **258,** 597–610.

Branchek, T. A., Kates, M., and Gershon, M. D. (1984a). Enteric receptors for 5-hydroxytryptamine. *Brain Research* **324,** 107–18.

Branchek, T., Rothman, T., and Gershon, M. D. (1984b). Serotonin receptors on the processes of intrinsic enteric neurons: Reduction in the aganglionic bowel of the *ls/ls* mouse. *Society for Neuroscience Abstracts* **10,** 1097.

Brownlee, G. and Johnson, E. S. (1963). The site of the 5-hydroxytryptamine receptor on the peristaltic reflex. *British Journal of Pharmacology* **13,** 444–57.

Buchheit, K.-H., Engel, G., Mutschler, E., and Richardson, B. (1985). Study of the contractile effect of 5-hydroxytryptamine (5-HT) in the isolated longitudinal muscle strip from guinea-pig ileum. Evidence for two district release mechanisms. *Naunyn-Schmiedeberg's Archives of Pharmacology* **332,** 8–15.

Bülbring, E. and Crema, A. (1958). Observations concerning the action of 5-hydroxytryptamine on the peristalti reflex. *British Journal of Pharmacology* **13,** 444–57.

Bülbring, E. and Crema, A. (1959). The release of 5-hydroxytryptamine in relation to pressure exerted on the intestinal mucosa. *Journal of Physiology (London)* **146,** 381–407.

Bülbring, E. and Gershon, M. D. (1967). 5-Hydroxytryptamine participation in the vagal inhibitory innervation of the stomach. *Journal of Physiology (London)* **192,** 823–46.

Bülbring, E. and Lin, R. C. Y. (1958). The effect of intraluminal application of 5-hydroxytryptamine and 5-hydroxytryptophan on peristalsis, the local production of 5-hydroxytryptamine and its release in relation to intraluminal pressure and propulsive activity. *Journal of Physiology (London)* **140,** 381–407.

Cassuto, J., Jodal, M., Tuttle, R., and Lundgren, O. (1982). 5-Hydroxytryptamine and cholera secretion. *Scandinavian Journal of Gastroenterology* **17,** 695–703.

Cohen, M. L., Mason, N., Wiley, K. S., and Fuller, R. W. (1983). Further evidence that vascular serotonin receptors are of the 5-HT_2 type. *Biochemistry and Pharmacology* **28,** 565–71.

Cooke, H. J. (1987). Neural and humoral regulation of small intestinal electrolyte transport. In *Physiology of the gastrointestinal tract* (ed. L. R. Johnson) Vol. 2, pp. 1307–50. Raven Press, New York.

Cooke, H. J. and Carey, H. V. (1985). Pharmacological analysis of 5-hydroxytryptamine actions on guinea pig ileal mucosa. *European Journal of Pharmacology* **111,** 329–37.

Cooke, H. J., Montakhab, M., Wade, P. R., and Wood, J. D. (1983). Transmural fluxes of 5-hydroxytryptamine in guinea pig ileum. *American Journal of Physiology* **244,** G421–5.

Costa, M. and Furness, J. B. (1976). The peristaltic reflex: an analysis of the nerve pathways and their pharmacology. *Naunyn-Schmiedeberg's Archives of Pharmacology* **294,** 47–60.

Costa, M. and Furness, J. B. (1979a). Commentary: On the possibility that an indoleamine is a neurotransmitter in the gastrointestinal tract. *Biochemical Pharmacology* **28,** 565–71.

Costa, M. and Furness, J. B. (1979b). The sites of action of 5-HT in nerve muscle preparations from guinea-pig small intestine and colon. *British Journal of Pharmacology* **65,** 237–48.

Costa, M., Furness, J. B., Cuello, A. C., Verhofstad, A. A. J., Steinbusch, H. W. J., and Elde, R. P. (1982). Neurons with 5-hydroxytryptamine-like immunoreactivity in the enteric nervous system: Their visualization and reactions to drug treatment. *Neuroscience* **7,** 351–63.

Costa, M., Furness, J. B., and Llewellyn-Smith, I. (1987). Histochemistry of the enteric nervous system. In *Physiology of the gastrointestinal tract* (ed. L. R. Johnson) Vol. 1, 2nd ed., pp. 1–41. Raven Press, New York.

Davidson, H. I. and Pilot, M. A. (1986). Does endogenous neuronal 5-hydroxytryptamine influence canine intestinal motility? *Journal of Physiology (London)* **376,** 49P.

Dingledine, R. and Goldstein, A. (1976). Effect of synaptic transmission blockade on morphine action in the guinea pig myenteric plexus. *Journal of Pharmacology and Experimental Therapeutics* **196,** 97–106.

Donowitz, M., Tai, Y.-H., and Asarkof, N. (1980). Effect of serotonin on active electrolyte transport in rabbit ileum, gall bladder, and colon. *American Journal of Physiology* **239,** G463–72.

Drakontides, A. B. and Gershon, M. D. (1968). 5-HT receptors in the mouse duodenum. *British Journal of Pharmacology* **33,** 480–92.

Dreyfus, C. F., Bornstein, M. B., and Gershon, M. D. (1977a). Synthesis of serotonin by neurons of the myenteric plexus *in situ* and in organotypic tissue culture. *Brain Research* **128,** 109–23.

Dreyfus, C. F., Sherman, D., and Gershon, M. D. (1977b). Uptake of serotonin by intrinsic neurons of the myenteric plexus grown in organotypic tissue culture. *Brain Research* **128,** 109–23.

Dubois, A. and Jacobowitz, D. M. (1974). Failure to demonstrate serotonergic neurons in the myenteric plexus of the rat. *Cell Tissue Research* **150,** 493–6.

Engel, G., Hoyer, D., Kalkman, H. O., and Wick, M. B. (1984). Identification of 5-HT_2-receptors on longitudinal muscle of the guinea-pig ileum. *Journal of Receptor Research* **4,** 113–26.

Erde, S., Sherman, D., and Gershon, M. D. (1985). Morphology of the serotonergic innervation of physiologically identified cells of the guinea pig myenteric plexus. *Journal of Neuroscience* **5,** 617–33.

Erspamer, V. (1966). Occurrence of indolealkylamines in nature. In *Handbook of experimental pharmacology,* Vol. 19, *5-Hydroxytryptamine and related indolealkylamines* (ed. V. Erspamer) pp. 132–81. Springer-Verlag, New York.

Feher, E. (1974). Effect of monoamine oxidase inhibition on the nerve elements of the isolated cat ileum. *Acta Morphologica Academy of Sciences of Hungary* **22,** 249–63.

Feher, E. (1975). Effects of monoamine oxidase inhibitors on the nerve elements of the isolated cat's ileum. *Verhandlungen der Anatomischen Gesellschaft* **69,** 477–82.

Feldberg, W. and Toh, C. C. (1953). Distribution of 5-hydroxytryptamine (serotonin, enteramine) in the wall of the digestive tract. *Journal of Physiology (London)* **119,** 352–62.

Fozard, J. R. (1984). Neuronal 5-HT receptors in the periphery. *Neuropharmacology* **23,** 1473–86.

Fozard, J. R. (1985). 5-Methoxytryptamine (5-MeOT) discriminates between excitatory neuronal 5-hydroxytryptamine (5-HT) receptors in the guinea-pig ileum. *Journale de Pharmacologie (Paris)* **16,** 498.

Fozard, J. R. and Mobarok Ali, A. T. M. (1978). Receptors for 5-hydroxytryptamine on sympathetic nerves of the rabbit heart. *Naunyn-Schmeideberg's Archives of Pharmacology* **301,** 224–35.

Furness, J. B. and Costa, M. (1973). The nervous release and the action of substances which affect intestinal muscle through neither adrenoreceptors nor cholinoreceptors. *Philosophical Transactions of the Royal Society* Series B **265,** 123–33.

Furness, J. B. and Costa, M. (1982). Neurons with 5-hydroxytryptamine-like immunoreactivity in the enteric nervous system: Their projections in the guinea pig small intestine. *Neuroscience* **7,** 341–50.

Furness, J. B. and Costa, M. (1987). *The enteric nervous system,* pp. 65–9. Churchill Livingston, New York.

Fuxe, K. and Jonsson, G. (1967). A modification of the histochemical fluorescence method for the improved localization of 5-hydroxytryptamine. *Histochemistry* **11,** 161–6.

Gaddum, J. H. and Picarelli, Z. P. (1957). Two kinds of tryptamine receptor. *British Journal of Pharmacology and Chemotherapy* **12,** 323–8.

Gaginella, T. S., Rimele, T. J., and Wietecha, M. (1983). Studies on rat intestinal epithelial cell receptors for serotonin and opiates. *Journal of Physiology (London)* **335,** 101–11.

Gershon, M. D. (1967). Effects of tetrodotoxin on innervated smooth muscle preparations. *British Journal of Pharmacology* **29,** 259–79.

Gershon, M. D. (1981). The enteric nervous system. *Annual Review of Neuroscience* **4,** 227–72.

Gershon, M. D. (1982). Enteric serotonergic neurons. In *Biology of serotonergic neurotransmission* (ed. N. Osborne) pp. 363–99. John Wiley, New York.

Gershon, M. D. and Altman, R. F. (1971). An analysis of the uptake of 5-hydroxytryptamine by the myenteric plexus of the small intestine of the guinea-pig. *Journal of Pharmacology and Experimental Therapeutics* **179,** 29–41.

Gershon, M. D. and Dreyfus, C. F. (1980). Stimulation of tryptophan uptake into enteric neurons by 5-hydroxytryptamine: a novel form of neuromodulation. *Brain Research* **184,** 229–33.

Gershon, M. D. and Jonakait (1979). Uptake and release of 5-hydroxytryptamine by enteric serotonergic neurons: effects of fluoxetine (Lilly 110140) and chlorimipramine. *British Journal of Pharmacology* **66,** 7–9.

Gershon, M. D. and Ross, L. L. (1966). Radioisotopic studies of the binding, exchange and distribution of 5-hydroxytryptamine synthesized from its radioactive precursor. *Journal of Physiology (London)* **186,** 451–76.

Gershon, M. D. and Sherman, D. L. (1982a). Selective demonstration of serotonergic neurons and terminals in electron micrographs: Loading with 5,7-dihydroxy-tryptamine and fixation with $NaMnO_4$. *Journal of Histochemistry and Cytochemistry* **30,** 769–73.

Gershon, M. D., and Sherman, D. L. (1982b). Identification of and interactions between noradrenergic and serotonergic neurites in the myenteric plexus. *Journal of Comparative Neurology* **204,** 407–21.

Gershon, M. D. and Sherman, D. L. (1987). Noradrenergic innervation of serotoninergic neurons in the myenteric plexus. *Journal of Comparative Neurology* **259,** 193–210.

Gershon, M. D. and Tamir, H. (1981a). Release of endogenous 5-hydroxytryptamine from resting and stimulated enteric neurons. *Neuroscience* **6,** 2277–86.

Gershon, M. D. and Tamir, H. (1981b). Serotonin binding protein: role in transmitter storage in central and peripheral serotonergic neurons. In *Serotonin* (eds. B. Haber, S. Gabay, M. Issidorides, and S. Alivisatos) pp. 37–50. Plenum, New York.

Gershon, M. D. and Tamir, H. (1985). Peripheral sources of serotonin binding proteins. In *Serotonin and the cardiovascular system* (ed. P. M. Vanhoute) pp. 15–26. Raven Press, New York.

Gershon, M. D., Drakontides, A. B., and Ross, L. L. (1965). Serotonin: synthesis and release from the myenteric plexus of the mouse intestine. *Science* **149,** 197–9.

Gershon, M. D., Robinson, R., and Ross, L. L. (1976). Serotonin accumulation in the guinea pig's myenteric plexus: ion dependence, structure activity relationship, and the effect of drugs. *Journal of Pharmacology and Experimental Therapeutics* **198,** 548–61.

Gershon, M. D., Sherman, D., and Dreyfus, C. F. (1980). Effects of indolic neurotoxins on enteric serotonergic neurons. *Journal of Comparative Neurology* **190,** 581–96.

Gershon, M. D., Liu, K. P., Karpiak, S. E., and Tamir, H. (1983). Storage of serotonin *in vivo* as a complex with serotonin-binding protein in central and peripheral serotonergic neurons. *Journal of Neuroscience* **3,** 1901–11.

Gershon, M. D., Takaki, M., Tamir, H., and Branchek, T. (1985). The enteric neural receptor for 5-hydroxytryptamine. *Experientia* **41,** 863–8.

Goodrich, J. T., Bernd, P., Sherman, D. L., and Gershon, M. D. (1980). Phylogeny of enteric serotonergic neurons. *Journal of Comparative Neurology* **190,** 15–28.

Grafstein, B. and Forman, D. (1980). Intracellular transport in neurons. *Physiological Reviews* **60,** 1167–283.

Grønstad, K. O., DeMagistris, L., Dahlström, A., Nilsson, O., Price, B., Zinner, M. J., Jaffe, B. M., and Ahlman, H. (1985). The effects of vagal nerve stimulation on endoluminal release of serotonin and substance P into the feline small intestine. *Scandinavian Journal of Gastroenterology* **20,** 163–9.

Gwee, M. C. E. and Yeoh, T. S. (1968). The release of 5-hydroxytryptamine from rabbit small intestine *in vitro*. Journal of Physiology (London) **194,** 817–25.

Gyermek, L. (1966). Drugs which antagonize 5-hydroxytryptamine and release indolealkylamines. In *Handbüch der experimentellen Pharmakologie*, vol. 19 (ed. V. Erspamer) pp. 471–528. Springer-Verlag, New York.

Hardcastle, J., Hardcastle, P. E. T. and Redfern, J. S. (1981). Action of 5-hydroxytryptamine on intestinal transport in the rat. *Journal of Physiology (London)* **320,** 41–55.

Harry, J. (1963). The action of drugs on the circular muscle strip from the guinea pig isolated ileum. *British Journal of Pharmacology and Chemotherapy* **20,** 399–417.

Heuring, R. E. and Peroutka, S. J. (1987). Characterization of a novel ^{3}H-5-hydroxytryptamine binding site subtypes in bovine brain membranes. *Journal of Neuroscience* **7,** 894–903.

Hirst, G. D. S., Holman, M. E., and Spence, I. (1974). Two types of neurones in the myenteric plexus of duodenum in the guinea-pig. *Journal of Physiology (London)* **236,** 303–26.

Holman, M. E., Hirst, G. D. S., and Spence, I. (1972). Preliminary studies of the neurones of Auerbach's plexus using intracellular microelectrodes. *Australian Journal of Experimental and Biological Medicine* **50,** 795–801.

Holzer, P. and Skofitsch, G. (1984). Release of endogenous 5-hydroxytryptamine from the myenteric plexus of the guinea-pig isolated small intestine. *British Journal of Pharmacology* **81,** 381–6.

Humphrey, P. P. A. (1983). Pharmacological characterization of cardiovascular 5-hydroxytryptamine receptors. In *Proceedings of the Fourth International Symposium on Vascular Neuroeffector Mechanisms* (eds. J. A. Bevan, R. A. Maxwell, S. Shibata, M. Fujiwara, K. Muhri, and N. Toda, pp. 237–42. Raven Press, New York.

Humphrey, P. P. A., Feniuk, W., and Watts, A. D. (1983). Prejunctional effects of 5-hydroxytryptamine on noradrenergic nerves in the cardiovascular system. *Federation Proceedings* **42,** 218–22.

Ireland, S. J. and Tyers, M. B. (1987). Pharmacological characterization of 5-hydroxytryptamine-induced depolarization of the rat isolated vagus nerve. *British Journal of Pharmacology* **90,** 229–38.

Johnson, S. M., Kayatama, Y., and North, R. A. (1980a). Slow synaptic potentials in neurones of the myenteric plexus. *Journal of Physiology (London)* **301,** 505–16.

Johnson, S. M., Katayama, Y., and North, R. A. (1980b). Multiple actions of 5-hydroxytryptamine on myenteric neurones of the guinea-pig ileum. *Journal of Physiology (London)* **304,** 459–79.

Johnson, S. M., Katayama, Y., Morita, K., and North, R. A. (1981). Mediators of slow synaptic potentials in the myenteric plexus of the guinea pig ileum. *Journal of Physiology (London)* **320,** 175–86.

Jonakait, G. M., Tamir, H., Rapport, M. M., and Gershon, M. D. (1977). Detection of a soluble serotonin binding protein in the mammalian myenteric plexus and other peripheral sites of serotonin storage. *Journal of Neurochemistry* **28,** 277–84.

Jonakait, G. M., Tamir, H., Gintzler, A. R., and Gershon, M. D. (1979). Release of [^{3}H]serotonin and its binding protein from enteric neurones. *Brain Research* **174,** 55–69.

Jonsson, G., Fuxe, K., Hamberger, B., and Hökfelt, T. (1969). 6-Hydroxytryptamine a new tool in monoamine fluoresence histochemistry. *Brain Research* **13,** 190–5.

Julé, Y. (1980). Nerve-mediated descending inhibition in the proximal colon of the rabbit. *Journal of Physiology (London)* **159,** 361–8.

Juorio, A. V. and Gabella, G. (1974). Noradrenaline in the guinea-pig alimentary canal: regional distribution and sensitivity to denervation and reserpine. *Journal of Neurochemistry* **221,** 851–8.

Kamikawa, Y. and Shimo, Y. (1983). Indirect action of 5-hydroxytryptamine on the isolated muscularis mucosa of the guinea pig oesophagus. *British Journal of Pharmacology* **78,** 103–10.

Katayama, Y. and North, R. A. (1978). Does substance P mediate slow synaptic excitation within the myenteric plexus? *Nature (London)* **274,** 387–8.

Katayama, Y., North, R. A., and Williams, J. T. (1979). The action of substance P on neurons of the myenteric plexus of the guinea-pig small intestine. *Proceedings of the Royal Society of London* **206,** 191–208.

Kurian, S. S., Feri, G. L., Demeg, J., and Polak, J. M. (1983). Immunocytochemistry of serotonin-containing nerves in the human gut. *Histochemistry* **78,** 523–9.

Langley, J. N. (1921). *The autonomic nervous system.* Part I. Heffer Press, Cambridge.

Le Douarin, N. M. (1982). *The neural crest.* Cambridge University Press, Cambridge.

Lees, G. J., Furness, J. B., and Costa, M. (1979). Apparent absence of tryptophan hydroxylase from the myenteric plexus of the guinea-pig ilium. *Proceedings of the Australian Physiology and Pharmacology Society* **10,** 304P.

Legay, C., Faudon, M., Héry, F., and Ternaux, J. P. (1983a). 5-HT metabolism in the intestinal wall of the rat. I. The mucosa. *Neurochemistry International* **5,** 721–7.

Legay, C., Faudon, M., Héry, F., and Ternaux, J. P. (1983b). 5-HT metabolism in the intestinal wall of the rat. II. The nerves plexuses – interactions between 5-HT containing cells. *Neurochemistry International* **5,** 571–7.

Legay, C., Saffrey, M. J., and Burnstock, G. (1984). Coexistence of immunoreactive substance P and serotonin in neurones of the gut. *Brain Research* **302,** 379–82.

Lew, W. Y. W. and Longhurst, J. C. (1986). Substance P, 5-hydroxytryptamine and bradykinin stimulate abdominal visceral afferents. *American Journal of Physiology* **250,** R465–73.

Leysen, J. E., Awouters, F., Kennis, L., Laduron, P. M. Vandenberk, J., and Janssen, P. A. J. (1982). Receptor binding profile of R41468, a novel anagonist at 5-HT_2 receptors. *Life Sciences* **28,** 1015–22.

Liu, K. P., Gershon, M. D., and Tamir, H. (1985). Identification, purification, and characterization of two forms of serotonin binding protein from rat brain. *Journal of Neurochemistry* **44,** 1289–301.

Mawe, G. M., Branchek, T., and Gershon, M. D. (1986). Peripheral neural serotonin receptors: Identification and characterization with specific agonists and antagonists. *Proceedings of the National Academy of Sciences (USA)* **83,** 9799–803.

Nada, O. and Toyohara, T. (1987). An immunohistochemical study of serotonin-containing nerves in the colon of rats. *Histochemistry* **86,** 229–32.

Nishi, S. and North, R. A. (1973). Intracellular recording from the myenteric plexus of the guinea pig ileum. *Journal of Physiology (London)* **231,** 471–91.

North, R. A. and Tokimasa, T. (1982). Muscarinic potentials in guinea-pig myenteric plexus neurones. *Journal of Physiology (London)* **333,** 151–6.

North, R. A., Henderson, C., Katayama, Y., and Johnson, S. M. (1980). Electrophysiological evidence of presynaptic inhibition of acetylcholine release by 5-hydroxytryptamine in the enteric nervous system. *Neuroscience* **5,** 581–6.

Ormsbee, H. S., Silver, D. A., and Hardy, F. E. (1984). Effects of 5-hydroxytryptamine on the migrating myoelectric complex in the canine intestine. *Journal of Pharmacology and Experimental Therapeutics* **231,** 436–40.

Paintal, A. S. (1954). The response of gastric stretch receptors and certain other abdominal and thoracic vagal receptors to some drugs. *Journal of Physiology (London)* **126,** 271–85.

Paintal, A. S. (1964). Effects of drugs on vertebrate mechanoreceptors. *Pharmacology Reviews* **16,** 341–80.

Paton, W. D. M. and Vane, J. R. (1963). An analysis of the response of the isolated stomach to electrical stimulation and to drugs. *Journal of Physiology (London)* **165,** 10–46.

Payette, R. F., Tennyson, V. M., Pham, T. D., Mawe, G. M., Pomeranz, H., Rothman, T. P., and Gershon, M. D. (1987). Origin and morphology of nerve fibres in the aganglionic colon of the lethal spotted (*ls*/*ls*) mutant mouse. *Journal of Comparative Neurology* **257,** 237–52.

Pazos, A., and Palacios, J. M. (1985a). Quantitative autoradiographic mapping of serotonin receptors in the rat brain. I. Serotonin-1 receptors. *Brain Research* **346,** 205–30.

Pazos, A., Cortes, R., and Palacios, J. M. (1985b). Quantitative autoradiographic mapping of serotonin receptors in the rat brain. II. Serotonin-2 receptors. *Brain Research* **346,** 231–49.

Pedigo, N. W. and Nelson, D. L. (1981). Discrimination of multiple [^{3}H]5-hydroxytryptamine binding sites by the neuroleptic spiperone in rat brain. *Journal of Neurochemistry* **36,** 220–6.

Peroutka, S. (1984). 5-HT, receptor sites and functional correlates. *Neuropharmacology* **23,** 1487–92.

Rattan, S. and Goyal, R. K. (1978). Evidence of 5-HT participation in vagal inhibitory pathway to opossum LES. *American Journal of Physiology* **234:** E273–6.

Richardson, B. P. and Engel, G. (1986). The pharmacology and functions of 5-HT_3 receptors. *Trends in Neuroscience* **9,** 424–8.

Richardson, B. P., Engel, G., Donatsch, P., and Stadler, P. A. (1985). Identification of serotonin M-receptor subtypes and their specific blockade by a new class of drugs. *Nature (London)* **316,** 216–31.

Robinson, R. and Gershon, M. D. (1971). Synthesis and uptake of 5-hydroxytryptamine by the myenteric plexus of the small intestine of the guinea pig. *Journal of Pharmacology and Experimental Therapeutics* **179,** 29–41.

Rothman, T. P. and Gershon, M. D. (1982). Phenotypic expression in the developing murine enteric nervous system. *Journal of Neuroscience* **2,** 381–93.

Rothman, T. P., Sherman, D., Cochard, P., and Gershon, M. D. (1986). Development of the monoaminergic innervation of the avian gut: transient and permanent expression of phenotypic markers. *Developmental Biology* **116,** 357–80.

Saffrey, M. J., Legay, C., and Burnstock, G. (1984). Development of 5-hydroxytryptamine-like immunoreactive neurones in cultures of the myenteric plexus from the guinea-pig caecum. *Brain Research* **304,** 105–16.

Salimova, N. (1978). Localization of biogenic monoamines in Amphioxus Brachiostroma lanceolatum. *Doklady (Proceedings) of the Academy of Sciences of the USSR* **242,** 939–941.

Sanger, G. J. (1985). Three different ways in which 5-hydroxytryptamine can affect choline activity in guinea-pig isolated ileum. *Journal of Pharmacy and Pharmacology* **37,** 584–6.

Schulz, R. and Cartwright, C. (1974). Effect of morphine on serotonin release from the myenteric plexus of the guinea pig. *Journal of Pharmacology and Experimental Therapeutics* **190,** 420–30.

Sheerin, H. E. (1979). Serotonin action on short circuit current and ion transport across isolated rabbit ileal mucosa. *Life Sciences* **24,** 1609–16.

Surprenant, A. and Crist, J. (1988). Electrophysiological characterization of functionally distinct 5-HT receptors on guinea-pig submucous plexus. *Neuroscience* **24:** 283–95.

Takaki, M., Branchek, T., Tamir, H., and Gershon, M. D. (1985a). Specific antagonism of enteric neural serotonin receptors by dipeptides of 5-hydroxytryptophan: Evidence that serotonin is a mediator of slow synaptic excitation in the myenteric plexus. *Journal of Neuroscience* **5,** 1769–80.

Takaki, M., Mawe, G. M., Barasch, J., and Gershon, M. D. (1985b). Physiological responses of guinea-pig myenteric neurons secondary to the release of endogeneous serotonin by tryptamine. *Neuroscience* **16,** 223–40.

Tamir, H. and Gershon, M. D. (1979). Storage of serotonin and serotonin binding protein in synaptic vesicles. *Journal of Neurochemistry* **33,** 35–44.

Tamir, H. and Huang, I. L. (1974). Binding of serotonin to soluble protein from synaptosomes. *Life Sciences* **14,** 83–93.

Tamir, H. and Kuhar, M. J. (1975). Association of serotonin-binding protein with projections of the midbrain raphe nuclei. *Brain Research* **83,** 164–72.

Tansy, M. F., Rothman, G., Bartlett, G., Farbert, P., and Hohenleitner, F. J. (1971). Vagal adrenergic degranulation of enterochromaffin cell system in guinea-pig duodenum. *Journal of Pharmacological Science* **60,** 81–5.

Vane, J. R. (1957). A sensitive method for the assay of 5-hydroxytryptamine. *British Journal of Pharmacology and Chemotherapy* **12,** 344–9.

Vizi, V. A. and Vizi, E. S. (1978). Direct evidence for acetylcholine releasing effect of serotonin in the Auerbach's plexus. *Journal of Neural Transmission* **42,** 127–38.

Wallis, D. I., Stansfeld, C. E., and Nash, H. L. (1982). Depolarizing responses recorded from nodose ganglion cells in the rabbit evoked by 5-hydroxytryptamine and other substances. *Neuropharmacology* **21,** 31–40.

Watson, A. H. D. (1979). Fluorescent histochemistry of the teleost gut: evidence for the presence of serotonergic neurons. *Cell Tissue Research* **197,** 155–64.

Wood, J. D. (1987). Physiology of enteric neurons. In *Physiology of the gastrointestinal tract.* (ed. L. R. Johnson) Vol. 1, 2nd ed., pp. 1–41. Raven Press, New York.

Wood, J. D. and Mayer, C. J. (1979). Serotonergic activation of tonic-type enteric neurons in guinea pig small bowel. *Journal of Neurophysiology* **422,** 582–93.

Yagaloff, K. A. and Hartig, P. R. (1985). ^{125}I-Lysergic acid diethylamide binds to a novel serotonergic site of rat choroid plexus epithelial cells. *Journal of Neuroscience* **5,** 3178–83.

Yntema, C. L. and Hammond, W. S. (1954). The origin of intrinsic ganglia of trunk viscera from vagal neural crest in the chick embryo. *Journal of Comparative Neurology* **101,** 515–41.

12

5-hydroxytryptamine and peripheral secretory mechanisms

BRIAN L. FURMAN and N. GRAHAM WATON

12.1 Introduction

5-HT is widely distributed in the gastro-intestinal tract (Feldberg and Toh 1953; Dalgleish *et al.* 1953; Erspamer 1954) and in various endocrine glands (Owman *et al.* 1973). The amine itself and its biosynthetic precursors, as well as various 5-HT receptor blocking drugs and monoamine oxidase inhibitors, have been reported to have marked effects on both endocrine and exocrine secretory functions. Either stimulatory or inhibitory effects of these substances have been found in different glands and, in some cases, there are conflicting data concerning the actions of 5-HT in any one secretory tissue. Nevertheless, attempts have been made to ascribe a physiological role to 5-HT in regulating various endocrine and exocrine secretory functions in the periphery. Such a role, in a given secretory structure, could be accepted if it could be demonstrated unequivocally that: (1) 5-HT was present in the tissue, together with appropriate biosynthetic enzymes and inactivation mechanisms; (2) 5-HT could produce a potent effect in modulating secretion; and (3) secretion was appropriately modulated by selective depletion of 5-HT and by its specific pharmacological antagonism. This chapter will review the actions of 5-HT on various exocrine and endocrine secretions, together with evidence for and against its physiological role in regulating, or modulating, the secretory activity of these endocrine and exocrine structures. Emphasis has been placed on the secretions of the pancreatic islets, the adrenal cortex and the stomach, as there is a considerable amount of data on these topics. Central 5-HT mechanisms may be important in regulating the secretions of the anterior pituitary and therefore, indirectly, the secretions of certain other endocrine glands (Tuomisto and Mannisto 1985). However, this important topic is beyond the scope of the book and we will concentrate on those effects of 5-HT exerted peripherally.

12.2 5-HT and endocrine secretion

Falck and Hellman (1963) demonstrated formaldehyde-induced monoamine fluorescence in the islets of Langerhans. Subsequently this was shown to be due to the presence of 5-HT and dopamine (Cegrell 1968). These monoamines have been found in other polypeptide-producing endocrine cells, such as the thyroid C cells (Owman *et al.* 1973). Although many polypeptide-secreting cells do not contain detectable monoamines, they can take up and decarboxylate (to the corresponding amines) the precursors 5-hydroxytryptophan and L-dopa (Ritzen *et al.* 1965; Hakanson *et al.* 1967a; Larson *et al.* 1966; Cegrell 1968). Pearse (1976) used the acronym APUD to denote cells capable of Amine Precursor Uptake and Decarboxylation and suggested their common embryological origin. At least 26 APUD cells have been suggested (see Pearse 1976) but studies on the *functional* significance of their ability to take up and decarboxylate 5-hydroxytryptophan and store 5-HT have been limited largely to the pancreatic islets. The islets may represent a useful model for other APUD cell systems.

12.2.1 5-HT and the function of the endocrine pancreas

Histochemical fluorescence studies have shown 5-HT to be present in the pancreatic islets of the cat, dog, pig, and guinea-pig but not of human fetus, golden hamster, or adult rabbit (Cegrell 1968; Lundquist *et al.* 1975). The failure to detect 5-HT in some species may reflect an insensitivity of the histochemical fluorescence technique. Using highly sensitive radioenzymatic or double isotope methods, high concentrations of 5-HT were found in hamster or mouse islets, in which 5-HT had not previously been found by histochemical fluorescence (Hansen and Hedeskov 1977; Bird *et al.* 1980). 5-HT has been demonstrated in the glucagon-secreting A-cells (α_2-cells) and the insulin-secreting B-cells (β-cells) but not in the somatostatin-secreting D-cells (α_1-cells) (Cegrell 1968; Lundquist *et al.* 1975).

12.2.1.1 *Subcellular location of 5-HT in the endocrine pancreas*

Studies using electron-microscopy histochemistry, autoradiography (following administration of ^{3}H-labelled precursors) or subcellular fractionation, show clearly that 5-HT is stored within the secretory granules of the islet B cells (Jaim-Etcheverry and Zieher 1968; Ekholm *et al.* 1971; Hellman *et al.* 1972). Although treatment with the B-cell toxin streptozotocin depleted guinea-pig islets of both insulin and 5-HT, selective depletion of 5-HT could be obtained by reserpine treatment (Lundquist *et al.* 1975). Gylfe (1978) showed a clear relationship between secretion of insulin and release of [^{3}H]5-HT by perifused islets pre-loaded with [^{3}H]5-hydroxytryptophan. He postulated that 5-HT release occurred by exocytosis, along with insulin. However, a second study (Hellman *et al.* 1972) showed no parallel between the release of 5-HT and that of insulin.

12.2.1.2 *Uptake of 5-HT by pancreatic islets*

Tritiated 5-HT is transported into islet cells and becomes associated with the insulin-secretory granules (Hellman *et al.* 1972). There appears to be a high affinity uptake mechanism saturable at 1–3 μM 5-HT and a second, non-saturable (up to 1 mM) uptake process (Lindstrom *et al.* 1980). The high affinity uptake process is impaired at low temperatures, by metabolic inhibitors and by sodium deficiency (Lindstrom *et al.* 1980). Moreover, it is sensitive to inhibition by imipramine (Lindstrom 1981a) and thus appears similar to uptake mechanisms into platelets or neurones (Stahl and Meltzer 1978). In rat insulinoma cells, however, uptake of 5-HT from solutions containing 0.1 μM to 5 mM appeared to be due largely to a simple diffusion process (Hutton *et al.* 1983). Isolated granules from these cells could concentrate 5-HT and uptake was favoured by alkaline media and blocked by a permeant base. Hutton *et al.* (1983) suggested that the accumulation of 5-HT in the B-cell *in vivo* could be accounted for by an uptake driven by proton gradients across the granular membrane.

12.2.1.3 *Uptake and decarboxylation of 5-hydroxytryptophan*

The origin of islet 5-HT is uncertain. There is no evidence that islets contain the enzyme tryptophan hydroxylase. Moreover, parachlorophenylalanine, an inhibitor of tryptophan hydroxylase, failed to deplete hamster islets of their 5-HT content (Bird *et al.* 1980). However, islets contain the enzyme, aromatic L-amino acid decarboxylase (Wilson *et al.* 1974). 5-Hydroxytryptophan administered *in vivo* to the intact animal, or incubated with islets *in vitro*, is taken up and decarboxylated, thus increasing intra-islet 5-HT content (Cegrell 1968; Ekholm *et al.* 1971; Gylfe *et al.* 1973; Wilson *et al.* 1974; Zern *et al.* 1980). This increase is prevented by inhibition of the decarboxylase enzyme (Pulido *et al.* 1978).

12.2.1.4 *Actions of 5-HT on insulin secretion*

Functional studies support the histochemical studies and indicate a role for 5-HT in the regulation of insulin secretion. The evidence has been obtained from studies using 5-HT, 5-hydroxytryptophan, monoamine oxidase inhibitors and antagonists at 5-HT receptors.

12.2.1.4.1 *Effects of 5-HT* Many *in vitro* studies using islets, perfused pancreas, or pancreatic pieces from the mouse, rabbit, rat, and golden hamster have shown that 5-HT inhibits insulin secretion in response to glucose (the main physiological secretagogue) without reducing basal insulin release (Telib *et al.* 1968; Feldman and Lebovitz 1970; Feldman *et al.* 1972; de Bold and Benscombe 1978; Pontiroli *et al.* 1978; Gagliardino *et al.* 1974; Furman *et al.* 1981; Lindstrom and Sehlin 1983; Lindstrom *et al.* 1984), and without any effect on synthesis (Gagliardino *et al.* 1974; Lin and Haist 1975). 5-HT also inhibits the insulinaemic response to intravenous glucose in dogs (Robertson and Guest 1978). A striking observation, however, is the low potency of 5-HT in exerting this inhibitory effect *in vitro*. In all studies 5-HT is active in the range 0.05–1 mM, whereas more classical actions of 5-HT are exerted in the nanomolar to micromolar range. Quickel *et al.* (1971a) showed that 5-HT was most active in the hamster (82 per cent inhibition of insulin secretion at 50 μM) and least active in the rabbit (38 per cent inhibition at 1 mM). Attenuation of the inhibitory effect by reserpine (Quickel *et al.* 1971a), together with the fact that hamster islets have a particularly dense noradrenergic innervation (Cegrell 1968), may indicate an indirect action of 5-HT through the release of noradrenaline (see, for example, Trendelenburg 1972; Doggrell and Woodruff 1978; Fozard 1984).

Although most studies have shown inhibition of glucose-induced insulin secretion, *enhancement* of *basal* secretion was shown using rabbit pancreatic slices (Telib *et al.* 1968) and mouse isolated islets (Lindstrom *et al.* 1984; Lindstrom and Sehlin 1983). Moreover, in *in vivo* studies in dogs and humans, 5-HT showed a stimulatory effect on insulin release (Federspil *et al.* 1974).

12.2.1.4.2 *Effects of increasing intra-islet 5-HT using 5-hydroxytryptophan and monoamine oxidase inhibitors* Wilson *et al.* (1974) showed inhibition of glucose-induced insulin secretion by pancreatic segments removed from rabbits infused with 5-hydroxytryptophan. Similarly, 5-hydroxytryptophan pre-treatment of hamsters inhibited glucose-induced insulin secretion in subsequently isolated islets, but only if the animals were additionally pre-treated with the monoamine oxidase inhibitor, tranylcypromine (Zern *et al.* 1980). Direct incubation of rat islets with 5-hydroxytryptophan inhibited glucose-induced insulin secretion (Furman *et al.* 1981). Using islets from ob/ob (obese hyperglycaemic) mice, inhibition of glucose induced insulin release was found only after *pre-incubation* with 5-hydroxytryptophan (Lindstrom and Sehlin 1983). *In vivo* experiments showed 5-hydroxytryptophan to inhibit insulin secretion stimulated by sulphonylureas or β-adrenoceptor agonists but not that stimulated by glucose (Lundquist *et al.* 1971). In some of these studies the inhibitory effect of 5-hydroxytryptophan was prevented by inhibition of aromatic L-amino acid decarboxylase, thus suggesting mediation by 5-HT (Wilson *et al.* 1974; Lundquist *et al.* 1971). Other work has shown 5-hydroxytryptophan to *stimulate* insulin secretion *in vitro*. The effect is seen as an increase in basal secretion (Pontiroli *et al.* 1978), but much more markedly on glucose induced insulin secretion (Lernmark 1971; Lindstrom and Sehlin 1983 and 1986). *In vivo* studies in mice and rats suggest a stimulatory action of large doses of 5-hydroxytryptophan (Jacoby and Bryce 1978 and 1979; Furman and Wilson 1980). The involvement of 5-HT in these effects of 5-hydroxytryptophan is suggested by their prevention by inhibition of aromatic L-amino acid decarboxylase (Furman and Wilson 1980; Lindstrom and Sehlin 1983 and 1986). However, Lindstrom and Sehlin (1986) suggested that activation of decarboxylase *per se* may be the stimulus to insulin secretion rather than the 5-HT formed from the precursor. This suggestion was made on the basis of the opposite effects of 5-HT and 5-hydroxytryptophan on glucose-induced insulin secretion, and the fact that other substrates for aromatic L-amino acid decarboxylase (*m*-tyrosine and *o*-tyrosine) also enhanced insulin release. On the other hand, L-dopa, a very good substrate for aromatic L-amino acid decarboxylase, *inhibited* insulin secretion. It was argued that dopamine, formed in the islets from L-dopa, was a more powerful inhibitor of insulin secretion than 5-HT. The role of decarboxylation in the stimulatory effects of 5-hydroxytryptophan clearly requires further study. It must be pointed out that the majority of the available inhibitors of aromatic L-amino acid decarboxylase are not entirely specific (Bender 1980; Fozard 1982a). One study showed benserazide pre-treatment to inhibit glucose-induced insulin release in the subsequently isolated and perfused rat pancreas (Pulido *et al.* 1978). Interestingly, the latter study showed that raising intra-islet 5-HT by pre-loading with 5-hydroxytryptophan did not modify glucose-induced insulin secretion. A similar lack of effect of raised 5-HT on glucose-induced insulin secretion was seen in hamster islets in which 5-HT was elevated 22-fold by pre-administration of 5-hydroxytryptophan (Bird *et al.* 1980).

12.2.1.4.3 *Effects of monoamine oxidase inhibitors* Pancreatic islets show significant activity of both the A and B forms of monoamine oxidase (Feldman and Chapman 1975a; Lenzen *et al.* 1983). Thus, if 5-HT is present in islets and has a modulatory role in regulating insulin secretion, inhibition of its oxidative deamination should modify hormone release. Several studies have shown that inhibition of monoamine oxidase, without administration of 5-hydroxytryptophan, inhibits glucose-induced insulin secretion (Aleyassine and Lee 1971 and 1972; Aleyassine and Gardiner 1975; Feldman and Chapman 1975a and b; Lenzen *et al.* 1983). *In vivo* studies showed inhibition of isoprenaline-stimulated insulin secretion following administration of monoamine oxidase inhibitors to mice (Lundquist *et al.* 1971). Although these observations support the idea that intra-islet 5-HT may inhibit insulin secretion, their interpretation is difficult. First, none of these studies included measurements of islet monoamine content following monoamine oxidase inhibitor treatment. Thus it is difficult to attribute the effects of the drugs to changes in 5-HT accumulation or, indeed, of other monoamines (dopamine) that may accumulate after monoamine oxidase inhibition. It may also be significant that the drugs have been studied in species whose islets may not normally contain 5-HT (rat, rabbit: Cegrell 1968). Second, there is no clear relationship between inhibition of monoamine oxidase and modification of insulin secretion. Some monoamine oxidase inhibitors *potentiate* glucose-stimulated insulin secretion (Feldman and Chapman 1975b). Moreover, the IC_{50} values for inhibition of islet monoamine oxidase, using 5-HT as a substrate, were

0.01, 0.9 and 0.1 μM for clorgyline, deprenyl, and tranylcypromine respectively. On the other hand, the corresponding IC_{50} values for inhibition of glucose-induced insulin were 0.85, 2.5 and 0.7 mM, concentrations at least 100-fold greater than those required for complete inhibition of islet monoamine oxidase (Lenzen *et al.* 1983). This casts doubt on the specificity of action of the drugs in the concentrations required to inhibit insulin release and suggests that inhibition of insulin secretion is not due to inhibition of monoamine oxidase. Finally, at least in rabbit pancreas, monoamine oxidase inhibitors, although potentiating the inhibitory action of dopamine on insulin secretion, did not potentiate 5-HT (Feldman and Chapman 1975b).

12.2.1.4.4 *Effect of antagonists at 5-HT receptors* A role for endogenous 5-HT in modulating insulin secretion would be supported if insulin secretion were altered by blocking 5-HT receptors. The 5-HT receptor antagonists, cyproheptadine, methysergide, and cinanserin, potentiated glucose-induced insulin secretion in rabbit pancreatic segments although methysergide failed to block the effect of 5-HT and the other two drugs were not examined for this latter property (Feldman *et al.* 1972). Methysergide, but not cyproheptadine or cinanserin, blocked the inhibitory action of dopamine and other catecholamines. Very high concentrations of metergoline (1 mM) enhanced basal and glucose-induced insulin secretion (Pontiroli *et al.* 1978) but no data were presented on the ability of the drug to inhibit the effect of 5-HT. A further complicating issue is the inhibitory action of cyproheptadine on insulin release and its ability to decrease islet insulin content, actions seemingly unrelated to its 5-HT receptor blocking properties (Halben *et al.* 1979). Human studies have shown metergoline to increase glucose-stimulated insulin secretion in normal subjects (Pontiroli *et al.* 1978) and methysergide and cyproheptadine to increase insulin secretion in acromegalics and/or diabetics during the oral glucose tolerance test (Quickel *et al.* 1971b; Feldman *et al.* 1975). On the other hand, another study in non-insulin-dependent diabetics revealed no effect of metergoline on arginine- or glucose-provoked elevations in the plasma insulin concentrations (Giugliano *et al.* 1981). Thus, to date, studies with 5-HT receptor blocking drugs have not produced conclusive data supporting, or refuting, a role for 5-HT in the regulation of insulin secretion.

12.2.1.4.5 *Effects of 5-HT depletion* There have been no systematic studies examining the effects of specific 5-HT depleting drugs on insulin secretion. *p*-Chlorophenylalanine, a selective inhibitor of tryptophan hydroxylase, neither modified glucose-induced insulin secretion from rabbit pancreatic pieces, nor prevented the enhancing effect of methysergide (Feldman *et al.* 1972). In human studies *p*-chlorophenylalanine augmented the hyperinsulinaemic response to intravenous glucose (de Leiva *et al.*

1978). No data have been presented on the effect of *p*-chlorophenylalanine on the 5-HT content of rabbit or human islets and, as mentioned previously, there is no evidence that 5-HT is normally present in the islets of these species. In the hamster, *p*-chlorophenylalanine did not deplete the islets of their 5-HT content (*Bird et al.* 1980). Destruction of rat islet 5-HT-containing nerves, using 5,7-dihydroxytryptamine, did not modify basal or glucose-induced insulin release from the subsequently isolated and perfused pancreas (Jansson *et al.* 1985). It is possible, of course, that the *in vivo* function of the islets was modified by the procedure.

12.2.1.4.6 *Mechanism of action of 5-HT on insulin secretion* The consensus of opinion is that 5-HT is inhibitory to insulin secretion stimulated by glucose but may augment basal release. The initiation of insulin secretion is thought to be due to the accumulation of Ca^{2+} in the B cell (Malaisse 1973; Herchuelz *et al.* 1980). Ca^{2+} enters the B cell, at least in part, via voltage-operated Ca^{2+} channels (Lebrun *et al.* 1982) which are opened following depolarization of the B cell membrane secondarily to a reduction in potassium permeability (Henquin 1978; Atwater *et al.* 1979). Although no electrophysiological studies have been made, 5-HT increased the potassium permeability of the membrane, as evidenced by the increased $^{86}Rb^{+}$ efflux (Lindstrom and Sehlin 1982a). Moreover 5-HT diminished the ability of glucose to reduce $^{86}Rb^{+}$ efflux (Lindstrom *et al.* 1984) which suggests a reduction in the ability of glucose to depolarize the islet B cell membrane.

Exposure of islets to 5-HT reduces their ability to accumulate Ca^{2+} both basally and in response to glucose (Lindstrom and Sehlin 1983). 5-HT also reduces glucose-induced increases in chloride efflux (Lindstrom and Sehlin 1982b) and diminishes the 'phosphate flush' in response to glucose (Lindstrom 1981b). Thus 5-HT appears to act at some point early in the stimulus-secretion coupling sequence. Although some of the inhibiting action of 5-HT on insulin secretion may be attributable to increased noradrenaline release, catecholamines probably inhibit secretion by an action at later stages (Morgan and Montague 1985).

12.2.1.5 *5-HT and the secretion of other islet hormones*

There are few data concerning the actions of 5-HT on the secretion of glucagon and none concerning its effect on the secretion of somatostatin or pancreatic polypeptide. Using isolated islets from various species, 5-HT has been shown either to inhibit (Pontiroli *et al.* 1978; Marco *et al.* 1977) or to stimulate (Matsumoto *et al.* 1984; Furman *et al.* 1981) glucagon secretion. Stimulation of glucagon secretion has also been shown *in vivo* in rats (Jacoby and Bryce 1978) whereas Marco *et al.*. (1976) provided evidence for an inhibitory role of 5-HT in modulating glucagon secretion in man; in normal volunteers, arginine- or hypoglycaemia-induced glucagon secretion was enhanced by the 5-HT receptor antagonists cyproheptadine or methysergide or by 5-HT depletion using *p*-chlorophenylalanine. Some of these discrepancies may reflect a 5-HT-induced release of catecholamines from noradrenergic varicosities. Catecholamines stimulate glucagon secretion (Itoh *et al.* 1982) and the stimulatory effects of 5-HT on glucagon secretion in the intact rat (Jacoby and Bryce 1978) and from hamster islets *in vitro* (Matsumoto *et al.* 1984) can be attenuated by adrenoceptor blocking drugs.

Although there are no data on *pancreatic* somatostatin secretion, 5-HT is a potent inhibitor of somatostatin secretion by the rat isolated perfused stomach (Koop and Arnold 1984). In this preparation maximal inhibition was obtained using μM concentrations of 5-HT and the response was prevented by methysergide or the 5-HT_2 receptor antagonist, ketanserin. The response was also blocked by atropine suggesting, paradoxically, the involvement of an inhibitory cholinergic pathway.

12.2.1.6 *The glycaemic effects of 5-HT*

5-Hydroxytrytophan produces hypoglycaemia in mice, an effect apparently due to the formation of 5-HT (Lundquist *et al.* 1971; Furman 1974; Furman and Wilson 1980; Wilson and Furman 1982). Although this is partly due to increased insulin secretion (Furman and Wilson 1980) other mechanisms are clearly involved (Lundquist *et al.* 1971; Furman and Wilson 1980; Wilson and Furman 1982). Recently, it was suggested that 5-hydroxytrytophan-induced hypoglycaemia was due to the hepatic accumulation of 5-HT (Endo 1987). On the other hand, 5-hydroxytryptophan produces *hyper*glycaemia in rats pre-treated with monoamine oxidase inhibitors (Wong and Tyce 1978). This may be related to the recent observations of hyperglycaemia and inhibition of glucose-induced insulin secretion in response to the selective 5-HT_{1A} receptor agonist, 8-OH-DPAT (Chaouloff and Jeanrenaud 1987a). However, these effects appear to be mediated via an effect in the central nervous system, at least in the genetically obese rat (Chaouloff and Jeanreneaud 1987b).

12.2.2 5-HT and steroid hormone secretion

Central 5-HT mechanisms controlling the secretion of adrenal corticosteroids via pituitary adrenocorticotrophic hormone have been investigated extensively (Tuomisto and Mannisto 1985). Direct effects of 5-HT on adrenal cortical function have received much less attention.

12.2.2.1 *5-HT and the adrenal cortex*

5-HT is a potent stimulator of corticosterone and aldosterone secretion from pieces of adrenal gland or suspensions of adrenal cortical cells *in vitro* (Muller and Ziegler 1968; Albano *et al.* 1974; Bing and Schulster, 1977; Fujita *et al.* 1979; Matsuoka *et al.* 1985). This effect is seen only in cell suspensions rich in zona glomerulosa cells, suggesting that 5-HT has little, if any, effect on steroidogenesis by the zona fasciculata (Haning *et al.* 1970). The potency of 5-HT in this respect is in accord with its action on smooth muscle, responses occurring at concentrations of 0.001-1 μM. Infusion of 5-HT increases plasma aldosterone in man (Mantero *et al.* 1982).

12.2.2.1.1 *Mechanism of action of 5-HT on the adrenal cortex* As 5-HT increases both corticosterone and aldosterone secretion it is likely that it stimulates one or more early steps in steroidogenesis (cholesterol to pregnenolone) but it may also increase the conversion of corticosterone to aldosterone (Muller and Ziegler 1968; Haning *et al.* 1970). The response involves 5-HT receptors as it can be prevented by cyproheptadine, methysergide, or metergoline (Mendelsohn and Kachel 1981; Williams *et al.* 1984; Matsuoka *et al.* 1985). Moreover, prevention of the response by ketanserin indicates the involvement of 5-HT_2 receptors (Williams *et al.* 1984; Matsuoka *et al.* 1985). Like adrenocorticotrophic hormone, 5-HT increases the formation of cyclic AMP in adrenal cortical cells *in vitro* (Albano *et al.* 1974; Matsuoka *et al.* 1985). However, higher concentrations were required to stimulate cyclic nucleotide production than were necessary to increase steroidogenesis. Nevertheless, 5-HT differs from angiotensin peptides which do not increase cyclic AMP. The effect of angiotensin II, but not that of 5-HT, appears to be linked to an increase in cytosolic Ca^{2+} (Williams *et al.* 1981), since angiotensin II, but not 5-HT stimulates $^{45}Ca^{2+}$ efflux from perifused adrenal cortical cells. Angiotensin II rapidly (within 5 s) decreased the ^{32}P content of phosphatidylinositol bisphosphate in rat zona glomerulosa cells (Farese *et al.* 1984). 5-HT did not do this but both 5-HT and angiotensin II increased the incorporation of ^{32}P into phosphatidylinositol and phosphatidic acid (Farese *et al.* 1983); however, this was confirmed only for angiotensin II by other workers (Whitley *et al.* 1984). The steroidogenic action of 5-HT appears to be unrelated to any action on adrenal Na^+, K^+– ATPase activity (Meuli and Muller 1982) and 5-HT does not modify the intracellular potassium content of adrenal zona glomerulosa cells (Mendelsohn and Mackie 1975).

The consensus is that 5-HT stimulates steroidogenesis via increasing cyclic AMP formation; in contrast, angiotensin II works via activation of phosphatidylinositol metabolism.

12.2.2.1.2 *Physiological significance of the effect of 5-HT on aldosterone secretion* The action of 5-HT on aldosterone secretion was discovered whilst studying the stimulating action of serum on steroidogenesis. 5-HT probably makes a major contribution to the aldosterone-releasing ability of serum (Muller and Ziegler 1968; Mendelsohn and Kachel 1981). The concentration of 5-HT in serum ultrafiltrates (0.3–0.8 μM) (Mendelsohn and Kachel 1981) is certainly adequate to stimulate aldosterone secretion.

This may be physiologically relevant as even healthy human *plasma* may contain around 0.1 μM of free 5-HT (Fozard 1982b).

5-HT receptor blocking drugs lower basal aldosterone secretion *in vitro* (Mendelsohn and Kachel 1981; Matsuoka *et al.* 1985) as well as that stimulated by 5-HT, K^+, adrenocorticotrophic hormone, and angiotensin II (Mendelsohn and Kachel 1981; Matsuoka *et al.* 1985). They also reduce aldosterone secretion from isolated adrenal cells derived from patients with Cushing's syndrome (Racz *et al.* 1979). The significance of the effect of these antagonists on basal and non-5-HT stimulated aldosterone secretion *in vitro* is unclear. No attempt has been made to relate this action to the presence of 5-HT in the cell preparations. The slight reduction in plasma aldosterone concentrations produced by ketanserin in patients with primary hyperaldosteronism (Mantero *et al.* 1985) may suggest a role for 5-HT in pathological conditions associated with hypersecretion of the hormone.

12.2.2.2 *Steroidogenic action of 5-HT on the ovary*

5-HT (1–100 μM) stimulated progesterone production from cow isolated corpus luteum cells, an effect which was inhibited by the 5-HT receptor blocking drug mianserin (Battista and Condon 1986). This observation, together with the presence of high concentrations of 5-HT in rat (Clausell and Soliman 1978) and cow (Battista and Condon 1986) ovarian tissue suggests a role for 5-HT in ovarian function. Moreover, Clausell and Soliman (1978) showed cyclical changes in the 5-HT content of rat ovary, with the highest value being found at oestrus.

12.2.2.3 *Activation of the renin–angiotensin system by 5-HT*

Subcutaneous injection of either 5-HT or 5-hydroxytryptophan produces a marked increase in plasma renin activity in the rat, accompanied by a dipsogenic response (Barney *et al.* 1981; Kikta *et al.* 1981). The increase in plasma renin activity due to 5-hydroxytryptophan could be prevented by a peripheral aromatic L-amino acid decarboxylase inhibitor (Barney *et al.* 1981). The physiological significance of this remains to be established but, clearly, activation of the renin–angiotensin system via peripheral 5-HT receptors could also result in an increase in aldosterone secretion.

12.2.3 Endocrine secretion–concluding remarks

Among the endocrine structures containing APUD cells, only the endocrine pancreas has been studied in detail, in relation to the possible role of 5-HT. 5-HT inhibits insulin secretion. This appears to involve a direct effect of 5-HT on islet B-cells but there may be a contribution from 5-HT-induced noradrenaline release. The high concentrations of 5-HT required to inhibit secretion raises questions about the physiological significance of the effect. However, the high concentration required may be explained by a predominantly intracellular action. On the other hand it appears that very high intracellular levels of 5-HT are also required to produce an effect. Although somatostatin from the islet D-cell (Taborsky 1983) or insulin itself (Draznin *et al.* 1986) may act as negative modulators of insulin secretion, one must also consider the possibility that 5-HT, stored within the secretory granules and co-secreted with insulin, could participate in feedback inhibition of secretion. It is also possible that altered intracellular levels of 5-HT could explain the altered insulin secretion seen under different situations. However, it is unlikely that this can explain the diminished secretion associated with fasting, as no difference was observed between fasted and fed values for intra-islet 5-HT in the mouse (Hansen and Hedeskov 1977). Studies are required using more selective 5-HT receptor agonists and antagonists and highly sensitive methods for measuring 5-HT both within islet cells and released from them. The role of 5-HT-containing nerves in modulating insulin release also remains to be investigated. 5-HT is a potent steroidogenic substance, especially on the zona glomerulosa of the adrenal cortex, although its importance as a physiological aldosterone-secretagogue remains to be established.

12.3 5-HT and exocrine secretion

12.3.1 Gastric acid secretion

Unlike histamine, neither 5-HT, nor its precursor, 5-hydroxytryptophan, stimulates gastric acid secretion in anaesthetized dogs or in conscious dogs with Heidenhain pouches (Black *et al.* 1956 and 1958; Haverback *et al.* 1957 and 1958), although 5-HT may produce an initial enhancement of histamine-stimulated secretion (Smith 1958). This initial enhancement was also seen occasionally in the cat (Corral-Saleta 1960). However, the main action of 5-HT and its precursor, 5-hydroxytryptophan, is to cause *inhibition* of stimulated acid secretion. This has been shown in dogs (Black *et al.* 1956 and 1958; Haverback *et al.* 1957 and 1958), cats (Corral-Saleta 1960), rats (Haverback and Davidson 1958; Hakanson *et al.* 1967a; Winship and Schuck 1971), and man (Haverback and Davidson 1958; Warner *et al.* 1959). Black *et al.* showed that the inhibition only occurred when the 5-HT was given at the same time as the stimulant to acid secretion (histamine). They also showed (Black *et al.* 1959) that 5-hydroxytryptophan mimicked the inhibitory action of 5-HT, but only after a longer latent period. In cats, the effect of 5-HT depends on the dose and on the stimulus, low doses augmenting and high doses inhibiting pentagastrin-stimulated secretion; in contrast, histamine-stimulated secretion is usually inhibited (Izzat and Waton, unpublished observations).

5-HT- and 5-hydroxytryptophan-induced inhibition of secretion was claimed to be partly dependent on vagal innervation, indicating the possible involvement of a central site for controlling gastric secretion (Black *et al.* 1958). Further support for a vagal involvement came from the observations that 5-HT inhibited secretion from a gastric fistula (vagally innervated) but not from a Heidenhain pouch (vagally denervated) (Jaffe *et al.* 1977). Although there may be a vagal component, this may not be essential as 5-HT inhibits bethanechol-stimulated secretion from Heidenhain pouches in dogs (Haverback *et al.* 1957 and 1958). Moreover, a direct inhibitory action of 5-HT on acid secretion by the rat isolated stomach has been demonstrated (Canfield and Spencer 1983). This also tends to negate the suggestion of Smith (1958) that the inhibitory action of 5-HT is due to vasoconstriction and increased mucus production. Beck and Andersson (1985) showed that this action of 5-HT was not mediated by changes in gastric mucosal blood flow.

It is not clear whether 5-HT exerts its inhibitory effect directly, or via the release of other inhibitory factors. In addition to being blocked by 5-HT-receptor blocking drugs, the effect is prevented by β-adrenoceptor blocking agents (Beck 1986a) and by inhibitors of cyclo-oxygenase (Canfield and Spencer 1983) suggesting the involvement of catecholamines and eicosenoids, respectively.

12.3.1.1 *5-HT as the putative mediator of the inhibition of acid secretion produced by duodenal acidification*

It is well known that duodenal acidification to a pH of 2.5 or less produces a marked inhibition of gastric acid secretion in response to histamine, gastrin, cholinomimetics, or feeding (Pincus *et al.* 1942; Code and Watkinson 1955; Sircus 1958; Jones and Harkins 1959; Andersson and Nilsson 1969; Harrison and Johnson 1970), although this inhibition is more controversial in the cat (Konturek *et al.* 1969; Emas *et al.* 1978). Moreover, it is generally accepted that this is largely mediated by a humoral mechanism (Johnston and Duthie 1964 and 1966; Konturek and Grossman 1965; Schuck *et al.* 1968; Andersson *et al.* 1973 and 1976). There is considerable evidence to suggest that 5-HT is the probable hormonal mediator.

12.3.1.1.1 *Release of 5-HT by duodenal acidification* The first real link between acid in the duodenum and the release of 5-HT came from the observation that degranulation of rat and guinea-pig duodenal enterochromaffin cells was induced by hydrochloric acid (Wermel and Kacharova 1948). Perfusion of the rat upper intestine *in vivo* with physiological amounts of 0.1N hydrochloric acid reduced duodenal tissue 5-HT content (Resnick and Gray 1962a and b). Moreover, incubation of duodenal segments *in vitro* at low pH also reduced 5-HT in the tissue and produced degranulation of enterochromaffin cells. Resnick and Gray (1962a and b) further showed that 5-HT was present in the luminal perfusate of loops of intestine perfused with hydrochloric acid. Early attempts failed to show the presence of 5-HT in portal or hepatic blood, probably due to the insensitivity of the methods (Resnick and Gray 1962a; Wise *et al.* 1968; Wise 1972). However, using a sensitive radioimmunoassay, elevation of 5-HT was detected in the portal vein within one minute after duodenal acidification in anaesthetized dogs (Kellum and Jaffe 1976). A similar observation was reported by Koren *et al.* (1976). Although it is clear that duodenal acidification increases portal venous 5-HT, it could be argued that insufficient 5-HT would escape hepatic and pulmonary inactivation to achieve an inhibitory concentration in the stomach. However, in conscious dogs with Heidenhain pouches and gastric fistulae, duodenal acidification resulted in significant inhibition of gastric output and elevated *peripheral* venous 5-HT concentrations (Jaffe *et al.* 1977). Furthermore, infusion of 5-HT, to produce similar blood concentrations, inhibited acid output from the gastric fistulae.

It is not yet clear how 5-HT is released from the duodenum in response to acid. Local anaesthesia of the duodenal mucosa abolishes acid-induced inhibition of gastric secretion, suggesting a neuronal involvement (Sircus 1958; Thompson 1964; Wise *et al.* 1968). However, it is not clear whether local anaesthesia blocks a short (duodeno-gastric) or a long reflex arc, or merely stabilizes the membrane of the hormone-containing cell. It is also possible that 5-HT release is coincidental with the release (from APUD cells) of duodenal peptide hormones having an inhibitory action on gastric secretion. Alternatively, 5-HT may mediate the release of these hormones (Thompson 1971). 5-HT can release gastro-intestinal hormones, although this has been shown only for gastrin (Koop and Arnold 1984), which would, of course, *stimulate* gastric secretion.

12.3.1.1.2 *Effects of 5-HT receptor blockade or 5-HT depletion* The inhibitory effect of 5-HT on gastric acid secretion is prevented by methysergide (Thompson 1964; Beck 1986a). Methysergide prevented the inhibitory gastric secretory response to duodenal acidification, thus supporting a role for 5-HT as the humoral mediator (Thompson 1964; Wise *et al.* 1968; Schuck *et al.* 1968; Wise 1972; Mate *et al.* 1985). Moreover, depletion of 5-HT, using reserpine, abolished the effects of duodenal acidification in dogs and in rats (Resnick and Gray 1962b; Thompson 1964; Wise 1972). These observations must be interpreted cautiously, in view of the lack of specificity of the drugs used. It is also possible that reserpine, in addition to depleting 5-HT, depletes APUD cells of their duodenal peptide hormone stores. However, this seems unlikely, by analogy with the selective depletion by reserpine of guinea-pig islet 5-HT without any effect on insulin (Lundquist *et al.* 1975) (see Section 12.2.1.1).

12.3.1.2 *Role of 5-HT in the 'fade' of stimulated gastric secretion*

When pentagastrin is infused, the stimulation of acid secretion shows a characteristic fade from the initial peak value. This fade was reduced by reserpine or by 5-HT receptor blockade (Sanyal and Waton 1976). Moreover, in the cat, pentagastrin infusion raised the plasma 5-HT concentration approximately four-fold (Izzat and Waton 1987). These data suggest a role for 5-HT in the fade phenomenon.

12.3.2 Pepsin secretion

Grechiskin (1967, quoted by Thompson 1971) reported that 5-HT increased pepsin output in dogs. Furthermore, the pepsin secreted in response to acetylcholine, carbachol, insulin hypoglycaemia, and feeding was correlated with the concentration of 5-HT in the gastric juice, suggesting a direct relationship (Grechiskin 1970, quoted by Thompson 1971). Wazna *et al.* (1977 and 1978) concluded from a study in dogs using reserpine and 5-HT antagonists that 5-HT stimulated pepsin secretion by a direct hormonal effect on the gastric chief cells. More recently Beck and Andersson (1985) showed that 5-HT had a dual role on pentagastrin-stimulated secretion; low and high doses facilitated the response whereas intermediate doses were inhibitory. The inhibition was non-competitive and could be counteracted by β-adrenoceptor blocking drugs; on this basis, Beck and Andersson concluded that the inhibitory effect was mediated via a release of noradrenaline from sympathetic nerves. These results and conclusions have been fully supported by further work on bethanecol-stimulated pepsin secretion (Beck 1986b).

12.3.3 Mucus secretion

12.3.3.1 *Gastric mucus*

In some species, 5-HT is a potent stimulant of gastric mucus secretion. Large amounts of mucus were secreted by the stomach and intestine of dogs following an intravenous infusion of 5-HT (Black *et al.* 1958; Smith 1958). Both the volume and total output of gastric mucus are increased (Menguy 1969). White and Magee (1958) found that atropine, but not hexamethonium, reduced 5-HT-induced mucin secretion. In man, mucus production is increased both by 5-HT and 5-hydroxytryptophan (Haverback and Davidson 1958; Warner *et al.* 1959) but in rats administration of 5-hydroxyptrytophan did not increase mucus output (Julkunen *et al.* 1970). It has been suggested that an increased mucus secretion may be a physiological action to protect the mucosa against the excessive action of acid and that 5-HT is the neurohumoral substance released at the nerve endings which stimulates the mucus secreting cells (Wise 1972).

12.3.3.2 *Colonic mucus*

5-HT also stimulates the secretion of colonic mucus when infused intravenously (Black *et al.* 1979; Bradbury *et al.* 1980). Unlike in the stomach, however, this was not prevented by atropine; methysergide was similarly ineffective (Bradbury *et al.* 1980). Although the effects of 5-HT on mucus secretion in other parts of the intestinal tract do not seem to have been investigated, it remains possible that a physiological role of 5-HT may be to stimulate the production of a protective mucus lining for the whole gastrointestinal tract.

12.3.4 Intestinal fluid secretion

There is good evidence that 5-HT causes a net water and electrolyte secretion in the small intestine *in vivo* (Kisloff and Moore 1976; Donowitz and Charney 1977; Donowitz *et al.* 1979). These changes do not involve adenylate cyclase, cyclic AMP, or cyclic GMP but rather an alteration in the enterocyte plasma membrane permeability to Ca^{2+} (Donowitz *et al.* 1980; Zimmerman and Binder 1984). The effect is on the crypt cells rather than on the villous cells (Hardcastle *et al.* 1981).

Cassuto *et al.* (1982) claim that their studies with 5-HT and cholera toxin in cats and rats point to a neuronal mechanism for the intestinal secretory action of 5-HT. A similar and more detailed conclusion was made by Keast *et al.* (1985) using isolated innervated mucosa–submucosa preparations of guinea-pigs mounted in Ussing chambers, in which the short circuit current was measured as an indication of net ion transport. Doses of 5-HT $>$ 0.5 μM and electrical stimulation produced two components with different latencies. Hyoscine abolished the first and significantly reduced the second component. At concentrations <0.5 μM, only the latter component was seen and this was unaffected by hyoscine. The results indicate that the early component is due to acetylcholine release from cholinergic nerves but the second component is only partially due to acetylcholine release. The hyoscine-resistant responses were partially blocked by a substance P antagonist, suggesting the involvement of non-cholinergic substance P fibres. A role for 5-HT in chloride secretion produced by stimulation of cholinergic neurones within the submucosal plexus of the guinea-pig was also suggested by Cooke and Carey (1985). Further evidence for a neuronal mechanism has resulted from the work of Baird and Cuthbert—see Baird and Cuthbert (1987) for references. In ileal and colonic epithelia from sensitized guinea-pigs, 5-HT or antigen produced a chloride-secretory response, which was blocked by tetrodotoxin, indicating a neuronal involvement (Baird and Cuthbert 1987). Moreover, the response to 5-HT was blocked by the selective 5-HT_3 receptor antagonist ICS 205-930, but not by the 5-HT_2 receptor blocking drug ketanserin or by cyproheptadine (Baird and Cuthbert 1987). On the other hand the response to antigen was blocked by ICS 205-930 *and* by the other 5-HT antagonists, suggesting the involvement of 5-HT in the hypersensitivity reaction, although the receptor type(s) involved remain to be clarified.

Some other work in rats (Beubler *et al.* 1984 and 1986) suggests that 5-HT induces intestinal fluid secretion by stimulating local prostaglandin formation, in particular PGE_2, which probably acts in physiological concentrations via facilitation of calcium entry.

Paradoxically, 5-HT receptor antagonists (methysergide or ketanserin), or 5-HT depletion using *p*-chlorophenylalanine, blocked the antisecretory effect of morphine against secretion induced by vasoactive intestinal peptide in anaesthetized rats (Couper and Taylor 1987). This suggests an *antisecretory* role for 5-HT in mediating the actions of morphine, although neither the 5-HT receptor blocking drugs, nor *p*-chlorophenylalanine, themselves modified vasoactive intestinal peptide-induced secretion.

12.3.5 Pancreatic enzyme secretion

Using very high doses (0.5–32 mg kg^{-1} intravenously) it has been claimed (Mori *et al.* 1979) that 5-HT decreases the rate of flow and of protein secretion from rat pancreas. In recent work, using more physiological doses, it has been found that zymogen granules of the rat concentrate 5-HT taken up by the acinar cells, which not only take up and store 5-HT themselves, but will also form it from 5-hydroxytryptophan. When dispersed acinar cells were stimulated with caerulein there was a synchronous increase in amylase activity and release of 5-HT, thus showing an involvement of 5-HT in pancreatic enzyme secretion, but not indicating specifically the role played by 5-HT (Stern *et al.* 1983). However, these authors suggest the possibility that 5-HT may influence the rate of flow of pancreatic fluid.

12.3.6 Bile secretion

Although not much work has been done on 5-HT and bile secretion, Kortz *et al.* (1985) have recently claimed to show that 5-HT inhibits fasting and meal stimulated bile formation in dogs, i.e. it has the opposite effect to secretin. This suppression occurred in the absence of endogenous intestinal hormone and was always short lasting, suggesting a direct mechanism of action.

12.3.7 Salivary enzyme secretion

It has been shown in guinea-pigs that 5-HT is a potent enzyme secretogogue of the submandibular salivary gland. However, as this effect was readily blocked by propranolol the authors concluded that it was mediated via the β-adrenoceptor (Carlson *et al.* 1974). In earlier work, using the parotid gland *in vivo* in rabbits, it was claimed that the stimulatory effect on enzyme secretion was a combination of a direct and an indirect effect mediated by noradrenaline (Kojima *et al.* 1973).

12.3.8 Nasal secretion

Tonnesen and Schaffalitzky de Muckadell (1987) showed that nasal challenge with 5-HT induced dose-related itching, sneezing, and secretion in normal subjects. The effect was partially blocked by atropine but was unaffected by methysergide. In addition unilateral nasal challenge induced bilateral secretion. From these observations the authors concluded that 5-HT may cause the effects by acting on sensory nerves and they suggested that 5-HT may thus play a role in the normal nasal secretory response. This suggestion is supported by the finding of 5-HT in nasal secretion produced by histamine, methacholine, and allergens. However, substance P was also found in all 5-HT stimulated secretions. This substance is present in the sensory nerves of the nose but, although highly active in animals, it appears to have no effect on nasal secretion or mucociliary function in humans. This relationship of 5-HT with substance P may be similar to that suggested in the case of intestinal secretions (Keast *et al.* 1985).

12.3.9 Exocrine secretion—concluding remarks

5-HT undoubtedly modifies exocrine secretory activity. At present, however, there is only strong evidence for a physiological role for 5-HT in modulating gastric acid secretion. However, recent work suggests its involvement in intestinal and, possibly nasal secretions, in which situations the effects of 5-HT appear to be mediated through activation of neuronal mechanisms.

References

Albano, J. D. M., Brown, B. L., Ekins, R. P., Tait, S. A. S., and Tait, J. F. (1974). The effects of potassium, 5-hydroxytryptamine, adrenocorticotrophin and angiotensin II on the concentration of adenosine 3′, 5′-cyclic monophosphate in suspensions of dispersed rat adrenal zone glomerulosa and zona fasciculata cells. *Biochemical Journal* **142,** 391–400.

Aleyassine, H. and Gardiner, R. J. (1975). Dual action of antidepressant drugs (MAO inhibitors) on insulin release. *Endocrinology* **96,** 702–10.

Aleyassine, H. and Lee, S. H. (1971). Inhibition by hydrazine, phenelzine and pargyline of insulin release from rat pancreas. *Endocrinology* **89,** 125–9.

Aleyassine, H. and Lee, S. H. (1972). Inhibition of insulin release by substrates and inhibitors of monoamine oxidase. *American Journal of Physiology* **222,** 565–9.

Andersson, S. (1973). Secretion of gastrointestinal hormones. *Annual Review of Physiology* **35,** 431–52.

Andersson, S. and Nilsson, G. (1969). pH dependence of the mechanism in the duodenal bulb inhibiting gastric acid responses to exogenous gastrin. *Acta Physiologica Scandinavica* **76,** 182–90.

Andersson, S., Nilsson, G., Sjodin, L., and Uvnas, B. (1973). Mechanism of duodenal inhibition of gastric acid secretion. In *Frontiers in gastrointestinal hormone research* (ed. S. Andersson), Nobel Symposium, Vol. 16, pp. 223–38. Almquist & Wiksell, Stockholm.

Andersson, S., Nilsson, G., and Sjodin, L. (1976). The significance of the duodenal bulb for gastric functions in the dog. *Scandinavian Journal of Gastroenterology* **11,** 49–54.

Atwater, I., Ribalet, B., and Rojas, E. (1979). Mouse pancreatic β cells: Tetraethylammonium blockage of the potassium permeability increase induced by depolarization. *Journal of Physiology* **288,** 561–4.

Baird, A. W. and Cuthbert, A. W. (1987). Neuronal involvement in Type 1 hypersensitivity reactions in gut epithelia. *British Journal of Pharmacology* **92,** 647–55.

Barney, C. C., Threatte, R. M., Kikta, D. C., and Fregly, M. J. (1981). Effects of serotonin and L-5-hydroxytryptophan on plasma renin activity in rats. *Pharmacology, Biochemistry and Behaviour* **14,** 895–900.

Battista, P. J. and Condon, W. A. (1986). Serotonin-induced stimulation of progesterone production by cow luteal cells *in vitro. Journal of Reproduction and Fertility* **76,** 231–8.

Beck, K. (1986a). Exogenous serotonin and histamine-stimulated gastric acid and pepsin secretion in dogs. *Scandinavian Journal of Gastroenterology* **21,** 1205–10.

Beck, K. (1986b). Effect of serotonin on bethanecol-stimulated gastric pepsin secretion in dogs. *Scandinavian Journal of Gastroenterology* **21,** 537–41.

Beck, K. and Andersson, D. (1985). Effect of serotonin on pentagastrin-stimulated gastric pepsin secretion in dogs with gastric fistula. Scandinavian Journal of Gastroenterology **20,** 1204–8.

Bender, D. A. (1980). Inhibition *in vitro* of the enzymes of the oxidative pathway of tryptophan metabolism and of nicotinamide nucleotide synthesis by benserazide, carbidopa and isoniazid. *Biochemical Pharmacology* **29,** 707–12.

Beubler, E., Bukhave, K., and Rask-Madsen, J. (1984). Colonic secretion mediated by prostaglandin E_2 and 5-hydroxytryptamine, may contribute to diarrhoea due to morphine withdrawal in the rat. *Gastroenterology* **87,** 1042–81.

Beubler, E., Bukhave, K., and Rask-Madsen, J. (1986). Involvement of calcium in prostaglandin E_2 mediated fluid secretion induced by 5-hydroxytryptamine in the rat jejunum *in vivo. Gastroenterology* **90,** 1345.

Bing, R. F. and Schulster, D. (1977). Steroidogenesis in isolated rat adrenal glomerulosa cells: response to physiological concentrations of angiotensin II and effects of potassium, serotonin and [$Sar^1.Ala^8$] angiotensin II. *Journal of Endocrinology* **74,** 261–72.

Bird, J. L., Wright, E. E., and Feldman, J. M. (1980). Pancreatic islets: A tissue rich in serotonin. *Diabetes* **29,** 305–8.

Black, J. W., Fisher, E. W., and Smith, A. N. (1956). The effects of 5-hydroxytryptamine on gastric acid secretion. *International Physiology Congress* **20,** 102.

Black, J. W., Fisher, E. W., and Smith, A. N. (1958). The effects of

5-hydroxytryptamine on gastric secretion in anaesthetized dogs. *Journal of Physiology (London)* **141,** 27–34.

Black, J. W., Fisher, E. W., and Smith, A. N. (1959). The effects of precursors of 5-hydroxytryptamine on gastric secretion in anaesthetized dogs. *Journal of Physiology (London)* **146,** 10–17.

Black, J. W., Bradbury, J. E., and Wyllie, J. H. (1979). Stimulation of colonic mucus output in the rat. *British Journal of Pharmacology* **66,** 456P–7P.

de Bold, M. L. and Benscombe, S. A. (1978). Effects of serotonin and its antagonists on the first phase of insulin release in normal and hypersomatotrophic rats. *Hormone and Metabolic Research* **10,** 500–5.

Bradbury, J. E., Black, J. W., and Wyllie, J. H. (1980). Stimulation of mucus output from rat colon *in vivo. European Journal of Pharmacology* **68,** 417–25.

Canfield, S. P. and Spencer, J. E. (1983). The inhibitory effects of 5-hydroxytryptamine on gastric acid secretion by the rat isolated stomach. *British Journal of Pharmacology* **78,** 123–9.

Carlson, B., Danielsson, A., Marklund, S., and Stigbrand, T. (1974). Effects of alpha- and beta-receptor blocking agents on catecholamine and 5-hydroxytryptamine induced peroxidase and amylase secretion from guinea-pig submandibular gland. *Acta Physiologica Scandinavica* **92,** 263–71.

Cassuto, J., Jodal, M., Tuttle, R., and Lundgren, O. (1982). 5-hydroxytryptamine and cholera secretion; physiological and pharmacological studies in cats and rats. *Scandinavian Journal of Gastroenterology* **17,** 695–703.

Cegrell, L. (1968). The occurrence of biogenic monoamines in the mammalian endocrine pancreas. *Acta Physiologica Scandinavica (Supplement)* **314,** 1–60.

Chaouloff, F. and Jeanrenaud, B. (1987a). 5-HT_{1A} and α_2-adrenergic receptors mediate the hyperglycaemic and hypoinsulinaemic effects of 8-hydroxy-2-(di-n-propylamino) tetralin (8-OH-DPAT) in the conscious rat. *Journal of Pharmacology and Experimental Therapeutics* **243,** 1159–68.

Chaouloff, F. and Jeanrenaud, B. (1987b). Hyperinsulinaemia of the genetically obese (fa/fa) rat is decreased by a low dose of the 5-HT_{1A} receptor agonist 8-hydroxy-2-(di-N-propylamino) tetraline (8-OH-DPAT). *European Journal of Pharmacology* **147,** 111–18.

Clausell, D. E. and Soliman, K. F. A. (1978). Ovarian serotonin content in relation to ovulation. *Experientia* **34,** 410–11.

Code, C. F. and Watkinson, G. (1955). Importance of vagal innervation in the regulatory effect of acid in the duodenum on gastric secretion of acid. *Journal of Physiology (London)* **130,** 233–52.

Cooke, H. J. and Carey, H. V. (1985). Pharmacological analysis of 5-hydroxytryptamine actions in guinea-pig ileal mucosa. *European Journal of Pharmacology* **111,** 329–37.

Corral-Saleta, J. M. de (1960). Accion de la serotinina sobre la secretion gastrica. *Revista Espanola de Fisiologia* **16,** 114–21.

Couper, I. M. and Taylor, D. A. (1987). Possible involvement of noradrenergic and tryptaminergic neurones in the intestinal antisecretory action of morphine. *British Journal of Pharmacology* **90,** 105P.

Dalgleish, C. E., Toh, C. C., and Work, T. S. (1953). Fractionation of the smooth muscle stimulants present in extracts of gastrointestinal tract. Identification of

5-hydroxytryptamine from substance P. *Journal of Physiology (London)* **120,** 298–310.

Doggrell, S. A. and Woodruff, G. N. (1978). Effects of drugs on the accumulation and spontaneous release of noradrenaline in the rat anococcygeus muscle. *British Journal of Pharmacology* **63,** 465–70.

Donowitz, M. and Charney, A. N. (1977). Effect of serotonin treatment on intestinal transport in the rabbit. *American Journal of Physiology* **232,** E85–E94.

Donowitz, M., Tai, Y.-H., and Asakoff, N. (1979). Serotonin-induced ileal secretion; active electrolyte secretion which is calcium dependent. *Gastroenterology* **76,** 1123.

Donowitz, M., Asakoff, N., and Pike, G. (1980). Serotonin-induced changes in rabbit ileal active electrolyte transport are calcium dependent associated with an increased ileal calcium uptake. *Journal of Clinical Investigation* **66,** 341–52.

Draznin, B., Goodman, M., Leitner, J. W., and Sussman, K. E. (1986). Feedback inhibition of insulin on insulin secretion in isolated pancreatic islets. *Endocrinology* **118,** 1054–8.

Ekholm, R., Ericson, L. E., and Lundquist, I. (1971). Monoamines in the pancreatic islets of the mouse. Subcellular localization of 5-hydroxytryptamine by electron microscopic autoradiography. *Diabetologia* **7,** 339–48.

Emas, S., Fernstrom, M., and Svensson, S.-O. (1978). Potentiation of the action of gastrin by a humoral factor released by duodenal feeding in cats. *Acta hepato-gastroenterologica* **25,** 503–9.

Endo, Y. (1987). Suppression and potentiation of 5-hydroxytryptophan-induced hypoglycaemia by α-monofluoromethyldopa: correlation with the accumulation of 5-hydroxytryptamine in the liver. *British Journal of Pharmacology* **90,** 161–5.

Erspamer, V. (1954). Pharmacology of indolealkylamines. *Pharmacological Reviews 6.* 425–87.

Falck, B. and Hellman, B. (1963). Evidence for the presence of biogenic amines in pancreatic islets. *Experientia* **19,** 139–40.

Farese, R. V., Larson, R. E., Sabir, M. A., and Gomez-Sanchez, C. E. (1983). Effects of angiotensin II, K^+, adrenocorticotrophin, serotonin, adenosine 3′5′—monophosphate, guanosine 3′5′—monophosphate, A23187 and EGTA on aldosterone synthesis and phospholipid metabolism in the rat adrenal zona glomerulosa. *Endocrinology* **113,** 1377–86.

Farese, R. V., Larson, R. E., and Davis, J. S. (1984). Rapid effects of angiotensin II on polyphosphoinositide metabolism in the rat adrenal glomerulosa. *Endocrinology* **114,** 302–4.

Federspil, B., Casara, D., Pedrazzoli, S., Sicolo, N., and Scandelari, L. (1974). In vivo studies on 5-hydroxytryptamine and insulin secretion in dogs and man. *Diabetologia* **10,** 13–17.

Feldberg, W. and Toh, C. (1953). Distribution of 5-hydroxytryptamine (Serotonin. Enteramine). in the wall of the digestive tract. *Journal of Physiology (London)* **119,** 352–62.

Feldman, J. M. and Chapman, B. (1975a). Characterization of pancreatic islet monoamine oxidase. *Metabolism,* **24,** 581–8.

Feldman, J. M. and Chapman, B. (1975b). Monoamine oxidase inhibitors: nature of interaction with rabbit pancreatic islets. *Diabetologia* **11,** 487–94.

Feldman, J. M. and Lebovitz, H. E. (1970). Specificity of serotonin inhibition of insulin release from golden hamster pancreas. *Diabetes* **19,** 475–9.

Feldman, J. M., Quickel, K. E. jun., and Lebovitz, H. E. (1972). Potentiation of insulin secretion in vitro by serotonin antagonists. *Diabetes* **21,** 779–88.

Feldman, J. M., Bivens, C. M., Skyler, J. S., and Lebovitz, H. E. (1975). Effect of cyproheptadine administration on insulin secretion in acromegalic, diabetic and normal subjects. *Hormone and Metabolic Research* **7,** 279–83.

Fozard, J. R. (1982a). Highly potent irreversible inhibitors of aromatic L-amino acid decarboxylase. *Trends in Pharmacological Sciences* **11,** 429.

Fozard, J. R. (1982b). Serotonin, migraine and platelets. In *Progress in pharmacology,* Vol. 414, pp. 135–46. Gustav Fischer Verlag, Stuttgart and New York.

Fozard, J. R. (1984). Neuronal 5-HT receptors in the periphery. *Neuropharmacology* **23,** 1473–86.

Fujita, K., Aguilara, G., and Catt, K. J. (1979). The role of cyclic AMP in aldosterone production by isolated zona glomerulosa cells. *Journal of Biological Chemistry* **254,** 8567–74.

Furman, B. L. (1974). The hypoglycaemic effect of 5-hydroxytryptophan. *British Journal of Pharmacology* **50,** 575–80.

Furman, B. L. and Wilson, G. A. (1980). Further studies on the effects of 5-hydroxytryptophan on plasma glucose and insulin in the mouse. *Diabetologia* **19,** 386–90.

Furman, B. L., Itoh, M., and Gerich, J. E. (1981). Effect of 5-hydroxytryptamine (5-HT) and its precursor 5-HTP on insulin and glucagon secretion in vitro. *Diabetologia* **20,** 669–70.

Gagliardino, J. J., Nierle, C., and Pfeiffer, E. F. (1974). The effect of serotonin on in vitro insulin secretion and biosynthesis in mice. *Diabetologia* **10,** 411–14.

Giugliano, D., Passariello, N., Santoro, L., Cerciello, T., and Gentile, S. (1981). Impaired insulin secretion in human diabetes mellitus III. The effect of the serotonin antagonist metergoline. *Diabete et Metabolisme* **7,** 231–4.

Gylfe, E. (1978). Association between 5-hydroxytryptamine release and insulin secretion. *Journal of Endocrinology* **78,** 239–48.

Gylfe, E., Hellman, B., Sehlin, J., and Taljedal, I-B. (1973). Amino acid conversion into 5-hydroxytryptamine in pancreatic B-cells. *Endocrinology* **93,** 932–7.

Hakanson, R., Lilja, B., and Owman, C. (1967a). Properties of a new system of amine-storage cells in the gastric mucosa of the rat. *European Journal of Pharmacology* **1,** 188–99.

Hakanson, R., Lilja, B., and Owman, C., and Thunell, S. (1967b). Changes in gastric secretion induced by certain amines, amine precursors and related enzyme inhibitors. *European Journal of Pharmacology* **1,** 425–33.

Halben, P. A., Wollheim, C. B., Blondel, B., Niesor, E., and Renold, A. E. (1979). Perturbation of hormone storage and release induced by cyproheptadine in rat pancreatic islets *in vitro. Endocrinology* **104,** 1096–106.

Haning, R., Tait, S. A. S., and Tait, J. F. (1970). *In vitro* effects of ACTH, angiotensins, serotonin and potassium on steroid output and conversion of corticosterone to aldosterone by isolated adrenal cells. *Endocrinology* **87,**1147–67.

Hansen, S. E. and Hedeskov, C. J. (1977). Simultaneous determination of the content of serotonin, dopamine, noradrenaline and adrenaline in pancreatic islets isolated from fed and starved mice. *Acta Endocrinologica* **86,** 820.

Hardcastle, J., Hardcastle, P. T., and Redfern, J. S. (1981). Action of 5-hydroxytryptamine on intestinal ion transport in the rat. *Journal of Physiology (London)* **320,** 41–55.

Harrison, L. A. and Johnson, L. R. (1970). Analysis of gastric secretory inhibition by duodenal acidification. *Gastroenterology* **58,** 1043.

Haverback, B. J. and Davidson, J. D. (1958). Serotonin and the gastrointestinal tract. *Gastroenterology* **35,** 570–8.

Haverback, B. J., Hogben, C. A. M., Moran, N. C., and Terry, L. L. (1957). Effect of serotonin and related compounds on gastric secretion and intestinal motility in the dog. *Gastroenterology* **32,** 1058–65.

Haverback, B. J., Bogdanski, D. F., and Hogben, C. A. M. (1958). Inhibition of gastric acid secretion in the dog by the precursor of serotonin, 5-hydroxytryptophan. *Gastroenterology* **34,** 188–95.

Hellman, B., Lernmark, A., Sehlin, J., and Taljedal, I-B. (1972). Transport and storage of 5-hydroxytryptamine in pancreatic B-cells. *Biochemical Pharmacology* **21,** 695–706.

Henquin, J. C. (1978). D-Glucose inhibits potassium efflux from pancreatic islet cells. *Nature (London)* **271,** 271–3.

Herchuelz, A., Couturier, E., and Malaisse, W. J. (1980). Regulation of calcium fluxes in pancreatic islets: glucose-induced calcium–calcium exchange. *American Journal of Physiology* **238,** E96–E103.

Hutton, J. C., Peshavaria, M., and Tooke, N. E. (1983). 5-Hydroxytryptamine transport in cells and secretory granules from a transplantable rat insulinoma. *Biochemical Journal* **210,** 803–10.

Itoh, M., Reach, G., Furman, B., and Gerich, J. (1982). Secretion of glucagon. In *The islets of Langerhans* (eds. S. J. Cooperstein and D. Watkins) pp. 225–255. Academic Press, London.

Izzat, A. and Waton, N. G. (1987). Release of 5-hydroxytryptamine by pentagastrin and its role in the fade of stimulated gastric secretion in cats. *Journal of Physiology (London)* **383,** 499–507.

Jacoby, J. H. and Bryce, G. F. (1978). The acute pharmacological effects of serotonin on the release of insulin and glucagon in the intact rat. *Archives Internationales de Pharmacodynamie et de Therapie* **235,** 254–70.

Jacoby, J. H. and Bryce, G. F. (1979). The acute effects of 5-HTP, fluoxetine and quipazine on insulin and glucagon release in the intact rat. *Hormone and Metabolic Research* **11,** 90–4.

Jaffe, B. M., Kopen, D. F., and Lazan, D. W. (1977). Endogenous serotonin in the control of gastric acid secretion. *Surgery* **82,** 156–63.

Jaim-Etcheverry, G. and Zieher, L. M. (1968). Electron microscopic cytochemistry of 5-hydroxytryptamine (5-HT) in the beta cells of guinea pig endocrine pancreas. *Endocrinology* **83,** 917–23.

Jansson, L., Leung, P. E., and McEvoy, R. C. (1985). Ablation of serotoninergic nerves in the rat pancreas. Lack of effects on hormone secretion and intrinsic blood flow. *Acta Endocrinologica* **110,** 515–21.

Johnston, D. and Duthie, H. L. (1964). Effect of acid in the duodenum on histamine-stimulated gastric secretion in man. *Gut* **5,** 573–80.

Johnston, D. and Duthie, H. L. (1966). Inhibition of histamine-stimulated gastric secretion by acid in the duodenum in man. *Gut* **7,** 58–68.

Jones, T. W. and Harkins, H. N. (1959). The mechanism of inhibition of gastric acid secretion by the duodenum. *Gastroenterology* **37,** 81–6.

Julkunen, R. J. K., Airaksinen, N. M., and Duro, K. (1970). Effects of reserpine and 5-HT on the secretion of mucopolysaccharides in gastric juice in rats. *Annals of Medicine and Experimental Biology, Fenn* **48,** 149–54.

Keast, J. R., Furness, J. B., and Costa, M. (1985). Investigations of nerve populations influencing ion transport that can be stimulated electrically, by serotonin and by a nicotinic agonist. *Naunyn-Schmiedeberg's Archives of Pharmacology* **331,** 260–6.

Kellum, J. M. and Jaffe, B. M. (1976). Release of immuno reactive serotonin following acid perfusion of the duodenum. *Annals of Surgery* **184,** 633–6.

Kikta, D. C., Threate, R. M., Barney, C. C., Fregly, M. J., and Greenleaf, J. E. (1981). Peripheral conversion of L-5-hydroxytryptophan to serotonin induces drinking in rats. *Pharmacology, Biochemistry and Behaviour* **14,** 889–93.

Kisloff, B. and Moore, E. W. (1976). Effect of serotonin on water and electrolyte transport in the *in vivo* rabbit small intestine. *Gastroenterology* **71,** 1033–8.

Kojima, S., Ikeda, M., and Tsujimoto (1973). Effects of serotonin, glucagon and other hormones on amylase secretion from rabbit 'parotid gland. *Japanese Journal of Pharmacology* **23,** 588–91.

Konturek, S. J. and Grossman, M. I. (1965). Effect of perfusion of intestinal loops with acid, fat or dextrose on gastric secretion. *Gastroenterology* **49,** 481–9.

Konturek, S. J., Dubiel, J., and Gabrys, B. (1969). Effect of acid infusion into various levels of the intestine on gastric and pancreatic secretions in the rat. *Gut* **10,** 749–53.

Koop, H. and Arnold, R. (1984). Serotoninergic control of somatostatin and gastrin release from the isolated rat stomach. *Regulatory Peptides* **9,** 101–8.

Koren, E., Wadnick, S., Solowiejczyk, M., Pfeiffer, Y., and Sulman, F. G. (1976). Serotonin in the portal vein after acidification. *International Surgery* **61,** 370–1.

Kortz, W. J., Schirmer, B. D., Nashold, J. R., Jones, R. S., and Meyers, W. O. (1985). Effects of serotonin on canine bile formation. *Surgery* **98,** 907–13.

Larson, B., Owman, C., and Sundler, F. (1966). Monoaminergic mechanisms in parafollicular cells of the mouse thyroid gland. *Endocrinology* **78,** 1109–14.

Lebrun, P., Malaisse, W. J., and Herchuelz, A. (1982). Evidence for two distinct modalities of Ca^{2+} influx into pancreatic B cell. *American Journal of Physiology* **242,** E59–E66.

de Leiva, A., Tanenberg, R. J., Anderson, B., Greenberg, B., Senske, B., and Goetz, F. C. (1978). Serotoninergic activation and inhibition: effects on carbohydrate tolerance and plasma insulin and glucagon. *Metabolism* **27,** 511–20.

Lenzen, S., Nahrstedt, H., and Panten, U. (1983). Monoamine oxidase in pancreatic islets, exocrine pancreas and liver from rats. Characterization with clorgyline, deprenyl, pargyline, tranylcypromine and amezinium. *Naunyn-Schmiedeberg's Archives of Pharmacology* **324,** 190–5.

Lernmark, A. (1971). The significance of 5-hydroxytryptamine for insulin secretion in the mouse. *Hormone and Metabolic Research* **3,** 305–9.

Lin, B. J. and Haist, R. E. (1975). Insulin biosynthesis: the monoaminergic mechanisms and the specificity of 'glucoreceptor'. *Endocrinology* **96,** 1247–53.

Lindstrom, P. (1981a). Further studies on 5-hydroxytryptamine transport in pancreatic islets and isolated B-cells. *British Journal of Pharmacology* **73,** 385–91.

Lindstrom, P. (1981b). 5-HT inhibits glucose-induced changes in fluxes of ^{86}Rb and ^{36}Cl and ^{32}P in pancreatic B-cells. *Diabetologia* **21,** 297–8.

Lindstrom, P. and Sehlin, J. (1982a). 5-hydroxytryptamine stimulates $^{86}Rb^{+}$ efflux from pancreatic B-cells. *Biochimica Biophysica Acta* **720,** 400–04.

Lindstrom, P. and Sehlin, J. (1982b). Action of 5-hydroxytryptamine on the ^{36}Cl efflux in pancreatic beta-cells. *Acta Physiologica Scandinavica* **116,** 101–2.

Lindstrom, P. and Sehlin, J. (1983). Opposite effects of 5-hydroxytryptophan and 5-hydroxytryptamine on the function of microdissected ob/ob—mouse pancreatic islets. *Diabetologia* **24,** 52–7.

Lindstrom, P. and Sehlin, J. (1986). Aromatic amino acids and pancreatic islet function: a comparison of 1-tryptophan and 1-5-hydroxytryptophan. *Molecular and Cellular Endocrinology* **48,** 121–6.

Lindstrom, P., Sehlin, J., and Taljedal, I-B. (1980). Characteristics of 5-hydroxytryptamine transport in pancreatic islets. *British Journal of Pharmacology* **68,** 773–8.

Lindstrom, P., Norlund, L., Sehlin, J., and Taljedal, I-B. (1984). Effects of 5-hydroxytryptamine on rubidium ion fluxes and insulin release in cultured pancreatic islets. *Endocrinology* **115,** 2121–5.

Lundquist, I., Ekholm, R., and Ericson, L. E. (1971). Monoamines in the pancreatic islets of the mouse. 5-hydroxytryptamine as an intracellular modifier of insulin secretion and the hypoglycaemic action of monoamine oxidase inhibitors. *Diabetologia* **7,** 414–22.

Lundquist, I., Sundler, F., Hakanson, R., Larsson L-I., and Heding, L. G. (1975). Differential changes in the 5-hydroxytryptamine and insulin content of guinea-pig B cells. *Endocrinology* **97,** 937–47.

Malaisse, W. J. (1973). Insulin secretion: multifactorial regulation for a single process of release. *Diabetologia* **9,** 167–73.

Mantero, F., Opocher, G., Boscaro, M., and Armanini, D. (1982). Effect of serotonin on plasma aldosterone in man. *Journal of Endocrinological Investigation* **5,** 97–9.

Mantero, F., Rocco, S., Opocher, G., Armanini, D., Boscaro, M., and Agostino, D. (1985). Effect of ketanserin in primary hyperaldosteronism. *Journal of Cardiovascular Pharmacology* **7,** (Supplement), S172–5.

Marco, J., Hedo, J. A., Martinell, J., Calle, C., and Villaneuva, M. L. (1976). Potentiation of glucagon secretion by serotonin antagonists in man. *Journal of Clinical Endocrinology and Metabolism,* **42,** 215–21.

Marco, J., Hedo, J. A., and Villaneuva, M. L. (1977). Inhibition of glucagon release by serotonin in mouse pancreatic islets. *Diabetologia* **13,** 585–8.

Mate, L., Sakamoto, T., Greeley, G. H., and Thompson, J. C. (1985). Regulation of gastric acid secretion by secretin and serotonin. *American Journal of Surgery* **149,** 40–5.

Matsumoto, P., Brown, A., and Dunbar, J. C. (1984). Biogenic amine regulation of glucagon secretion. *Acta Diabetologica Latina* **21,** 333–40.

Matsuoka, H., Ishii, M., Goto, A., and Sugimoto, T. (1985). Role of serotonin type 2 receptors in regulation of aldosterone production. *American Journal of Physiology* **249,** E234–8.

Mendelsohn, F. A. O. and Kachel, C. D. (1981). Stimulation by serum of aldosterone production by rat zona glomerulosa cells in vitro: relationships to K^+, serotonin and angiotensin II. *Acta Endocrinologica* **97,** 231–42.

Mendelsohn, F. A. and Mackie, C. (1975). Relation of intracellular K^+ and steroidogenesis in isolated adrenal zona glomerulosa and fasciculata cells. *Clinical Science and Molecular Medicine* **49,** 13–26.

Menguy, R. (1969). Gastric mucus and the gastric mucosal barrier. A review. *American Journal of Surgery* **117,** 806–12.

Meuli, C. and Muller, J. (1982). Characterization of rat capsular adrenal (zona glomerulosa) (Na, K)—ATPase activity. *Journal of Steroid Biochemistry* **16,** 129–32.

Morgan, N. and Montague, W. (1985). Studies on the mechanism of inhibition of glucose-stimulated insulin secretion by noradrenaline in rat islets of Langerhans. *Biochemical Journal* **226,** 571–6.

Mori, J., Satoh, H., Satoh, Y. A., Honda, F. (1979). Amines and the rat exocrine pancreas: (3) Effects of amines on pancreatic secretions. *Japanese Journal of Pharmacology* **29,** 923–33.

Muller, J. and Ziegler, W. H. (1968). Stimulation of aldosterone biosynthesis in vitro by serotonin. *Acta Endocrinologica* **59,** 23–35.

Owman, C., Hakanson, R., and Sundler, F. (1973). Occurrence and function of amines in endocrine cells producing polypeptide hormones. *Federation Proceedings* **32,** 1785–91.

Pearse, A. G. E. (1976). Evolutionary and developmental relationships among the cells producing peptide hormones. In *Peptide hormones* (ed. J. A. Parsons). Macmillan Press, London.

Pincus, I. J., Thomas, J. E., and Rehfuss, A. (1942). A study of gastric secretion as influenced by changes in duodenal acidity. *Proceedings of the Society for Experimental Biology and Medicine* **51,** 367–8.

Pober, Z. (1976). Effect of pylorotomy on inhibition of gastric secretion by intestinal acidification. *Digestive Diseases* **21,** 448–52.

Pontiroli, A. E., Viberti, G. C., Tognetti, A., and Puzza, G. (1975). Effect of metergoline, a powerful and long acting antiserotoninergic agent, on insulin secretion in normal subjects and in patients with chemical diabetes. *Diabetologia* **11,** 165–7.

Pontiroli, A. E., Micossi, P., and Foa, P. P. (1978). Effects of serotonin, of its biosynthetic precursors and of the anti-serotonin agent metergoline on the release of glucagon and insulin from the rat pancreas. *Hormone and Metabolic Research* **10,** 177–262.

Pulido, O. M., Benscombe, S. A., de Bold, M. L., and de Bold, A. J. (1978). Intracellular pancreatic B cell serotonin and the dynamics of insulin release. *Diabetologia* **15,** 197–204.

Quickel, K. E., Feldman, J. M., and Lebovitz, H. E. (1971a). Inhibition of insulin secretion by serotonin and dopamine: species variation. *Endocrinology* **89,** 1295–302.

Quickel, K. E., Feldman, J. M., and Lebovitz, H. E. (1971b). Enhancement of insulin secretion in adult onset diabetes by methysergide maleate: evidence for an endogenous biogenic monoamine mechanism as a factor in the impaired insulin secretion of diabetes mellitus. *Journal of Clinical Endocrinology* **33,** 877–81.

Racz, K., Wolf, I., Kiss, R., Lada, G., Vida, S., and Glaz, E. (1979). Corticosteroidogenesis by isolated human adrenal cells: effect of serotonin and serotonin antagonists. *Experientia* **35,** 1532–4.

Resnick, R. H. and Gray, S. J. (1962a). Serotonin release by hydrochloric acid. *Journal of Laboratory and Clinical Medicine* **59,** 462–8.

Resnick, R. H. and Gray, S. J. (1962b). Chemical and histological demonstration of hydrochloric acid induced release of serotonin from intestinal mucosa. *Gastroenterology* **42,** 48–55.

Ritzen, M., Hammarstrom, L., and Ullberg, S. (1965). Autoradiographic distribution of 5-hydroxytryptamine and 5-hydroxytryptophan in the mouse. *Biochemical Pharmacology* **14,** 313–21.

Robertson, R. P. and Guest, R. J. (1978). Reversal by methysergide of inhibition of insulin secretion by prostaglandin E in the dog. *Journal of Clinical Investigation* **62,** 1014–9.

Rocco, S., Boscaro, M., de Agostino, D., Armanini, D, Opocher, G., and Mantero, F. (1986). Effect of ketanserin on the in vitro regulation of aldosterone. *Journal of Hypertension* **4** (Supplement), S51–4.

Sanyal, A. K. and Waton, N. G. (1976). The possible mechanism for the 'fade' in acid gastric secretion during continuous infusions of pentagastrin. *Journal of Physiology (London)* **261,** 445–51.

Schuck, M., Santillana, M., Wise, L., and Ballinger, W. F. (1968). Studies on the inhibition of gastric secretion by the small intestine. *Surgery, Gynaecology and Obstetrics* **127,** 1295–9.

Sircus, W. (1958). Studies on the mechanisms in the duodenum inhibiting gastric secretion. *Quarterly Journal of Experimental Physiology* **43,** 114–33.

Smith, A. N. (1958). The effect of 5-hydroxytryptamine on acid and gastric secretion. In *5-Hydroxytryptamine* (ed. G. P. Lewis) pp. 183–92. Pergamon, London.

Stahl, S. M. and Meltzer, H. Y. (1978). A kinetic and pharmacologic analysis of 5-hydroxytryptamine transport by human platelets and platelet storage granules; comparison with central serotonergic neurones. *Journal of Pharmacology and Experimental Therapeutics* **205,** 118–32.

Stern, L., Tenenhouse, A., and Yu, E. W. (1983). Uptake storage and secretion of 5-hydroxytryptamine and its amino acid precursor by dispersed rat pancreas acinar cells. *Journal of Physiology (London)* **340,** 555–67.

Taborsky, G. T. (1983). Evidence for a paracrine role of pancreatic somatostatin in vivo. *American Journal of Physiology* **245,** E598–E603.

Telib, M., Raptis, S., Schroder, K. E., and Pfeiffer, E. F. (1968). Serotonin and insulin release in vitro. *Diabetologia* **4,** 253–6.

Thompson, J. H. (1964). The possible role of 5-hydroxytryptamine in the control of gastric acid secretion. A new concept. *Irish Journal of Medicinal Science* **461,** 191–201.

Thompson, J. H. (1971). Serotonin and the alimentary tract. *Research Communications in Chemical Pathology and Pharmacology* **2,** 687–781.

Tonnesen, P. and Schaffalitzky de Muckadell, O. B. (1987). Substance P and vasoactive intestinal peptide in serotonin-induced nasal secretions in normal subjects. *Allergy* **42,** 146–50.

Trendelenburg, U. (1972). Classification of sympathomimetic amines. In *Catecholamines* (eds. H. Blaschko and E. Muscholl) pp. 356. Springer Verlag, Berlin, Heidelberg and New York.

Tuomisto, J. and Mannisto, P. (1985). Neurotransmitter regulation of anterior pituitary hormones. *Pharmacological Reviews* **37,** 249–332.

Warner, R. R. P., Janowitz, H. D., and Dreiling, D. A. (1959). Increased mucus production in man by 5-HT and 5-HTP. *Clinical Research* **7,** 32.

Wazna, M. F., Stein, T., and Wise, L. (1977). The effect of serotonin on pepsin inhibition by duodenal fat. *Annals of Surgery* **186,** 130–5.

Wazna, M. F., Stein, T., and Wise, L. (1978). Studies on the role of serotonin in gastric pepsin reaction. *Surgery, Gynaecology and Obstetrics* **141,** 929–32.

Wermel, E. M. and Kacharova, E. A. (1948). The role of the basal granular cells of the mucosa of the small intestine in the production of secretin. *Anatomical Record,* **101,** 595–604.

White, T. T. and Magee, D. F. (1958). The influence of serotonin on gastric mucus production. *Gastroenterology* **35,** 289–91.

Whitley, G. S., Bell, J. B., Chu, F. W., Tait, J. F., and Tait, S. A. (1984). The effects of ACTH, serotonin, K^+ and angiotensin analogues on ^{32}P incorporation into phospholipids of the rat adrenal cortex: basis for an assay method using zona glomerulosa cells. *Proceedings of the Royal Society of London (Biology)* **222,** 273–94.

Williams, B. C., McDougall, J. G., Tait, J. F., and Tait, S. A. (1981). Calcium efflux and steroid output from superfused rat adrenal cells: effect of potassium, adrenocorticotrophic hormone, 5-hydroxytryptamine, adenosine 3′5′-cyclic monophosphate and angiotensins II and III. *Clinical Science* **61,** 541–51.

Williams, B. C., Shaikh, S., and Edwards, C. R. (1984). The specificity of ketanserin in the inhibition of serotonin-induced steroidogenesis in the rat adrenal zona glomerulosa. *Journal of Hypertension* (Supplement) **2,** S559–61.

Wilson, G. A. and Furman, B. L. (1982). Effects of inhibitors of 5-hydroxytryptamine uptake on plasma glucose and their interaction with 5-hydroxytryptophan in producing hypoglycaemia in mice. *European Journal of Pharmacology* **78,** 263–70.

Wilson, J. P., Downs, R. W. jun., Feldman, J. M., and Lebovitz, H. E. (1974). Beta cell monoamines: further evidence for their role in modulating insulin secretion. *American Journal of Physiology* **227,** 305–12.

Winship, D. H. and Schuck, T. J. (1971). Serotonin inhibition of gastric induced gastric acid secretion. *Clinical Research* **19,** 406.

Wise, L. (1972). Peptic ulcer: A re-appraisal of its aetiology. *Annals of the Royal College of Surgeons of England* **50,** 145–63.

Wise, L., Burkholder, J., Zagaloff, A., Schuck, M., and Ballinger, W. F. (1968). Studies on the role of serotonin in the inhibition of gastric acid secretion by the duodenum. *Annals of Surgery* **168,** 824–30.

Wong, K. L. and Tyce, G. M. (1978). Effect of the administration of L-5-hydroxytryptophan and a monoamine oxidase inhibitor on glucose metabolism in the rat brain. *Journal of Neurochemistry* **31,** 613–20.

Zern, R. T., Bird, J. L., and Feldman, J. M. (1980). Effect in increased pancreatic islet norepinephrine, dopamine and serotonin concentration on insulin secretion in the golden hamster. *Diabetologia* **18,** 341–6.

Zimmerman, T. W. and Binder, H. J. (1984). Serotonin-induced alteration in colonic electrolyte transport in the rat. *Gastroenterology* **86,** 310–7.

13

5-Hydroxytryptamine and cardiopulmonary and carotid body reflex mechanisms

D. S. McQUEEN and A. K. MIR

13.1 Introduction

5-HT can evoke various reflexes by activating sensory receptors in the heart, lungs, and carotid body. In this chapter we shall review the reflexes initiated by exogenous 5-HT, examine the mechanism(s) whereby sensory receptors are affected, and consider whether these actions of 5-HT are merely pharmacological curiosities or whether they have physiological and/or pathophysiological significance.

13.2 Reflexes in general: the role of 5-HT

During the study of cardiopulmonary and carotid body reflexes various physiological and pharmacological stimuli have been utilized, including agents such as 5-HT which are used to evoke reflexes via particular types of sensory nerve (e.g. unmyelinated as opposed to myelinated nerve fibres). Much data have accumulated concerning reflexes initiated by actions of 5-HT on sensory nerves, but variable results from individual species and gross interspecies differences make it difficult to assess the significance of the information obtained, particularly when attempting to extrapolate it to either the normal or the diseased state in man. An additional complication is that stimulation of certain thoracic sensory receptors in man gives rise to sensations (e.g. dyspnoea–breathlessness, or the feeling of inability to get enough air into the lungs) that cannot be reported by animals.

Recently, rather than simply using 5-HT as a probe for activating particular sensors, attention has been given to characterizing the 5-HT receptors associated with these sensory nerves (e.g. thoracic vagal afferents) and to considering what implication the presence of 5-HT and 5-HT receptors in tissues close to the sensory receptors has for normal physiology and for pathophysiological states such as bronchial asthma, chronic obstructive pulmonary disease, and sudden infant death syndrome (cot death). The cause of these disorders remains to be established, so any endogeneous substance whose role is unknown is very likely to be investigated with the aim of establishing its function in the tissues. Results from the experiments may indicate whether or not the substance is involved in a particular disorder, and might also reveal its function.

Similarly, the mechanism in the arterial chemoreceptors of the carotid body responsible for transducing a physiological stimulus (hypoxia) into increased chemosensory neural discharge, and hence reflex hyperventilation and cardiovascular changes, is unknown and has led to studies on substances, including 5-HT, which are found within the carotid body. The logical basis is that, because they are present, they must have a function, and that function may be concerned with chemoreception. The outcome of such studies provides information relating to the actions of 5-HT on specific pharmacological receptors associated with particular chemosensory nerves, as well as evidence concerning the question of whether or not 5-HT is directly or indirectly involved in chemoreception.

It will be appreciated that, because of species variability, lack of substantive evidence, and the existence of conflicting evidence, it is not possible to classify definitively either the sensory receptors, their natural stimuli, or the reflexes evoked. Consequently, the summary of reflexes associated with cardiopulmonary and carotid body sensory receptors presented in Table 13.1 is undoubtedly rudimentary and imperfect, but it does serve to show which of the reflexes can be activated by 5-HT. Although 5-HT is particularly implicated in reflexes associated with unmyelinated (C) sensory fibres, it does not seem to activate all unmyelinated afferents, and may be able to affect certain sensors associated with myelinated (A) afferents (e.g. carotid body chemoreceptors). However, much of the evidence is equivocal, and several types of sensory receptor (e.g. rapidly adapting pulmonary receptors, carotid and aortic baroreceptors) have been omitted from detailed consideration because it is likely that they are indirectly affected by 5-HT secondary to actions of the amine on smooth muscle and other tissues surrounding the sensory element. Only those receptors for which a direct action of 5-HT on sensory fibres appears to have been established will be considered in the following sections, but more detailed information and discussion of cardiopulmonary receptors can be found in reviews by Donald and Shepherd (1978), Thóren (1979), Barnes (1986), and Widdicombe (1986).

13.3 Cardiac receptors

In common with many other sensory receptors (e.g. nociceptors in skin), histological studies have not so far provided evidence concerning the detailed structure of the *epicardial receptors* associated with C fibre afferents which are mainly located on and in the left ventricle. However, it is well established that 5-HT and phenylbiguanide can affect these vagally innervated

Table 13.1 Commonly encountered reflexes associated with activation of particular sensory nerves in different structures indicating whether or not the reflex can be evoked by 5-HT

Type of receptor	Location	Sensory fibres	Stimuli	Major reflexes evoked	Activated by 5-HT
CARDIAC					
Atrial A + B	Atria (L+R) Junction of atrial walls	Vagal myelinated	*A TYPE* Tension of atrial muscle (active in systole). Veratridine	Tachycardia Vasodilatation	NO
			B TYPE Atrial filling (volume and rate of change of volume—active in diastole) Veratridine		NO
Ventricular pressure	Ventricle (L+R) Subendocardium	Vagal myelinated Some unmyelinated	Increased ventricular pressure (discharge during systole) Veratridine (irregular discharge, L ventricle) (Major component of Bezold-Jarisch effect)	Bradycardia, Hypotension Apnoea	NO ?YES (unmyelinated)
Epicardial	Ventricle (mainly L) Surface/walls	Vagal unmyelinated	Increased atrial or ventricular pressure Veratridine, Phenylbiguanide Acetylstrophanthidin Unaffected by hypoxia	Bradycardia, Hypotension Nausea Vomiting	YES
Sympathetic afferents	Atria and ventricles	Sympathetic myelinated and unmyelinated	Coronary occlusion Increased ventricular filling Increased systemic arterial pressure	Tachycardia Hypertension Pain (of myocardial ischaemia)	NOT KNOWN

Table 13.1 Continued.

Type of receptor	Location	Sensory fibres	Stimuli	Major reflexes evoked	Activated by 5-HT
PULMONARY					
Slowly adapting (stretch)	Smooth muscle/ mucosa of conducting airways	Vagal myelinated	Change in tension across airway wall (as in inflation/ deflation of lungs)	Inhibition of inspiration prolongation of expirations (Hering-Breuer lung inflation reflex) Bronchodilatation Tachycardia Relief of dysponeic sensations	NO
Rapidly adapting (irritant)	Larynx trachea (cough)	Vagal myelinated	Mechanical stimuli Dust, Histamine Cigarette smoke	Cough Bronchoconstriction	NOT DIRECTLY
	Intrapulmonary	Vagal myelinated	Mechanical, Irritants (ammonia, sulphur dioxide, cigarette smoke) Substantial lung inflation Histamine, $PGF_{2\alpha}$, Pulmonary congestion (as in chronic obstructive pulmonary disease), Anaphylaxis, Inflammation (as with influenza)	Hyperpnoea Bronchoconstriction Dyspnoea	NOT DIRECTLY

Unmyelinated	Airway epithelium	Vagal unmyelinated	Pulmonary congestion, Capsaicin, Bradykinin Histamine, $PGF_{2\alpha}$, PGE_2, PGI_2 Local oedema.	Cough Transient bronchoconstriction	YES
Type-J (Juxta-pulmonary capillary)	Collagen tissue of alveolar walls	Vagal unmyelinated	Lung deflation Increased pulmonary capillary pressure—giving pumonary oedema (as in severe exercise, especially at high altitude); secondary to sudden left heart failure, anaphylaxis, pulmonary embolism; volatile anaesthetics, Capsaicin, Histamine (not directly), Nicotine, Lobeline, Phenylbiguanide	Rapid shallow breathing Initial apnoea if intense stimulus. Pain. Bronchoconstriction Laryngeal constriction Dyspnoea Bradycardia, Hypotension. Inhibition of somatic muscles.	YES
CAROTID BODY					
Arterial chemoreceptors	Carotid body cells (Type I, Type II, nerves)	Glossopharyngeal (IX cranial) afferents—Carotid sinus nerve branch Myelinated and unmyelinated	Hypoxia, Hypercapnia Acidaemia, Reduced Blood flow (as following a substantial fall in B.P.) Nicotine, Acetylcholine, sodium cyanide, phenylbiguanide, lobeline.	Hyperventilation Bradycardia Hypertension Reflex vasoconstriction Release of adrenal medullary catecholamines	YES (some)

receptors to evoke reflex bradycardia, hypotension, and apnoea, although the latter response seems not to occur in dogs or man (Paintal 1973). It is worth noting that phenylbiguanide, whose chemical structure is shown in comparison with that of 5-HT (Fig. 13.1), has been extensively used as a probe for activating 5-HT-sensitive receptors—more so than 5-HT itself. This may have arisen because 5-HT, when injected, tends to cause smooth muscle contraction, changes in systemic arterial pressure, and tachyphylaxis—which complicates analysis —whereas phenylbiguanide selectively affects the 5-HT-sensitive cardiopulmonary receptors (e.g. Dawes and Mott 1950; Paintal, 1955; Fastier *et al.* 1959) with fewer 'side effects'.

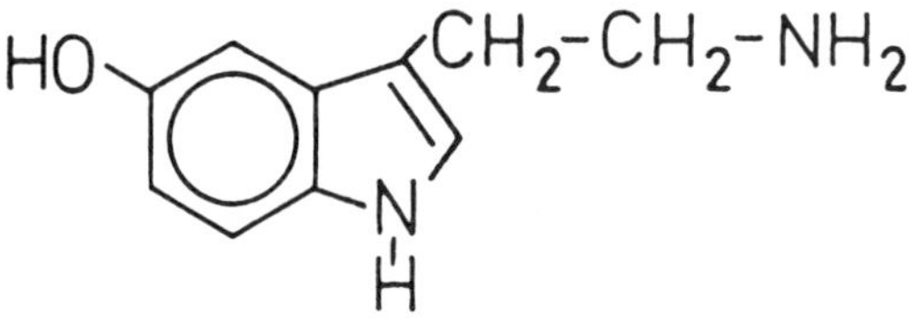

5–HT

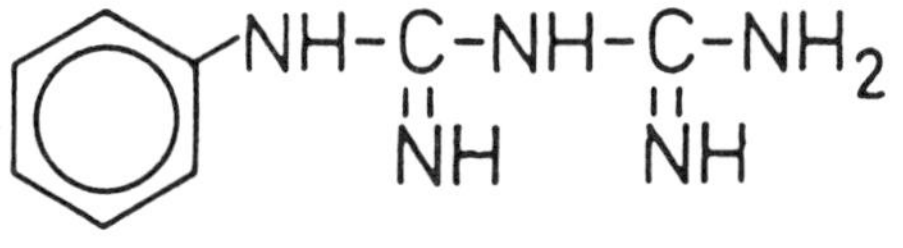

phenylbiguanide

Fig. 13.1 Structure of 5-HT in comparison with that of phenylbiguanide which is also known as phenyldiguanide

In a study of 5-HT and phenylbiguanide-evoked depolarization in rat isolated vagus nerves, Fortune *et al.* (1983) showed that both agonists were equipotent at depolarizing the nerve, and both were antagonized by metoclopramide, suggesting that the same 5-HT receptor subtype is involved in the response; it is not possible in this type of *in vitro* study to establish that sensory modality is normally associated with the afferent fibres being depolarized. Collins and Fortune (1983) demonstrated in anaesthetized rats that phenylbiguanide and 5-HT both caused abrupt bradycardia and hypotension when injected intravenously, which is a 'Bezold–Jarisch-like' effect. It is worth recalling at this point that the Bezold–Jarisch effect or reflex is, by definition, evoked by veratrum alkaloids (which include veratridine) and involves more types of sensory receptor than just those affected by 5-HT. Again, metoclopramide antagonized the vagal reflex, and the suggestion was made that the effects of 5-HT and phenylbiguanide are mediated through the same 5-HT receptor subtype.

A Bezold–Jarisch-like effect of 5-HT in anaesthetized rats was also studied by Fozard and co-workers (see Fozard 1984; and Chapter 15), and it was found that the selective 5-HT_3 receptor antagonist, MDL 72222, caused a marked reduction in the response. Richardson *et al.* (1985) found that the 5-HT_3 receptor antagonist, ICS 205-930, in doses as low as 1 μg/kg antagonized the Bezold–Jarisch effect in anaesthetized rats. They also showed that ICS 205-930 antagonized 5-HT-induced reduction of the C fibre component of the compound action potential in the desheathed rat vagus.

All the evidence outlined above suggests that the 5-HT_3 receptor (Bradley *et al.* 1986) is associated with vagally-innervated epicardial receptors. That being the case, one can envisage studies being performed with selective 5-HT_3 receptor agonists and antagonists (Richardson and Engel 1986; Fozard 1987) in order to characterize these receptors further. It would be of considerable interest to establish the role of epicardial ventricular receptors in the reflex control of the cardiovascular system under normal and pathophysiological (e.g. acute/chronic heart failure) conditions. Available evidence suggests that ventricular C fibre afferents are normally tonically active, discharging 1–2 impulses per second, and may inhibit vasomotor tone (see Thóren, 1979).

Injection of 5-HT into the left atrium or close to the proximal branches of the coronary artery in anaesthetized dogs results in a hypertensive coronary chemoreflex (James *et al.* 1975). The reflex is characterized by intense vasoconstriction, transient bradycardia, hyperpnoea, and changes in cardiac contractility and impulse conduction. In dogs the afferent limb of this reflex is reported to be carried by the vagi and the efferent limb by both the vagi and sympathetic nerves (James *et al.* 1975; Urthaler *et al.* 1978). Recently Richardson *et al.* (1987) have shown that ICS 205–930 blocks potently and reversibly the hypertensive coronary chemoreflex evoked by left intra-atrial injection of 5-HT in anaesthetized dogs, suggesting the involvement of 5-HT_3 receptors. In conscious dogs in direct contrast to the situation in anaesthetized dogs, the 5-HT-induced coronary chemoreflex appears to be markedly altered, such that no consistent hypertension is observed; rather, hypotension associated with vagal bradycardia and respiratory stimulation were the predominant features (Zucker and Cornish 1980).

The nature of the 5-HT-induced hypertensive chemoreflex in the conscious monkey shows certain similarities to that observed in the dog (Cornish and Zucker 1983). However, unlike in the dog, injections into the ascending aorta or the carotid arteries produced essentially similar responses; moreover, sino-aortic chemoreceptor denervation abolished the reflex despite the vagal pathways being intact. Therefore, the presence of a specific 5-HT receptor in the primate coronary circulation mediating the hypertensive chemoreflex is by no means certain; rather the reflex appears to be primarily associated with respiratory control. In this context it is interesting that the cardiovascular responses to nicotine, a chemoreceptor stimulant, when injected at the same sites were qualitatively similar to those obtained with 5-HT (Cornish and Zucker 1983). In view of the foregoing, the proposed benefit of this reflex to humans with respect to the clearance of thrombi by reflex hypertension evoked by 5-HT released from the thrombi locally in the coronary circulation (James *et al.* 1976) remains to be established.

13.4 Pulmonary receptors

5-HT, in common with various other chemicals (e.g. capsaicin, bradykinin, prostaglandin $F_{2\alpha}$, prostaglandin E_2), can activate unmyelinated vagal afferents in airway epithelium (Coleridge and Coleridge 1977). However, much less is known about these sensory receptors than is known about unmyelinated vagal afferents associated with type J receptors that have been described by Paintal (1973 and 1977) and are located in collagenous alveolar tissue, as illustrated in Fig. 13.2. Electron microscopy has provided evidence for unmyelinated nerve endings and 'free nerve endings' in the interstitium close to pulmonary capillaries where electrophysiological evidence predicted these sensory receptors would be found, so it is probable that these nerve terminals and fibres are part of the type J receptor sensory system.

J receptors are stimulated by pulmonary oedema and by various chemicals including volatile anaesthetics (e.g. halothane), nicotine, capsaicin, phenylbiguanide (Fig. 13.2), and 5-HT (see Fig. 10 in Paintal 1955). Although there may be species differences in responsiveness to some of the chemicals, and 'pulmonary' C fibre (i.e. unmyelinated) vagal afferents may differ somewhat from 'bronchial' C fibres, evidence suggests that J receptors may function as a type of protective nociceptor giving rise to dyspnoeic sensations as well as causing reflex bronchoconstriction, bradycardia, rapid shallow breathing, and inhibition of voluntary muscles in response to pulmonary oedema. More intense stimulation causes transient apnoea, followed by rapid shallow breathing, and the respiratory effects are more prominent than the cardiovascular effects (cf. cardiac receptors).

There is evidence linking activation of J receptors with the distressing dyspnoeic sensation associated with intense exercise, especially at high altitude, with dyspnoea of chronic obstructive pulmonary disease, and possibly with the sensation of suffocation experienced during acute left ventricular failure (Paintal 1977). Victims of the Bhopal gas tragedy died of pulmonary oedema, and those who survived complained of severe muscle weakness in addition to cough and breathlessness; all symptoms can be attributed to activation of J receptors by the oedema (Paintal 1986).

Experiments in anaesthetized rabbits demonstrated that 5-HT and phenylbiguanide injected into the right atrium caused rapid shallow breathing, bradycardia, and hypotension, the typical reflex pattern evoked by stimulating J receptors (Armstrong and Kay 1985). The response was antagonized by MDL 72222, which suggests involvement of 5-HT_3 receptors, as in the case of epicardial sensory receptors with unmyelinated vagal afferents. The observa-

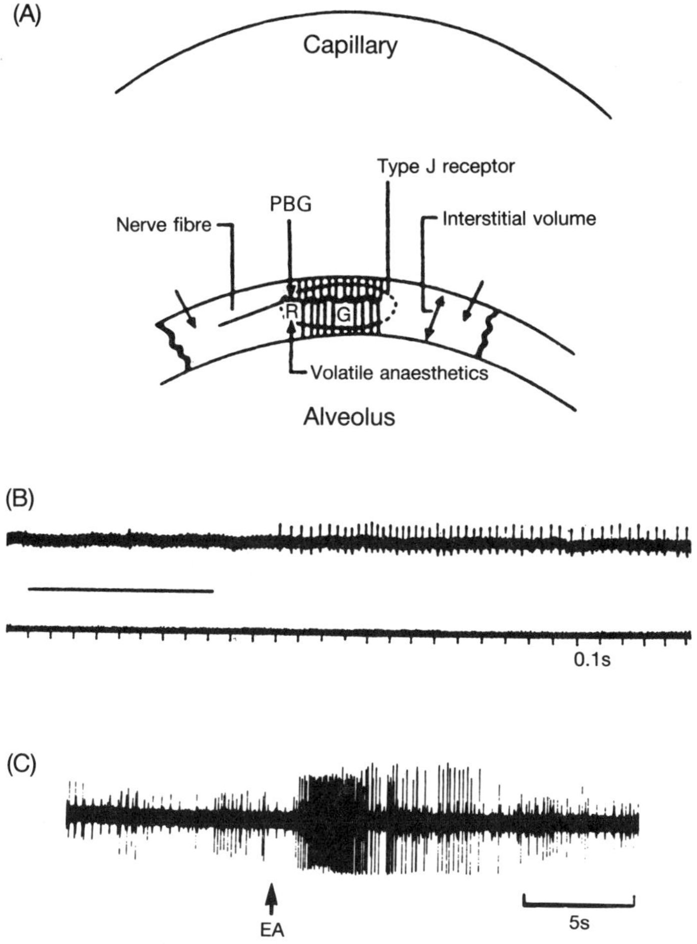

Fig. 13.2 A. Schematic representation of a type J receptor lying in interstitial tissue, between capillary and alveolus, containing collagen fibrils (vertical lines). The nerve-ending is stimulated by an increase in interstitial volume (or pressure) produced when inflow of fluid into interstitial tissue (which acts like a sponge) exceeds removal of fluid. Volatile anaesthetics and other substances, e.g. phenylbiguanide (PBG), act on the regenerative region (R) of the receptor, while an increase in interstitial volume acts on the generator region (G) of the ending (modified from Paintal 1977). B. Neurogram showing typical excitatory response of a type J receptor (cat) to intra-arterial injection of 1.3 μmoles phenylbiguanide at signal (modified from Paintal 1969). C. Pulmonary J receptor fibre activity in a thin filament of the caudal end of the left vagus (rat). Note sparse activity without cardiac or respiratory pattern before the arrow. Injection at arrow of an enkephalin analog [Dmet^2, Dpro^5]enkephalinamide (EA, 500 μg/kg right atrium) produced a burst of activity, as did phenyldiguanide (40 μg/kg) and capsaicin (0.2 μg/kg) (modified from Sapru *et al.* 1981)

tion that veratridine gave a similar response, but which was unaffected by MDL 72222, could mean either that veratridine activates J receptors, or alternatively epicardial receptors were being stimulated. In subsequent experiments on rabbits, Armstrong *et al.* (1986a) concluded that the MDL 72222-sensitive reflex tachypnoea evoked by phenylbiguanide emanated from pulmonary vagal afferents. In these latter experiments phenylbiguanide was about eight times more potent than 1,4-diguanadinobenzene, which evidently activates the same receptors as phenylbiguanide because it was also antagonized by MDL 72222. The authors refer to their 1,4 compound as phenyldiguanide which is potentially confusing because many workers (Dawes and Mott 1950; Black *et al.* 1972; Paintal 1977), used phenyldiguanide as an alternative name for what is now commonly called phenylbiguanide. Since phenylbiguanide and phenyldiguanide are alternative names used for the structure shown in Fig. 13.1, it is probably best to avoid using phenyldiguanide when referring to 1,4 diguanadinobenzene.

Armstrong *et al.* (1986b) have reported that pre-treatment of anaesthetized rabbits with MDL 72222 reduces the vagally-mediated reflex increase in respiratory frequency evoked by pulmonary embolism, which supports the hypothesis that stimulation of intrathoracic vagal afferent nerve fibres by 5-HT intitiates post-embolic tachypnoea. Further experiments are needed to determine the extent to which direct or indirect actions of 5-HT on pulmonary J receptors contribute to the reflexes involved in the rabbit experiments. Selective 5-HT_3 receptor antagonists should be very helpful in establishing the role of endogenous 5-HT at J receptors, and could also be relevant for the therapy of certain pathophysiological states such as pulmonary embolism.

An interesting report by Sapru *et al.* (1981), based on experiments in decerebrate rats, shows that an enkephalin analogue [Dmet^2,Dpro^5] enkephalinamide injected into the right atrium at a dose of 10 μg/kg can evoke the typical J receptor reflex of tachypnoea following transient apnoea, bradycardia, and hypotension. The authors confirmed by electrophysiological evidence (see Fig. 13.2) that the discharge pattern in vagal afferents stimulated by the opioid was consistent with that of J receptors and discharge increased rapidly on inhalation of halothane (a test for J receptors). The opioid-induced reflex was abolished by the opioid receptor antagonist, naloxone, without affecting responses to capsaicin or phenylbiguanide. It is interesting to speculate whether the sensory receptor/nerve complex has different pharmacological receptors (e.g. in the regenerator or generator regions—see Fig. 13.2), or whether different substances may act indirectly to release the same endogenous material, which in turn affects the sensors. In this context it would be interesting to establish whether or not a 5-HT_3 receptor antagonist, such as MDL 72222, would prevent the action of opioids. If it did, it would suggest that opioids act to release 5-HT, perhaps in a way analogous to their histamine-releasing action. Sapru *et al.* (1981) found that methysergide did not affect the reflex response to opioids in their experiments, but this drug is only active at '5-HT_1-like' and 5-HT_2 receptors, being ineffective at the 5-HT_3 sites (see Bradley *et al.* 1986).

Neuroendocrine cells which lie close to unmyelinated nerve fibres are also found in the lung; they are present in the foetus and tend to disappear as the animal matures, suggesting that they are particularly concerned with functioning during lung development. These cells may detect hypoxia (Pack and Widdicombe 1984) and there are fewer of them in infants with respiratory distress syndrome. Since neuroendocrine cells contain 5-HT (as well as bombesin, calcitonin, and [Leu]enkephalin—see Cutz 1982), it is possible that 5-HT, perhaps in conjunction with one or more of the other substances, acts as a neurotransmitter to activate sensory afferents. Further studies are required to establish whether or not 5-HT does act as a transmitter, and to explore the possibility that neuroendocrine cells may have a function in controlling respiration via reflex mechanisms. It will also be interesting to compare the mechanism in these cells with that operating in the carotid body, and to determine how they interact functionally.

13.5 Carotid body chemoreceptors

The arterial chemoreceptors of the carotid body have been more extensively investigated than the cardiopulmonary receptors described above, probably because they are more accessible and associated with various powerful reflexes that are readily elicited and easy to study (e.g. see Heymans and Neil 1958; Anichkov and Belen'kii 1963; Daly 1985). Arterial chemoreceptors are also found in the aortic bodies, but these are more difficult to study for various reasons, including their diffuse anatomy and the fact that they do not appear to be functional in many species. Where the actions of 5-HT on aortic chemoreceptors have been investigated neuropharmacologically, they are found to be qualitatively similar to those obtained from carotid body chemoreceptors (e.g. Paintal 1967). It is worth noting that the aortic bodies receive a blood supply from the coronary circulation, which means that coronary chemoreflexes (i.e. reflex cardiovascular and respiratory changes evoked by intra-coronary injection of particular chemicals—see above) have to be interpreted cautiously since they may be partly or entirely the consequence of activating aortic chemoreceptors rather than cardiac sensory mechanoreceptors.

The sensory nerves from the carotid body run in the carotid (sinus) branch of the glossopharyngeal (IX cranial) nerve (Fig. 13.3) and are a mixture of baroreceptors, which discharge in phase with the rise in arterial pressure following systole, and chemoreceptors, which do not discharge in a regular manner but do, unlike baroreceptors, increase their discharge in response to hypoxia or hypercapnia. There is some evidence that baroreceptors can be activated by 5-HT (e.g. Nishi 1975), but Paintal (1955) found that 5-HT does not stimulate or sensitize aortic baroreceptors. Since the effect on baroreceptors is not well studied, and could well be secondary to smooth muscle contraction, it will not be considered in this review.

The carotid nerve comprises myelinated and unmyelinated fibres, with two thirds of the former and one third of the latter being chemoreceptors in cat (Fidone and Sato 1969). However, there are major technical problems involved in obtaining reliable measurements of conduction velocity in the relatively short length of carotid nerve and most workers do not try measuring it. Consequently, it is not generally possible to state whether the actions of 5-HT on carotid chemoreceptors are associated with fast (myelinated) or slow (unmyelinated) conducting afferents, in contrast to the situation with aortic chemoreceptors or cardiopulmonary afferents where longer lengths of nerve enable reasonably accurate measurements of conduction velocity to be made, or carotid baroreceptors where the discharge pattern in response to physiological stimuli is quite characteristic for the different types of fibre (Fidone and Sato 1969). The evidence available suggests that 5HT can act at myelinated as well as unmyelinated chemoreceptor afferents. The carotid body also receives efferent fibres via the carotid nerve and the sympathetic (ganglioglomerular) nerves from the superior cervical ganglion.

13.5.1 Presence of 5-HT in the carotid body

Relatively large amounts of 5-HT have been found in the carotid body of various species, including man, using techniques such as fluorimetry (Chiocchio *et al.* 1967), high pressure liquid chromatography (McQueen and Mir 1984; Leitner *et al.* 1986) and immunohistochemistry (Gronblad *et al.* 1983; Perrin *et al.* 1986). Fig. 13.3 illustrates the gross morphology of the carotid body and also demonstrates that 5-HT is located in some, but not all, glomus or type I cells, being present in dense-cored vesicles. Steele and Hinterberger (1972) measured 5-HT levels in human carotid bodies and found that they were raised in patients who had died from hypertension or chronic obstructive pulmonary disease, as opposed to 'control' carotid bodies from people dying of other causes (e.g. accidents). Recent work by Perrin *et al.* (1986) indicates that 5-HT is the main amine (cf. dopamine, noradrenaline) in the carotid bodies of infants.

13.5.2 5-HT-Induced reflexes from carotid body chemoreceptors

Cardiovascular and respiratory reflexes are evoked by the action of injected 5-HT on chemoreceptors in the carotid body. Responses are more variable than those involving cardiopulmonary sensory receptors, and it has been suggested that species differences, dose levels, or tachyphylaxis may be responsible for the variability. A common respiratory response to 5-HT is hyperventilation, which is sometimes preceded by apnoea, particularly in cats. Phenylbiguanide produces similar effects (Black *et al.* 1972), and most authors conclude that there is evidence for a biphasic action of

5-HT. Data are, rather unusually, also available from experiments in man. Skinner and Whelan (1962) showed that a rapid injection of 5-HT into the carotid artery caused transient apnoea followed by hyperventilation, whereas infusions of 5-HT had no effect. Surgical removal of the carotid body abolished the hyperventilation without affecting the initial apnoea, suggesting that 5-HT was stimulating arterial chemoreceptors in the carotid body and thereby reflexly increasing ventilation. Inhibition of ventilation probably resulted from stimulation of vagal afferent fibres, or perhaps cell bodies in the nodose ganglion, possibly associated with pulmonary receptors (Black *et al.* 1972).

The existence of biphasic reflex effects, coupled with interactions between reflexes (e.g. hyperventilation activates a lung inflation reflex which counteracts the primary response) and complications of tachyphylaxis to 5-HT, made it difficult to establish the primary action of 5-HT on the chemoreceptors and led to the use of neurophysiological techniques for studying more directly the effect of 5-HT on the chemoreceptors.

13.5.3 Neuropharmacological studies of the effect of 5-HT on carotid body chemoreceptors

Neural recording from the carotid (sinus) nerve established that injection of 5-HT into the carotid artery caused an increase in chemosensory discharge in about 20 per cent of recordings in the cat and dog (Black *et al.* 1972; and see Figs. 13.3 and 13.4). In others there was no change in frequency, or a decrease was obtained. Phenylbiguanide was qualitatively similar to 5-HT in causing a brief burst of activity (see Fig. 13.5). Aortic chemoreceptors are also stimulated by 5-HT (Paintal 1967). More detailed investigation by Bisgard *et al.* (1979) demonstrated that in dogs low doses of 5-HT injected into the carotid artery depressed chemosensory discharge, whereas higher doses caused an initial increase followed by depression of discharge. Intracarotid injection of 5-HT to the contralateral intact carotid body evoked no change in ventilation following low doses that caused chemodepression, whereas higher doses that caused chemoexcitation resulted in an increase in ventilation. It was found that chemoexcitation was antagonized by D-tubocurarine in a dose lower than that required for neuromuscular block, without affecting 5-HT-induced chemodepression. The latter response could be blocked by dihydroergotamine, without affecting the excitatory component. Like 5-HT, dopamine- and noradrenaline-induced chemoexcitation was also antagonized by D-tubocurarine and the chemodepression was blocked by dihydroergotamine, whereas haloperidol only antagonized dopamine-induced chemodepression. These results point to the existence of two separate responses (excitation, inhibition) of the chemoreceptors to 5-HT, but it is not clear what particular pharmacological receptors are involved. D-Tubocurarine can

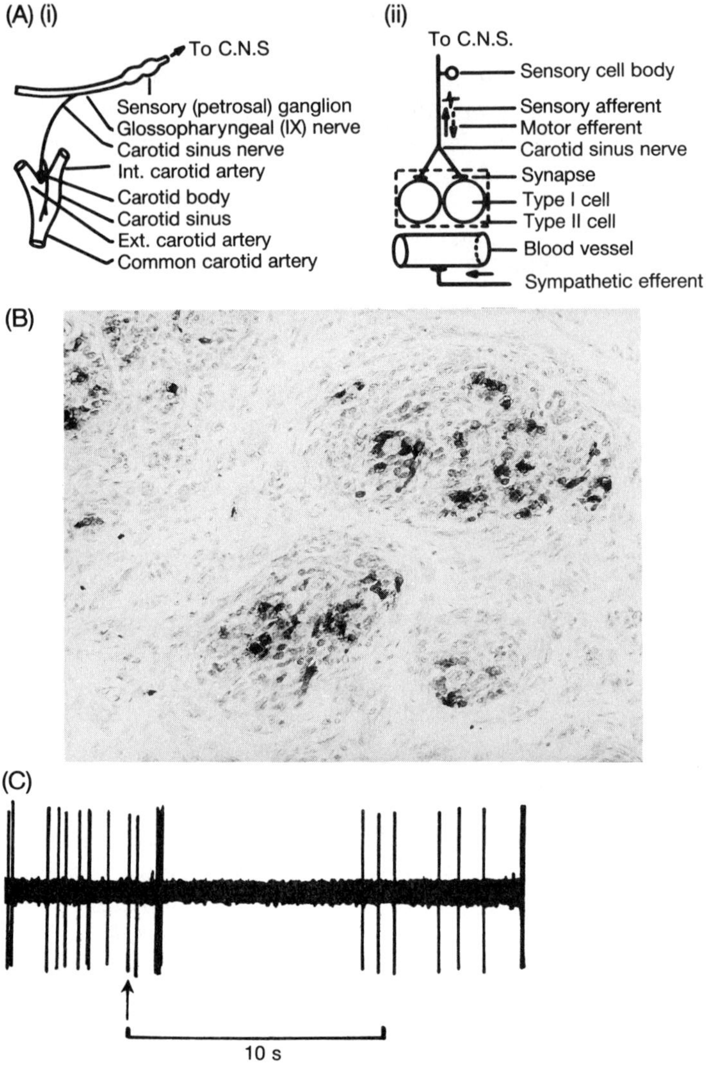

Fig. 13.3 A. Carotid body: (*i*) gross anatomy, (*ii*) simplified schematic representation of the receptor complex. Individual nerve fibres typically course to 10–20 type 1 (glomus) cells. Not all type 1 cells receive a nerve supply, and the surrounding type II (sustentacular) cells do not appear to be innervated. The carotid sinus nerve contains both myelinated and unmyelinated sensory (afferent) fibres as well as motor (parasympathetic efferent) fibres. The presence of vesicles and putative neurotransmitters in the type 1 cells suggests that chemical transmission may occur at the cleft or 'synapse' between the cell and the nerve-ending. Parasympathetic efferents could release transmitter(s) to affect the cells or nearby sensory terminals, providing efferent control or modulation of the chemoreceptors in addition to that achieved by sympathetic nerves; the latter can influence

block certain molluscan dopamine receptors (Ascher 1972), but the fact that it affects the action of all three amines implies either that they all act directly/indirectly at the same receptor, or that they act at different receptors which share a common pathway that can be blocked by D-tubocurarine (e.g. by blocking a particular ion channel). The results with dihydroergotamine are also difficult to interpret because it is not very selective. However, Bisgard *et al.* (1979) also used phenoxybenzamine (10 mg/kg) and found it had no effect on 5-HT-induced chemodepression, which indicates that α-adrenoceptor antagonism is unlikely to explain the actions of dihydroergotamine.

Nishi (1975) noted a biphasic response from cat chemoreceptors to injected 5-HT, with transient excitation preceding depression of discharge. Chemoexcitation was not affected by large doses of atropine, hexamethonium, LSD, or methysergide, which suggests that cholinergic receptors are not involved in the response, and that the 5-HT receptor involved is not affected by LSD or by methysergide (which affect '5-HT_1-like', and 5-HT_2 receptors).

Sapru and Krieger (1977) showed that 5-HT induced an increase in chemosensory discharge in rats and noted that this response was unaffected by treatment with methysergide, but could be reduced by the ganglion blocking agent, mecamylamine. Docherty and McQueen (1978) obtained transient chemoexcitation in response to 5-HT followed by depression of discharge (see Fig. 13.3) and noted that high doses of the dopamine antagonist, α-flupenthixol, could prevent the excitation and the delayed depression of chemosensory discharge. A similar pattern was also observed in rabbits (Docherty and McQueen 1979) where responses to 5-HT were unaffected by mecamylamine or atropine but were substantially reduced by high doses of α-flupenthixol. Thus, neither nicotine nor muscarine receptors seem to be involved in chemoreceptor responses to 5-HT, and although *high* doses of α-flupenthixol were able to block the effect, this observation cannot be taken as evidence for dopamine D_2 receptor involvement because domperidone (a selective dopamine D_2 receptor antagonist) antagonizes the chemodepressant action of dopamine without affecting

---

discharge by altering blood flow through the organ. The question of whether the cells or the afferent nerve-endings are the primary sensory element in the receptor complex has yet to be answered (from McQueen 1985). B. Immunoperoxidase staining for 5-HT in the carotid body of a three-month-old infant who died from cold exposure. Note that both single and clusters of chief (type 1, glomus) cells within the glomic area stain positively. The surrounding stroma is negative ×38 (from Perrin *et al.* 1986). C. Response of a single cat carotid chemoreceptor afferent fibre to an intracarotid injection of 25nmoles 5-HT (arrow) showing transient excitation followed by longer-lasting depression of discharge; ganglioglomerular (sympathetic) nerves cut (modified from Docherty and McQueen 1978) the response to 5-HT (Kirby and McQueen 1985) suggesting that 5-HT-induced chemodepression is not due to release of dopamine within the carotid body. A logical though unproven explanation would be that α-flupenthixol in high doses may be a 5-HT receptor antagonist.

The finding *in vivo* that excitatory and/or inhibitory actions can be obtained from the carotid body chemosensors that do respond to 5-HT leads one to speculate as to whether some or all these effects may be secondary to changes in carotid body blood flow (which can alter chemoreceptor activity) resulting from vascular actions of 5-HT. For the rapid excitation (see Fig. 13.4) and the early chemodepression this seems very improbable because it takes a finite time for changes in vascular tone to cause changes in blood flow, and a further delay for such changes to influence the chemosensors and thereby affect discharge. The effects occur too quickly for this to be a likely mechanism; however, vascular changes cannot be excluded from involvement in the more delayed aspects of the response to 5-HT.

Removing the carotid body from the animal and studying its chemoreceptor discharge *in vitro* should resolve this problem because *in vitro* there is no blood supply in the superfused preparation to be affected by 5-HT. However, very little information is available concerning the effects of 5-HT *in vitro*. Eyzaguirre and Koyano (1965) reported that injections of 5-HT into the fluid superfusing the isolated cat carotid body caused a slight decrease in discharge at low doses, but high doses (200 μg) increased chemosensory discharge.

The available evidence thus suggests that 5-HT can increase or decrease chemosensory discharge under conditions where vascular effects are not able to influence chemoreceptor activity. It is probable that a high local concentration of 5-HT is needed at the chemoreceptors in order to activate the sensory nerves, and this is difficult to achieve by superfusion *in vitro*—hence the need to use high doses. Injection into a superfusing fluid *in vitro* is probably more akin to an infusion *in vivo* rather than an injection and, as observed by Skinner and Whelan (1962), infusions do not activate the chemoreceptors.

A further potential complication to interpreting the action of 5-HT on carotid chemoreceptors lies in the fact that 5-HT can stimulate sympathetic ganglia (see Chapter 10) and might thereby affect chemosensory discharge secondary to changes in local blood flow. However, several workers have demonstrated that responses to 5-HT *in vivo* are unaffected by cutting the ganglioglomerular (sympathetic) nerves (e.g. Nishi 1975; Docherty and McQueen 1978; Fig. 13.3C).

13.5.4 Classification of carotid body 5-HT receptors

None of the evidence described above enables conclusions to be reached regarding the type(s) of receptor involved in chemoexcitation or chemodepression evoked by injected 5-HT. The advent of MDL 72222 (see Chapter 15) provided a selective tool for investigating neuronal actions of 5-HT and it was used by Kirby and McQueen (1984a) to help characterize the 5-HT receptors in the cat carotid body. MDL 72222 virtually abolished the transient chemoexcitation obtained during injection of 5-HT in 56 per cent of recordings (see Fig. 13.4) and greatly reduced the chemodepression which was obtained in all recordings, regardless of whether or not transient chemoexcitation occurred before (i.e. depression is not simply a secondary consequence of excitation). These effects were dose-related; the dose of 5-HT injected into the carotid artery was 2.6–129 nmoles with the half-maximum response obtained following 15 nmoles. A rather variable delayed chemoexcitation was observed in many experiments which was concurrent with the hypotension associated with injection of 5-HT. This delayed increase in discharge was either unaffected by MDL 72222, or was increased in magnitude. Ketanserin, a 5-HT_2 receptor antagonist, substantially reduced both the fall in blood pressure and the delayed chemoexcitation without affecting the early excitation or subsequent depression of discharge. The combination of MDL 72222 and ketanserin abolished most of the effects of injected 5-HT, but had no appreciable effect on the responsiveness of the chemoreceptors to hypoxia. MDL 72222 itself caused transient chemodepression followed by transient (30s) slight excitation.

In view of the possibility that 5-HT_3 receptors were responsible for both the initial chemoexcitation and for the chemodepression evoked by 5-HT, while 5-HT_2 receptors may have contributed to the delayed excitatory response, further experiments were performed by Kirby and McQueen (1985) in which 'selective' 5-HT receptor agonists were used. 2-Methyl

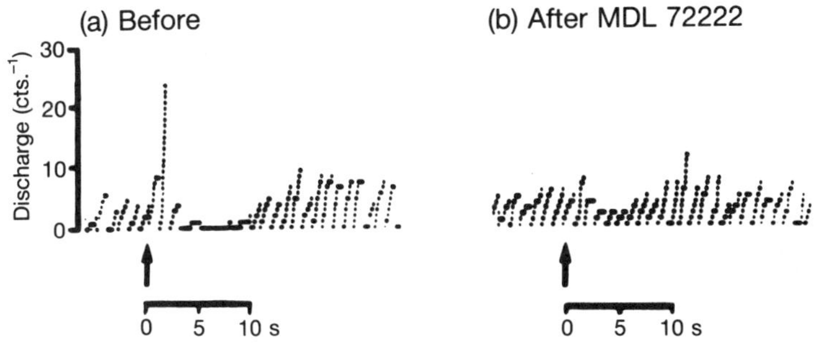

Fig. 13.4 Response of cat chemoreceptors (multi-unit recording) to intracarotid injection of 5-HT (13nmoles) at arrows before and after the 5-HT_3 receptor antagonist MDL 72222 (100 μg/kg i.c.). The initial transient chemoexcitation and the subsequent depression of chemosensory discharge were both virtually abolished by the antagonist. Delayed chemoexcitation was not greatly affected by MDL 72222, but was substantially reduced by the 5-HT_2 receptor antagonist ketanserin. The ramps show the number of action potentials counted cumulatively in 1s intervals (from Kirby and McQueen 1984a)

5-HT, a selective 5-HT_3 receptor agonist (Richardson and Engel 1986), caused chemodepression equal to or greater than that seen with 5-HT, but was less active than 5-HT in causing transient excitation. However, considering that both the initial chemoexcitation and the chemodepression induced by 5-HT are sensitive to blockade with MDL 72222 and that 2-methyl 5-HT does not discriminate between the putative 5-HT_3 receptor subtypes (see Richardson and Engel 1986; and Chapter 10) it is not readily explicable why 2-methyl 5-HT is less active in evoking the initial chemoexcitation. In contrast, 5-methoxytryptamine, which does not stimulate 5-HT_3 receptors (Fozard 1986) caused only slight changes in chemosensory discharge although its effects on blood pressure were similar to those of 5-HT. Taken together, these results point to the existence of receptors in the carotid body which are sensitive to MDL 72222 and are activated by 2-methyl 5-HT. There is probably sufficient evidence to suggest that 5-HT_3 receptors are involved in chemoexcitation and chemodepression evoked by exogenous 5-HT, whereas delayed chemoexcitation may involve 5-HT_2 receptors in the carotid body. Moreover, in view of the 5-HT_3 receptor subtype selectivity of MDL 72222 (see Fozard 1986; Richardson and Engel 1986) it seems unlikely that the 5-HT_3 receptor subtype of the guinea-pig ileum is involved in the responses to 5-HT. Further studies are needed to confirm these observations and, in addition, to localize the receptors to particular elements of the chemosensory complex. The fact that physiological stimuli, such as hypoxia, appear to be unaffected by doses of MDL 72222 which adequately antagonize exogenous 5-HT suggests that endogenous 5-HT does not have a crucial function in chemoreception, although it may be a modulator. By analogy with actions of 5-HT at autonomic ganglia (e.g. Wallis and Woodward 1975; see Chapter 10) it is possible that 5-HT is acting to increase sodium conductance in some, but not all, chemosensory fibres. Matsumoto (1981) provided indirect evidence favouring this when he found that tetrodotoxin abolished 5-HT-induced respiratory changes when injected close-arterially to the carotid body in anaesthetized rabbits, but had no significant effect on the response to cyanide.

Some preliminary experiments have shown that 5-HT-induced chemodepression is potentiated during an infusion of substance P (7 nmoles/min) into the ipsilateral common carotid artery, as was the delayed chemoexcitation (Kirby and McQueen 1984b). It will be interesting to examine in detail how 5-HT, substance P and other peptides interact—given that substance P-like material, opioids, and vasoactive intestinal peptide are present in cells and nerves in the cat carotid body (Cuello and McQueen 1980; Wharton *et al.* 1980). Moreover, capsaicin treatment, which results in a destruction of substance P and other polypeptide containing unmyelinated sensory fibres, causes a reduction in respiratory responses to peripheral chemoreceptor stimulation following neonatal treatment in rats (Bond *et al.* 1982) and is accompanied by marked increases in catechol-

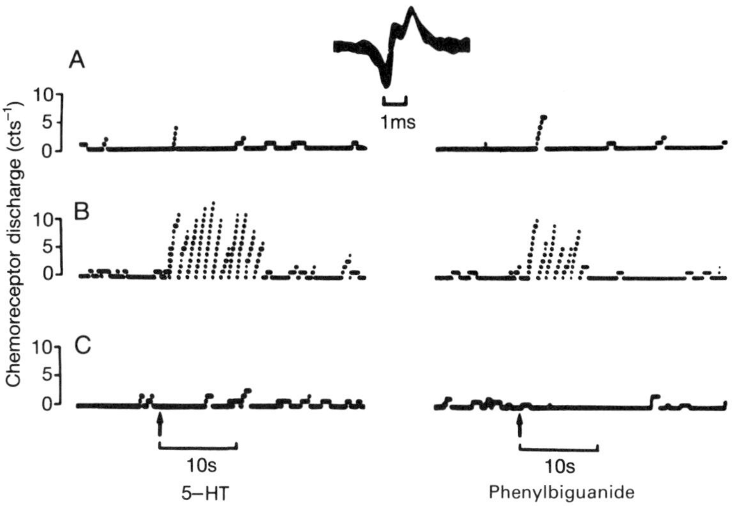

Fig. 13.5 Responses of a single carotid chemoreceptor fibre (cat, 25 successive superimposed sweeps in neurogram) to intracarotid injections of 5-HT (65 nmoles) or phenylbiguanide (283 nmoles) at arrows: A. control; B. after the chemoreceptor stimulant almitrine (0.5 mg/kg i.v.); C. after the 5-HT_3 antagonist MDL 72222 (100 μg/kg i.c.). Both agonists caused a transient chemoexcitation which was potentiated by almitrine (mechanism unknown) and antagonized by MDL 72222. The latter evidence suggests that both 5-HT and phenylbiguanide cause chemoexcitation via actions at 5-HT_3 receptors. The ramps show the number of action potentials counted cumulatively in 1s intervals (McQueen and Mir 1988, unpublished observations)

amine and 5-HT concentrations in the carotid body (McQueen and Mir 1984). In this respect it is interesting that surgical transection of the carotid sinus nerve, which contains the sensory fibres, also results in an increase in carotid body 5-HT concentration in the rabbit (Leitner *et al.* 1986).

13.5.5 Actions of 5-HT and phenylbiguanide on carotid body chemoreceptors

It is known that phenylbiguanide can stimulate carotid body chemoreceptors in a manner qualitatively similar to that of 5-HT (Black *et al.* 1972). In view of the evidence that phenylbiguanide activates cardiopulmonary afferents via an action that can be blocked by MDL 72222, it was considered of interest to investigate whether the action of phenylbiguanide on carotid body chemoreceptors would, in common with that of 5-HT, be antagonized by the 5-HT_3 receptor antagonist. Accordingly, during preparation of this review, experiments were performed on cats and the results from four animals were consistent in showing that MDL 72222 does indeed antagonize phenylbiguanide-induced chemoexcitation and chemodepression, as well as blocking the effects of 5-HT (McQueen and Mir 1988; see Fig. 13.5). This can be interpreted as evidence that both 5-HT and phenylbiguanide act at 5-HT_3 receptors; the fact that there was no delayed chemoexcitation with phenylbiguanide, nor any marked changes in blood pressure, suggests that the latter effects must be associated with some other action of 5-HT that is not shared by phenylbiguanide (i.e. at 5-HT_2 receptors). Note that responses to phenylbiguanide and 5-HT were potentiated by the peripheral chemoreceptor stimulant drug, almitrine, *but* excitation evoked by almitrine was largely unaffected by MDL 72222, making it seem unlikely that almitrine was acting by release of endogenous 5-HT or by directly activating 5-HT_3 receptors in the carotid body. Regarding the selective action of phenylbiguanide on 5-HT_3 receptors, Ireland and Tyers (1987) have recently reported that phenylbiguanide potently mimics the 5-HT-induced depolarization responses on the rat isolated cervical vagus nerve and that these effects were effectively antagonized by metoclopramide and, albeit non-competitively, by low concentrations of MDL 72222 and ICS 205-930.

13.5.6 Physiological significance of 5-HT and 5-HT receptors in the carotid body

What is the function of the 5-HT which is present in carotid body cells? The finding that exogenous 5-HT affects discharge when injected close to the carotid body (causing chemodepression in all fibres, but transient excitation in only some) may be evidence for a transmitter or modulator role, but the fact that these actions can be antagonized by the 5-HT_3 antagonist, MDL 72222, without any obvious effect on the response to a physiological stimulus (hypoxia) raises certain questions about the involvement of endogenous 5-HT and 5-HT_3 receptors in normal functioning of the chemoreceptors. It may be that the time course (seconds, minutes) studied is not appropriate, or the antagonist or dose of antagonist are inadequate for antagonizing high local concentrations of endogenous 5-HT in the carotid body; however, if endogenous material is producing the same effect at the same receptors as affected by injected 5-HT, then it is difficult to sustain this argument. An alternative would be that endogenous 5-HT is not concerned with chemoreception but has another role in the carotid body (e.g. in vascular control, including permeability, or in efferent neural control) or is not *normally* involved in chemoreception but becomes involved only under particular circumstances (e.g. pathophysiological states).

Large amounts of 5-HT are present in the carotid body glomus cells of normal infants (Perrin *et al.* 1986; see Fig. 13.3B), which may indicate involvement of the amine in respiratory regulation in young animals, with perhaps less contribution in the mature normal animal: in general most experiments are performed on healthy adult animals. High levels of dopamine are found in the carotid bodies of infants dying from sudden infant death syndrome (Perrin *et al.* 1984)—is the level of 5-HT also raised? Do increased levels indicate decreased or increased release of amine? Levels of 5-HT in the carotid body are raised in patients dying with chronic obstructive pulmonary disease or hypertension, but not in bronchial asthma (Steele and Hinterberger 1972). The carotid body has been implicated in essential hypertension (Przybylski 1981) particularly at the very early stages; others (Honig *et al.* 1981) have questioned whether changes within the carotid body vasculature and alterations in chemoreceptor activity may be a consequence, rather than a cause, of essential hypertension. Thus, the answer to the question posed above is that we do not know what role 5-HT plays in the carotid body. However, evidence is accumulating and the availability of selective 5-HT receptor agonists and antagonists (see Fozard, 1987) and the partial characterization of 5-HT receptors that has already taken place suggests that rapid progress can be envisaged. The recent availability of suitable 5-HT_3 receptor radioligands (see Chapter 4) will help to localize the receptors to particular types of nerve fibre or cell, and patch-clamp techniques on cultured cells or dissociated cells, or standard electrophysiological techniques in carotid body slice preparations *in vitro*, coupled with local application of selective 5-HT receptor agonists and antagonists, should advance knowledge considerably. Furthermore, it will be challenging to try and mimic pathophysiological disorders in animal models—e.g. does neonatal capsaicin treatment of rats, which raises 5-HT levels in the adult carotid body and reduces chemosensitivity, represent a useful model for disorders such as chronic obstructive pulmonary disease or sudden infant death syndrome (see McQueen 1985)?

13.6 Concluding remarks

The observations that exogenous 5-HT can affect various sensory receptors and evoke quite potent reflexes may initially have been considered as a pharmacological curiosity. However, the finding that 5-HT is present in the vicinity of sensory nerves (often in considerable quantities) and that levels of 5-HT can be altered in several pathophysiological states does strongly suggest that 5-HT and 5-HT receptors do have a physiological or pathophysiological role.

A feature common to the major sensory nerves (cardiac, pulmonary, and carotid chemoreceptor) reviewed in this chapter is the ability of MDL 72222 to antagonize the action of 5-HT on neural discharge, and it may be that these different sensory receptors share a common type of 5-HT receptor which may not be present at all the sensory complexes. Does 5-HT activate the sensors directly, or is it acting to affect the release/activity of another endogenous substance? Some of the actions are very rapid (e.g. chemoexcitation—Fig. 13.4), but release of endogenous material may occur within tens of milliseconds which is quite rapid compared with the normal time-course of responses—generally occurring over several seconds. Is it relevant to study the actions of 5-HT alone? Might it not be co-stored with one or more polypeptides (see Johannson *et al.* 1981), and also co-released, such that investigating 5-HT receptor agonists or antagonists alone will provide only limited information regarding the true function of the amine? Could 5-HT be a co-transmitter, or involved in axon reflexes, e.g. via peptidergic neurones, leading to abnormal control of smooth muscle in pulmonary airways (Barnes 1986), or be acting to alter local blood flow or vascular permeability?

It is interesting that phenylbiguanide appears, from the available evidence, to be a selective 5-HT_3 receptor agonist. It has, historically, been used as a probe for evoking particular cardiovascular reflexes from cardiopulmonary vagal afferents (Fastier *et al.* 1959) and has been considered to affect unmyelinated fibres (e.g. Paintal 1977). However, since it seems largely free of actions at other 5-HT receptors, it may represent a specific probe for 5-HT_3 receptors associated with particular sensory receptors, and could prove a useful tool in further studies.

The role of the 5-HT-sensitive sensory receptors will need to be investigated in more detail, and this is problematic because of technical difficulties in reaching the receptors, identifying them, controlling the local drug concentrations, and avoiding vascular complications when working *in vivo*. Studies on isolated tissue may prove helpful in resolving some of these problems, but it will be necessary to ensure that local concentrations of 5-HT or other drugs and the time over which they are free to act reflect conditions *in vivo*. How the various reflexes described in this chapter are processed and integrated centrally is also a complex area that will require careful investigation—e.g. does activation of sensory nerves in the periphery affect central 5-HT neurotransmission in regions concerned with, for example, respiratory control (Mueller *et al.* 1982)?

Since sensory reflexes are undoubtedly important for homeostatic regulation in normal or pathophysiological states, the study of 5-HT receptors associated with sensory systems should not only advance our knowledge concerning the role of 5-HT in normal physiological processes, but could also be very important in helping to understand, and perhaps treat, various disorders (e.g. cardiopulmonary disturbance following pulmonary embolism, coronary thrombosis, or anaphylactic reactions) that may result from local or reflex release of 5-HT. Given the difficulty of studying cardiopulmonary receptors, further information may well be obtained by using the more accessible carotid body chemoreceptors as a general model for studying sensory 5-HT receptors in normal and pathophysiological states. We are only just beginning to make major advances in this field.

References

Anichkov, S. V. and Belen'kii, M. L. (1963). *Pharmacology of the carotid body chemoreceptors*. Pergamon Press, London.

Armstrong, D. J. and Kay, I. S. (1985). MDL 72222 (a 5-HT antagonist) antagonizes the pulmonary depressor and respiratory chemoreflexes evoked by phenylbiguanide in anaesthetized rabbits. *Journal of Physiology (London)* **365,** 104P.

Armstrong, D. J., Kay, I. S., and Russell, N. J. W. (1986a). The pulmonary chemoreflexes evoked by phenylbiguanide, diguanide and guanidine in anaesthetized rabbits. *Journal of Physiology (London)* **371,** 115P.

Armstrong, D. J., Kay, I. S., and Russell, N. J. W. (1986b). MDL 72222 antagonizes the reflex tachypnoeic response to miliary pulmonary embolism in anaesthetized rabbits. *Journal of Physiology (London)* **381,** 13P.

Ascher, P. (1972). Inhibitory and excitatory effects of dopamine on aplysia neurones. *Journal of Physiology (London)* **225,** 173–209.

Barnes, P. J. (1986). Neural control of human airways in health and disease. *American Review of Respiratory Disease* **134,** 1289–314.

Bisgard, G. E., Mitchell, R. A., and Herbert, D. A. (1979). Effects of dopamine, norepinephrine and 5-hydroxytryptamine on the carotid body of the dog. *Respiration Physiology* **37,** 61–80.

Bond, S. M., Cervero, F., and McQueen, D. S. (1982). Influence of neonatally administered capsaicin on baroreceptor and chemoreceptor reflexes in the adult rat. *British Journal of Pharmacology* **77,** 517–24.

Bradley, P. B., Engel, G., Feniuk, W., Fozard, J. R., Humphrey, P. P. A., Middlemiss, D. N., Mylecharane, E. J., Richardson, B. P., and Saxena, P. R. (1986). Proposals for the classification and nomenclature of functional receptors for 5-hydroxytryptamine. *Neuropharmacology* **25,** 563–76.

Black, A. M. S., Comroe, J. H., and Jacobs, L. (1972). Species difference in carotid body response of cat and dog to dopamine and serotonin. *American Journal of Physiology* **223,** 1097–102.

Chiocchio, S. R., Biscardi, A. M., and Tramezzani, J. H. (1967). 5-Hydroxytryptamine in the carotid body of the cat. *Science* **158,** 790–1.

Coleridge, H. M. and Coleridge, J. C. G. (1977). Impulse activity in afferent vagal C-fibres with endings in the intrapulmonary airways of dogs. *Respiration Physiology* **29,** 125–42.

Collins, D. P. and Fortune, D. H. (1983). Phenylbiguanide mimics the Bezold–Jarisch effect of 5-HT in the rat. *British Journal of Pharmacology* **80,** 570P.

Cornish, K. G. and Zucker, I. H. (1983). Is there a serotonin-induced hypertensive coronary chemoreflex in the nonhuman primate? *Circulation Research* **52,** 312–18.

Cuello, A. C. and McQueen, D. S. (1980). Substance P: a carotid body peptide. *Neuroscience Letters* **17,** 215–9.

Cutz, E. (1982). Neuroendocrine cells of the lung. An overview of morphologic characteristics and development. *Experimental Lung Research* **3,** 185–208.

Daly, M. de B. (1985). Chemoreceptor reflexes and cardiovascular control. *Acta Physiologica Polonica* **36,** 4–20.

Dawes, G. S. and Mott, J. C. (1950). Circulatory and respiratory reflexes caused by aromatic guanidines. *British Journal of Pharmacology* **5,** 65–76.

Docherty, R. J. and McQueen, D. S. (1978). Inhibitory action of dopamine on cat carotid chemoreceptors. *Journal of Physiology (London)* **279,** 425–36.

Docherty, R. J. and McQueen, D. S. (1979). The effects of acetylcholine and dopamine on carotid chemosensory activity in the rabbit. *Journal of Physiology (London)* **288,** 411–23.

Donald, D. E. and Shepherd, J. T. (1978). Reflexes from the heart and lungs: physiological curiosities or important regulatory mechanisms. *Cardiovascular Research* **12,** 446–69.

Eyzaguirre, C. and Koyano, H. (1965). Effects of some pharmacological agents on chemoreceptor discharges. *Journal of Physiology (London)* **178,** 410–37.

Fastier, F. N., McDowall, M. A., and Waal, H. (1959). Pharmacological properties of phenyldiguanide and other amidine derivates in relation to those of 5-hydroxytryptamine. *British Journal of Pharmacology* **14,** 527–35.

Fidone, S. J. and Sato, A. (1969). A study of chemoreceptor and baroreceptor A and C fibres in the cat carotid nerve. *Journal of Physiology (London)* **205,** 527–48.

Fortune, D. H., Ireland, S. J., and Tyers, M. B. (1983). Phenylbiguanide mimics the effects of 5-hydroxytryptamine on the rat isolated vagus nerve and superior cervical ganglion. *British Journal of Pharmacology* **79,** 298P.

Fozard, J. R. (1984). MDL 72222: a potent and highly selective antagonist at neuronal 5-hydroxytryptamine receptors. *Naunyn-Schmiedeberg's Archives of Pharmacology* **326,** 36–44.

Fozard, J. R. (1986). Peripheral neuronal 5-hydroxytryptamine receptors and their biological significance. In *Proceedings of the Third SCI-RSC Medicinal Chemistry Symposium* (ed. R. W. Lambert) pp. 37–56. Royal Society of Chemistry, London.

Fozard, J. R. (1987). 5-HT: The enigma variations. *Trends in Pharmacological Sciences* **8,** 501–5.

Green, F. J., Schmidt, N. D., Schultz, H. D., Roberts, A. M., Coleridge, H. M., and Coleridge, J. C. G. (1984). Pulmonary C fibres evoke both apnoea and tachypnea of pulmonary chemoreflex. *Journal of Applied Physiology* **57,** 562–7.

Gronblad, M., Liesi, P., and Rechardt, L. (1983). Serotonin-like immunoreactivity in rat carotid body. *Brain Research* **276,** 348–50.

Heymans, C. and Neil, E. (1958). *Reflexogenic areas of the cardiovascular system.* Churchill, London.

Honig, A., Habeck, J. O., Pfeiffer, C., Schmidt, M., Huckstorf, C., Rotter, H., and Eckermann, P. (1981). The carotid bodies of spontaneously hypertensive rats (SHR): a functional and morphologic study. *Acta Biologica et Medica Germanica* **40,** 1021–30.

Ireland, S. J. and Tyers, M. B. (1987). Pharmacological characterisation of 5-hydroxytryptamine-induced depolarization of the rat isolated vagus nerve. *British Journal of Pharmacology* **90,** 229–38.

James, T. N., Isobe, J. H., and Urthaler, F. (1975). Analysis of components in a cardiogenic hypertensive chemoreflex. *Circulation* **52,** 179–92.

James, T. N., Urthaler, F., and Hageman, G. R. (1976). Chemoreceptors of the heart. *Transactions of the American Clinical Climatology Association* **88,** 18–31.

Johansson, O., Hokfelt, T., Pernow, B., Jeffcoate, S. L., White, N., Steinbusch, H. W. M., Verhofstad, A. A. J., Emson, P. C., and Spindel, E. (1981). Immunohistochemical support for three putative transmitters in one neuron: coexistence of 5-hydroxytryptamine, substance P- and thyrotropin releasing hormone-like immunoreactivity in medullary neurones projecting to the spinal cord. *Neuroscience* **6,** 1857–81.

Kirby, G. C. and McQueen, D. S. (1984a). Effects of the antagonists MDL 72222 and ketanserin on responses of cat carotid body chemoreceptors to 5-hydroxytryptamine. *British Journal of Pharmacology* **83,** 259–69.

Kirby, G. C. and McQueen, D. S. (1984b). Effects of substance P on responses of cat carotid body chemoreceptors to dopamine, noradrenaline and 5-hydroxytryptamine. *British Journal of Pharmacology* **83,** 455P.

Kirby, G. C. and McQueen, D. S. (1985). Effects of selective 5-hydroxytryptamine agonists on carotid body chemoreceptor discharge in anaesthetized cats. *British Journal of Pharmacology* **86,** 733P.

Leitner, L. M., Roumy, M., Ruckebusch, M., and Sutra, J. F. (1986). Monoamines and their catabolites in the rabbit carotid body. Effects of reserpine, sympathectomy and carotid sinus nerve section. *Pflügers Archiv* **406,** 552–6.

Matsumoto, S. (1981). Effects of carotid body chemoreceptor stimulation by 5-HT on phrenic nerve activity and ventilation in the rabbit. *Archives Internationales de Pharmacodynamie et de Therapie* **254,** 282–92.

McQueen, D. S. (1985). Carotid body chemoreceptors as a pharmacological model. *Trends in Pharmacological Sciences* **6,** 167–70.

McQueen, D. S. and Mir, A. K. (1984). Changes in carotid body amine levels and effects of dopamine on respiration in rats treated neonatally with capsaicin. *British Journal of Pharmacology* **83,** 909–18.

McQueen, D. S. and Mir, A. K. (1988). Involvement of 5-HT_3 receptors in responses of cat carotid body chemoreceptors to phenylbiguanide. *Journal of Physiology (London)* **401,**86P.

Mueller, R. A., Lundberg, D. B. A., Breese, G. R., Hedner, J., Hedner, T., and Jonason, J. (1982). The neuropharmacology of respiratory control. *Pharmacological Reviews* **34,** 255–85.

Nishi, K. (1975). The action of 5-hydroxytryptamine on chemoreceptor discharges of the cat's carotid body. *British Journal of Pharmacology* **55,** 27–40.

Pack, R. J. and Widdicombe, J. C. (1984). Amine containing cells of the lung. *European Journal of Respiratory Disease* **65,** 559–78.

Paintal, A. S. (1955). Impulses in vagal afferent fibres from specific pulmonary deflation receptors. The response of these receptors to phenyldiguanide, potato starch, 5-hydroxytryptamine, and nicotine and their role in respiratory and cardiovascular reflexes. *Quarterly Journal of Experimental Physiology* **40,** 89–111.

Paintal, A. S. (1967). Mechanism of stimulation of aortic chemoreceptors by natural stimuli and chemical substances. *Journal of Physiology (London)* **189,** 63–84.

Paintal, A. S. (1969). Mechanism of stimulation of type J pulmonary receptors. *Journal of Physiology (London)* **203,** 511–32.

Paintal, A. S. (1973). Sensory mechanisms involved in the Bezold–Jarisch effect. *Australian Journal of Biological and Medical Science* **51,** 3–15.

Paintal, A. S. (1977). Thoracic receptors connected with sensation. *British Medical Bulletin* **33,** 169–74.

Paintal, A. S. (1986). The significance of dry cough, breathlessness and muscle weakness. *Indian Journal of Tuberculosis* **33,** 51–5.

Perrin, D. G., Cutz, E., Becker, L. E., Bryan, A. C., Madapallimatum, A., and Sole, M. J. (1984). Sudden infant death syndrome: increased carotid body dopamine and noradrenaline content. *Lancet* **ii,** 535–7.

Perrin, D. G., Chan, W., Cutz, E., Madapallimattam, A., and Sole, M. J. (1986). Serotonin in human infant carotid body. *Experientia* **42,** 562–4.

Przybylski, J. (1981). Do arterial chemoreceptors play a role in the pathogenesis of hypertension? *Medical Hypotheses* **7,** 127–31.

Richardson, B. P. and Engel, G. (1986). The pharmacology and function of 5-HT_3 receptors. *Trends in Neurosciences* **9,** 424–8.

Richardson, B. P., Engel, G., Donatsch, P., and Stadler, P. A. (1985). Identification of serotonin M-receptor subtypes and their specific blockade by a new class of drugs. *Nature (London)* **316,** 126–31.

Richardson, B. P., Engel, G., Donatsch, P., and Buchheit, K-H. (1987). 5-HT receptors on afferent neurones. In *Serotonin: Actions, Receptors, Pathophysiology. Proceedings of the 1987 IUPHAR Congress Satellite Meeting, Heron Island, Australia* (eds E. J. Mylecharane, J. A. Angus, I. S. de la Lande, and P. P. A. Humphrey). Macmillan, London. (In press).

Sapru, H. N. and Krieger, A. J. (1977). Effect of 5-hydroxytryptamine on the peripheral chemoreceptors in the rat. *Research Communications in Chemical Pathology and Pharmacology* **16,** 245–50.

Sapru, H. N., Willette, R. N., and Krieger, A. J. (1981). Stimulation of pulmonary J receptors by an enkephalin-analog. *Journal of Pharmacology and Experimental Therapeutics* **217,** 228–34.

Skinner, S. L. and Whelan, R. F. (1962). Carotid body stimulation by 5-hydroxytryptamine in man. *Journal of Physiology (London)* **162,** 35–43.

Steele, R. H. and Hinterberger, H. (1972). Catecholamines and 5-hydroxytryptamine in the carotid body in vascular, respiratory and other diseases. *Journal of Laboratory and Clinical Medicine* **80,** 63–70.

Thóren, P. (1979). Role of cardiac vagal C fibres in cardiovascular control. *Review of Physiology and Biochemical Pharmacology* **86,** 1–94.

Urthaler, F., Hageman, G. R., and James, T. N. (1978). Haemodynamic components of a cardiogenic hypertensive chemoreflex in dogs. *Circulation Research* **42,** 135–42.

Wallis, D. I. and Woodward, B. (1975). Membrane potential changes induced by 5-hydroxytryptamine in the rabbit superior cervical ganglion. *British Journal of Pharmacology* **55,** 199–212.

Wharton, J., Polak, J. M., Pearse, A. G. E., McGregor, G. P., Bryant, M. G., Bloom, S. R., Emson, P. C., Bisgard, G. E., and Will, J. A. (1980). Enkephalin-, VIP- and substance P-like immunoreactivity in the carotid body. *Nature (London)* **284,** 269–71.

Widdicombe, J. (1986). The neural reflexes in the airways. *European Journal of Respiratory Disease* (Supplement) **144,** 1–33.

Zucker, I. H. and Cornish, K. G. (1980). Reflex cardiovascular and respiratory effects of serotonin in the conscious and anaesthetized dogs. *Circulation Research* **47,** 509–15.

14

Vascular responsiveness to 5-hydroxytryptamine in experimental hypertension

MILA B. TURLA and R. CLINTON WEBB

14.1 Introduction

Vascular reactivity, defined as the magnitude of the response to a constrictor agent, is increased in clinical and experimental hypertension. The response may be measured as a pressor effect in the intact animal, as the magnitude of a resistance change in a perfused vascular bed, or as a contractile event in an isolated vascular smooth muscle segment. The increased vascular reactivity in hypertension may be due to: (1) a functional change in the vascular smooth muscle which results in a greater contractile response to a given concentration of a constrictor agonist; (2) a decreased ability of blood vessel wall components (e.g. endothelium, prejunctional modulation of transmitter release, etc.) to limit contractile responses to normal stimuli; and/or (3) a structural change in which the wall-to-lumen ratio is increased, resulting in a greater decrease in vessel radius for a given vascular smooth muscle shortening. The importance of the structural basis for the increase in vascular reactivity has been emphasized by Folkow (1978) and is generally accepted. Structural changes in the blood vessel wall of hypertensive subjects are thought to be an adaptive phenomenon in which the wall of the resistance artery thickens in response to the increase in arterial pressure.

This chapter reviews experimental observations which characterize the abnormal vascular reactivity to 5-HT in hypertension. As a preface, we will provide a brief statement about the sources of the monoamine in hypertension. Experimental evidence will be presented indicating that the mechanism of abnormal contractile responsiveness probably relates to cellular events that occur following activation of receptors specific for 5-HT. Aspects of the inhibitory and potentiating actions of the monoamine in hypertension will be considered and, finally, experimental observations relating to the effects of 5-HT on the endothelium and the adrenergic nerve endings in hypertension will be summarized. For an extensive overview of vascular changes in hypertension the reader is directed to more complete publications (Folkow 1982; Webb 1984). Other aspects of the possible pathophysiological role of 5-HT in hypertension have been reviewed recently (Vanhoutte 1982a and b, 1983, and 1985; see also Chapter 16).

14.2 Sources of 5-HT

5-HT can be found in significant quantities in several tissues of the body: (1) enterochromaffin cells of the gastro-intestinal tract; (2) blood platelets; and (3) neurones of the central nervous system and the gastro-intestinal wall. The monoamine is synthesized in certain neurones of the central nervous system and some of these neurones are located in brain regions that play a role in cardiovascular regulation (Kuhn *et al.* 1980). Peripherally, 5-HT is synthesized within and released from the enterochromaffin cells (Thompson 1971). Following release into the circulation, 5-HT is taken up by platelets through a saturable uptake process (Lingjaerde 1977) and this source of 5-HT can be liberated upon platelet aggregation. 5-HT uptake also occurs at adrenergic nerve endings and this may be an important source of the monoamine that could influence vascular function (Verbeuren *et al.* 1983). Recent evidence has documented that pial arterioles in rats and rabbits are innervated by 5-HT-containing neurones (Scatton *et al.* 1985).

Abnormalities in the uptake and/or metabolism of 5-HT have been reported in hypertension (Kamal *et al.* 1983; Koulu *et al.* 1986; Ozaki *et al.* 1986; Roth and Wallace 1980; Vanhoutte 1985). The results of different studies are highly variable and some work is consistent with the hypothesis that alterations in both central regulatory centres and peripheral storage sites may contribute to altered vascular tone in hypertension. It should be recognized, however, that some studies with selective antagonists of 5-HT receptors do not provide strong support for this hypothesis (see Section 14.3.3). For a more extensive consideration regarding the sources of 5-HT and its altered metabolism in disease states, the reader is directed to Chapters 1 and 16.

14.3 Vasoconstrictor responses to 5-HT

14.3.1 Activation of 5-HT receptors

Responses to 5-HT have been used to demonstrate the large differences in vascular smooth muscle reactivity between normotensive and hypertensive animals. In 1970, McGregor and Smirk described the blood pressure responses to 5-HT in genetic and renal hypertensive rats. They observed that the pressor response to 5-HT in hypertensive rats was about twice that in normotensive controls. In the perfused mesenteric vasculature of both types of hypertensive animals, the pressor responses to noradrenaline and angiotensin II were also increased. However, the magnitude of the enhanced pressor responses to noradrenaline and angiotensin II (approximately 50 per cent above control) was greatly exceeded by the enhanced response to 5-HT (approximately 400–700 per cent; McGregor and Smirk 1970). This exaggerated responsiveness to 5-HT in hypertension (compared with that for other vasoconstrictors) is often cited as evidence that structural adaptations in the blood vessels cannot solely account for increased vascular reactivity in this disease process (McGregor and Smirk 1970; Webb and Bohr 1984).

Greater enhancement of the response to 5-HT in hypertension than to other vasoconstrictor agents has been demonstrated by several laboratories (Berecek *et al.* 1980; Collis and Vanhoutte 1977; De Mey and Vanhoutte 1981). Additionally, the sensitivity of the vasculature (threshold for constrictor effect) to 5-HT is augmented in hypertensive animals compared with normotensive animals, although the degree of increase varies from one vascular bed to another and from one type of hypertension to another (Skolasinska *et al.* 1982; Webb 1982; Webb and Bohr 1984; Winquist and Bohr 1983). In the renal vasculature of spontaneously hypertensive rats, constrictor responses to 5-HT are potentiated to a greater degree than those to noradrenaline, angiotensin II, or barium (Collis and Vanhoutte 1977). Furthermore, the relative changes in sensitivity (threshold response) observed in the hypertensive rats were greatest for 5-HT (Fig. 14.1). In the perfused mesenteric vascular bed of renal, deoxycorticosterone acetate and spontaneously hypertensive rats, thresholds to 5-HT are lower, dose-response curves to the monoamine are steeper, and maximal constrictor responses to 5-HT are greater than in normotensive rats. In contrast, the threshold for constriction in hind quarter preparations of hypertensive rats does not differ from that in normotensive rats (Haeusler and Finch 1972). The maximal pressor responses to 5-HT are augmented in the hind quarters of renal and deoxycorticosterone acetate hypertensive rats but not in the hind quarters nor hind limbs of spontaneously hypertensive rats (Ahlund *et al.* 1977; Haeusler and Finch 1972). The aorta from spontaneously hypertensive rats is about ten times more sensitive to 5-HT, but the portal vein less than twice as sensitive as the corresponding blood vessels from normotensive rats (Ahlund *et al.* 1977). In the isolated perfused heart (Langendorff technique) of normotensive rats, 5-HT causes concentration-dependent increases in coronary flow (see below for vasodilator properties of 5-HT), whereas in hearts from spontaneously hypertensive rats, the monoamine causes concentration-dependent coronary vasoconstriction (Lüscher *et al.* 1986). Thus, there is a marked variability in sensitivity to 5-HT between different vascular preparations from hypertensive animals.

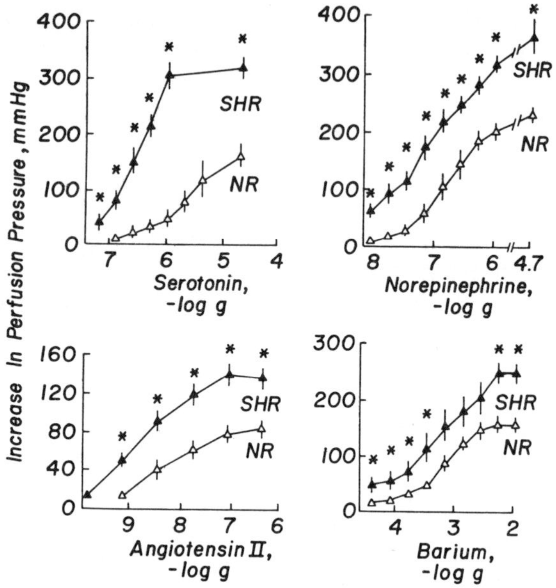

Fig. 14.1 Vascular reactivity to various constrictor agonists in the perfused renal vascular bed of spontaneously hypertensive rats (SHR) and normotensive controls (NR). The threshold response to serotonin (5-HT) in the SHR is greater than that of the response to noradrenaline (norepinephrine), angiotensin II, or barium. (From Collis and Vanhoutte 1977 by permission)

14.3.1.1 *Affinity and number of 5-HT receptors*

Enhanced vascular reactivity in hypertensive animals to different agonists may reflect either alterations in specific membrane receptor sites or changes in the sequence of events initiated by the different receptors. Insight into the mechanisms that play a role in the exaggerated potentiation of 5-HT responses in hypertension has been gained from information regarding the 5-HT receptors in vascular smooth muscle. Increased affinity or number of 5-HT receptors may contribute to the augmented vascular responses to 5-HT in hypertension. Relative affinities of 5-HT receptors in normotensive and hypertensive rat femoral artery were assessed using cumulative concentration-response curves to 5-HT in the absence and presence of two different 5-HT receptor antagonists, methysergide and ketanserin (Mecca and Webb 1984a). Schild analysis (pA_2 values) revealed that the competitive antagonism of 5-HT-induced contractions by methysergide or ketanserin is similar in arteries from normotensive and deoxycorticosterone acetate hypertensive rats. These results suggest that the increased sensitivity to 5-HT does not result from a change in receptor affinity in this model of mineralocorticoid hypertension.

It has been proposed that there are more receptors than necessary for a maximal functional response and that the maximal response may be obtained with only a small fraction of receptors occupied (Nickerson 1956; Stephenson 1956). Suzuki *et al.* (1986) studied this spare receptor concept for 5-HT receptors in the blood vessels from stroke-prone spontaneously hypertensive and Wistar-Kyoto normotensive rats. Dibenamine treatment caused a shift to the right in dose-response curves for 5-HT in the aorta and superior mesenteric artery from both hypertensive and normotensive rats. There was no difference between the two rat strains and the magnitude of the shift following treatment with dibenamine was less than that for dose-response curves to noradrenaline. Dose-response curves to 5-HT for the femoral artery, tail artery and portal vein from both normotensive and hypertensive rats were depressed but not shifted to the right. The authors concluded that the receptor reserve for 5-HT is very small compared with that for noradrenaline, and that there is no change in this pharmacological parameter in blood vessels from stroke-prone spontaneously hypertensive rats. Thus, the abnormal responses to 5-HT in hypertension do not appear to be influenced by the affinity or the number of 5-HT receptors which take part in the contractile response.

14.3.1.2 *Calcium mobilization*

Several cellular mechanisms involved in cell calcium regulation may contribute to the increased sensitivity to 5-HT in hypertension. In an attempt to elucidate the underlying mechanisms, Mecca and Webb (1984a) assessed: (1) transmembrane movement of calcium following receptor activation; and (2) mobilization of activator calcium from cellular stores by 5-HT. Following depletion of cellular calcium with EGTA, contractile responses to added calcium were measured. The dose–response relationships to calcium in the presence of a maximal dose of 5-HT are not significantly different in arteries from deoxycorticosterone acetate hypertensive rats compared with those from normotensive rats. In contrast, contractile responses to 5-HT in calcium-free solution following loading of cellular stores of calcium are 50 per cent greater in the arteries from the hypertensive rats (Fig. 14.2). Thus, the enhanced vascular sensitivity to 5-HT in deoxycorticosterone acetate hypertension is not related to an alteration in the transmembrane movement of calcium folowing receptor activation, but rather to an altered mobilization of calcium from cellular stores.

14.3.1.3 *5-HT and phosphoinositide metabolism*

In the rat aorta, 5-HT elicits smooth muscle contraction and increased phosphoinositide hydrolysis (Conn and Sanders-Bush 1986; Nakaki *et al.* 1985; Roth *et al.* 1984; see also Chapter 4). Agonist-induced hydrolysis of membrane phosphoinositides is thought to serve as a multifunctional transducing mechanism which results in the liberation of two second messengers, membrane-associated diacylglycerol and water-soluble inositol trisphosphate. Diacylglycerol activates protein kinase C, whereas inositol trisphosphate liberates calcium from intracellular stores (Berridge 1984). Activation of protein kinase C and mobilization of activator calcium have been shown to be involved in vascular smooth muscle contraction (Forder *et al.* 1985; Nakaki *et al.* 1985; Turla and Webb 1987a).

Recent awareness of these post-receptor events unveiled the possibility that increased responsiveness to 5-HT in hypertension may be due to an abnormal phosphoinositide metabolism. Koutouzov *et al.* (1983) compared the basal turnover of phosphoinositides of erythrocytic membranes from Wistar-Kyoto normotensive, spontaneously hypertensive and stroke-prone spontaneously hypertensive rats. They observed an impaired basal phosphoinositide metabolism in the hypertensive animals. The impairment was also present in young rats before they became hypertensive, suggesting that the defect in phosphoinositide metabolism was not a consequence of blood pressure elevation and could be related to the pathogenesis of hypertension. The ^{32}P-labelling of phosphoinositides, an index of phospholipid turnover, was investigated in kidney cortex of spontaneously hypertensive, Wistar-Kyoto normotensive, Sabra hypertension-prone and -resistant rats under various dietary conditions known to affect their blood pressure (Marche *et al.* 1986). The results indicated that the basal turnover of phosphoinositides in hypertensive rats is less than in their corresponding normotensive controls.

In contrast to these experiments showing depressed basal activity of the phosphoinositide pathway, a recent study by Turla and Webb (unpublished observations) indicates that, following receptor stimulation with 5-HT, there is augmented release of [^{3}H]inositol phosphates in aortae from stroke-prone spontaneously hypertensive rats compared with that in aortae from Wistar-Kyoto normotensive rats. In these experiments, the release of [^{3}H]inositol phosphates was increased four-fold above the basal state by 5-HT (100 μM) in hypertensive aorta compared with two-fold in normotensive aorta. These data suggest that the generation of these second messengers is augmented in hypertensive rats and that this enhanced activity may account for the exaggerated phasic response to 5-HT in arteries from hypertensive rats (see Fig. 14.2). It is known that inositol trisphosphate participates in the release of membrane-bound calcium in several cell systems (Berridge 1984) and it seems likely that abnormal responsiveness to 5-HT in hypertension is linked to this cellular mechanism.

The role of protein kinase C in hypertension has been studied through the use of protein kinase C activators. Turla and Webb (1987a and b) demonstrated an augmented vascular reactivity to the protein kinase C activators, 12-O-tetradecanoylphorbol-13-acetate and mezerein, in arteries from deoxycorticosterone acetate and stroke-prone spontaneously hypertensive rats. The augmentation may be a result of an enhanced activity of protein kinase C in vascular smooth muscle from hypertensive animals. A selective inhibitor of protein kinase C (Hidaka *et al.* 1984), H-7, was also employed to investigate the role of protein kinase C in 5-HT-mediated aortic smooth muscle contraction in stroke-prone spontaneously hypertensive and Wistar-Kyoto normotensive rats (Fig. 14.3; Turla and Webb, unpublished observations). In the presence of H-7, maximal responses to 5-HT were depressed in both groups of rats. In addition, H-7 shifted the dose-response curves to 5-HT to the right of the dose-response curves

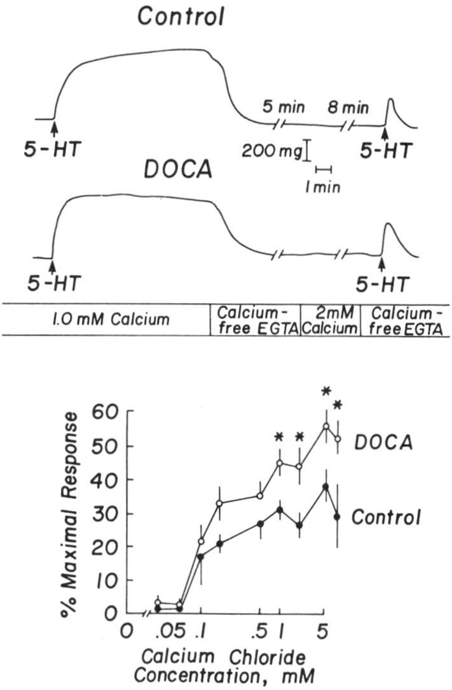

Fig. 14.2 Demonstration of a greater membrane store of calcium in femoral arteries of deoxycorticosterone acetate hypertensive (DOCA) rats. Upper: Femoral arteries were made to contract to 5-HT (0.57 μM) in solution containing 1.6 mM calcium; the strips were then depleted of calcium by placement in EGTA-calcium-free buffer for 10 min. Following this calcium depletion the strips were 'loaded' for 5 min in buffer with a known concentration of calcium, placed in EGTA-calcium-free buffer, and then challenged with 5-HT (0.57 μM). Lower: Data are expressed as a percentage of contraction in 1.6 mM calcium buffer and represent the mean (± SEM) of responses from six femoral artery strips. *The contractile responses to 5-HT in femoral arteries from DOCA rats were significantly greater ($p < 0.05$) than responses from control rats at 'loading' concentrations of calcium above 0.5 mM. (From Mecca and Webb 1984a by permission)

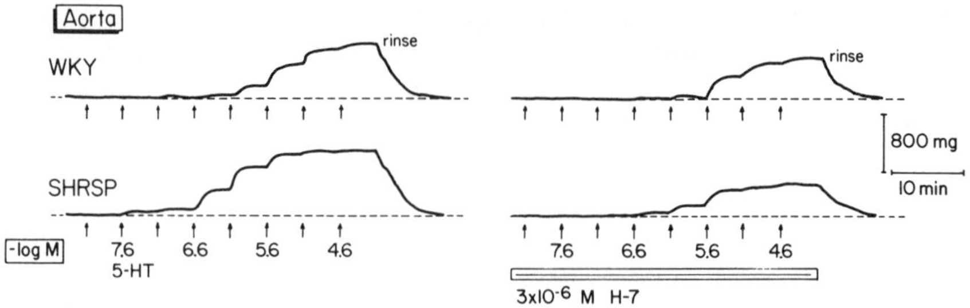

Fig. 14.3 Effects of protein kinase C inhibition by H-7 on 5-HT-induced contractions in aorta from normotensive Wistar-Kyoto (WKY) and stroke-prone spontaneously hypertensive (SHRSP) rats. In the presence of H-7, contractile responses to 5-HT were depressed in both groups of rats. (From Turla and Webb unpublished observations)

generated in the absence of the protein kinase C inhibitor. Since protein kinase C is involved in the regulation of the calcium-calmodulin regulatory system and in protein phosphorylation (Berridge 1984; Hidaka and Kawamoto 1985), it is possible that alterations in the activity of this enzyme may contribute to the increased vascular responsiveness in hypertension.

14.3.1.4 *Activation of receptor-operated ionic channels*

A difference in the activation of receptor-operated ionic channels by 5-HT may also account for the altered sensitivity to the monoamine in hypertension. Noradrenaline, angiotensin II, 5-HT, and vasopressin induce an increased turnover of radiolabelled K^+ in aortic segments from deoxycorticosterone acetate hypertensive and normotensive rats (Jones *et al.* 1981). With the exception of vasopressin, the dose of each agonist which causes a 50 per cent change in K^+ turnover rate is lower in arteries from hypertensive than in those from normotensive rats (Fig. 14.4). The difference between the dose-response relationships for K^+ turnover for normotensive and hypertensive rats is greater for 5-HT than for noradrenaline and angiotensin II. Thus, the specific increase in vascular responsiveness to 5-HT in hypertension may result from an altered relationship in the receptor activation of ionic channels important for membrane excitability.

14.3.1.5 *5-HT and prostaglandins*

Several recent studies have demonstrated that prostaglandins are generated in response to 5-HT receptor activation and that these prostaglandins contribute to the overall vascular response to the monoamine (Blackshear *et al.* 1986; McGiff *et al.* 1972; Sametz and Juan 1982). As mentioned earlier, 5-HT causes a dose-dependent constriction of the coronary vasculature in spontaneously hypertensive rats, whereas the monoamine

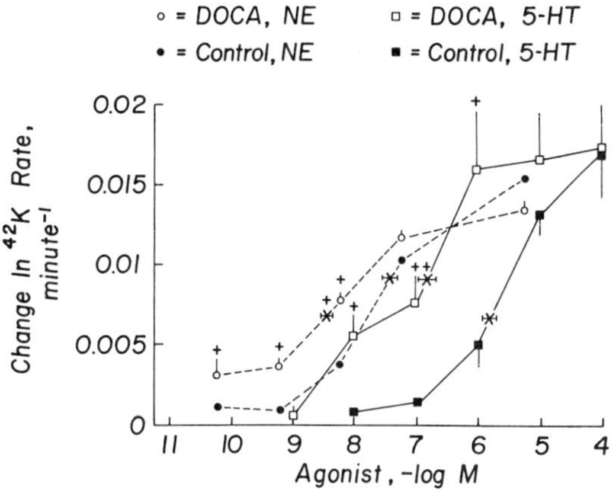

Fig. 14.4 Dose-dependent effect of noradrenaline (NE) and 5-HT on the potassium efflux (^{42}K rate, min^{-1}) from aortic strips isolated from deoxycorticosterone acetate hypertensive (DOCA) and normotensive rats. The ED_{50} values are indicated by (x). The responses for hypertensive rats which are significantly different from the corresponding controls are identified by (+). Values are the means ± SEM. (From Jones *et al.* 1981 by permission)

produces vasodilatation in the hearts from normotensive rats (Lüscher *et al.* 1986). Indomethacin, a cyclo-oxygenase inhibitor, blocked the constrictor response to 5-HT in hypertensive rats but did not alter the vasodilator effect to 5-HT in normotensive rats (Fig. 14.5). The authors conclude that the mediator of the vasoconstriction in hearts from spontaneously hypertensive rats is a product of cyclo-oxygenase.

In the femoral artery of mineralocorticoid hypertensive rats, the sensitivity to 5-HT is increased and cyclo-oxygenase inhibitors (indomethacin, aspirin) shift dose-response curves to 5-HT to the right. However, the magnitude of the shift produced by cyclo-oxygenase inhibitors in the contractile effect to 5-HT in mineralocorticoid hypertensive rats does not differ from that in normotensive rats (Mecca and Webb 1983). Thus, cyclo-oxygenase products contribute to the constrictor response to 5-HT but it does not appear that the change in sensitivity to the monoamine can be explained by an altered production of prostaglandins in this model of hypertension.

14.3.2 Partial 5-HT receptor agonists

Hauesler and Finch (1972) observed that methysergide, at doses which completely blocked the pressor effects of 5-HT in the pithed rat, failed to lower blood pressure in the intact hypertensive animal. They concluded that 5-HT plays no role in the maintenance of the elevated blood pressure

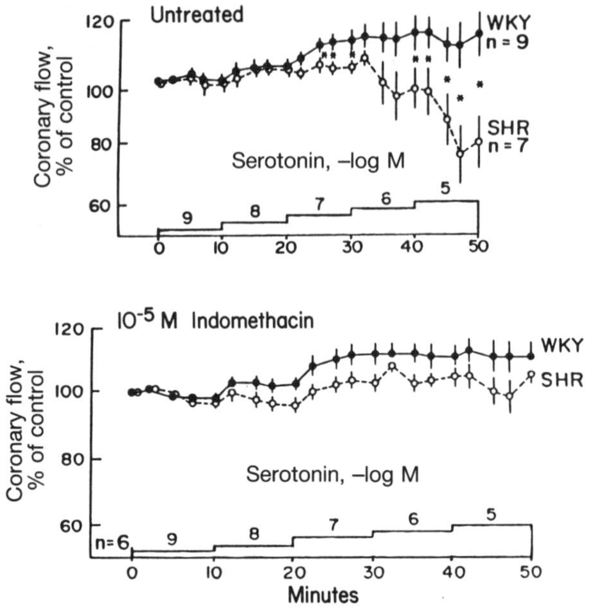

Fig. 14.5 Effects of 5-HT in isolated hearts from normotensive Wistar-Kyoto (WKY) rats and spontaneously hypertensive rats (SHR). The hearts were paced and exposed to increasing concentrations of 5-HT at 10 min intervals. The effects of the monoamine on coronary flow and tension are given for experiments done in the absence (upper) or in the presence (lower) of indomethacin (10 μM). The data are given as means ± SEM. $^*p<0.05$, statistically significant difference between WKY rats and SHR. (From Lüscher *et al.* 1986 by permission)

in experimental hypertension. In the light of more recent studies on the agonistic nature of methysergide (Mecca and Webb 1984a; Webb 1982) it is perhaps not surprising that attempts to lower blood pressure with methysergide have failed in hypertensive animals.

In hypertensive rats, the vascular responsiveness to the partial agonistic properties of methysergide is augmented (Mecca and Webb 1984a; Webb and Bohr 1984) and this effect is inhibited by the 5-HT_2 receptor antagonist, ketanserin. Isolated arterial segments from mineralocorticoid hypertensive rats develop greater force and are more sensitive to methysergide than those from normotensive rats. The relative change in responsiveness to methysergide is greatest in the aorta, intermediate in the mesenteric artery, and least in the tail artery (Fig. 14.6). This sequence is the reverse of that describing the density of innervation of these three blood vessels. The significance, if any, of this interesting correlation between responsiveness to methysergide and adrenergic innervation is as yet unknown.

The altered vascular responsiveness to methysergide parallels the

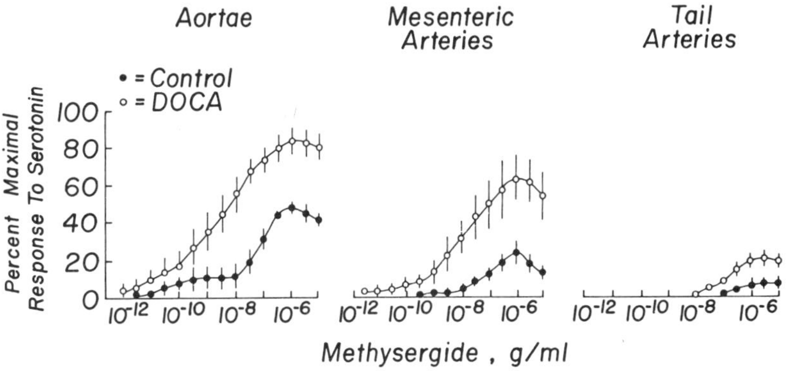

Fig. 14.6 Dose–response curves to methysergide in arteries from deoxycorticosterone acetate hypertensive (DOCA) and normotensive rats. Cumulative addition of methysergide to the muscle bath produced contraction. The contractile responses to methysergide are expressed as a percentage of the maximal response to 5-HT since the agonistic properties of the 5-HT antagonist are mediated by 5-HT receptors. Arterial preparations from DOCA rats are more sensitive to methysergide than those from control rats. Values are means ± SEM for six control and six hypertensive rats. (From Webb and Bohr 1984 by permission)

development of the increase in arterial blood pressure in deoxycorticosterone acetate hypertensive rats (Webb 1982). Placement of these hypertensive rats on a low-sodium diet reverses both the elevated blood pressure and the enhanced sensitivity to methysergide to near normal levels. An increased responsiveness to the partial agonistic properties of methysergide is also greater in genetic and renal models of hypertension in rats (Webb 1982) and in mineralocorticoid hypertensive sheep (Mecca *et al.* 1985). Thus, the hypertensive process may be associated with an alteration in the sensitivity of the receptors that mediate the partial agonistic properties of methysergide.

14.3.3 5-HT receptor antagonists

Studies on isolated blood vessels clearly show that 5-HT receptor antagonists can cause relaxation of vascular smooth muscle by interfering with the agonist–receptor interaction (De Clerck and Vanhoutte 1982). However, the inability of 5-HT receptor antagonists to lower blood pressure of hypertensive animals (Vanhoutte *et al.* 1981; Cohen (H. O.) *et al.* 1983; Cohen (M. L.) *et al.* 1983) does not support a role for 5-HT in the maintenance of hypertension. Recent evidence suggests that several 5-HT receptor antagonists are partial agonists and thus may produce a direct vasoconstriction that could mask potentially beneficial effects of 5-HT receptor blockade (Webb 1982; Saxena 1974). Ketanserin, a selective antagonist of the $5\text{-}HT_2$ receptor, does not possess these agonistic actions and it lowers blood pressure in hypertensive animals and humans (Wenting *et al.* 1982; Van Nueten *et al.* 1981; Vanhoutte 1982b). However, it should be noted that a portion of the blood pressure lowering effect of ketanserin is probably related to blockade of α_1-adrenoceptor receptors (Mecca *et al.* 1985; Robertson *et al.* 1986). Studies with more selective antagonists that do not exhibit agonist properties have yielded conflicting findings regarding the possible role of 5-HT in hypertension (see Chapter 16).

14.3.4 5-HT metabolites, 5-hydroxykynuramine and kynuramine

Recent work (Thompson and Webb 1987) has examined vascular responsiveness to the major degradative products of 5-HT in mineralocorticoid hypertensive rats (Fig. 14.7). Contractile responses to 5-hydroxyindoleacetic acid and 5-hydroxytryptophol were greater in arteries from hypertensive rats, whereas reactivity to melatonin in arteries from hypertensive rats did not differ from that in arteries from normotensive rats. Mesenteric arteries from both rat groups were unresponsive to the 5-HT metabolite, N-acetyl-5-HT. The contractile properties of the 5-HT metabolites were inhibited by ketanserin. These experiments with 5-HT metabolites may have particular importance to studies demonstrating that the metabolism of 5-HT is altered in hypertension (Guicheney *et al.* 1985). It seems possible that the degradative products of 5-HT at the local level may contribute to increased vascular responsiveness; however, the pathophysiological role of these metabolites in hypertension remains unknown.

In addition to the major degradative products of 5-HT, there are other related metabolic compounds that have been shown to have vasoactive properties. The major class of these compounds with biological actions similar to 5-HT are the kynuramines (Charlton *et al.* 1984 and 1985; Murakami and Makino 1971). These aromatic diamines are believed to be formed from precursor amino acids (L-tryptophan, L-5-hydroxytryptophan) and indoleamines (tryptamine, 5-HT, melatonin). Vascular sensitivity to 5-hydroxykynuramine and kynuramine in mesenteric arteries from deoxycorticosterone acetate hypertensive rats is increased compared with that in arteries from normotensive rats (Gilbert and Webb, unpublished observations). The increased sensitivity to 5-hydroxykynuramine is similar to that to 5-HT and contractile responses to 5-hydroxykynuramine are inhibited by ketanserin. In contrast, contractile responses to kynuramine are not inhibited by the 5-HT_2 receptor antagonist (Fig. 14.8). These observations suggest that several metabolically related compounds may contribute to the complex vascular actions of 5-HT.

14.3.5 Potentiation of constrictor responses to other stimuli by 5-HT

In addition to its vasoconstrictor and vasodilator actions, 5-HT interacts with other vasoactive agents in a complex manner (Van Nueten *et al.* 1981;

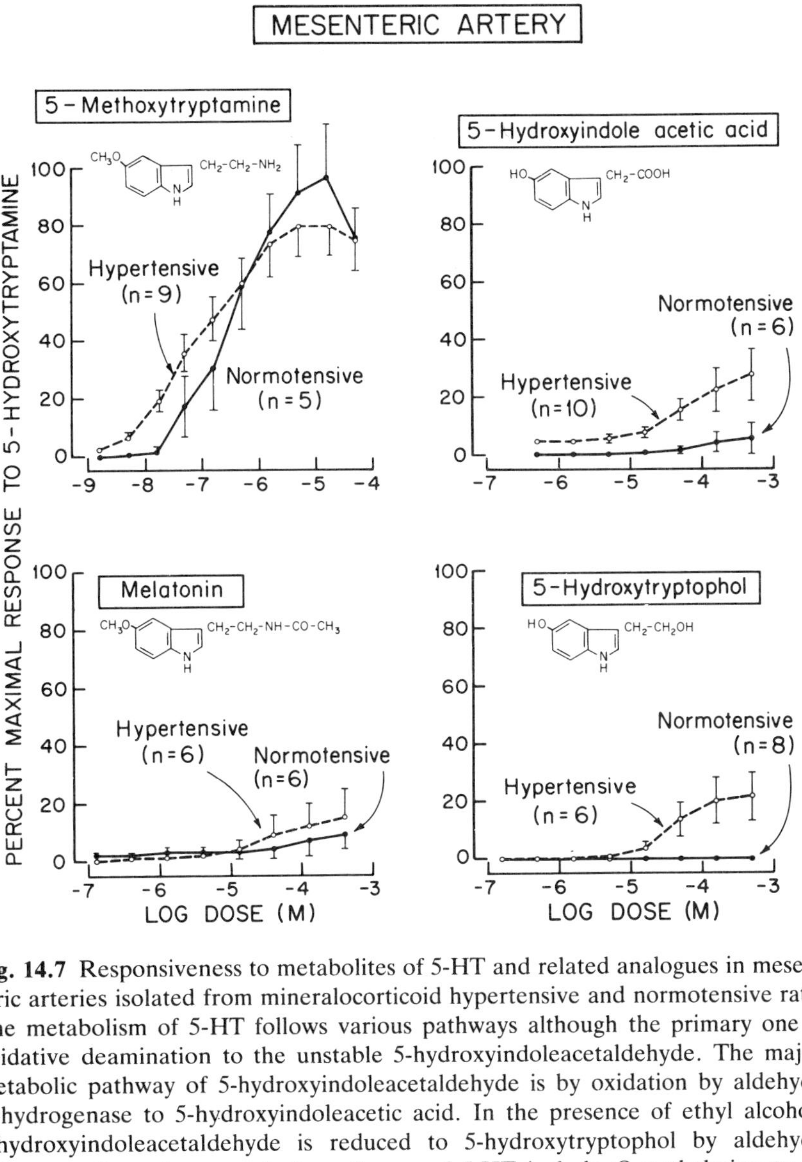

Fig. 14.7 Responsiveness to metabolites of 5-HT and related analogues in mesenteric arteries isolated from mineralocorticoid hypertensive and normotensive rats. The metabolism of 5-HT follows various pathways although the primary one is oxidative deamination to the unstable 5-hydroxyindoleacetaldehyde. The major metabolic pathway of 5-hydroxyindoleacetaldehyde is by oxidation by aldehyde dehydrogenase to 5-hydroxyindoleacetic acid. In the presence of ethyl alcohol, 5-hydroxyindoleacetaldehyde is reduced to 5-hydroxytryptophol by aldehyde reductase. Secondary metabolic pathways of 5-HT include *O*-methylation to 5-methoxytryptamine and *N*-acetylation to N-acetyl-5-HT leading to the formation of melatonin. Contractile responses are expressed as a percentage of the maximal force to 5-HT. Values are means ± SEM. The number of rats in each experimental group is indicated in parentheses. (From Thompson and Webb 1987 by permission)

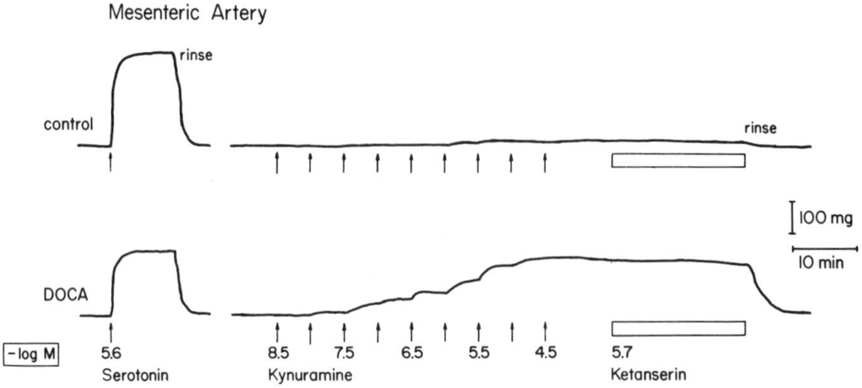

Fig. 14.8 Dose–response curves to kynuramine in mesenteric arteries from deoxycorticosterone acetate hypertensive (DOCA) and normotensive rats. Maximal force development to 5-HT in arteries from hypertensive rats was similar to that in arteries from normotensive rats, whereas cumulative addition of kynuramine caused greater contractions in arteries from hypertensive rats compared to controls. Ketanserin did not antagonize the contractile responses to kynuramine. (From Gilbert and Webb unpublished observations)

Van Nueten 1982). Bohr and Elliot (1962) demonstrated that subthreshold doses of 5-HT potentiate contractile responses to adrenaline in isolated canine mesenteric arteries. Several other investigators have also observed this phenomenon (see Chapter 6) and it has been demonstrated that this potentiating or amplifying effect of 5-HT is blocked by ketanserin (Van Nueten *et al.* 1981). The potentiating effect of 5-HT is not limited to the catecholamines, but occurs with many other agonists (Van Nueten *et al.* 1981; Van Nueten 1982). In the mesenteric vasculature of spontaneously hypertensive rats, the potentiation by 5-HT of pressor responses to exogenous noradrenaline is more pronounced than that in the mesentery of normotensive rats (Kubo and Su 1983). The potentiating action of 5-HT on pressor responses to noradrenaline is also augmented in the renal vasculature of spontaneously hypertensive rats (Janssens and Van Nueten 1986) and in the intact mineralocorticoid hypertensive sheep (Myers *et al.* 1985). In femoral arteries from two kidney-one clip hypertensive rabbits (Fig. 14.9), the absolute magnitude of potentiation by 5-HT of noradrenaline-induced contractile activity is greater at low doses of the monoamine than that in arteries from normotensive rabbits (Pott *et al.* 1986). At high doses of 5-HT, the magnitude of potentiation is similar for arteries from both hypertensive and normotensive rabbits. Although it is difficult to determine if this potentiating effect of 5-HT is important in hypertension, it seems certain that this effect contributes to the complex actions of the monoamine when injected into the intact animal.

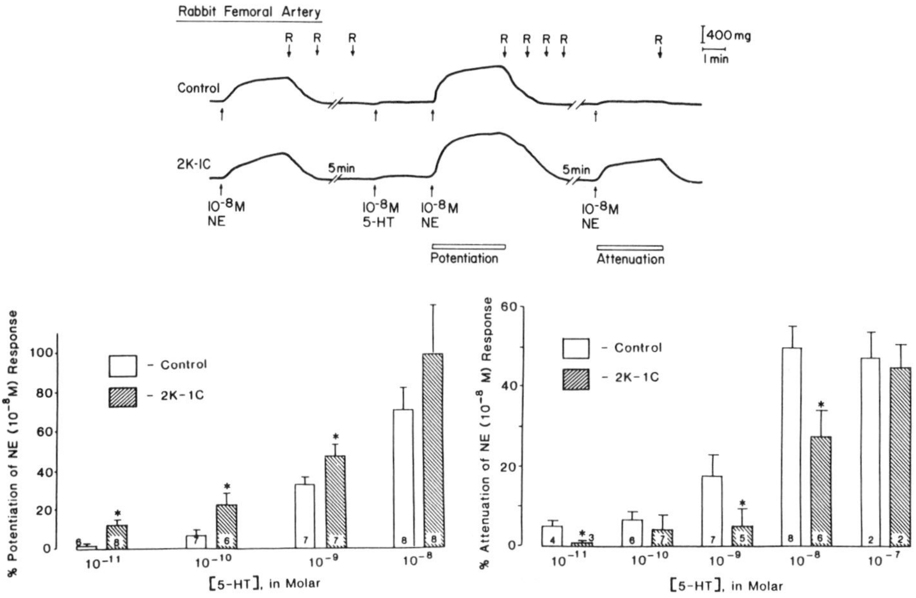

Fig. 14.9 The influence of 5-HT on noradrenaline (NE) responses in helical strips of femoral artery from control and two kidney-one clip (2K-1C) hypertensive rabbits. Upper: Shown are representative tracings depicting the potentiation by the presence of 5-HT and attenuation by previous exposure to 5-HT of a NE response. (R) indicates rinse. Note the greater potentiation and lesser attenuation of the NE response in vessels from hypertensive rabbits than in those from control rabbits. Lower left: Potentiation of NE responses by the presence of 5-HT in femoral artery strips from control and hypertensive rabbits. This figure shows the results of several experiments of the type described above with varying concentrations of 5-HT. The NE concentration was 0.01 μM. Data presented are expressed as a percentage increase in NE response in the presence of 5-HT compared with a NE response alone. The force development to 5-HT was not included in the calculation of the potentiation. The NE response of arterial strips from 2K-1C rabbits was potentiated more by lower concentrations of 5-HT than that for strips from control rabbits. Maximal potentiation was similar for arterial strips from both sources. Lower right: Attentuation of NE responses by previous exposure to 5-HT in femoral artery strips from control and hypertensive rabbits. This figure depicts the results of several experiments of the type described above with varying concentrations of 5-HT. The NE concentration was 0.01 μM. Data presented are expressed as a percentage decrease in NE response after previous exposure to 5-HT compared with NE responses before 5-HT exposure. The attenuation of NE is less in arterial strips from 2K-1C than in strips from control rabbits. Maximal attenuation is the same for arterial strips from both sources. Values are means ± SEM for the number of animals denoted (range shown where $n = 2$). $^*p < 0.05$, hypertensive compared with control animals. (From Pott *et al.* 1986 by permission)

14.3.6 Tachyphylaxis to 5-HT

The responsiveness of certain vascular preparations to repeated doses of 5-HT decreases with time (Collis and Vanhoutte 1977). This tachyphylaxis to 5-HT is decreased in the renal vasculature of spontaneously hypertensive rats (Collis and Vanhoutte 1977; De Mey and Vanhoutte 1981) and it is more pronounced in the male than in the female spontaneously hypertensive rat (Fig. 14.10). Tachyphylaxis to 5-HT is less in older than in younger genetically hypertensive rats (De Mey and Vanhoutte 1981). The cellular mechanism responsible for the delayed tachyphylaxis in the renal vasculature of the spontaneously hypertensive rat is unknown. De Mey and Vanhoutte (1981) have shown that the effect is unrelated to the presence of activator calcium in the extracellular fluid.

14.3.7 Indirect sympathomimetic effect of 5-HT

5-HT can be actively transported into sympathetic neurones by the neuronal uptake mechanism in the blood vessel wall (Vanhoutte *et al.* 1981; Verbeuren *et al.* 1983). In the nerve endings, 5-HT can displace noradrenaline, causing vasoconstriction via an indirect sympathomimetic effect (Collis *et al.* 1979). The monoamine can also be stored and released

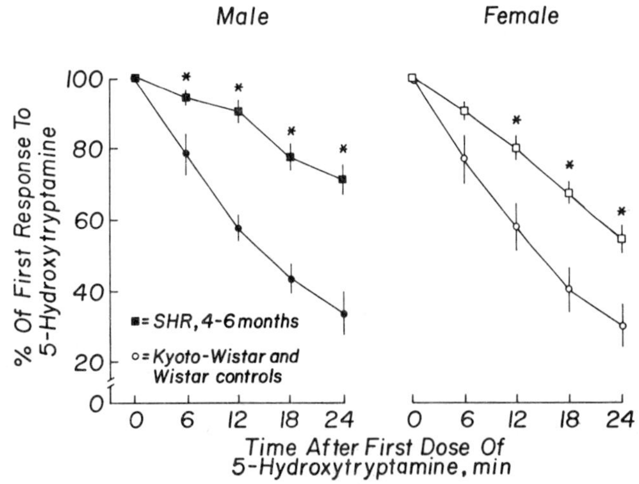

Fig. 14.10 Rate of development of tachyphylaxis to 5-HT in kidneys from male and female spontaneously hypertensive and sex-matched control rats. The graph shows mean percentage decrease in response amplitude to repeated doses (20 μg) of 5-HT in kidneys from female control and spontaneously hypertensive rats (right), and from male control and spontaneously hypertensive rats (left). Significant differences between kidneys from spontaneously hypertensive and control rats are indicated by asterisks. (From Collis and Vanhoutte 1977 by permission)

together with noradrenaline upon activation of the nerve endings (Verbeuren *et al.* 1983; see Chapter 1).

In innervated kidneys of young spontaneously hypertensive rats, the vasoconstrictor responses to 5-HT are increased compared with normotensive values (Collis *et al.* 1979). This difference between the two animal groups is absent following sympathetic denervation with 6-hydroxydopamine. Thus, 5-HT has a more pronounced indirect sympathomimetic effect in the young spontaneously hypertensive rats.

14.4 Vascular relaxation in response to 5-HT

14.4.1 Partial 5-HT receptor agonists

14.4.1.1 *Direct vasodilator effect*

5-HT can produce vasodilatation by direct action on the vascular smooth muscle cells (Lamb and Webb 1987; Mecca and Webb 1984b; Thompson *et al.* 1987). This direct inhibitory action is more prominent in isolated veins than in arteries (Mecca and Webb 1984b) and is not antagonized by adrenoceptor, muscarine, or histamine receptor blockade nor by inhibition of prostaglandin synthesis (for review, see Lamb and Webb 1987). The vasodilator action is mediated by a receptor that is distinct from the 5-HT_2 subclass, and it may be a '5-HT_1-like' receptor-mediated response (Feniuk *et al.* 1981a; see Chapter 7). The cellular mechanism that is responsible for the dilator effect of 5-HT may relate to either of the following: (1) elevations in cyclic AMP (Trevithick *et al.* 1984 and 1986); or (2) enhancement of potassium conductance (Lamb and Webb 1987).

5-HT causes relaxation of femoral veins from renal hypertensive rabbits and the magnitude of this relaxation response does not differ from that in veins from normotensive rabbits (Webb and Vanhoutte 1985). In deoxycorticosterone acetate hypertensive rats, the selective '5-HT_1-like' receptor agonist, 5-carboxamidotryptamine, produces greater depressor responses compared with that in normotensive rats (Dalton *et al.* 1986).

14.4.1.2 *Attenuation of constrictor responses*

Exposure to and subsequent removal of 5-HT attenuates contractile responses to adrenaline (Bohr and Elliot 1962). This attenuation is blocked by methysergide and may involve activation of the electrogenic sodium pump and depression of calcium influx during adrenergic activation (Moreland and Bohr 1983; Pott *et al.* 1986). The attenuation of noradrenaline responsiveness that follows the exposure to 5-HT is less in arteries from hypertensive than in those from normotensive rabbits (see Fig. 14.9; Pott *et al.* 1986). These changes induced by 5-HT may be related to a greater influx of calcium into the smooth muscle cells from hypertensive rabbits during exposure to the adrenoceptor agonist (Moreland *et al.* 1985).

14.4.2 Endothelium

The endothelium produces relaxing substances that may be important in the local control of the circulation (Furchgott and Zawadzki 1980). Several investigators have observed that EDRF depresses constrictor responses to 5-HT (Angus and Cocks 1984; Cohen 1986; Cohen *et al.* 1984; see Chapter 8). Pressor responses to 5-HT in the perfused rat mesenteric bed are potentiated following endothelial disruption by free radicals generated by electrolysis of the perfusate (Thompson *et al.* 1987). However, this potentiation did not eliminate the difference in vascular sensitivity to 5-HT in perfused mesenteric beds from deoxycorticosterone acetate hypertensive and control normotensive rats (Fig. 14.11; King and Webb unpublished observations). In aortic segments from Dahl hypertensive rats (salt-sensitive and salt-resistant strains), 5-HT-induced contractions are potentiated following removal of the endothelium and this effect is greatest in aortae obtained from rats that are maintained on a low sodium intake (Lüscher *et al.* 1987). Contractions to 5-HT in aortic segments with endothelium were greater in rats maintained on a high sodium diet. Since blood pressures were elevated only in the salt-sensitive substrain maintained on a high salt diet, these findings suggest a differential effect of salt and blood pressure on endothelium-dependent responses.

Lüscher and Vanhoutte (1986) investigated the endothelium-dependent responses to 5-HT in aortic rings (contracted with prostaglandin $F_{2\alpha}$) from spontaneously hypertensive and Wistar-Kyoto normotensive rats. Lower concentrations of 5-HT induced a slight relaxation of aortic rings with endothelium, but higher concentrations induced contraction. In rings without endothelium, 5-HT caused only contractions. In the spontaneously

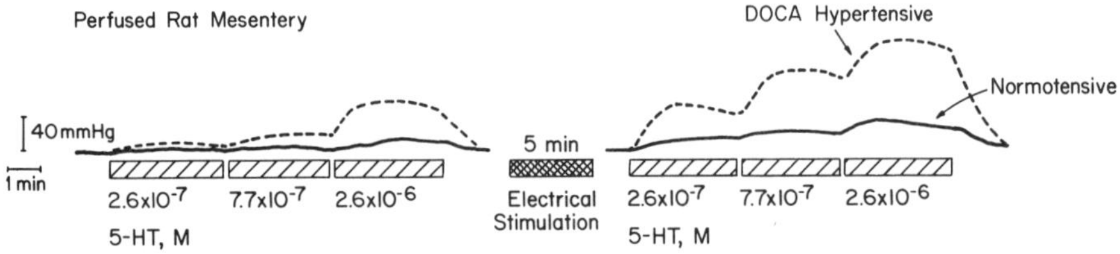

Fig. 14.11 5-HT and the endothelium in the perfused rat mesentery from deoxycorticosterone acetate (DOCA) hypertensive (dashed line) and normotensive (solid line) rats. The mesentery was perfused with physiological salt solution (PSS) and perfusion pressure was measured (constant flow). Destruction of endothelial cells by perfusion with oxygen free radicals generated in the PSS by an electrical stimulus (Lamb *et al.* 1987) caused a potentiation of 5-HT-induced vasoconstriction in both rat strains. However, the degree of potentiation was not different between DOCA hypertensive and normotensive rats. (From King and Webb unpublished observations)

hypertensive rat aorta, the contractions caused by higher concentrations of 5-HT were larger in the presence than in the absence of endothelium, suggesting that a vasoconstrictor factor may be released from the endothelium in response to 5-HT. This effect was absent in the normotensive rat (Fig. 14.12). These studies suggest that in the aorta of spontaneously hypertensive rats, there is a release of one or more endothelium-derived vasoactive factors.

14.4.3 Prejunctional component

In the intact animal, 5-HT infusion inhibits vasoconstriction in response to sympathetic nerve stimulation (Feniuk *et al.* 1981b). This inhibitory response occurs as a consequence of reduced transmitter release after interaction of 5-HT at the level of the adrenergic nerve ending (Bevan *et al.* 1975; Feniuk *et al.* 1979; Kimura and Satoh 1983; Martinez and Lokhandwala 1980; McGrath 1977; Watts *et al.* 1981). The 5-HT-induced inhibition of adrenergic transmitter release is not blocked by ketanserin (Van Nueten *et*

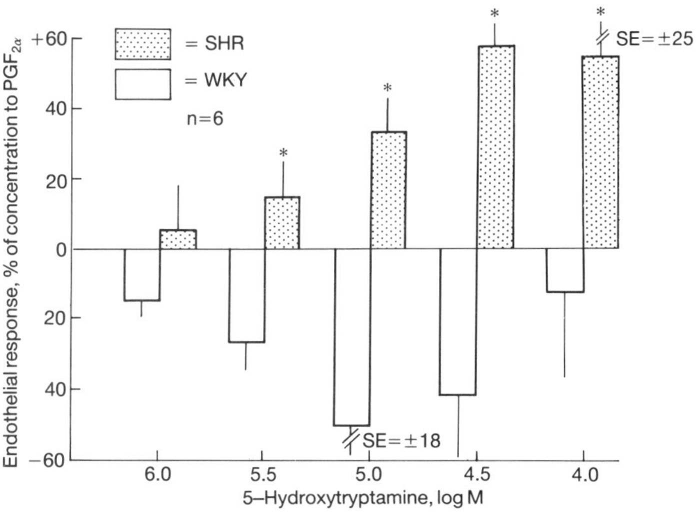

Fig. 14.12 Endothelial response to 5-HT in aortic rings from normotensive Wistar Kyoto (WKY) and spontaneously hypertensive (SHR) rats contracted with prostaglandin $F_{2\alpha}$). In all experiments, the 5-HT_2 receptor antagonist, ketanserin (0.1 μM), was present to minimize the direct contractile effects of 5-HT on vascular smooth muscle. The endothelial response was calculated as the tension of rings with endothelium minus the tension of rings without endothelium and expressed as a percentage of the contractile response to $PGF_{2\alpha}$. The data are given as means ± SEM for 6 experiments. Asterisks denote statistically significant differences between WKY and SHR ($p<0.05$). (From Lüscher and Vanhoutte 1986 by permission)

al. 1981) but it is antagonized by methiothepin (Cohen 1985). Kubo and Su (1983) have demonstrated that 5-HT exerts a prejunctional inhibition in mesenteric arteries from Wistar-Kyoto normotensive but not in arteries from spontaneously hypertensive rats (Fig. 14.13). Furthermore, 5-HT reduces the stimulated overflow of tritium-labelled products from normotensive mesenteric arteries previously incubated with [^{3}H]noradrenaline but not in arteries from spontaneously hypertensive rats (Su and Uruno 1984). Thus, prejunctional modulation by 5-HT is depressed in hypertension and this effect may contribute to the elevated vascular resistance characteristic of this disease process.

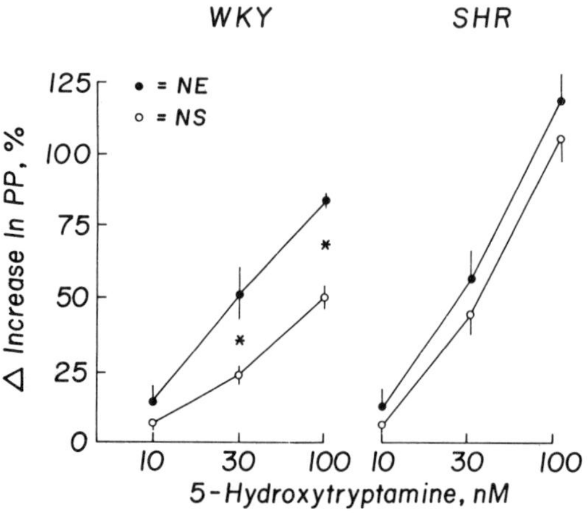

Fig. 14.13 Dose–response curves for the effects of 5-HT on the pressor response to nerve stimulation (NS) and noradrenaline (NE) infusion in perfused mesenteric arteries from normotensive Wistar-Kyoto (WKY) (left) and spontaneously hypertensive (SHR) rats (right). The data for five experiments are expressed as means ± SEM. The asterisks indicate statistically significant differences between responses to norepinephrine and nerve stimulation. (From Kubo and Su 1983 by permission)

14.5 Concluding remarks

5-HT exerts complex and varying actions on the cardiovascular system and thus it has been difficult to characterize a role for the monoamine in pathophysiological states. In hypertension, vasoconstrictor sensitivity to 5-HT is augmented, whereas vasodilator properties of 5-HT are either reduced or unchanged compared with that in the normal animal. The enhanced ability of 5-HT to exert contractile effects in isolated blood vessels from hypertensive animals probably relates to alterations in calcium metabolism via the phosphoinositide pathway. It is also likely that changes in specific receptor-operated ion channels and in the synthesis of cyclo-oxygenase products contribute to the enhanced contractile effect of 5-HT in this disease process. Several compounds that are metabolically related to 5-HT cause greater contraction in blood vessels from hypertensive animals. The ability of 5-HT to potentiate the actions of other vasoactive agents is increased in hypertension and tachyphylaxis to 5-HT is depressed in some vascular beds of hypertensive animals. The direct dilator effect of 5-HT on smooth muscle does not appear to be altered in hypertension but its dilator action mediated via the endothelium is either depressed or offset by a constricting factor. Finally, the inhibitory effect of 5-HT on adrenergic neurotransmitter release is reduced in hypertensive animals. These observations are consistent with the hypothesis that abnormal responsiveness to 5-HT characterizes the hypertensive state.

Acknowledgement

Work reported in this manuscript was supported by National Institutes of Health grant HL-27020.

References

Ahlund, L., Lundgren, Y., Sjoberg, B., and Weiss, L. (1977). Vascular reactivity to 5-hydroxytryptamine (5-HT) in hindquarter vascular beds, aortic strips and portal veins from spontaneously hypertensive and normotensive rats. *Acta Physiologica Scandinavica* **101,** 489–92.

Angus, J. A. and Cocks, T. M. (1984). Role of endothelium in vascular responses to norepinephrine, serotonin, and acetylcholine. In *Vasodilator mechanisms* (eds. P. M. Vanhoutte and S. F. Vatner), pp. 43–52. S. Karger, Basel.

Berecek, K. H., Schwertschlag, U., and Gross, F. (1980). Alterations in renal vascular resistance and reactivity in spontaneous hypertension of rats. *American Journal of Physiology* **238,** H287–93.

Berridge, M. J. (1984). Inositol trisphosphate and diacylglycerol as second messengers. *Biochemical Journal* **220,** 345–60.

Bevan, J. A., Duckles, S. P., and Lee, T. J.-F. (1975). Histamine potentiation of nerve and drug-induced responses of a rabbit cerebral artery. *Circulation Research* **36,** 647–53.

Blackshear, J. L., Orlandi, C., and Hollenberg, N. K. (1986). Serotonin and the renal blood supply: Role of prostaglandins and the 5HT-2 receptor. *Kidney International* **30,** 304–10.

Bohr, D. F. and Elliot, J. (1962). 5-Hydroxytryptamine and the contraction of smooth muscle from resistance vessels. In *Metabolismus parietis vasorum* (eds. B. Prusik, Z. Reinis, and O. Riedl), pp. 337–45. Statni Zdravotnicke Nakladatelstvi, Prague.

Charlton, K. G., Johnson, T. D., and Clarke, D. E. (1984). Vasoconstrictor and norepinephrine potentiating action of 5-hydroxykynuramine in the isolated perfused rat kidney: Involvement of serotonin receptors and alpha$_1$-adrenoceptors. *Naunyn-Schmiedeberg's Archives of Pharmacology* **328,** 154–9.

Charlton, K. G., Johnson, T. D., Maurice, R. W., and Clarke, D. E. (1985). Kynuramine: High affinity for [^{3}H]tryptamine binding sites. *European Journal of Pharmacology* **106,** 661–4.

Cohen, H. O., Harms, Y. M., van Gelderen, E. M., Batnik, H. D., Timmermans, P. B. M. W. M., and van Zwieten, P. A. (1983). Hypotensive activity of serotonin antagonists; correlation with α_1-adrenoceptor and serotonin receptor blockade. *Life Sciences* **32,** 1499–505.

Cohen, M. L., Fuller, R. W., and Kurz, K. D. (1983). Evidence that blood pressure reduction by serotonin antagonists is related to alpha receptor blockade in spontaneous hypertensive rats. *Hypertension* **5,** 676–81.

Cohen, R. A. (1985). Serotonergic prejunctional inhibition of canine coronary adrenergic nerves. *Journal of Pharmacology and Experimental Therapeutics* **235,** 76–80.

Cohen, R. A. (1986). Adenine nucleotides and 5-hydroxytryptamine released by aggregating platelets inhibit adrenergic neurotransmission in canine coronary artery. *Journal of Clinical Investigation* **77,** 369–75.

Cohen, R. A., Shepherd, J. T., and Vanhoutte, P. M. (1984). Endothelium and asymmetrical responses of the coronary arterial wall. *American Journal of Physiology* **247,** H403–8.

Collis, M. G. and Vanhoutte, P. M. (1977). Vascular reactivity of isolated perfused kidneys from male and female spontaneously hypertensive rats. *Circulation Research* **41,** 759–67.

Collis, M. G., De Mey, C., and Vanhoutte, P. M. (1979). Renal vascular reactivity in the young spontaneously hypertensive rat. *Hypertension* **2,** 45–52.

Conn, J. P. and Sanders-Bush, E. (1986). Biochemical characterization of serotonin stimulated phosphoinositide turnover. *Life Sciences* **38,** 663–9.

Dalton, D. W., Feniuk, W., and Humphrey, P. P. A. (1986). An investigation into the mechanisms of the cardiovascular effects of 5-hydroxytryptamine in conscious normotensive and DOCA-salt hypertensive rats. *Journal of Autonomic Pharmacology* **6,** 219–28.

De Clerck, F. and Vanhoutte, P. M. (1982). *5-Hydroxytryptamine in peripheral reactions.* Raven Press, New York.

De Mey, C. and Vanhoutte, P. M. (1981). Effect of age and spontaneous hypertension on the tachyphylaxis to 5-hydroxytryptamine and angiotensin II in the isolated rat kidney. *Hypertension* **3,** 718–24.

Feniuk, W., Humphrey, P. P. A., and Watts, A. D. (1979). Presynaptic inhibitory action of 5-hydroxytryptamine in dog isolated saphenous vein. *British Journal of Pharmacology* **67,** 247–54.

Feniuk, W., Humphrey, P. P. A., and Watts, A. D. (1981a). An analysis of the receptors involved in 5-hydroxytryptamine-induced smooth muscle relaxation. *British Journal of Pharmacology* **74,** 850P–1P.

Feniuk, W., Humphrey, P. P. A., and Watts, A. D. (1981b). Modifications of the vasomotor actions of methysergide in the femoral artery bed of the anaesthetized dog by changes in sympathetic nerve activity. *Journal of Autonomic Pharmacology* **1,** 127–32.

Folkow, B. (1978). Cardiovascular structural adaptation: Its role in the initiation and maintenance of primary hypertension. *Clinical Science and Molecular Medicine* **55,** 3s-22s.

Folkow, B. (1982). Physiological aspects of primary hypertension. *Physiological Reviews* **62,** 347–504.

Forder, J., Scriabine, A., and Rasmussen, H. (1985). Plasma membrane calcium flux, protein kinase C activation and smooth muscle contraction. *Journal of Pharmacology and Experimental Therapeutics* **235,** 267–73.

Furchgott, R. F. and Zawadzki, J. V. (1980). The obligatory role of endothelial cells in the relaxation of arterial smooth muscle by acetylcholine. *Nature (London)* **288,** 373–6.

Guicheney, P., Baudouin-Egros, M., Garnier, J. P., Roques, P., Dreux, C., and Meyer, P. (1985). Platelet serotonin and blood tryptophan in spontaneously hypertensive and normotensive Wistar-Kyoto rats. *Journal of Cardiovascular Pharmacology* **7,** 515–7.

Hauesler, G. and Finch, L. (1972). Vascular reactivity to 5-hydroxytryptamine and hypertension in the rat. *Naunyn-Schmiedeberg's Archives Pharmacology* **272,** 101–16.

Hidaka, H. and Kawamoto, S. (1985). Protein kinase C-mediated phosphorylation in smooth muscle and platelet myosins. In *Phospholipids and cellular regulation* (ed. J. F. Kuo), pp. 171–88. CRC Press, New York.

Hidaka, H., Inagaki, M., Kawamoto, S., and Sasaki, Y. (1984). Isoquinolinesulfonamides, novel and potent inhibitors of cyclic nucleotide dependent protein kinase and protein kinase C. *Biochemistry* **23,** 5036–41.

Janssens, W. J. and Van Nueten, J. M. (1986). The direct and amplifying effects of serotonin are increased with age in the isolated perfused kidney of Wistar and spontaneously hypertensive rats. *Naunyn-Schmiedeberg's Archives of Pharmacology* **334,** 327–32.

Jones, A. W., Heidlage, J. F., Meyer, R., Day, B., and Freeland, A. (1981). Non-specific supersensitivity of aortic ^{42}K effluxes during DOCA hypertension in the rat, and the effects of antihypertensive therapy. In *New trends in arterial hypertension* (ed. M. Worcel *et al.*), pp. 163–73. Elsevier/North Holland Biomedical Press, Amsterdam.

Kamal, L. A., Cliox, J. F., Devynck, M. A., and Meyer, P. (1983). Reduced ^{3}H-serotonin uptake in human blood platelets by ouabain and endogenous digitalis-like inhibitors at Na^{+}/K^{+}-ATPase. *European Journal of Pharmacology* **92,** 167–8.

Kimura, R. and Satoh, S. (1983). Presynaptic inhibition by serotonin of cardiac sympathetic transmission in dogs. *Clinical and Experimental Pharmacology and Physiology* **10,** 535–42.

Koulu, M., Savedra, J. M., Bjelogrlic, N., Niwa, M., Agren, H., and Linnoila, M. (1986). Serotonin turnover in discrete hypothalamic nuclei and mesencephalic raphe nuclei of young and adult spontaneously hypertensive rats. *Brain Research* **379,** 257–63.

Koutouzov, S., Marche, P., Girard, A., and Meyer, P. (1983). Altered turnover of polyphosphoinositides in the erythrocyte membrane of the spontaneously hypertensive rat. *Hypertension* **5,** 409–14.

Kubo, T. and Su, C. (1983). Effects of serotonin and some other neurohumoral agents on adrenergic neurotransmission in spontaneously hypertensive rat vasculature. *Clinical and Experimental Hypertension* **5,** 1501–10.

Kuhn, D. M., Wolf, W. A., and Lovenberg, W. (1980). Review of the role of the central serotonergic neuronal system in blood pressure regulation. *Hypertension* **2,** 243–55.

Lamb, F. S. and Webb, R. C. (1987). Inhibitory actions of serotonin on vascular smooth muscle. In *Neuronal messengers in vascular function,* Fernstrom Series No. 10 (eds. A. Nobin, C. Owman, and B. Arneklo-Nobin), pp. 45–56. Elsevier Science Publishers, Amsterdam.

Lamb, F. S., King, C., Herrel, K., Burkel, W., and Webb, R. C. (1987). Free radical-mediated endothelial damage in blood vessels after electrical stimulation. *American Journal of Physiology* **252,** H1041–6.

Lingjaerde, O. (1977). Platelet uptake and storage of serotonin. In *Serotonin in health and disease* (ed. W. B. Essman), pp. 139–99. Spectrum, New York.

Lüscher, T. F. and Vanhoutte, P. M. (1986). Endothelium-dependent responses to platelets and serotonin in spontaneously hypertensive rats. *Hypertension* **8,** 55–60.

Lüscher, T. F., Rubanyi, G. M., Aarhus, L. K., Edoute, Y., and Vanhoutte, P. M. (1986). Serotonin reduces coronary flow in the isolated heart of the spontaneously hypertensive rat. *Journal of Hypertension* **4,** S148–50.

Lüscher, T. F., Raij, L., and Vanhoutte, P. M. (1987). Endothelium-dependent vascular responses in normotensive and hypertensive Dahl rats. *Hypertension* **9,** 157–63.

McGiff, J. C., Crowshaw, K., Terango, N. A., Malik, K. U., and Lonigno, A. J. (1972). Differential effect of noradrenaline and renal nerve stimulation on vascular resistance in the dog kidney and the release of a prostaglandin E-like substance. *Clinical Science* **42,** 223–33.

McGrath, M. A. (1977). 5-Hydroxytryptamine and neurotransmitter release in canine blood vessels. *Circulation Research* **41,** 428–35.

McGregor, D. D. and Smirk, F. H. (1970). Vascular responses to 5-hydroxytryptamine in genetic and renal hypertensive rats. *American Journal of Physiology* **219,** 687–90.

Marche, P., Koutouzov, S., Girard, A., Elghozi, J.-L., Meyer, P., and Ben-Ishay, D. (1986). Phosphoinositide metabolism in primary hypertension: An intrinsic membrane deficiency. *Journal of Hypertension* **4,** S267–9.

Martinez, A. A. and Lokhandwala, M. F. (1980). Evidence for a presynaptic inhibitory action of 5-hydroxytryptamine on sympathetic neurotransmission to the myocardium. *European Journal of Pharmacology* **63,** 303–11.

Mecca, T. E. and Webb, R. C. (1983). Vascular effects of arachidonic acid in deoxycorticosterone acetate (DOCA)-treated rats. *The Pharmacologist* **38,** 268.

Mecca, T. E. and Webb, R. C. (1984a). Vascular responses to serotonin in steroid hypertensive rats. *Hypertension* **6,** 887–92.

Mecca, T. E. and Webb, R. C. (1984b). Serotonin and vasodilatation. In *Vasodilator mechanisms* (eds. P. M. Vanhoutte and S. F. Vatner) pp. 81–90. S. Karger, Basel.

Mecca, T. E., Mitchell, J., Bohr, D. F., and Webb, R. C. (1985). Effects of serotonin antagonists on blood pressure in mineralocorticoid hypertensive sheep. *Journal of Cardiovascular Pharmacology* **7,** 660–5.

Moreland, R. S. and Bohr, D. F. (1983). Prior exposure to serotonin attenuates norepinephrine contraction in dog mesenteric artery. *Blood Vessels* **20,** 203.

Moreland, R. S., van Breemen, C., and Bohr, D. F. (1985). Mechanism by which serotonin attenuates contractile response of canine mesenteric arterial smooth muscle. *Journal of Pharmacology and Experimental Therapeutics* **232,** 322–9.

Murakami, Y. and Makino, K. (1971). Identification of 5-hydroxykynuramine in the urine of kynurenine administered mice. *Journal of Biochemistry* **70,** 533–5.

Myers, J. H., Mecca, T. E., and Webb, R. C. (1985). Direct and sensitizing effects of serotonin agonists and antagonists on vascular smooth muscle. *Journal of Cardiovascular Pharmacology* **7,** 544–8.

Nakaki, T., Roth, B. L., Chuang, D.-M., and Costa, E. (1985). Phasic and tonic components in 5-HT_2 receptor-mediated rat aorta contraction: Participation of Ca^{++} channels and phospholipase C. *Journal of Pharmacology and Experimental Therapeutics* **234,** 442–6.

Ozaki, M., Himero, A., Uchida, S., Ohta, H., and Niwa, M. (1986). Accelerated uptake of serotonin in mesenteric arteries of young spontaneously hypertensive rats. *Journal of Hypertension* **4** (Suppl. 3), S227–8.

Nickerson, M. (1956). Receptor occupancy and tissue response. *Nature (London)* **178,** 697–8.

Page, I. H. (1987). *Hypertension mechanisms.* Grune & Stratton, Orlando, Florida.

Pott, J. W. R., Moreland, R. S., Gantzos, R. D., Babcock, C., and Bohr, D. F. (1986). Vascular interactions of serotonin and norepinephrine in renal hypertensive rabbits. *Journal of Hypertension* **4,** 207–13.

Robertson, J. I. S., Scott, D. J., and Ball, S. G. (1986). The serotonin antagonist ketanserin in the treatment of clinical hypertension: A short review. *Journal of Hypertension* **4** (Suppl.), S119–21.

Roth, R. A. and Wallace, K. B. (1980). Disposition of biogenic amines and angiotensin I by lungs of spontaneously hypertensive rats. *American Journal of Physiology* **239,** H736–61.

Roth, B. L., Nakaki, T., Chuang, D.-M., and Costa, E. (1984). Aortic recognition sites for serotonin (5-HT) are coupled to phospholipase C and modulate phosphatidylinositol turnover. *Neuropharmacology* **23,** 1223–5.

Sametz, W. and Juan, H. (1982). Release of different prostaglandins from vascular tissue by different stimulators. *Prostaglandins and Leukotrienes in Medicine* **9,** 593.

Saxena, P. R. (1974). Selective vasoconstriction in carotid vascular bed by methysergide: possible relevance to its antimigraine effect. *European Journal of Pharmacology* **27,** 99–105.

Scatton, B., Duverger, D., l'Heureux, R., Serrano, A., Fage, D., Nowicki, J. P., and MacKenzie, E. T. (1985). Neurochemical studies on the existence, origin and characteristics of the serotonergic innervation of small pial vessels. *Brain Research* **345,** 219–29.

Skolasinska, K., Kihara, M., Horie, R., and Yamori, Y. (1982). CBF response to l-dopa influenced by peripheral decarboxylase inhibitor and serotonin in conscious SHR-SP, SHR-SR and WK rats. *Clinical and Experimental Pharmacology and Physiology* **9,** 631–7.

Stephenson, R. P. (1956). A modification of receptor theory. *British Journal of Pharmacology* **11,** 376–93.

Su, C. and Uruno, T. (1984). Excitatory and inhibitory effects of 5-hydroxytryptamine in mesenteric arteries of spontaneously hypertensive rats. *European Journal of Pharmacology* **106,** 283–90.

Suzuki, A., Hideaki, H., Shigetoshi, K., Yamanishi, Y., Yoshimura, Y., and Niwa, A. (1986). Spare receptors in the blood vessels from stroke-prone spontaneously hypertensive rats. *Journal of Hypertension* **4,** S77–9.

Thompson, J. H. (1971). Serotonin and the alimentary tract. *Research Communications in Chemical Pathology and Pharmacology* **2,** 687–782.

Thompson, L. P. and Webb, R. C. (1987). Vascular responsiveness to serotonin metabolites in mineralocorticoid hypertension. *Hypertension* **9,** 277–81.

Thompson, L. P., Lamb, F. S., King, C. M., and Webb, R. C. (1988). Serotonin: Vasodilator properties and interaction with atriopeptin III. In *Mechanisms of vasodilatation* (ed. P. M. Vanhoutte), pp. 223–8 Raven Press, New York.

Trevethick, M. A., Feniuk, W., and Humphrey, P. P. A. (1984). 5-Hydroxytryptamine-induced relaxation of neonatal porcine vena cava *in vitro*. *Life Sciences* **35,** 477–86.

Trevethick, M. A., Feniuk, W., and Humphrey, P. P. A. (1986). 5-Carboxamidotryptamine: A potent agonist mediating relaxation and elevation of cyclic AMP in the isolated neonatal porcine vena cava. *Life Sciences* **38,** 1521–8.

Turla, M. B. and Webb, R. C. (1987a). Enhanced vascular reactivity to protein kinase C activators in genetically hypertensive rats. *Hypertension* **9 (suppl III),** 150–4.

Turla, M. B. and Webb, R. C. (1987b). Increased vascular reactivity to phorbol ester in the mineralocorticoid hypertensive rat. *Federation Proceedings* **46,** 1289.

Van Nueten, J. M. (1982). Serotonergic amplification mechanisms in vascular tissues. In *5-Hydroxytryptamine in peripheral reactions* (eds. F. De Clerck and P. M. Vanhoutte), pp. 77–82. Raven Press, New York.

Van Nueten, J. M., Janssen, P. A. J., Van Beek, J., Xhonneux, R., Verbeuren, T. J., and Vanhoutte, P. M. (1981). Vascular effects of ketanserin (R 41 468), a novel antagonist of 5-HT_2 serotonergic receptors. *Journal of Pharmacology and Experimental Therapeutics* **218,** 217–30.

Vanhoutte, P. M. (1982a). Does 5-hydroxytryptamine play a role in hypertension? *Trends in Pharmacological Sciences* **3,** 370–3.

Vanhoutte, P. M. (1982b). 5-Hydroxytryptamine, vasospasm and hypertension. In *5-Hydroxytryptamine in peripheral reactions* (ed. P. M. Vanhoutte), pp. 163–74. Raven Press, New York.

Vanhoutte, P. M. (1983). Introductory remarks: Why 5-hydroxytryptamine? *Federation Proceedings* **42,** 211–12.

Vanhoutte, P. M. (1985). Peripheral serotonergic receptors and hypertension. In *Serotonin and the cardiovascular system* (ed. P. M. Vanhoutte), pp. 123–33. Raven Press, New York.

Vanhoutte, P. M., Verbeuren, T. J., and Webb, R. C. (1981). Local modulation of the adrenergic neuroeffector interaction in the blood vessel wall. *Physiological Reviews* **61,** 151–247.

Verbeuren, T. J., Jordaens, F. H., and Herman, A. G. (1983). Accumulation and release of [^{3}H]5-hydroxytryptamine in saphenous veins and cerebral arteries of the dog. *Journal of Pharmacology and Experimental Therapeutics* **226,** 579–88.

Watts, A. D., Feniuk, W., and Humphrey, P. P. A. (1981). A prejunctional action of 5-hydroxytryptamine and methysergide on noradrenergic nerves in dog isolated saphenous vein. *Journal of Pharmacy and Pharmacology* **33,** 515–20.

Webb, R. C. (1982). Increased vascular sensitivity to serotonin and methysergide in hypertension in rats. *Clinical Science* **63,** 73–5s.

Webb, R. C. (1984). Vascular changes in hypertension. In *Cardiovascular pharmacology* (ed. M. J. Antonaccio), pp. 215–55. Raven Press, New York.

Webb, R. C. and Bohr, D. F. (1984). The membrane of the vascular smooth muscle cell in experimental hypertension and its response to serotonin. In *Smooth muscle contractions* (ed. N. Stephens), pp. 485–508. Marcel Dekker, New York.

Webb, R. C. and Vanhoutte, P. M. (1985). Responsiveness of the hypertensive blood vessel wall to serotonergic stimuli. In *Serotonin and the cardiovascular system* (ed. P. M. Vanhoutte), pp. 113–21. Raven Press, New York.

Wenting, G. J., Man in't Veld, A. J., Wottiez Boomsma, F., and Schalekamp, M. A. O. M. (1982). Treatment of hypertension with ketanserin, a new selective 5-HT_2 receptor antagonist. *Clinical Research* **284,** 537–9.

Winquist, R. J. and Bohr, D. F. (1983). Structural and functional changes in cerebral arteries from spontaneously hypertensive rats. *Hypertension* **5,** 292–7.

15

The development and early clinical evaluation of selective 5-HT_3 receptor antagonists

J. R. FOZARD

15.1 Introduction

The remarkable recent advances in our understanding of 5-HT mechanisms and function (Fozard 1987a) reflect in large part the increasing availability of compounds with selectivity for individual 5-HT receptor subtypes (Bradley *et al.* 1986; Fozard 1987a). Of the latter, it is perhaps the 5-HT_3 receptor category (Bradley *et al.* 1986) where the most spectacular developments have occurred. Potent and selective antagonists at 5-HT_3 receptors have been developed by Merrell Dow (e.g. MDL 72222, MDL 72422), Sandoz (e.g. ICS 205-930), Beecham (e.g. BRL 24924, BRL 43694), and Glaxo (e.g. GR38032) (Table 15.1). Despite a genuine concern on the part of the present author to relate without bias the development of *all* these agents, for obvious reasons the article will deal mainly with the development of the Merrell Dow 5-HT_3 receptor antagonists and, in particular, with MDL 72222. Bias will be less evident in the section dealing with the clinical evaluation of these compounds since information is available for all the principal molecules in this context.

A brief account of the development of 5-HT_3 receptor antagonists has appeared (Fozard 1989); a certain overlap of the first part of the present chapter with this article has been inevitable.

15.2 The development of MDL 72222

15.2.1 5-HT and the isolated rabbit heart: definition of an atypical 5-HT receptor

The author's interest in 5-HT_3 receptors and their antagonists has its origin in a long-standing preoccupation with 5-HT and its cardiovascular effects (Fozard 1968) but is indisputably linked with a post-doctoral period spent in the laboratory of Professor Erich Muscholl of the University of Mainz. The introduction whilst in Mainz to the rabbit heart as a particularly useful model for the evaluation of drugs acting prejunctionally at the autonomic neuroeffector junction (Fozard and Muscholl 1972) led directly to a

Table 15.1 The principal selective 5-HT_3 receptor antagonists and their activities at the excitatory neuronal 5-HT receptors on afferent, autonomic and enteric neurones

Compound	Afferent (rabbit vagus) pA_2	Autonomic (rabbit heart) pA_2	Enteric (guinea-pig ileum) pA_2
MDL 72222	7.9[a]	9.3[b]	<6[b]
ICS 205–930	10.2[c]	10.6[c]	7.9[c]
BRL 24924	8.5[d]	8.9[e]	7.6[e]
BRL 43694	9.9[d]	10.7[f]	8.1[f]
GR38032	9.2[d]	10.1[d]	7.5[d]

Data sources: [a]Donatsch *et al.* (1984); [b]Fozard (1984a); [c]Richardson *et al.* (1985); [d]Buchheit *et al.* (personal communication); [e] Sanger (1987a); [f]Fake *et al.* (1987)

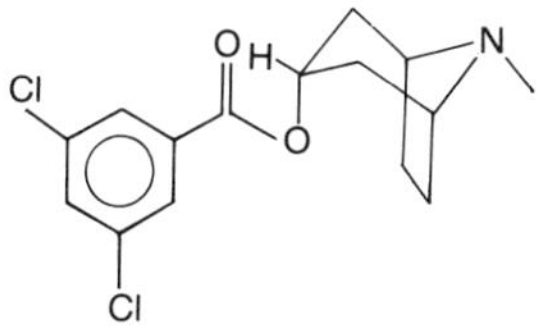

MDL 72222

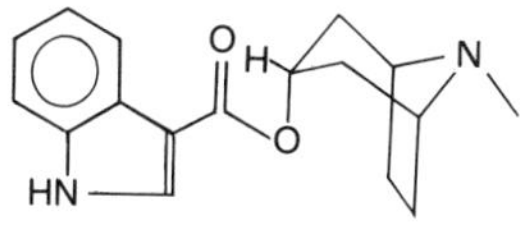

ICS 205–930

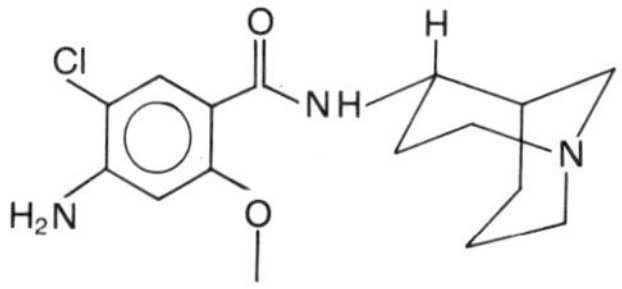

BRL 24924

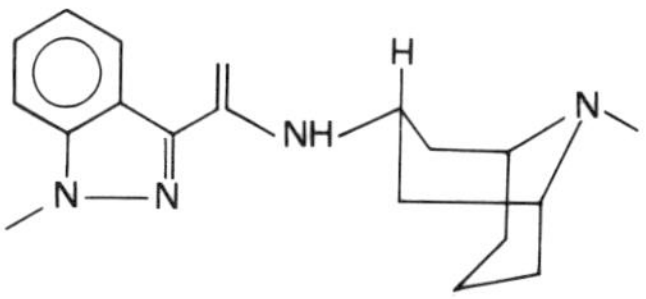

BRL 43694

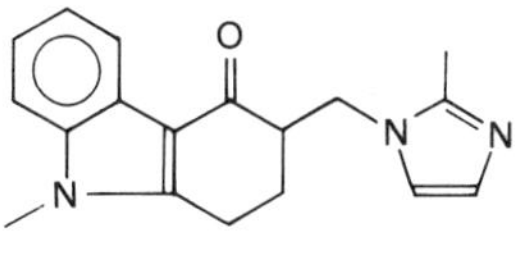

GR38032

detailed analysis of the indirect sympathomimetic effects of 5-HT on this preparation. This work, which formed the Ph.D. project of Dr Gabriel Mwaluko at the University of Manchester, confirmed the entirely indirect sympathomimetic effect of 5-HT, established the terminals of the sympathetic nerves as the exclusive source of the noradrenaline released, and provided strong evidence for receptor-mediated depolarization being the principle mechanism of transmitter release (Fozard and Mwaluko 1976). However, the putative receptor could not be defined as either D or M according to the accepted reference classification of the day (Gaddum and

Picarelli 1957) since neither high concentrations of morphine nor methysergide inhibited selectively the response to 5-HT (Fozard and Mwaluko 1975; Table 15.2).

In 1975, Dr Abu T. M. Mobarok Ali took over as his Ph.D. project the task of identifying and defining the atypical neuronal 5-HT receptor of the rabbit heart. Since no leads for putative antagonists were available, the initial strategy was to characterize the receptor in terms of the relative potency of selective agonists. For reference, the guinea-pig ileum was used, set up in either its D receptor mode (i.e. with the M receptor responses suppressed by the continual presence of atropine, 33 μM, and morphine, 23 μM) or the M receptor mode (in the presence of methysergide, 2.8 μM). The results of these studies established the total dissimilarity between the neuronal receptor of the rabbit heart and the D receptor; moreover, although certain similarities to the M receptor were evident there were sufficient major discrepancies to conclude with some confidence that the sites were not identical (Fozard and Mobarok Ali 1978a). It was nevertheless accepted that convincing evidence of a novel 5-HT receptor would require the development of a selective and surmountable antagonist and efforts were therefore directed towards this goal.

15.2.2 Antagonist lead finding: (−)-cocaine and metoclopramide

An early strategy in the search for selective antagonists at the 5-HT receptor present on the sympathetic neurones of the rabbit heart was to screen a number of compounds with activities at a variety of 5-HT receptor or acceptor (uptake, enzyme) sites. Potencies for blockade of 5-HT were defined and selectivity was assessed by measuring potency to inhibit transmitter release evoked by the nicotine receptor agonist, dimethylphenylpiperazinium. Full details of the methodology appear in Fozard *et al.* (1979). As can be seen from Table 15.2, the strategy did prove successful; several compounds had activity (albeit weak) against 5-HT *and* were selective for 5-HT *vis-à-vis* dimethylphenylpiperazinium. Of these, it was (−)-cocaine which eventually proved the most fruitful lead.

It has to be admitted that the decision to use (−)-cocaine had been taken with no great enthusiasm. Although it had been used by Gaddum and Picarelli in their original work on M receptors, it was tested by us more on the basis of rounding out a Ph.D. project than with any real expectation of achieving *selective* antagonism of 5-HT. The compound was, after all, a potent local anaesthetic; moreover, it had earlier been shown to be a powerful *potentiator* of 5-HT on the cardiovascular system of the rat (Fozard and Leach 1969)! In the event, (−)-cocaine not only proved to be selective as an antagonist of 5-HT but also surmountable and with a potency (pA_2 6.24) incompatible with its acting as a local anaesthetic (concentrations > mmol/1 required) (Fozard *et al.* 1979). The great advan-

Table 15.2A The route to the development of MDL 72222[a]

Compound	Antagonist potency[b] vs 5-HT (A)	*n*	DMPP (B)	*n*	Selectivity[c] (A–B)
'Classical' antagonists					
Morphine	4.7±0.1	3	4.4±0.1	3	0.33
Methysergide	4.7±0.1	3	4.7±0.1	3	0.05
LSD	6.4±0.1	4	6.2±0.2	3	0.24
Picrotoxin	5.5±0.1	4	4.5±0.1	3	1.04
Phenylbiguanide	5.5±0.1	3	4.1±0.2	3	1.40
Atropine	4.8±0.1	3	4.6±0.1	3	0.20
(−) Cocaine	6.2±0.1	4	4.9±0.1	3	1.29
'Acceptor' site ligands					
Desipramine	5.2±0.2	3	6.1±0.1	3	−0.91
Amitriptyline	5.3±0.1	3	5.4±0.1	3	−0.16
Mianserin	5.9±0.1	3	5.3±0.1	3	−0.65
Femoxetine	5.6±0.2	4	5.6±0.1	3	−0.02
Clorgyline	6.9±0.1	5	<4.8	3	>2.10
Cocaine analogues					
Nor-(−)-cocaine	7.7±0.1	3	5.0	1	2.68
(+)-Pseudococaine	6.9±0.1	3	5.0±0.1	3	1.88
Tropine benzoate	7.2±0.1	4	5.1±0.1	3	2.14
Pseudotropine benzoate	7.0±0.1	4	4.8±0.1	3	2.13
Local anaesthetics					
Lignocaine	3.9±0.1	3	4.2±0.1	3	−0.29
Procaine	5.6±0.1	3	4.1±0.1	3	1.45
Procainamide	5.0±0.1	3	4.1±0.1	3	0.86
Metoclopramide	7.3±0.1	3	4.5±0.1	3	2.75
MDL 72222 and analogues					
Tropine 3-chlorobenzoate	8.6±0.1	4			
Tropine 3,5-dichlorobenzoate (MDL 72222)	9.3±0.1	7	6.1±0.1	6	3.13
Tropine 3-methylbenzoate	8.2±0.4	3			
Tropine 3,5-dimethylbenzoate (MDL 72422)	9.0±0.1	3	6.1±0.2	4	2.94

[a]Screening data from the Langendorff-perfused rabbit heart model; for details of methodology, see Fozard *et al.* (1979). Details of other analogues of (−)-cocaine and MDL 72222 can be found in Fozard *et al.* (1979), Fozard and Gittos (1983) and Fig. 15.2.
[b]Antagonist potency estimated by the method of Schild (1947). The figures represent the negative logarithm of the molar concentration of antagonist which reduced the effects of twice the ED50 of agonist to that of the ED50 in the absence of antagonist. Mean values (±SEM) of *n* individual experiments are presented.
[c]Selectivity is represented by the difference between the antagonist potency against 5-HT and that against dimethylphenylpiperazinium (DMPP).

Table 15.2B Structural formulae of compounds listed in Table 15.2A

'Classical' antagonists

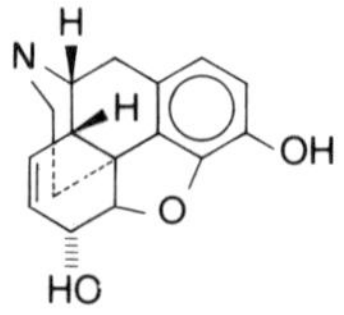

Morphine

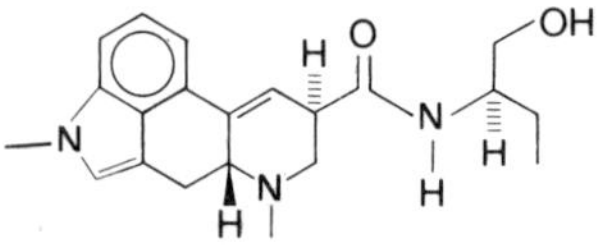

Methysergide

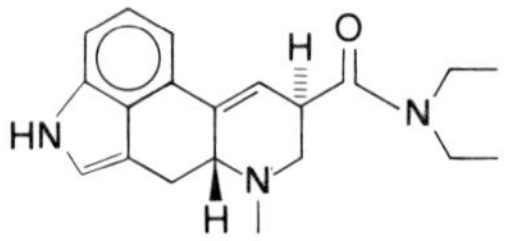

LSD

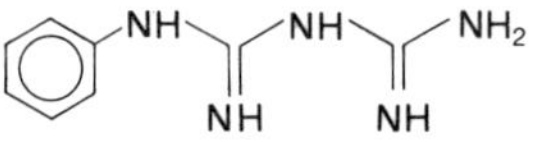

Phenylbiguanide

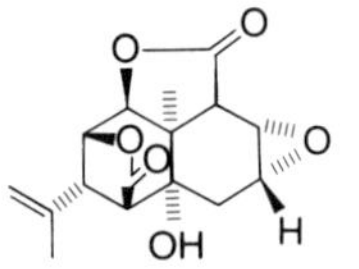

Picrotixinin

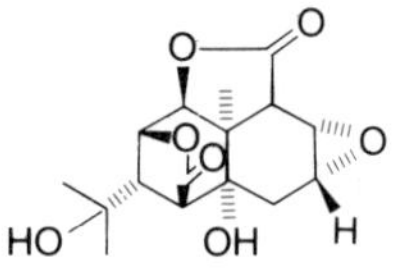

Picrotin

Picrotoxinin + *Picrotin* = *Picrotoxin*

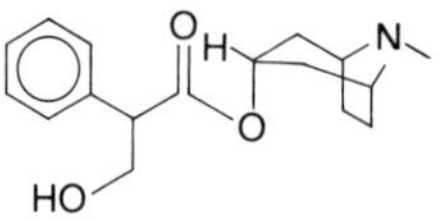

Atropine

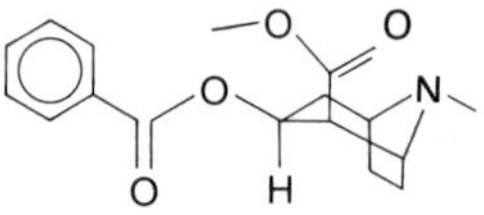

(–)–Cocaine

'Acceptor' site ligands

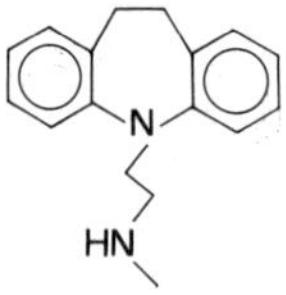

Desipramine

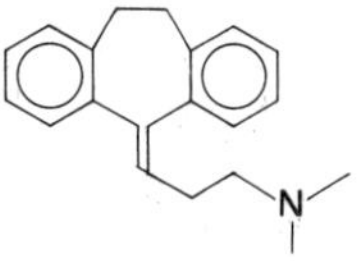

Amitriptyline

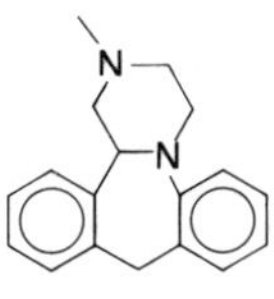

Mianserin

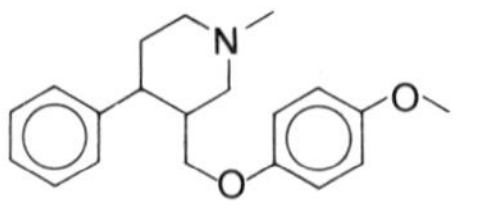

Femoxetine

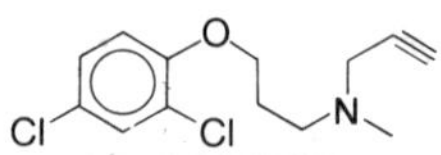

Clorgyline

Cocaine analogues

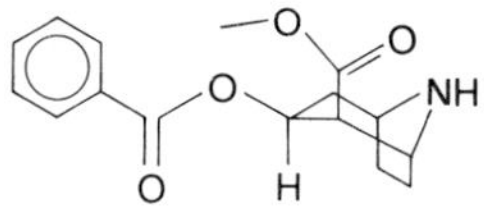

Nor-(−)-cocaine

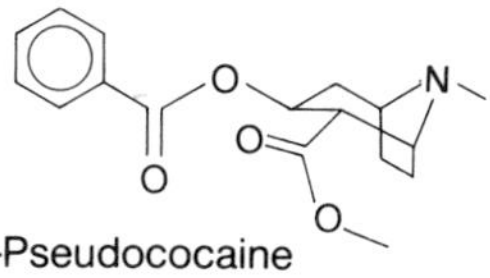

(+)-Pseudococaine

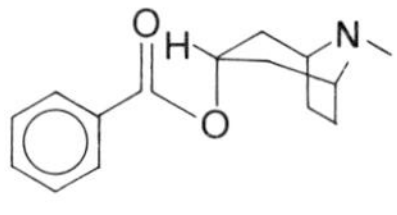

Tropine benzoate

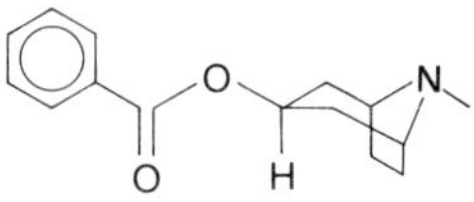

Pseudotropine benzoate

Local anaesthetics

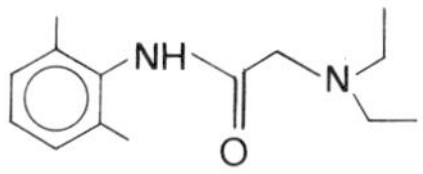

Lignocaine

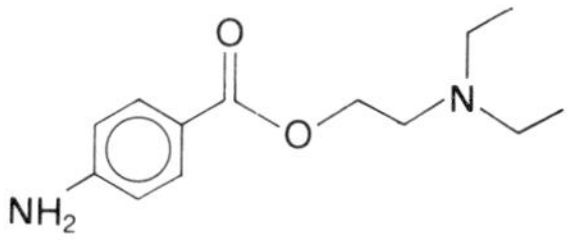

Procaine

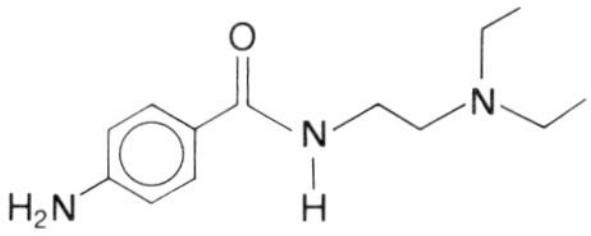

Procainamide

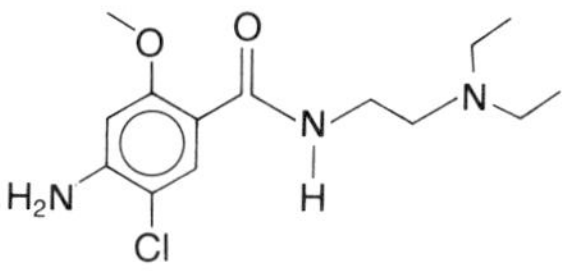

Metoclopramide

MDL 72222 and analogues

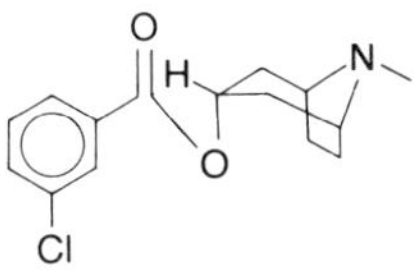

Tropine 3-chlorobenzoate

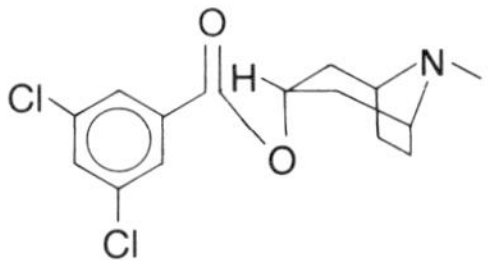

Tropine 3,5-dichlorobenzoate
(MDL 72222)

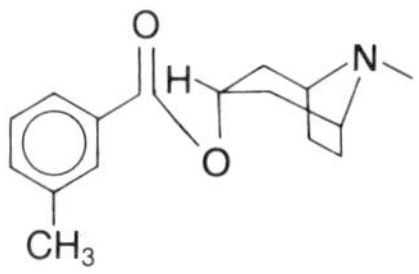

Tropine 3-methylbenzoate

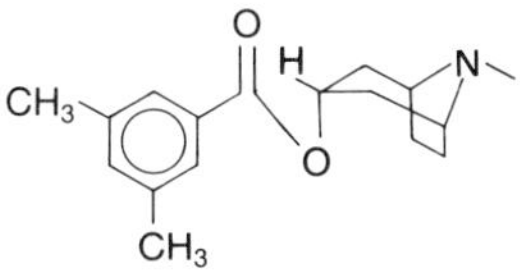

Tropine 3,5-dimethylbenzoate

tage of (−)-cocaine as a lead (over picrotoxin or clorgyline, for example) to an academic pharmacology laboratory was that sufficient related compounds could be obtained from commercial sources or as gifts to allow meaningful structure–activity relationship studies to be carried out. Such studies established the key structural features of a series of tropine or pseudotropine derivatives for potency and selectivity at the sympathetic neuronal 5-HT_3 receptors of the rabbit heart (Table 15.2; Fozard *et al.* 1979). Nor (−)-cocaine proved an especially potent compound (Table 15.2) and was used to provide the first *antagonist* evidence for the 5-HT_3 receptor of the rabbit heart (pA_2 7.79) being different from the M receptor of the ileum (pA_2 6.49) (Fozard 1980 and 1983).

From the control experiments carried out to verify that local anaesthesia was indeed playing no role in the mechanism of action of (−)-cocaine came a second important advance. Certain local anaesthetics (e.g. lignocaine, tetracaine, benzocaine and butacaine) were active only at high concentrations and showed selectivity for dimethylphenylpiperazinium (Fozard *et al.* 1979); others (e.g. procaine and procainamide) were active at concentrations well below those causing local anaesthesia and showed selectivity for 5-HT (Table 15.2). These data were presented at an internal seminar at the University of Manchester in 1976 and during discussion the similarity between procainamide and metoclopramide, a substituted benzamide with dopamine D_2 receptor blocking activity, was pointed out. Subsequent testing of metoclopramide revealed it to be a potent, surmountable, and entirely silent antagonist of the sympathetic neuronal 5-HT_3 receptor of the rabbit heart (Fozard and Mobarok Ali 1978b; Table 15.2).

15.2.3 The synthesis of MDL 72222 and its pharmacological evaluation

The present author joined the Merrell International Research Centre (now the Merrell Dow Research Institute) at Strasbourg in July 1977. The emerging pharmacophore based on data obtained with the analogues of cocaine and structures similar to metoclopramide was discussed and chemical synthesis initiated under the direction of Dr Maurice Gittos. The criteria set at the outset were for a compound with a pA_2 on the rabbit heart of >9.0, a selectivity of at least 1000 *vis-à-vis* dimethylphenylpiperazinium and, of course, good oral bioavailability. The projected clinical utility was the treatment of migraine (see Section 15.4.1). It was also recognized that the availability of a compound meeting the above criteria would afford a unique opportunity of exploring the physiological and possibly the pathophysiological role(s) of 5-HT_3 receptors. MDL 72222 was one of several substituted benzoic acid esters of tropine with high affinity and selectivity for 5-HT_3 receptors *in vitro* (Fozard and Gittos 1983) (see Fig. 15.1 and Table 15.2).

The development of any drug requires the availability of reliable models to quantify its principal activity *in vivo*. Since 5-HT induces a multitude of reflex effects *in vivo*, it was logical to seek to exploit this activity in the design of suitable tests for 5-HT$_3$ receptor antagonists. The von Bezold-Jarisch effect of 5-HT in the rat provided the basis for a simple and reproducible animal model (Fozard 1982; Fozard and Host 1982; Fozard 1984a; Fig. 15.1) and there was a good correlation between the potency of a number of 5-HT$_3$ receptor antagonists to inhibit the chronotropic effects of 5-HT on the rabbit heart and the von Bezold-Jarisch effect *in vivo* (Fig. 15.2). MDL 72222 proved a selective and particularly potent antagonist of the von Bezold-Jarisch effect of 5-HT, responses being suppressed by 0.1–0.3 mg/kg i.v. (Fozard 1984a) and for >4 h following 1.25 mg/kg p.o. (Fig. 15.1). Although at the time it was both surprising and somewhat disappointing to find that complete blockade of 5-HT$_3$ receptors did not result in overt pharmacological activity, MDL 72222 more than adequately satisfied the pre-set criteria and was chosen for clinical development.

The test developed for the evaluation of 5-HT$_3$ receptor blockade in man was based on the fact that intradermal injections of 5-HT induce a flare response which is believed to reflect axon-reflex vasodilatation (Greaves and Schuster 1967; *vide infra*). The response is dose-related, easily quantified, and allowed confirmation in man that a dose of approximately 0.3 mg/kg of MDL 72222 given intravenously was sufficient to markedly suppress the consequences of 5-HT$_3$ receptor activation (Orwin and Fozard 1986). Since the present author was one of the subjects in this study, confirmation was obtained on a highly personal basis that substantial blockade of peripheral 5-HT$_3$ receptors does not result in overt pharmacological activity!

15.3 The development of other 5-HT$_3$ receptor antagonists

Richardson *et al.* (1985) and Richardson and Engel (1986) have described in some detail the approach which led to the development of ICS 205-930 and its analogues at Sandoz. The starting point in this case was the 5-HT molecule itself, by analogy with the approach used for the development of β-adrenoceptor and histamine H$_2$ antagonists from derivation of the respective natural agonists. Systematic methyl substitutions in the indole nucleus and ethylamine side chain of 5-HT led first to 2-methyl-5-HT and S(+)α-methyl-5-HT, agonists with selectivity for 5-HT$_3$ and 5-HT$_2$ receptors, respectively. Extending the ethylamine side chain and incorporating the terminal nitrogen in a tropine or homotropine ring system, yielded a series of potent, selective, and surmountable antagonists at 5-HT$_3$ receptors, including ICS 205-930 (Richardson *et al.* 1985; Table 15.1).

Little information is available on the development of the Beecham and

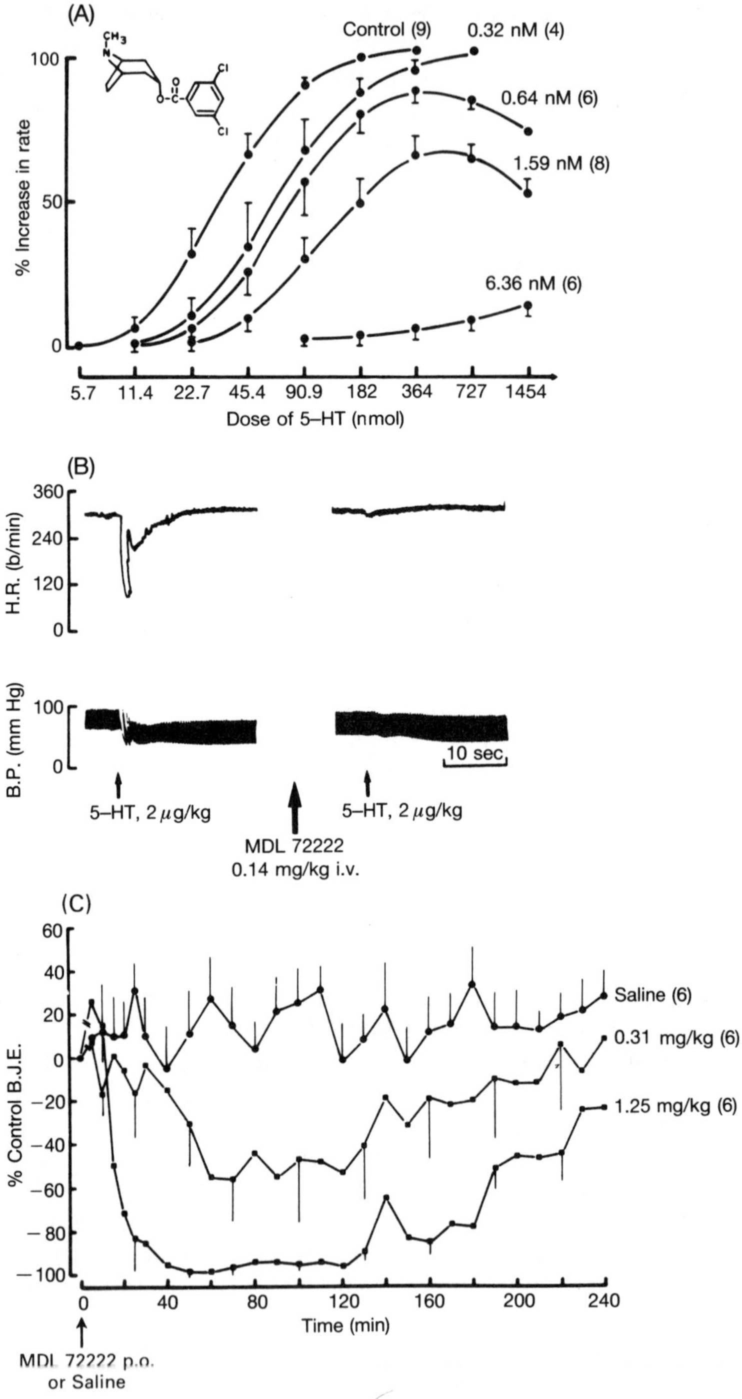
(A)
CH_3
Cl
Cl
Control (9)
0.32 nM (4)
0.64 nM (6)
1.59 nM (8)
6.36 nM (6)
% Increase in rate
100
50
0
5.7
11.4
22.7
45.4
90.9
182
364
727
1454
Dose of 5–HT (nmol)
(B)
H.R. (b/min)
360
240
120
0
B.P. (mm Hg)
100
0
10 sec
5–HT, 2 µg/kg
MDL 72222
0.14 mg/kg i.v.
5–HT, 2 µg/kg
(C)
% Control B.J.E.
60
40
20
0
−20
−40
−60
−80
−100
Saline (6)
0.31 mg/kg (6)
1.25 mg/kg (6)
0
40
80
120
160
200
240
Time (min)
MDL 72222 p.o.
or Saline

Glaxo 5-HT_3 receptor antagonists. Chemically, however, BRL 24924 clearly reflects the heritage of a company with a long-standing interest in the mechanism of action of metoclopramide, and BRL 43694 retains the amide bridge of metoclopramide in a molecule having structural similarities to both the Merrell Dow and Sandoz compounds. GR38032 is structurally quite dissimilar to any of the other classes of 5-HT_3 receptor antagonists (Table 15.1).

15.4 Actions of 5-HT_3 receptor antagonists in the periphery: the basis of clinical utility

Details of the location of 5-HT_3 receptors on peripheral autonomic, afferent, and enteric neurones, and the excitatory responses generated in those neurones when 5-HT_3 receptors are activated, have been repeatedly reviewed (Wallis 1981; Fozard 1984b and 1986; Richardson and Engel 1986; Richardson and Buchheit 1988; see also Chapters 10 and 11). Thus, only the properties of these sites which can be considered the basis of the early clinical evaluation of these compounds will be emphasized here; these are afferent neuronal 5-HT_3 receptor stimulation as the basis both of pathophysiological pain (e.g. migraine) and vomiting induced by cancer chemotherapy and radiotherapy and gastro-intestinal 5-HT_3 receptors as the basis of gastrokinesis and certain features of the carcinoid syndrome.

15.4.1 5-HT_3 receptors mediating pain in man: rationale for the use of 5-HT_3 receptor antagonists in migraine

5-HT is *per se* a potent pain-producing substance in man either when applied to a blister base or injected intradermally (Armstrong *et al.* 1953;

Fig. 15.1 The pharmacological evaluation of MDL 72222. A: The effects of MDL 72222 on the changes in rate of rabbit isolated hearts evoked by bolus injections of 5-HT. Dose response curves were established (control) and repeated after 15 min perfusion with MDL 72222 at the concentrations shown. The points represent the mean values with SEM of the number of individual determinations shown in parentheses (from Fozard 1984a with permission). B: Selective blockade of the Bezold-Jarisch component of the cardiovascular response to 5-HT in the anaesthetized rat. The second 5-HT injection was given 8 min after the first and 3 min after MDL 72222 (from Fozard 1984a with permission). C: Potency and duration of action of MDL 72222 given orally as an inhibitor of the Bezold-Jarisch effect of 5-HT in the conscious rat. A single submaximal dose of 5-HT (usually 2 μg/kg) was injected every 5 min. Saline or MDL 72222 at the doses indicated were administered as indicated. Points represent the mean values with SEM of the number of individual values shown in parentheses

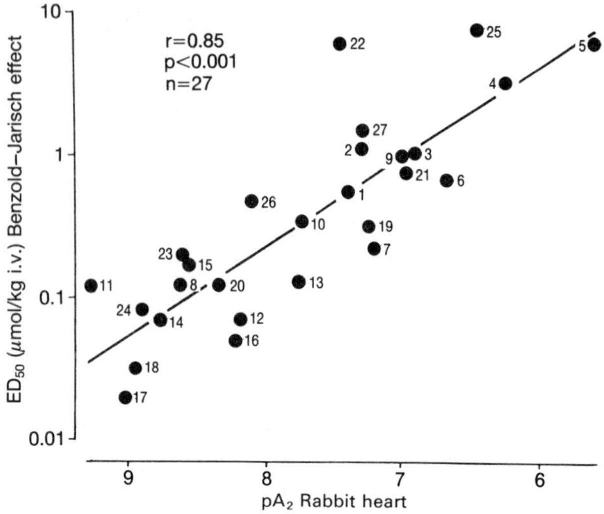

Fig. 15.2 Correlation between potency of 5-HT_3 receptor antagonists to inhibit the chronotropic effects of 5-HT on rabbit isolated heart and the Bezold-Jarisch effect induced by 5-HT in the anaesthetized rat. Points represent the means of 3–7 individual observations. pA_2 values obtained by the method described in detail in Fozard (1984a)

1. Metoclopramide methiodide
2. Metoclopramide
3. (+)-Cocaine
4. (−)-Cocaine
5. Procaine
6. Nortropacocaine
7. Tropine benzoate
8. Tropine 3-chlorobenzoate
9. Tropine 4-chlorobenzoate
10. Tropine 3,4-dichlorobenzoate
11. Tropine 3,5-dichlorobenzoate (MDL 72222)
12. Tropine 3-methylbenzoate
13. Tropine 4-methylbenzoate
14. Tropine 2,4-dimethylbenzoate
15. Tropine 2,5-dimethylbenzoate
16. Tropine 3,4-dimethylbenzoate
17. Tropine 3,5-dimethylbenzoate (MDL 72422)
18. Tropine 3,4,5-trimethylbenzoate
19. Tropine 3,4-dimethoxybenzoate
20. Tropine 3,5-dimethoxybenzoate
21. Pseudotropine benzoate
22. Pseudotropine 3,4-dichlorobenzoate
23. Pseudotropine 3,5-dichlorobenzoate
24. Pseudotropine 3,5-dimethylbenzoate
25. Pseudotropine 4-fluorobenzoate
26. N-Methyl 4-hydroxypiperidine 3,5-dimethylbenzoate
27. N-Methyl 5-hydroxypiperidine 3,5-dimethylbenzoate

Lindahl 1961; Sicuteri *et al.* 1965; Greaves and Schuster 1967; Richardson *et al.* 1985). In addition, 5-HT powerfully enhances the algogenic response to nociceptive agents such as bradykinin (Sicuteri *et al.* 1965; Bleehen and Keele 1977; Richardson *et al.* 1985). Pain is accompanied by a cutaneous flare response indicative of axon reflex vasodilatation via the release of neurotransmitter(s) from the cutaneous sensory nerve endings.

Recent experiments have established that 5-HT$_3$ receptors mediate both the pain-producing and the pain-enhancing effects of 5-HT and the axon reflex vasodilatation. For instance, in the studies of Richardson *et al.* (1985), pain arising from application of 5-HT to the blister base was blocked surmountably and selectively by local application of ICS 205-930; in contrast, the combined '5-HT$_1$-like'/5-HT$_2$ receptor antagonist, methysergide (Bradley *et al.* 1986), was ineffective at the relatively high concentration of 10^{-6} M. Consistent with a role for 5-HT$_3$ receptors, the highly selective 5-HT$_3$ receptor agonists, phenylbiguanide (Fastier *et al.* 1959) and 2-methyl-5-HT (Richardson *et al.* 1985) are potent algogens in this model. That 5-HT$_3$ receptors are also involved in the potentiation of bradykinin by 5-HT is indicated by the fact that the response is abolished in the presence of ICS 205-930 (Richardson *et al.* 1985). With respect to axon reflex vasodilatation, Orwin and Fozard (1986) showed, in a double-blind study, that a low dose (20 mg) of MDL 72222 given intravenously attenuated the flare response to 5-HT injected intradermally in the forearm. Essentially similar results were obtained by Heavey *et al.* (1986) who gave MDL 72422 orally at doses of 20, 40, and 80 mg and observed marked, dose-dependent suppression of the flare induced by 5-HT injected intradermally in the back. Cooper *et al.* (1988) infused BRL 43694, 40 μg/kg in 30 min, into eight volunteers and observed significant reductions of the axon reflex flare induced in the back for up to 24 hours.

The fact that 5-HT could induce pain and axon reflex vasodilatation was the basis of a hypothesis implicating 5-HT in the pathophysiology of migraine headache (Fozard 1982 and 1985). In brief, it was suggested that 5-HT released locally in the blood vessels of the head would contribute both to the putative 'sterile' inflammatory response of migraine and, in particular, to the headache, by activating 5-HT$_3$ receptors on pain fibres innervating key elements of the cranial vasculature. The rationale for an antagonist at these sites being an effective treatment for the acute attack is obvious and was the basis for the clinical evaluation of MDL 72222 in the condition (Fozard *et al.* 1985; Loisy *et al.* 1985).

In the event, MDL 72222 given by intravenous injection proved an effective acute treatment both in a pilot open study and under placebo-controlled double-blind conditions. In the open trial (Fozard *et al.* 1985), 13 out of 21 patients claimed complete relief of symptoms following 20 mg of MDL 72222, and the response was significantly greater than that obtained following 10 mg (4 out of 16). In the placebo-controlled, double-blind trial (Loisy *et al.* 1985), doses of 20–40 mg of MDL 72222 brought more than 75 per cent relief of symptoms in 15 out of 24 patients, a significantly better result than was obtained with placebo (more than 75 per cent relief in 5 out of 23 patients). Unfortunately, toxicity in animals, unconnected to 5-HT$_3$ receptor blockade, led to suspension of the development of MDL 72222 in migraine and prevented consolidation of these promising findings.

The evidence for 5-HT_3 receptor blockade being the basis of the beneficial effect of MDL 72222 in migraine has been discussed in detail previously (Fozard *et al.* 1985). It bears emphasis that the doses used (10–40 mg) are close to that (20 mg) which markedly suppressed the consequences of peripheral 5-HT_3 receptor activation in man (Orwin and Fozard 1986; see Section 15.2.3). Clearly, however, the results from the clinical evaluation of other selective 5-HT_3 receptor antagonists in the condition (see Fozard 1987a) will be important in determining the potential or otherwise of this particular mechanistic approach to the treatment of migraine.

15.4.2 5-HT_3 receptors mediating vagal afferent neuronal stimulation: role in cytotoxic drug and radiation-induced vomiting

It should be made clear at the outset that knowledge of the presence of 5-HT_3 receptors on peripheral vagal afferent neurones was not *per se* the basis of the discovery of the anti-emetic effects of selective 5-HT_3 receptor antagonists. The key observations, made independently by scientists at Beecham (Miner and Sanger 1986) and the University of Bradford (Costall *et al.* 1986a), arose from experiments aimed at clarifying aspects of the anti-emetic action of metoclopramide. Metoclopramide is one of several dopamine D_2 receptor antagonists which are used clinically to control nausea and vomiting from a variety of causes (Harrington *et al.* 1983; McRitchie *et al.* 1984; Ison and Peroutka 1986). However, metoclopramide appears to be exceptional amongst such agents in affording protection against vomiting induced by cancer chemotherapeutic agents, albeit at high intravenous doses (Gralla 1983; see also Miner *et al.* 1987). On this basis, McRitchie *et al.* (1984) speculated that factors other than dopamine D_2 receptor blockade might be involved in suppressing vomiting to the strongly emetic cytotoxic agents.

One obvious candidate for an additional anti-emetic mechanism of metoclopramide was 5-HT_3 receptor blockade which had been clearly demonstrated both *in vitro* on the isolated rabbit heart (Fozard and Mobarok Ali 1978b; see Table 15.2) and *in vivo* as blockade of the von Bezold-Jarisch effect (Fozard 1982; Fozard and Host 1982). Moreover, 5-HT_3 receptor blockade had already been implicated in the enhancement of caudal-directed gastro-intestinal motility (Buchheit *et al.* 1985a) which could, in theory at least, oppose the retroperistalsis evoked by chemotherapeutic agents (Akwari 1983). The availability of MDL 72222 (Fozard 1984a) and ICS 205-930 (Richardson *et al.* 1985), with their exceptional potency and selectivity for 5-HT_3 receptors, provided the means by which the role of these sites might be explored. In the event, both MDL 72222 (Miner and Sanger 1986) and ICS 205-930 (Costall *et al.* 1986a and 1987a) proved potent inhibitors of emesis induced by cisplatin in the ferret, thus clearly implicating 5-HT$_3$ receptor blockade in the mechanism of action of metoclopramide (see also Fozard 1987b). Significantly, selective 5-HT$_3$ receptor antagonists were not effective inhibitors of vomiting induced by dopamine receptor agonists such as 2-di-*n*-propylamino-5,6-dihydrotetralin or apomorphine (Costall *et al.* 1987a; Miner *et al.* 1987).

The capacity of all the principal 5-HT$_3$ receptor antagonists to suppress cisplatin vomiting in the ferret has now been documented (Table 15.3). In general, their relative potency as anti-emetic agents accords with their

Table 15.3 Effects of 5-HT$_3$ receptor antagonists on vomiting induced by cisplatin (10 mg/kg i.v.) in the ferret

	Dose range for suppression of vomiting induced by cisplatin (mg/kg i.v.)	Reference
Metoclopramide	1 – 2	a,c,e,f
MDL 72222	0.1 – 1	d
ICS 205–930	0.01 – 0.1	a,f
BRL 24924	1.3 – 2.5	c
BRL 43694	<0.01	b
GR38032	0.01 – 1	e

[a]Costall *et al.*; [b]Boyle *et al.* (1987); [c]Miner *et al.* (1987); [d]Miner and Sanger (1986); [e]Stables *et al.* (1987); [f]Costall *et al.*

5-HT$_3$ receptor antagonist activities (see Table 15.1) which reinforces the conclusion that the latter is the cause of the former. Moreover, the principle seems to hold for other severe emetic stimuli. Thus, both BRL 24924 and BRL 43694 suppress vomiting induced by a doxorubicin/cyclophosphamide combination and X-radiation (Andrews and Hawthorn 1987; Boyle *et al.* 1987; Miner *et al.* 1987). Similarly, GR38032 inhibits emesis evoked by cyclophosphamide or X-radiation (Andrews *et al.* 1986). It bears emphasis, too, that the effects are not species selective since MDL 72222 is a potent inhibitor of cisplatin emesis in the dog (Berga *et al.* 1986).

These results established a key role for 5-HT$_3$ receptors in certain severe emetogenic stimuli and pointed clearly to selective antagonism at these sites as a novel and potentially more selective approach to the control of emesis associated with cancer chemotherapy and radiotherapy (see Fozard 1987b). The first clinical results with such compounds have now appeared and they appear to bear out fully the preclinical promise. In an open pilot study in 11 patients, ICS 205-930 had a protective effect against a range of cytotoxic compounds when given intravenously at the time of chemotherapy; in 31 out of 47 courses of chemotherapy no vomiting or retching occurred following total daily doses of 20 or 40 mg (Leibundgut and Lancranjan 1987). In the study of Cunningham *et al.* (1987), 15 patients with a history of chemotherapy-induced vomiting unresponsive to anti-emetic treatment, were treated with GR38032, 4 mg intravenously and 4 mg orally at the time of chemotherapy, the oral dose being repeated 5 and 10 h later. Only one patient vomited during a total of 32 courses of chemotherapy. Clearly, these preliminary findings must be confirmed under double-blind conditions but the data do suggest a key role for 5-HT_3 receptors in emesis induced by cytotoxic drugs in man.

Although the exact site(s) of action of 5-HT_3 receptor antagonists as anti-emetic agents remains to be established, a role for the 5-HT_3 receptors on vagal afferent neurones seems likely. Thus, the effects of 5-HT_3 receptor antagonists on cytotoxic drug and radiation-induced emesis in the ferret are mimicked closely by bilateral abdominal bivagotomy with or without splanchnectomy (Andrews *et al.* 1986; Andrews and Hawthorn 1987). The similarity extends to the fact that neither surgery nor drug treatment affects the ability of the animal to vomit in response to apomorphine (Andrews and Hawthorn 1987; Miner *et al.* 1987). Hence, a major factor in the vomiting induced by radiation and chemotherapy may be activation of vagal afferent fibres by 5-HT released from intestinal stores. Consistent with this, X-radiation is a powerful stimulus to 5-HT release from the small intestine of a number of species (Matsuoka *et al.* 1962). Further, clear increases in 5-HT and 5-hydroxyindoleacetic acid concentrations have been reported in the ileal mucosa of ferrets following treatment with cisplatin (Gunning *et al.* 1987). However, direct evidence for the presence of 5-HT_3 receptors in brain has recently been obtained (Kilpatrick *et al.* 1987; Peroutka and Hamik 1988; Watling *et al.* 1988) and it remains possible that activity at a central site contributes to anti-emetic efficacy. However, since 5-HT_3 receptor blockade does not suppress emesis evoked by apomorphine any such site must clearly be proximal to the outputs from the chemoreceptor trigger zone responsible for vomiting.

15.4.3 5-HT_3 receptors in the gastro-intestinal tract: role in gastrokinesis

5-HT activates a number of different receptor subtypes in the gastro-intestinal tracts of animals and man (Garattini and Valzelli 1965; Fozard 1984b and 1988; Bradley *et al.* 1986; Richardson and Buchheit 1988; see also Chapters 9 and 11). Applying the criteria of Bradley *et al.* (1986), 5-HT_3 receptors have at this stage only been conclusively demonstrated in guinea-pig myenteric plexus (Buchheit *et al.* 1985b) where they induce the release of substance P from (presumably) the afferent sensory nerve terminals (Chahl 1983; Buchheit *et al.* 1985b). The 5-HT_3 receptor should be clearly distinguished from a second excitatory neuronal 5-HT receptor, activation of which results in the release of acetylcholine from cholinergic nerve terminals within the myenteric plexus. These latter sites show particularly high affinity for 5-HT (Buchheit *et al.* 1985b), mediate the sustained increase in response of the electrical field-stimulated guinea-pig ileum in response to low concentrations of 5-HT (Kilbinger and Pfeuffer-Friedrich 1985; Sanger 1985, 1987a and b), and can neither be blocked nor activated by ICS 205-930 (Buchheit *et al.* 1985b) or MDL 72222 (Fozard 1984a). Metoclopramide and BRL 24924, in contrast, are potent *agonists* at the non-5-HT_3 excitatory neuronal 5-HT receptor, which accounts for why they and not compounds such as ICS 205-930 and MDL 72222 are able to enhance field-stimulation-evoked contractile responses of the guinea-pig ileum (Sanger 1987a and b).

In contrast to the situation in guinea-pig ileum, electrical field stimulation of guinea-pig isolated stomach muscle can be enhanced both by ICS 205-930 and (albeit to a lesser extent) by MDL 72222 at concentrations which block 5-HT_3 receptors in other test systems (Buchheit *et al.* 1985a; Gunning *et al.* 1986). Although the precise mechanism remains to be established, it is suggested that 5-HT released by electrical field stimulation would, by activating 5-HT_3 receptors, cause liberation of a transmitter inhibitory to the nerve-evoked response; disinhibition, as a result of 5-HT_3 receptor blockade, would result in an increase in twitch amplitude (Buchheit *et al.* 1985a).

From the above, it would perhaps be expected that 5-HT_3 receptor antagonists would increase gastric emptying *in vivo* and this is indeed the case. Low doses of metoclopramide, ICS 205-930, and GR38032 have been shown to increase gastric emptying in a dose-dependent manner in the guinea-pig (Buchheit *et al.* 1985a; Costall *et al.* 1987c) and both metoclopramide and BRL 24924 are potent and effective stimulants of gastric emptying in the rat (Cooper *et al.* 1986). Whilst these observations clearly implicate 5-HT_3 receptors in the control of gastric motility *in vivo*, the extent to which the capacity to stimulate the non-5-HT_3 excitatory neuronal receptors contributes to the effects of agents such as metoclopramide and BRL 24924 remains to be defined. Also to be defined is the extent to which 5-HT_3 receptors of the brain contribute to the gastrokinetic effects of such compounds. In the experiments of Costall *et al.* (1986b) lower doses of ICS 205-930 than were effective systemically, when given directly into the hypothalamus, increased gastric emptying in the guinea-pig.

There are, of course, many clinical data documenting the gastrokinetic properties of metoclopramide (see Harrington *et al.* 1983); by contrast, data on the newer, more selective 5-HT_3 receptor antagonists are just beginning to emerge. For instance, single doses of 1–5 mg BRL 24924 enhanced gastric emptying and reduced the oro-caecal transit time in normal volunteers (Staniforth and Corbett 1988). Doses of 1 mg/kg of BRL 24924 given orally twice daily for seven days to healthy volunteers increased liquid meal emptying (Rapeport and Thompson 1988). Of particular interest is the first clinical report in this context of the effects of a 5-HT_3 receptor antagonist lacking the interaction with the non-5-HT_3 excitatory neuronal receptor of the guinea-pig ileum. Akkermans *et al.* (1988) administered ICS 205-930 intravenously at doses of 10 and 20 mg to 12 normal subjects and monitored gastric emptying of a solid meal; both the lag time prior to the onset of gastric emptying and the time taken for the stomach to empty by 50 per cent were significantly reduced by ICS 205-930 treatment.

Although these data are preliminary, it does appear that selective 5-HT_3 receptor antagonists, with or without agonist activity at non-5-HT_3 excitatory neuronal 5-HT receptors, may have potential in clinical conditions associated with gastric stasis. Since two quite separate gastrokinetic principles may be involved, qualitatively different effects on gastro-intestinal motility might be expected from a differential activation of the two mechanisms. That BRL 24924 enhances liquid meal emptying in subjects with slow and fast emptying profiles (Rapeport and Thompson 1988), whereas metoclopramide mainly enhances emptying in subjects with a slow emptying profile (McClelland and Sutton 1986), could well reflect basically different mechanism(s) of action. One challenge both for the pharmaceutical industry and for clinical pharmacology will be to exploit to the full these differences.

15.4.4 5-HT_3 receptors in the gastro-intestinal tract: role in the symptomatology of the carcinoid syndrome

The principal disabling symptoms of the carcinoid syndrome are flushing and persistent diarrhoea which respond poorly, if at all, to treatment with classical 5-HT receptor antagonists such as methysergide or cyproheptadine (Welch and Malt 1977). With the advent of selective 5-HT_3 receptor antagonists, it was entirely reasonable that such compounds should be tried in a condition characterized by a massive and sustained overproduction of 5-HT. In a preliminary report, Anderson *et al.* (1987) treated three patients with carcinoid syndrome and two patients with metastatic vasoactive intestinal peptide secreting tumours (vipomas) with ICS 205-930, 20–30 mg given daily by slow intravenous injection for up to three days. Watery diarrhoea improved considerably in each of the three patients with carcinoid syndrome but not in the two patients with vipomas.

The now substantial evidence that 5-HT is capable of increasing intestinal fluid secretion and may play a physiological role in the process is discussed in detail in Chapter 12. Of particular interest is the observation that the secretory response is, to a large extent, neuronally mediated (Cassuto *et al.* 1982; Keast *et al.* 1985; Baird and Cuthbert 1987) and, in guinea-pig ileal and colonic epithelia at least, can be suppressed selectively by ICS 205-930

(Baird and Cuthbert 1987). The findings provide an experimental basis for the encouraging preliminary observations of Anderson *et al.* (1987) and suggest that 5-HT_3 receptor antagonist drugs will have an important role in the symptomatic treatment of watery diarrhoea in the carcinoid syndrome.

15.5 Concluding remarks

This review has described the development of a number of antagonists with high potency and quite remarkable selectivity for 5-HT_3 receptors and their progression from pharmacological curiosities to experimental drugs. The availability of these compounds has allowed definition of the 5-HT_3 receptor (Bradley *et al.* 1986), revealed the likely existence of 5-HT_3 receptor subtypes (Fozard 1986; Richardson and Engel 1986; see Chapter 10), and provided experimental data to justify clinical development in a variety of important indications (see above and Fozard 1987b). It bears emphasis that the exciting and diverse therapeutic potential of selective 5-HT_3 receptor antagonists became evident only when such compounds were available for evaluation. Certainly their clinical significance could not have been foreseen some 15 years ago, when the only stimulus to the development of such compounds was a largely academic interest in the mechanism of action of 5-HT on the rabbit isolated heart.

References

Akkermans, L. M. A., Vos, A., Hoekstra, A., Roelofs, J. M. M., and Horowitz, M. (1988). The effect of ICS 205-930 (a specific 5-HT_3 receptor antagonist) on gastric emptying of a solid meal in normal subjects. *Gut* **29,** 1249–52.

Akwari, O. E. (1983). The gastrointestinal tract in chemotherapy-induced emesis. A final common pathway. *Drugs* **25** (Suppl. 1), 18–34.

Anderson, J. V., Coupe, M. O., Morris, J. A., Hodgson, H. J. F., and Bloom, S. R. (1987). Remission of symptoms in carcinoid syndrome with a new 5-hydroxytryptamine M receptor antagonist. *British Medical Journal* **294,** 1129.

Andrews, P. L. R. and Hawthorn, J. (1987). Evidence for an extra-abdominal site of action for the 5-HT_3 receptor antagonist BRL 24924 in the inhibition of radiation-evoked emesis in the ferret. *Neuropharmacology* **26,** 1367–70.

Andrews, P. L. R., Hawthorn, J., and Sanger, G. J. (1986). The effect of abdominal visceral nerve lesions and a novel 5-HT-M receptor antagonist on cytotoxic and radiation induced emesis in the ferret. *Journal of Physiology (London)* **382,** 47P.

Armstrong, D., Dry, R. M. L., Keele, C. A., and Markham, J. W. (1953). Observations on chemical excitants of cutaneous pain in man. *Journal of Physiology (London)* **120,** 326–51.

Baird, A. W. and Cuthbert, A. W. (1987). Neuronal involvement in Type 1 hypersensitivity reactions in gut epithelia. *British Journal of Pharmacology* **92,** 647–55.

Berga, P., Bou, J., Llenas, J., Llupiá, J., and Massingham, R. (1986). Relacion entre los effectos antiemeticos de metoclopramida (MTC) y clebopride (CLB) y el antagonismo de los receptores de 5-HT. *Revista de Farmacologia Clinica y Experimental* **3,** 139.

Bleehen, T. and Keele, C. A. (1977). Observations on the algogenic actions of adenosine compounds on the human blister base preparation. *Pain* **3,** 367–77.

Boyle, E. A., Miner, W. D., and Sanger, G. J. (1987). Different anti-cancer therapies evoke emesis by mechanisms that can be blocked by the 5-HT_3 receptor antagonist BRL 43694. *British Journal of Pharmacology* **91,** 418P.

Bradley, P. B., Engel, G., Feniuk, W., Fozard, J. R., Humphrey, P. P. A., Middlemiss, D. N., Mylecharane, E. J., Richardson, B. P., and Saxena, P. R. (1986). Proposals for the classification and nomenclature of functional receptors for 5-hydroxytryptamine. *Neuropharmacology* **25,** 563–76.

Buchheit, K. H., Costall, B., Engel, G., Gunning, S. J., Naylor, R. J., and Richardson, B. P. (1985a). 5-Hydroxytryptamine receptor antagonism by metoclopramide and ICS 205-930 in the guinea-pig leads to enhancement of contractions of stomach muscle strips induced by electrical field stimulation and facilitation of gastric emptying in vivo. *Journal of Pharmacy and Pharmacology* **37,** 664–7.

Buchheit, K. H., Engel, G., Mutschler, E., and Richardson, B. (1985b). Study of the contractile effect of 5-hydroxytryptamine (5-HT) in the isolated longitudinal muscle strip from guinea-pig ileum. Evidence for two distinct release mechanisms. *Naunyn-Schmiedeberg's Archives of Pharmacology* **332,** 8–15.

Cassuto, J., Jodal, M., Tuttle, R., and Lundgren, O. (1982). 5-Hydroxytryptamine and cholera secretion: physiological and pharmacological studies in cats and rats. *Scandinavian Journal of Gastroenterology* **17,** 695–703.

Chahl, L. A. (1983). Substance P mediates atropine-sensitive response of guinea-pig ileum to serotonin. *European Journal of Pharmacology* **87,** 485–9.

Cooper, S. M., McClelland, C. M., McRitchie, B., and Turner, D. H. (1986). BRL 24924: A new and potent gastric motility stimulant. *British Journal of Pharmacology* **88,** 383P.

Cooper, S. M., Arnold, B. D. C., and Rapeport, W. G. (1988). Inhibition of 5-HT induced axon-reflex flares by BRL 43694 a novel 5-HT_3 receptor antagonist. *British Journal of Clinical Pharmacology* **25,** 106P–7P.

Costall, B., Domeney, A. M., Naylor, R. J., and Tattersall, F. D. (1986a). 5-Hydroxytryptamine M-receptor antagonism to prevent cisplatin-induced emesis. *Neuropharmacology* **25,** 959–61.

Costall, B., Kelly, M. E., Naylor, R. J., Tan, C. C. W., and Tattersall, F. D. (1986b). 5-Hydroxytryptamine M-receptor antagonism in the hypothalamus facilitates gastric emptying in the guinea-pig. *Neuropharmacology* **25,** 1293–6.

Costall, B., Domeney, A. M., Naylor, R. J., and Tattersall, F. D. (1987a). Emesis induced by cisplatin in the ferret as a model for the detection of anti-emetic drugs. *Neuropharmacology* **26,** 1321–6.

Costall, B., Domeney, A. M., Gunning, S. J., Kelly, M. E., Naylor, R. J., Nohria, V., Owera-Atepo, J. B., Simpson, K. M., Tan, C. C. W., and Tattersall, D. (1987b). The action of dazopride to enhance gastric emptying and block emesis. *Neuropharmacology* **26,** 669–77.

Costall, B., Gunning, S. J., Naylor, R. J., and Tyers, M. B. (1987c). The effect of GR38032F, a novel 5-HT_3-receptor antagonist on gastric emptying in the guinea-pig. *British Journal of Pharmacology* **91,** 963–4.

Cunningham, D., Hawthorn, J., Pople, A., Gazet, J. C., Ford, H. T., Challoner, T., and Coombes, R. C. (1987). Prevention of emesis in patients receiving cytotoxic drugs by GR38032F, a selective 5-HT_3 receptor antagonist. *Lancet* **i,** 1461–3.

Donatsch, P., Engel, G., Richardson, B. P., and Stadler, P. A. (1984). ICS 205-930: a highly selective and potent antagonist at peripheral neuronal 5-hydroxytryptamine receptors. *British Journal of Pharmacology* **81,** 34P–5P.

Fake, C. S., King, F. D., and Sanger, G. J. (1987). BRL 43694: A potent and novel 5-HT_3 receptor antagonist. *British Journal of Pharmacology* **91,** 335P.

Fastier, F. N., McDowall, M. A., and Waal, H. (1959). Pharmacological properties of phenylbiguanide and other amidine derivatives in relation to those of 5-hydroxytryptamine. *British Journal of Pharmacology* **14,** 527–35.

Fozard, J. R. (1968). Studies on the cardiovascular reactivity to 5-HT in the rat. Unpublished Ph.D. Thesis, University of Bradford.

Fozard, J. R. (1980). Differences between receptors for 5-hydroxytryptamine on autonomic neurones. *British Journal of Pharmacology* **70,** 143P–4P.

Fozard, J. R. (1982). Basic mechanisms of antimigraine drugs. In *Headache: physiopathological and clinical concepts* (eds. M. Critchly, A. Friedman, S. Gorini, and F. Sicuteri). *Advances in Neurology*, Vol. 33, pp. 295–307. Raven Press, New York.

Fozard, J. R. (1983). Differences between receptors for 5-hydroxytryptamine on autonomic neurones revealed by nor-(-)-cocaine. *Journal of Autonomic Pharmacology* **3,** 21–6.

Fozard, J. R. (1984a). MDL 72222, a potent and highly selective antagonist at neuronal 5-hydroxytryptamine receptors. *Naunyn-Schmiedeberg's Archives of Pharmacology* **326,** 36–44.

Fozard, J. R. (1984b). Neuronal 5-HT receptors in the periphery. *Neuropharmacology* **23,** 1473–86.

Fozard, J. R. (1985). 5-Hydroxytryptamine in the pathophysiology of migraine. In *Vascular neuroeffector mechanisms* (eds. J. A. Bevan, T. Godfraind, R. A. Maxwell, J. C. Stoclet, and M. Worcel) pp. 321–8. Elsevier, Amsterdam.

Fozard, J. R. (1986). Peripheral neuronal 5-hydroxytryptamine receptors and their biological significance. In *Proceedings of the third SCI/RSC medicinal chemistry symposium* (ed. R. W. Lambert) pp. 37–56. The Royal Society of Chemistry, London.

Fozard, J. R. (1987a). 5-HT: The enigma variations. *Trends in Pharmacological Sciences* **8,** 501–6.

Fozard, J. R. (1987b). 5-HT_3 receptors and cytotoxic drug-induced vomiting. *Trends in Pharmacological Sciences* **8,** 44–5.

Fozard, J. R. (1989). The development of 5-HT_3 receptor antagonists. In *Serotonin: Actions, Receptors, Pathophysiology. Proceedings of the 1987 IUPHAR Congress Satellite Meeting, Heron Island, Australia* (ed. E. J. Mylecharane, J. A. Angus, I. S. De La Lande, P. P. A, Humphrey) Macmillan, London (in press).

Fozard, J. R. and Gittos, M. W. (1983). Selective blockade of 5-hydroxytryptamine neuronal receptors by benzoic acid esters of tropine. *British Journal of Pharmacology* **80,** 511P.

Fozard, J. R. and Host, M. (1982). Selective inhibition of the Bezold-Jarisch effect of 5-HT in the rat by antagonists at neuronal 5-HT receptors. *British Journal of Pharmacology* **77,** 520P.

Fozard, J. R. and Leach, G. D. H. (1969). 5-HT autopotentiation and drug induced changes in the 5-HT vasomotor reponse in the pithed rat. *European Journal of Pharmacology* **7,** 14–22.

Fozard, J. R. and Mobarok Ali, A. T. M. (1978a). Receptors for 5-hydroxytryptamine on the sympathetic nerves of the rabbit heart. *Naunyn-Schmiedeberg's Archives of Pharmacology* **301,** 223–35.

Fozard, J. R. and Mobarok Ali, A. T. M. (1978b) Blockade of neuronal tryptamine receptors by metoclopramide. *European Journal of Pharmacology* **49,** 109–12.

Fozard, J. R. and Muscholl, E. (1972). Effects of several muscarinic agonists on cardiac performance and the release of noradrenaline from sympathetic nerves of the perfused rabbit heart. *British Journal of Pharmacology* **45,** 616–29.

Fozard, J. R. and Mwaluko, G. M. P. (1975). The stimulant response of 5-hydroxytryptamine on cardiac sympathetic nerves: resistance to blockade by classical tryptamine receptor antagonists. In *Proceedings of the Sixth International Congress of Pharmacology,* July 1975, Helsinki, Finland, Abstract 697, p. 299.

Fozard, J. R. and Mwaluko, G. M. P. (1976). Mechanism of the indirect sympathomimetic effect of 5-hydroxytryptamine on the isolated heart of the rabbit. *British Journal of Pharmacology* **57,** 115–25.

Fozard, J. R. and Mobarok Ali, A. T. M., and Newgrosh, G. (1979). Blockade of serotonin receptors on autonomic neurones by (-)-cocaine and some related compounds. *European Journal of Pharmacology* **59,** 195–210.

Fozard, J. R., Loisy, C., and Tell, G. P. (1985). Blockade of neuronal 5-hydroxytryptamine receptors with MDL 72222. A novel approach to the symptomatic treatment of migraine. In *Proceedings of the 5th Migraine Symposium, London 1984* (ed. F. Clifford Rose), pp. 264–272. Karger, Basel.

Gaddum, J. H. and Picarelli, Z. P. (1957). Two kinds of tryptamine receptor. *British Journal of Pharmacology* **12,** 323–8.

Garattini, S. and Valzelli, L. (1965). *Serotonin.* Elsevier, London.

Gralla, R. J. (1983). Metoclopramide: a review of antiemetic trials. *Drugs* **25** (Suppl. 1), 63–73.

Greaves, M. W. and Schuster, S. (1967). Responses of skin blood vessels to bradykinin, histamine and 5-hydroxytryptamine. *Journal of Physiology (London)* **193,** 255–67.

Gunning, S. J., Bradbury, A. J., Costall, B., and Naylor, R. J. (1986). Evidence that 5-hydroxytryptamine may exert both facilitatory and inhibitory control of electrical field stimulation-evoked contractions in longitudinal muscle taken from the body of guinea-pig stomach. *Journal of Pharmacy and Pharmacology* **38,** 182–7.

Gunning, S. J., Hagan, R. M., and Tyers, M. B. (1987). Cisplatin induces biochemical and histochemical changes in the small intestine of the ferret. *British Journal of Pharmacology* **90,** 135P.

Harrington, R. A., Hamilton, C. W., Brogden, R. N., Linkewich, J. A., Romankiewicz, J. A., and Heel, R. C. (1983). Metoclopramide—an updated review of its pharmacological properties and clinical use. *Drugs* **25,** 451–94.

Heavey, D. J., Barnes, V. F., Orwin, J., and Brown, M. J. (1986). Inhibition of 5-HT induced axon flares by MDL 72422. *British Journal of Clinical Pharmacology* **21,** 558P.

Ison, P. J. and Peroutka, S. J. (1986). Neurotransmitter receptor binding studies predict antiemetic efficacy and side effects. *Cancer Treatment Reports* **70,** 637–41.

Keast, J. R., Furness, J. B., and Costa, M. (1985). Investigations of nerve populations influencing ion transport that can be stimulated electrically by serotonin and by a micotinic agonist. *Naunyn-Schmiedeberg's Archives of Pharmacology* **331,** 260–6.

Kilbinger, H. and Pfeuffer-Freiderich, I. (1985). Two types of receptor for 5-hydroxytryptamine on the cholinergic nerves of the guinea-pig myenteric plexus. *British Journal of Pharmacology* **85,** 529–39.

Kilpatrick, G. J., Jones, B. J., and Tyers, M. B. (1987). The identification and distribution of 5-HT_3 receptors in rat brain using radioligand binding. *Nature (London)* **330,** 746–8.

Leibundgut, V. and Lancranjan, I. (1987). First results with ICS 205-930 (5-HT receptor antagonist) in prevention of chemotherapy-induced emesis. *Lancet* **i,** 1198.

Lindahl, O. (1961). Experimental skin pain produced by injection of water-soluble substances in humans. *Acta Physiologica Scandinavica* **51,** 1–90.

Loisy, C., Beorchia, S., Centonze, V., Fozard, J. R., Schechter, P. J., and Tell, G. P. (1985). Effects on migraine headache of MDL 72222, an antagonist at neuronal 5-HT receptors. Double-blind, placebo-controlled study. *Cephalalgia* **5,** 79–82.

McClelland, C. M. and Sutton, J. A. (1986). Epigastric impedence: a non-invasive method for the assessment of gastric emptying and motility. *Gut* **26,** 607–14.

McRitchie, B., McClelland, C. M., Cooper, S. M., Turner, D. H., and Sanger, G. J. (1984). Dopamine antagonists as anti-emetics and as stimulants of gastric motility. In *Mechanisms of gastro-intestinal motility and secretion* (eds. A. Bennett and G. Velo) pp. 287–301. Plenum Press, New York.

Matsuoka, O., Tsuchiya, T., and Furukawa, Y. (1962). The effect of X-radiation on 5-hydroxytryptamine (serotonin) contents in the small intestines of experimental animals. *Journal of Radiation Research* **32,** 104–9.

Miner, W. and Sanger, G. J. (1986). Inhibition of cis-platin-induced vomiting by selective 5-hydroxytryptamine M-receptor antagonism. *British Journal of Pharmacology* **88,** 497–9.

Miner, W. D., Sanger, G. J., and Turner, D. H. (1987). Evidence that 5-hydroxytryptamine$_3$ receptors mediate cytotoxic drug and radiation-evoked emesis. *British Journal of Cancer* **56,** 159–62.

Orwin, J. M. and Fozard, J. R. (1986). Blockade of the flare response to intradermal 5-hydroxytryptamine in man by MDL 72222, a selective antagonist at neuronal 5-hydroxytryptamine receptors. *European Journal of Clinical Pharmacology* **30,** 209–12.

Peroutka, S. J. and Hamik, A. (1988). ^{3}H-quipazine labels 5-HT_3 recognition sites in rat cortical membranes. *European Journal of Pharmacology* **148,** 297–9.

Rapeport, W. G. and Thompson, S. (1988). Effects of oral BRL 24924 on gastric emptying measured by epigastric impedence. *British Journal of Clinical Pharmacology* **24,** 263P.

Richardson, B. P. and Buchheit, K. H. (1988). The pharmacology, distribution and function of 5-HT_3 receptors. In *Neuronal serotonin*, (eds. N. N. Osborne and M. Hamon) pp. 465–504. Wiley, Chichester.

Richardson, B. P. and Engel, G. (1986). The pharmacology and function of 5-HT_3 receptors. *Trends in Neurosciences* **9,** 424–8.

Richardson, B. P., Engel, G., Donatsch, P., and Stadler, P. A. (1985). Identification of serotonin M-receptor subtypes and their specific blockade by a new class of drugs. *Nature (London)* **316,** 126–31.

Sanger, G. J. (1985). Three different ways in which 5-hydroxytryptamine can affect cholinergic activity in guinea-pig isolated ileum. *Journal of Pharmacy and Pharmacology* **37,** 584–6.

Sanger, G. J. (1987a). Increased gut cholinergic activity and antagonism of 5-hydroxytryptamine M-receptors by BRL 24924: potential clinical importance of BRL 24924. *British Journal of Pharmacology* **91,** 77–87.

Sanger, G. J. (1987b). Activation by metoclopramide of a myenteric 5-hydroxytryptamine-like receptor. *Journal of Pharmacy and Pharmacology* **39,** 449–53.

Schild, H. O. (1947). pA, a new scale for the measurement of drug anagonism. *British Journal of Pharmacology* **2,** 189–206.

Sicuteri, F., Fanciullacci, M., Franchi, G., and Del Bianco, P. L. (1965). Serotonin-bradykinin potentiation on the pain receptors in man. *Life Sciences* **4,** 309–16.

Stables, R., Andrews, P. L. R., Bailey, H. E., Costall, B., Gunning, SJ., Hawthorne, J., Naylor, R. J., and Tyers, M. B. (1987). Antiemetic properties of the 5-HT_3 receptor antagonist, GR38032F. *Cancer Treatment Reviews* **14,** 333–6.

Staniforth, D. H. and Corbett, R. (1988). The effect of BRL 24924 on upper gastro-intestinal tract activity. *British Journal of Clinical Pharmacology* **24,** 263P–4P.

Wallis, D. I. (1981). Neuronal 5-hydroxytryptamine receptors outside the central nervous system. *Life Sciences* **29,** 2345–55.

Watling, K. J., Aspley, S., Swain, C. J., and Saunders, J. (1988). [^{3}H]-Quaternised ICS 205-930 labels 5-HT_3 receptor binding sites in rat brain. *European Journal of Pharmacology* **149,** 397–8.

Welch, J. P. and Malt, R. A. (1977). Management of carcinoid tumours of the gastrointestinal tract. *Surgery, Gynecology and Obstetrics* **145,** 223–7.

16

Pathophysiological significance of 5-hydroxytryptamine in the periphery

DONALD S. HOUSTON and PAUL M. VANHOUTTE

16.1 Introduction

5-HT has excited interest as a possible mediator in a wide variety of pathological processes. New information about the pharmacology of the amine in diverse tissues is allowing more precise testing of some of the hypotheses concerning its involvement, though for the most part significant roles of 5-HT remain speculative.

The discussion in this chapter will remain restricted to the areas of expertise of the authors, the vascular system in which 5-HT can cause either constriction (Fig. 16.1) or dilatation (Fig. 16.2). Most of the injurious processes attributed to 5-HT involve spasm of blood vessels, and so spasm in different segments of the vasculature (coronary, digital, and cerebral arteries) will be discussed first in this review. This will be put in the context of vasospasm as a beneficial element of the normal haemostasic process. Hypertension and peripheral vascular disease may also involve abnormal vascular contractility, but on a more diffuse scale, and the possible role of 5-HT in these conditions will thus be discussed separately. Finally, we will consider carcinoid syndrome, which is instructive because it is the naturally occurring model of 5-HT excess.

Throughout this review will run the recurring theme that the major source of 5-HT available to the blood vessel wall is that released from aggregating platelets. Exceptions are that 5-HT-containing neurones appear to innervate cerebral blood vessels (Edvinsson *et al.* 1983; Marco *et al.* 1985), and that 5-HT-containing nerves may play a significant role in the regulation by the central nervous system of the vasculature (for review see Kuhn *et al.* 1980). In evaluating whether 5-HT may play a role in any pathological process, we will attempt to answer (so far as evidence is available) the questions: (1) Is 5-HT likely to be available at the site of pathology? (2) Is there a plausible mechanism by which 5-HT could cause observed events? (3) What are the effects of 5-HT or 5-HT receptor agonists? (4) Do 5-HT receptor antagonists prevent or modify the pathology? (see also Vanhoutte and Cohen 1983; Vanhoutte *et al.* 1984; Cohen and Vanhoutte 1985; Vanhoutte 1985a and b, 1986, and 1987a and b; Vanhoutte and Houston 1985; Webb and Vanhoutte 1985; Houston and Vanhoutte 1986; Vanhoutte and Shepherd 1986).

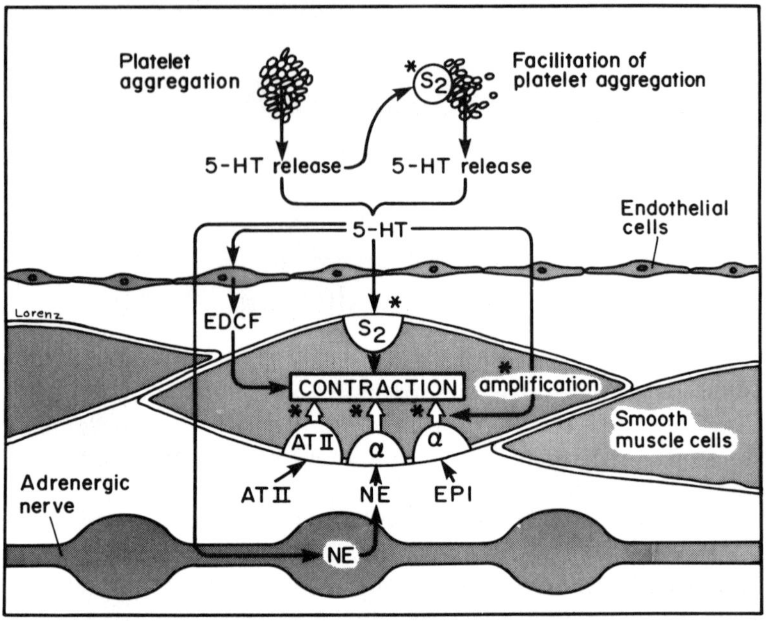

Fig. 16.1 Direct and indirect vasoconstrictor effects of 5-HT. α = α-adrenoceptor; EDCF = endothelium-derived contracting factor; AT II = angiotensin II; EPI = adrenaline; NE = noradrenaline; S_2 = 5-HT_2 receptor. *inhibited by 5-HT_2 receptor antagonists. (From Vanhoutte 1987b by permission)

16.2 Vasospasm

16.2.1 Haemostasis

5-HT is one of several vasoconstrictors which are released or formed during the process of platelet aggregation (De Clerck and Van Nueten 1982). Others include thromboxane A_2, catecholamines, adenine nucleotides, vasopressin (which is concentrated by platelets—Preibisz *et al.* 1983), and thrombin (which is formed upon activation of the coagulation cascade). 5-HT in its own right is a powerful vasoconstrictor in many vessels—indeed its alternate name, serotonin, was coined in recognition of that property. It has a further action of amplifying the reaction of blood vessels to other vasoconstrictors, including thromboxane A_2 (Mullane *et al.* 1982; see Chapter 6). Since the amplifying effect can occur at low concentrations of 5-HT, augmentation of the contraction evoked by the other platelet-released vasoconstrictors or of neurogenic tone may be even more important than the direct constrictor effect of 5-HT. Most blood vessels so far characterized have 5-HT_2 receptors mediating their response to 5-HT, but some vessels clearly express other subtypes of receptors (see Chapter 5). The question of whether amplification (which has been ascribed to 5-HT_2 receptors) occurs when the 5-HT response is mediated by another type of receptor has not yet been answered.

An indirect form of amplification occurs at the level of the platelets: 5-HT, thromboxane A_2, and adenine nucleotides released during platelet activation recruit other platelets from the blood stream and encourage their incorporation into the developing thrombus. 5-HT not only causes weak aggregation of human platelets, but amplifies the aggregatory response to other agonists (De Clerck *et al.* 1982). The platelet thrombus forms the primary haemostatic plug, but the contraction of the smooth muscle in the injured vessel wall as a response to 5-HT and the various other platelet-released vasoconstrictors, would help to further reduce luminal diameter and thus limit blood loss until complete occlusion from the clot can occur (Fig. 16.2).

Experimental evidence that 5-HT is important in haemostasis is that the Fawn-Hooded rat, which has markedly lower platelet 5-HT content than the control Wistar rat, has prolonged bleeding time and increased blood loss from a standardized tail wound (De Clerck *et al.* 1984); however, these animals also have platelet dysfunction. Wistar rats treated with 5-HT receptor antagonists such as ketanserin, or depleted of platelet 5-HT by prior treatment with reserpine or *p*-chlorophenylalanine, have significantly prolonged bleeding times and increased blood loss (De Clerck *et al.* 1984). A combination of ketanserin plus thromboxane synthetase inhibitor synergistically prolonged bleeding time.

Interestingly, 5-HT may be the critical agent responsible for the ability of platelets to maintain microvascular integrity, since 5-HT receptor agonists can maintain microvascular integrity in thrombocytopenic hamsters (Sweetman *et al.* 1981).

In considering vasospasm, it is of interest to note that the responses of different vascular beds to 5-HT are quite variable. For example, in the coronary artery of the dog, 5-HT evokes a maximal contraction that is only about half of the maximal active tone that the vessel can produce (as determined by contraction with potassium chloride). 5-HT does not appear to exert its effects through 5-HT_2 receptors in this tissue (Cohen 1986). In the femoral artery from the same species, where 5-HT appears to act via 5-HT_2 receptors, it evokes a maximal contraction. Furthermore, 5-HT-induced, endothelium-dependent relaxation has been noted with the greatest prominence in the coronary artery, and is minimal or non-existent in most other vessels so far studied (Vanhoutte and Houston 1986). In peripheral vessels from the dog studied *in vitro*, the net response to aggregating platelets is contraction. In coronary vessels, however, the net response is an endothelium-dependent relaxation (Cohen *et al.* 1983a and b), in dogs due principally to the release from the platelets of ADP and

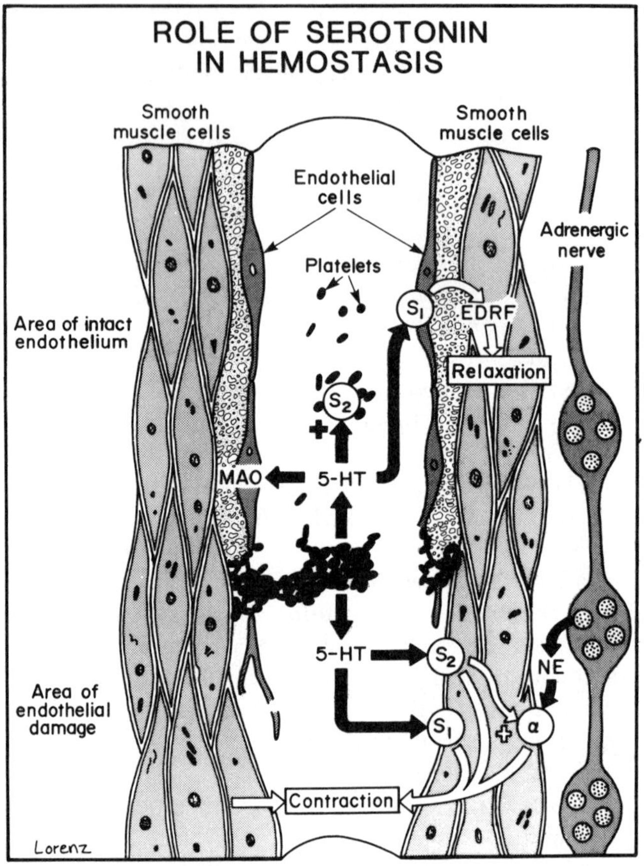

Fig. 16.2 Possible roles of 5-HT in haemostasis. 5-HT released from aggregating platelets amplifies further platelet aggregation; it is, however, avidly taken up and cleared from plasma by unaggregated platelets. If the endothelium is intact, 5-HT is also taken up and degraded by monoamine oxidase (MAO). Endothelial cells of some vessels can release EDRF in response to 5-HT and other platelet products. If the fragile endothelium is damaged during trauma, and 5-HT gains access to the underlying smooth muscle, it causes contraction directly (the receptor subtype depending on the blood vessel) and amplifies contraction to other vasoconstrictors such as noradrenaline (NE). Thus, the absence of endothelium favours not only platelet aggregation but also vasospasm and hence control of blood loss [s_1 = '5-HT_1-like' receptor; S_2 = 5-HT_2 receptor; α = α-adrenoceptor; + = amplification]. (From Houston and Vanhoutte 1986 by permission)

ATP (Houston *et al.* 1985 and 1986), but in the pig due largely to 5-HT (Shimokawa *et al.* 1987).

There may be a teleological explanation for these findings. During the course of our ancestors' evolution, vasospasm and a prompt haemostatic response when a peripheral vessel was injured would have been of obvious survival benefit. In the case of coronary artery laceration, however, the benefit of a haemostatic response would be moot. Platelet-evoked coronary spasm with myocardial ischaemia or infarction in the absence of trauma, on the other hand, could be expected to have had a pronounced negative selective effect.

Thus it appears that vasospasm in response to 5-HT and other platelet-released vasoconstrictors has evolved as a protective mechanism to aid haemostasis and hence limit blood loss. In vascular beds where vasospasm would exert a deleterious effect, the endothelium has developed an inhibitory role (Fig. 16.3).

16.2.2 Coronary spasm

16.2.2.1 *Clinical data*

Prinzmetal's or variant angina occurs because of active contraction of the smooth muscle of epicardial coronary arteries (Luchi *et al.* 1979). It can occur in arteries which are anatomically normal as determined by angiography (Maseri *et al.* 1977). Classic (exertional) angina and particularly unstable angina may also be caused or exacerbated by local vasospasm in the region of atheromatous plaque (Chahine *et al.* 1975). In patients with spasm, if infarction occurs it occurs in the territory subserved by the spastic vessel (Maseri *et al.* 1978). However, there is little doubt that deposition of platelet thrombus also contributes to vessel occlusion, since there is evidence of thrombus routinely both on angiography and at autopsy (Maseri *et al.* 1986; Davies and Thomas 1984).

16.2.2.2 *Evidence of platelet activation*

In patients with atherosclerotic disease, platelets are activated in transit through the coronary circulation. For example, levels of the platelet products, thromboxane B_2 (the stable metabolite of the short-lived vasoconstrictor, thromboxane A_2) and beta-thromboglobulin, were elevated in blood sampled from the coronary sinuses of patients with stable angina (Lewy *et al.* 1980; Rubenstein *et al.* 1981; Tada *et al.* 1981; Cortellaro *et al.* 1983). This is not too surprising, since loss of endothelial integrity and exposure of subendothelial layers in the diseased vessels would lead to activation of platelets and release of thromboxane. Perhaps more surprising is the fact that coronary sinus thromboxane B_2 levels were also elevated in patients with Prinzmetal's angina, most of whom had no angiographic evidence of atherosclerosis (Tada *et al.* 1981). During acute coronary

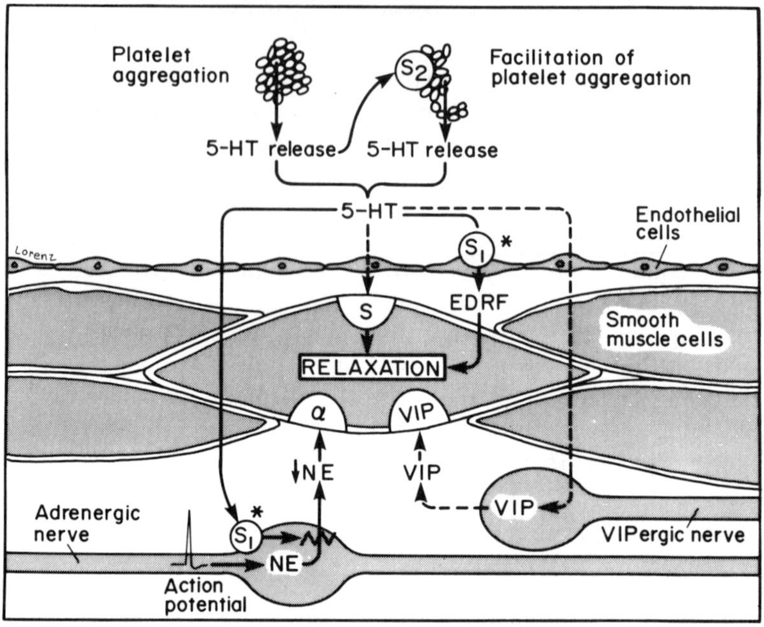

Fig. 16.3 Direct and indirect vasodilator effects of 5-HT. α = α-adrenoceptor; NE = noradrenaline; VIP = vasoactive intestinal peptide; S = 5-HT receptor; *inhibited by '5-HT_1-like' receptor antagonists; jagged line = inhibition of adrenergic neurotransmission. (From Vanhoutte 1987b by permission)

spasm, numbers of platelet aggregates and thromboxane B_2 levels are elevated (Robertson *et al.* 1980; Rubenstein *et al.* 1981; Hirsh *et al.* 1981). These results must be interpreted with some caution, however, because of the possibility of experimental artifact due to *ex vivo* platelet activation during the sampling or processing of blood. Nonetheless, with this evidence of platelet activation in the coronary bed, it would seem likely that the release of vasoactive products such as 5-HT and thromboxane A_2 would contribute to spasm and vessel narrowing.

16.2.2.3 *Animal models: a role for 5-HT*

5-HT has been shown to be a potent constrictor of canine coronary arteries *in vivo* (Bove and Dewey 1983). It has been demonstrated that (at least in the dog) ergonovine activates 5-HT receptors rather than adrenoceptors on coronary smooth muscle (Brazenor and Angus 1981; Holtz *et al.* 1982). Ergonovine has been used clinically to evoke coronary vasospasm during diagnostic coronary angiography (Schroeder *et al.* 1977). The spasm evoked by ergonovine mimics the episodes of spontaneous spasm witnessed in patients with Prinzmetal's angina (Curry *et al.* 1979), which supports the concept that 5-HT may be responsible for spontaneous vasospasm.

16.2.2.4 *Mechanisms of protective endothelial effect*

The endothelium may play an important protective role against such a postulated platelet- and 5-HT-evoked vasospasm by a number of mechanisms. Firstly, the endothelium would inhibit platelet activation, through its inherent non-thrombogenic surface characteristics and through the production of the potent platelet inhibitor, prostacyclin. Secondly, the endothelium would act as a diffusion barrier against platelet-released vasoconstrictor materials from reaching the underlying smooth muscle. Thirdly, the endothelium is metabolically active. Endothelial cells actively take up and destroy 5-HT via the monoamine oxidase enzyme they contain; furthermore, ectoenzymes on the endothelial cell surface degrade the proaggregatory adenine nucleotides, ATP and ADP, to the platelet inhibitor, adenosine (Pearson and Gordon 1985). Fourthly, the endothelium can, in response to a variety of vasoactive materials including 5-HT (Cocks and Angus 1983; Cohen *et al.* 1983a and b), adenine nucleotides (De Mey and Vanhoutte 1982; Houston and Vanhoutte 1988; Houston *et al.* 1987), and thrombin (De Mey *et al.* 1982; Ku 1982), produce the potent vasodilator substance known as EDRF (see Furchgott 1983; Vanhoutte *et al.* 1986). Prostacyclin, the other vasodilator known to be produced by endothelial cells, is also released in response to at least some of the same stimuli, such as thrombin (Weksler *et al.* 1978). It is of interest that EDRF has, like prostacyclin, been shown to be able to inhibit platelet aggregation, at least *in vitro* (e.g. Ayuma *et al.* 1986; Furlong *et al.* 1987), thus further amplifying the first protective function of the endothelium listed above.

16.2.2.5 *Animal models: artificial endothelial damage*

It is tempting to imagine that ergonovine, and 5-HT, evoke spasm in patients at locations where the endothelium is damaged or dysfunctional. Several investigators have performed experiments in animal models to look at the effect of endothelial damage on the responses to platelets and 5-HT receptor agonists. For example, removal of the canine coronary endothelium by balloon catheter *in vivo* augments the vasoconstrictor response to intracoronary injection of 5-HT (Brum *et al.* 1984; Lamping *et al.* 1985). In a more complex model, an isolated rabbit heart perfused with a perfluorocarbon artificial blood solution, constriction of epicardial coronary arteries occurred following addition of platelets to the perfusion solution only if the endothelium had first been removed (Bing *et al.* 1985). Although the authors did not directly address the issue, it seems that in this model 5-HT would undoubtedly have contributed to the coronary constriction.

16.2.2.6 *Animal models: atherosclerosis*

The effect of endothelial damage secondary to atherosclerosis has also been investigated in a number of animal models. The coronary arteries and aortas of genetically hyperlipidaemic rabbits are hypersensitive to 5-HT and ergonovine (Yokoyama *et al.* 1983) as are aortas made atherosclerotic by high cholesterol diets (Henry and Yokoyama 1980; Heric and Tackett 1985). Balloon endothelial denudation in conjunction with high cholesterol diet produces intimal changes (without lipid deposition) in canine coronary arteries. In these vessels, intravenously administered ergonovine evokes spasm at the site of denudation, and after sacrifice of the animals, the contractile responses to ergonovine and 5-HT of arterial muscle strips are augmented while responses to the α_1-adrenoceptor agonist, phenylephrine (for which no endothelial response has been reported) are unchanged (Kawachi *et al.* 1984). Perfused large limb arteries of cynomolgus monkeys made atherosclerotic by high-cholesterol diets show greatly enhanced resistance to blood flow when infused with 5-HT, compared to controls (Heistad *et al.* 1984).

Atherosclerosis may of course affect responses to 5-HT receptor agonists in a number of ways, including direct alterations of the characteristics of vascular smooth muscle, but it is tempting to attribute much of the changed response to alterations of the several functions of the intima alluded to above (impaired diffusion barrier, decreased metabolism of 5-HT, and decreased production of relaxing factors). For example, endothelial production of prostacyclin is inhibited in experimental atherosclerosis (Gryglewski *et al.* 1978).

Atherosclerosis may not even be required for loss of endothelial function. In pigs, regrowth of endothelium following balloon-denudation results in an endothelium which is morphologically intact but fails to exhibit the normal endothelium-dependent relaxation to 5-HT (Shimkawa *et al.* 1987).

In humans, although such prospective experiments are impossible to perform, it has been found that the coronary arteries from patients who have died of coronary artery disease are more responsive to 5-HT than those taken from controls (Kalsner and Richards 1984).

16.2.2.7 *Animal models: coronary occlusion*

If a canine coronary artery is occluded partially with an external band, the flow through the artery varies cyclically due to build-up and then dislodgement of platelet thrombi at the site of partial occlusion. Active spasm is not a contribution to reduction to flow in this model, presumably because the constricting ring compresses the smooth muscle of the underlying vessel to an extent that no further significant contraction can occur. However, this does indicate that local platelet aggregation does not trigger generalized vasoconstriction downstream. The cyclical obstruction is reduced by platelet inhibition with aspirin and other cyclo-oxygenase inhibitors (Folts *et al.* 1976 and 1982). It is also reduced by 5-HT receptor antagonists such as ketanserin, which reflects the platelet-inhibitory action of ketanserin in this crucial setting (Ashton *et al.* 1986; Bush 1987). It is of interest that endothelial injury is produced in such models of vascular occlusion, and might well contribute to a positive feedback cycle (Gertz *et al.* 1981).

16.2.2.8 *Human data: 5-HT antagonism*

The coronary sinus blood of patients with coronary artery disease contains a vasoconstrictor substance, the action of which can be prevented by the non-selective antagonist, methiothepin (Rubanyi *et al.* 1987). From this and the above arguments, it would be expected that 5-HT receptor antagonists might be beneficial in the treatment of coronary ischaemia. To test this hypothesis, De Caterina *et al.* (1984) administered ketanserin to five patients with variant angina, but found no significant benefit from the drug in preventing anginal attacks at doses that inhibited platelet aggregation to 5-HT. Similarly, coronary spasm induced by the intravenous administration of ergonovine in man was not blocked (though there was a trend towards inhibition) by a systemic dose of ketanserin that was sufficient to prevent contraction of hand veins evoked by local injection of 5-HT (Freedman *et al.* 1984). These findings have been replicated by Mata-Bouchart *et al.* (1985).

One explanation for the failure of these experiments to demonstrate a protective effect of 5-HT_2 receptor antagonists could be that human coronary smooth muscle bears a population of 5-HT receptors different from the conventional 5-HT_2 receptor. This appears to be the case in the dog, where ketanserin is a weak and non-competitive antagonist of the contractile effect of 5-HT (Van Nueten *et al.* 1981; Cohen 1986; Houston and Vanhoutte 1988), although in the pig ketanserin is an effective antagonist (Cocks and Angus 1983; Shimokawa *et al.* 1987). Unfortunately, many of the other 5-HT receptor antagonists (e.g. methysergide) which block 5-HT receptors other than 5-HT_2 receptors, act as partial agonists, causing contraction of coronary smooth muscle. This obviously would make them inappropriate for such a study.

Nonetheless, the platelet inhibiting action of ketanserin might be expected to offer some benefit in unstable angina, as does aspirin (Lewis *et al.* 1983); it is possible that more sensitive large-scale clinical trials might be able to demonstrate such a benefit.

16.2.3 Raynaud's phenomenon

The characteristically episodic nature of digital ischaemia in patients with Raynaud's phenomenon is suggestive of an active element of vascular spasm (see Cohen and Coffman 1984). Some data suggest that 5-HT may be involved in its pathogenesis. Infusion of 5-HT into the brachial artery of normal human volunteers produces digital cyanosis reminiscent of Raynaud's phenomenon (Roddie *et al.* 1955). 5-HT is a powerful constrictor of isolated human digital arteries, and is a major contributor to platelet-evoked contraction in these vessels (Jauernig and Moulds 1978; Moulds *et al.* 1984). The contractions in isolated canine cutaneous vessels are augmented by cooling (Vanhoutte and Shepherd 1970; Lindblad *et al.* 1984), which would be consistent with the fact that attacks of Raynaud's phenomenon are commonly cold-induced. Of particular interest, dermal vessels from patients with scleroderma and Raynaud's phenomenon are hypersensitive to 5-HT compared with vessels from normal subjects (Winkelmann *et al.* 1976). Platelets from subjects with Raynaud's phenomenon are hyperaggregable to adrenaline and less sensitive to the inhibitory effects of prostacyclin than controls. Furthermore, these subjects have higher plasma levels of beta-thromboglobulin, indicating increased platelet activation (Hutton *et al.* 1984).

More direct evidence for a role of 5-HT in Raynaud's phenomenon is the clinical efficacy of 5-HT receptor antagonists. Acute infusion of an ergot derivative can attenuate the fall in digital blood flow brought on by cold (Halpern *et al.* 1960). Intravenous infusion of ketanserin can acutely increase blood flow in the fingers of patients with Raynaud's syndrome (Stranden *et al.* 1982), though unfortunately no control group was included in this study. In an open study, orally administered ketanserin reduced the severity of symptoms and improved digital flow, particularly among patients with Raynaud's phenomenon as a manifestation of scleroderma (Seibold and Jageneau 1984). A cross-over placebo-controlled study subsequently did not confirm any benefit of oral ketanserin on finger blood flow, nor on haemorrheology or beta-thromboglobulin level (Longstaff *et al.* 1985). Nevertheless, some subgroups, particularly those with scleroderma, may benefit, perhaps because patients with scleroderma have intimal abnormalities that provoke platelet aggregation (Seibold 1985). It must be kept in mind, however, that part of whatever efficacy ketanserin may have may be related to blockade by ketanserin of α_1-adrenoceptors, since the specific α_1-adrenoceptor antagonist, prazosin, has also been reported to be of some therapeutic benefit in Raynaud's phenomenon (Nielsen *et al.* 1983).

16.2.4 Cerebral vasospasm

Spasm of the intracranial arteries is a frequent subacute complication of subarachnoid haemorrhage, and contributes to the morbidity and mortality of this ailment. Although structural changes eventually occur in such vessels (Mohan 1980), many workers have attempted to identify a mediator in the extravasated blood that may initiate the process of spasm. Haemoglobin, thrombin, prostaglandins, and K^+ have all been considered, as has 5-HT liberated from the platelets in the extravasated blood. 5-HT can contract human and canine cerebral vessels *in vitro* (Allen *et al.* 1974a and b, and 1976; Rusch *et al.* 1985), and furthermore rabbit basilar arteries have been found to become hypersensitive to the contractile action of 5-HT following experimental subarachnoid haemorrhage (Young *et al.* 1986). Depletion of platelet 5-HT stores with kanamycin or reserpine was found in one study to reduce the incidence of ischaemic symptoms following subarachnoid haemorrhage (Zervas *et al.* 1979). On the other hand, the concentration of 5-HT has not been found to be elevated in the cerebrospinal fluid of patients with subarachnoid haemorrhage (Voldby *et al.* 1982). Studies with the selective 5-HT receptor antagonist, methysergide, demonstrated that early but not late spasm in an experimental model could be reversed by 5-HT receptor blockade, perhaps reflecting the onset of irreversible pathological changes in 'late' spasm (Nagai *et al.* 1975). Phenoxybenzamine, which can irreversibly bind to 5-HT receptors, was found in a dog model to prevent or reverse the induction of spasm by injection of blood or 5-HT into the subarachnoid space (Allen *et al.* 1974a and b).

The results of these experiments are far from conclusive. We await the result of clinical trials, which will have to employ a 5-HT receptor antagonist which can penetrate the blood–brain barrier and which can antagonize selectively the 5-HT receptors located on the cerebral vascular smooth muscle which may be of the 5-HT_{1A} type (Peroutka *et al.* 1986; Taylor *et al.* 1986).

16.2.5 Migraine

Migraine is felt to occur in two phases, although both processes may be concurrent: constriction of intracranial arteries, which gives rise to the ischaemic neurological prodrome in classic migraine, and dilatation of extracranial arteries, which is associated with the throbbing pain (Raskin 1981). 5-HT could be involved in initiating either the vasoconstriction or the vasodilatation. Most evidence available indicates that 5-HT evokes contraction of isolated large cerebral arteries, but 5-HT-induced vasodilatation, either by direct action on the smooth muscle (e.g. Feniuk *et al.* 1983) or indirectly via the endothelium (e.g. Cocks and Angus 1983) has been reported in a number of blood vessels. 5-HT might have indirect effects as well, such as increased kinin formation or the sensitization of pain afferents (Fanchamps 1982). Support of the notion that 5-HT stimulation of nociceptive fibres may contribute to the pain of migraine is lent by the beneficial effect of the selective neuronal 5-HT_3 receptor antagonist, MDL 72222, administered acutely during migraine attacks (Loisy *et al.* 1985).

The classical observation made more than 25 years ago linking 5-HT to migraine was the finding that urinary excretion of the metabolite of 5-HT, 5-hydroxyindoleacetic acid, is increased following a migraine attack (Sicuteri *et al.* 1961). This may not, however, be a consistent finding (Fozard 1985). Decreased platelet content of 5-HT also occurs, which explains the decreased blood 5-HT concentration during attacks (Anthony *et al.* 1967; Rydzewski 1976; Somerville 1976), which returns to normal following the attack. These findings suggest that there is a transient increased release of 5-HT from platelets during the migraine attack. Furthermore, the blood of patients during a migraine attack has been reported to contain a factor which enhances release of 5-HT from platelets (Muck-Seler *et al.* 1979). This excess of liberated 5-HT would be available to trigger the vascular events above.

Fozard (1985) has argued the case that the most significant source of 5-HT is actually from perivascular nerves with their origin in the raphe nucleus of the brainstem. The 5-HT released from such nerves could cause vasomotion but may also trigger production of inflammatory mediators and stimulate nociceptive afferents.

Evidence that 5-HT does play a role is provided by the response to agents which enhance or suppress the effects of 5-HT in the management of migraine. Inhibitors of 5-HT uptake, such as zimelidine, may trigger migraine-like headache during initial administration when increased free 5-HT would be available, but establish resistance to migraine during long-term use when 5-HT is depleted (Syvälahti *et al.* 1979). 5-HT receptor antagonists, such as methysergide, pizotifen, and cyproheptadine have long been used for the prophylaxis of migraine. Even the β-adrenoceptor antagonist, propranolol, which has more recently been approved for use in migraine prophylaxis, has reasonably high affinity for 5-HT_1 recognition sites (Middlemiss 1984; see also Chapter 4). As noted above, the 5-HT_{1A} receptor subtype appears to be present on the vascular smooth muscle of the cerebral arteries. The exact mechanism of action of these antagonists remains far from clear. They most likely block the vasoconstriction that initiates the migraine sequence, but it is possible that their effect is the opposite: to prevent painful 5-HT-induced vasodilatation. Methysergide, at least, may be efficacious because of agonist rather than antagonist effects (Saxena 1974). 5-HT receptor agonists (usually ergotamine and ergonovine) are used to abort acute attacks. They most likely work by opposing the extracerebral vasodilatation. Somewhat perversely, 5-HT itself has shown beneficial effect when infused during the acute attack (Sjaastad 1975), perhaps through this mechanism. In this context, it is of interest that the novel and highly selective '5-HT_1-like' receptor agonists, AH 25086 and GR 43175, which selectively constrict elements of the cerebral circulation (Humphrey *et al.* 1987 and 1988; see Chapter 5) may be effective in the treatment of migraine (Doenicke *et al.* 1987).

Despite these data, it is unlikely that a simple excess of 5-HT is responsible for attacks of migraine. Patients with carcinoid syndrome, who have markedly elevated plasma 5-HT levels, do not characteristically suffer from migraine (Salmon *et al.* 1982).

A concluding speculation is that migraine and primary Raynaud's phenomenon might represent aspects of a 5-HT-related vasospastic diathesis; patients with migraine have an increased prevalence of Raynaud's symptoms (Zahavi *et al.* 1984).

16.3 Hypertension

16.3.1 Essential hypertension

Disposition mechanisms for 5-HT are multiple (clearance by hepatic and endothelial monoamine oxidase; uptake into platelets) and are efficient enough that 5-HT normally would exert significant vasoactive effects only in local areas of increased release. In other words, there is little to suggest that 5-HT normally acts as a circulating hormone. If 5-HT contributes to the pathogenesis of essential hypertension by diffusely altering vascular tone, one or more terms of the equation dictating the vascular response to 5-HT must be altered. Those terms are the production or release of 5-HT; the rate of uptake and disposition of 5-HT; the sensitivity of the blood vessel wall to vasoconstrictor effects of the monoamine; and vasodilator mechanisms of 5-HT. Alternatively, 5-HT could contribute to the hypertensive process by interference with the normal regulation of blood volume, for example through the renin–angiotensin–aldosterone system. There are suggestions from clinical observations and animal experiments that each of these could be abnormal in hypertension (Fig. 16.4); discussion here will be restricted to data from human studies.

16.3.1.1 *5-HT availability*

Circulating levels of beta-thromboglobulin, a marker of platelet aggregation, are elevated in hypertensive patients and normalize after antihypertensive therapy (Mehta and Mehta 1981). The increase in platelet activation may be due to microvascular damage caused by hypertension, to increased shear stresses associated with hypertension, or perhaps to alterations in the non-thrombogenic qualities of the endothelium. Furthermore, there may be increased aggregability of platelets from hypertensive patients. Plasma 5-HT concentrations unfortunately are difficult to measure reliably, but an indirect reflection of the elevated turnover of 5-HT in hypertensive patients is that the urinary excretion of 5-hydroxyindoleacetic acid can be lowered by antihypertensive therapy with ketanserin (Dzurik *et al.* 1985).

A decrease in one of the disposition mechanisms for 5-HT is reflected in the fact that platelet uptake of 5-HT is decreased in hypertensive patients (Kamal *et al.* 1984).

16.3.1.2 *Altered vascular responsiveness*

Most of the work demonstrating an increased sensitivity of the vascular wall to 5-HT has been done in animals (see Chapter 14).

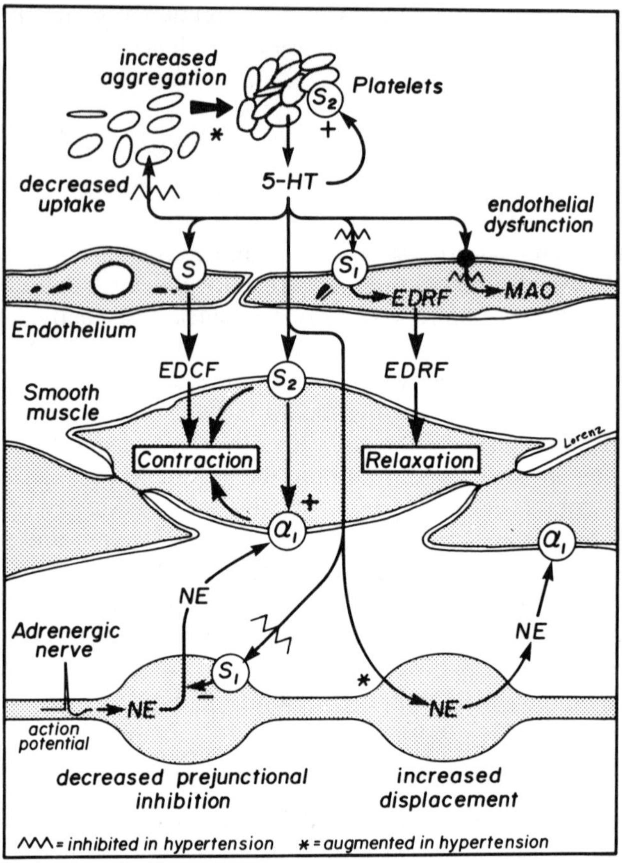

Fig. 16.4 Possible mechanisms by which 5-HT may contribute to hypertension. 5-HT uptake by platelets and elimination by the endothelium are diminished; coupled with increased aggregability of platelets; this may cause increased free 5-HT levels in plasma. Contractile effects of the amine are enhanced by increased sensitivity to the direct contractile action (mediated through 5-HT_2 receptors); reduced tachyphylaxis; amplification of other vasoconstrictor responses, e.g. noradrenaline and angiotensin II; facilitated displacement of noradrenaline from adrenergic nerve varicosities; and release of vasoconstrictor substances from the endothelium. The indirect relaxant effects of 5-HT (by presynaptic inhibition of noradrenaline release from adrenergic nerve varicosities and by release of EDRF) are diminished. Increased 5-HT concentrations, increased contractile effects, and decreased relaxant effects would all favour vasoconstriction and, if generalized, hypertension. MAO = monoamine oxidase; S_1 = '5-HT_1-like' receptor; S_2 = 5-HT_2 receptor; NE = noradrenaline; EDCF = endothelium-derived contracting factor; α = α-adrenoceptor; + = facilitation. (From Vanhoutte 1987b by permission)

The endothelium shows increased turnover and morphological abnormalities in several models of hypertension (Huttner and Gabbiani 1983). Impaired endothelium-dependent relaxation may occur in hypertension, but has been studied to date only in animal models. Likewise, the inhibitory presynaptic action of 5-HT on sympathetic nerve endings may be reduced in hypertension, but this has been shown only in experiments with animals.

16.3.1.3 *Renin–angiotensin–aldosterone system*

Renin secretion, enhanced by the administration of the diuretic furosemide, is decreased by the 5-HT receptor antagonist, cyproheptadine, in normal subjects (Epstein and Hamilton 1977). Other investigators have not found either cyproheptadine or ketanserin to affect renin or aldosterone levels in resting subjects (Gross *et al.* 1981; Degli Uberti *et al.* 1982; Fagard *et al.* 1984). However, 5-HT plays a role in the regulation of aldosterone secretion under certain circumstances (see Chapter 12). During high level exercise in normal men, the rise in plasma aldosterone concentration was attenuated by ketanserin infusion, as was the rise in arterial pressure (Fagard *et al.* 1984). In patients with idiopathic hyperaldosteronism due to adrenal hyperplasia, cyproheptadine suppressed aldosterone levels (Gross *et al.* 1981). Blood pressure, however, was apparently unchanged; hence a role for 5-HT-stimulated aldosterone secretion as a mechanism of hypertension must remain speculative.

16.3.1.4 *Effect of 5-HT receptor antagonists*

Certainly the most impressive argument for a causative role for 5-HT in the pathogenesis or maintenance of hypertension is the repeatedly demonstrated antihypertensive efficacy of the 5-HT_2 receptor antagonist, ketanserin (De Cree *et al.* 1981; see Schalekamp *et al.* 1986). This drug lowers blood pressure both acutely and over the long term. Animal studies suggest that the α_1-adrenoceptor blocking properties of ketanserin may explain its antihypertensive effect (e.g. Fozard 1982; Marwood and Stokes 1984), and an acute study in healthy volunteers has shown that ketanserin diminishes the pressor response to the α_1-adrenoceptor agonist, phenylephrine (Gasic *et al.* 1985). These conclusions are not supported by patient studies, which show that ketanserin can exert an acute antihypertensive effect without inhibiting the pressor response to phenylephrine. Furthermore, ketanserin lowers blood pressure even in patients with severe autonomic dysfunction, in whom adrenergic mechanisms for maintenance of blood pressure are less prominent (Wenting *et al.* 1984).

A specific instance of hypertension in which ketanserin appears to be valuable is that occurring following coronary bypass surgery. Patients who developed hypertension in the early post-operative period had significant reduction of pressure during infusion of ketanserin without any detrimental influence on heart rate or indices of cardiac oxygen demand (van der Starre *et al.* 1983). This could reflect augmented 5-HT release secondary to platelet injury during extracorporeal circulation; however, another study of patients undergoing bypass sugery demonstrated a pronounced antihypertensive effect of ketanserin administered *before* extracorporeal circulation was begun (Richter *et al.* 1986).

Against the 5-HT_2 receptor antagonist effect of ketanserin being responsible for the antihypertensive effect is the finding that other 5-HT_2 receptor antagonists do not display convincing antihypertensive properties, although this may be due in part to the fact that many of them are partial agonists. Another possibility is that the antihypertensive action of ketanserin occurs through effects on central 5-HT systems (Kuhn *et al.* 1980).

16.3.2 Pulmonary hypertension

One of the hypotheses for the cause of idiopathic pulmonary hypertension is that recurrent microscopic pulmonary embolization leads to loss of capillary cross-sectional area. 5-HT, released from such platelet aggregates, may contribute to the increase in resistance by causing constriction of resistance vessels. Alternatively, intimal changes associated with pulmonary hypertension could provoke ongoing platelet activation and 5-HT release.

5-HT contracts isolated canine and human pulmonary arteries (Chand and Altura 1980; Freeman *et al.* 1981). Aggregating platelets cause contractions of isolated canine pulmonary arteries, which are due to the platelet-released 5-HT (McGoon and Vanhoutte 1984). Pulmonary artery pressure increases during infusion of 5-HT into the pulmonary artery (Todd *et al.* 1983). However, experimental pulmonary hypertension induced by monocrotaline pyrrole is not affected by ketanserin or the less selective 5-HT receptor antagonist, metergoline (Ganey *et al.* 1986).

Plasma 5-HT concentrations rise in experimental pulmonary embolism. There is an associated rise in pulmonary arterial pressure, which is reduced by cyproheptadine, ketanserin, or methysergide (Utsunomiya *et al.* 1981; Huval *et al.* 1983; Todd *et al.* 1983). Thus, at least in the case of large pulmonary emboli, 5-HT does clearly contribute to the elevation in pulmonary vascular resistance. The benefit of 5-HT receptor antagonists presumably is due principally to inhibition of vasoconstriction although antiplatelet effects may play a role.

Sepsis-related adult respiratory distress syndrome, although different from primary pulmonary hypertension, is also associated with a rise in pulmonary vascular resistance. A positive correlation exists between plasma 5-HT levels and pulmonary vascular pressure gradient, suggesting that the amine may contribute to the pulmonary hypertension (Sibbald *et al.* 1980). On the other hand, the correlation may merely reflect pulmonary endothelial damage (and hence decreased endothelial 5-HT uptake) as a concomitant of the lung pathology. In early (less than five days) respiratory failure, transpulmonary platelet entrapment occurs, and ketanserin infusion has a beneficial effect on pulmonary pressures and physiological shunt (Huval *et al.* 1984). Endotoxin administration causes respiratory failure and pulmonary hypertension in several animal models, which are associated with an increase in the 5-HT content of lung lymph. Ketanserin reduces both the increased pulmonary artery pressure and the hypoxia (Meuleman *et al.* 1983; Demling *et al.* 1985; Olson 1987).

16.3.3 Pre-eclamptic hypertension

Pre-eclamptic toxaemia is characterized by hypertension, proteinuria, and oedema. The hypertension is believed to result from arteriolar vasoconstriction, but the mediator of this vasoconstriction is unknown. Several lines of evidence point to a possible role for 5-HT. 5-HT was observed over 30 years ago to cause vasoconstriction in the human placental circulation (Astrom and Samelius 1957); secondary vasoactive mediators may be released in response to placental ischaemia. The 5-HT content of platelets from patients with severe pre-eclampsia is reduced, and these platelets have decreased aggregability (Whigham *et al.* 1978), suggesting that platelet activation and 5-HT release may be ongoing. Further support for a role of 5-HT comes from two recent studies which have demonstrated that ketanserin has a potent antihypertensive effect in pre-eclamptic women (Weiner *et al.* 1984; Montenegro *et al.* 1985).

16.4 Peripheral vascular disease

As noted above, atherosclerotic blood vessels in animal models are hyperresponsive to 5-HT (Henry and Yokoyama 1980; Heistad *et al.* 1984), which might lead one to predict that there is an active element to limb artery stenosis. Nevertheless, clinical trials with many different vasodilators in human subjects with large vessel occlusive disease have been disappointing, and it probably is true that at the sites of stenosis such vessels are so extensively diseased and calcified that no vasoactive function remains. This does not mean, however, that there may be no role for 5-HT receptor antagonists in this disease.

Ongoing platelet activation occurs in such patients, as indicated by elevated levels in circulating blood of platelet factor IV and beta-thromboglobulin (Baele *et al.* 1981). This most likely occurs at the sites of endothelial disruption by atheroma. Ketanserin and other 5-HT receptor antagonists will not only inhibit platelet aggregation—thus inhibiting any tendency for platelet thrombus to occlude the vessel as discussed above—but will block downstream vasoconstrictor effects of the platelet-released 5-HT. This may be of particular importance in regard to collateral vessels, which have increased sensitivity to 5-HT (Verheyen *et al.* 1984). In animal models, the reduction of collateral blood flow secondary to a clot or to 5-HT infusion can be reversed by 5-HT depletion or 5-HT receptor blockade (Schaub *et al.* 1977; Nevelsteen *et al.* 1984).

An additional characteristic of ketanserin is that it increases red blood cell deformability (which is diminished in claudication), facilitating passage of red blood cells through the microcirculation (Dormandy 1987). Pentoxifylline, which is the only effective drug currently marketed for this indication in North America, also is believed to work by such a haemorrheologic effect (Taylor and Porter 1986).

A drug such as ketanserin might thus be ideal for the treatment of intermittent claudication. However, laboratory studies have shown improvement only of rest flow after intravenous administration of ketanserin (Bounameaux *et al.* 1985a), which may indicate that during exercise, maximal metabolic vasodilatation occurs. Nonetheless, some studies have shown significant increases in walking distance (e.g. De Cree *et al.* 1984) although not all authors have been able to replicate these findings (Bounameaux *et al.* 1985b).

Again presumably because of its effects on microcirculatory flow, ketanserin caused a marked increase in rest flow and in reactive hyperaemia in the feet of diabetics (Schneider *et al.* 1985). Likewise, a single case report suggests that ketanserin may aid healing in severe digital frostbite (Vayssairat *et al.* 1986).

16.5 Carcinoid syndrome

Enterochromaffin cells of the gastro-intestinal tract are the source of the great majority of the 5-HT synthesized in the body outside of the central nervous system. Neoplasms of these cells, or carcinoid tumours, can secrete large amounts of 5-HT into the bloodstream. Other vasoactive materials are typically also produced, however, including substance P, making it difficult to determine which of the clinical features of carcinoid syndrome are due to 5-HT (Ahlman 1985). Bowel hypermotility, which is characteristic of the syndrome, was ameliorated by the 5-HT receptor antagonists, methysergide and cyproheptadine (Brown 1977) or ketanserin (Ahlman 1985), indicating that 5-HT probably is responsible. The vascular effects (flushing and hypotension), though not improved in the latter two studies, were improved in five of seven patients treated with ketanserin (Gustafsen *et al.* 1986). In a recent study, the selective 5-HT_3 receptor antagonist, ICS 205-930, markedly attenuated the watery diarrhoea associated with the carcinoid syndrome (Anderson *et al.* 1987; see also Chapter 15). Whilst these data suggest that 5-HT may play some role in these responses, vasoactive peptides are probably also of importance. It is not known if the endocardial fibrosis seen in carcinoid syndrome is due to 5-HT, but it may not be coincidental that prolonged administration of methysergide, a 5-HT receptor antagonist with partial agonist activity, can result in a somewhat similar pathology (Graham 1967; Thompson 1977; Spatz 1979).

16.6 Concluding remarks

This chapter does not undertake to establish concrete proof for the involvement of 5-HT in any form of vascular pathology, for no such proof is currently available. Instead, we have tried to present what evidence exists that 5-HT could be involved, and to develop a schema for pathophysiology of the amine in a variety of diseases.

References

Ahlman, H. (1985). Serotonin and carcinoid tumors. *Journal of Cardiovascular Pharmacology* **7** (Suppl. 7), S79–S85.

Allen, G. S., Henderson, L. M., Chou, S. N., and French, L. A. (1974a). Cerebral arterial spasm. Part 1: in vitro contractile activity of vasoactive agents on canine basilar and middle cerebral arteries. *Journal of Neurosurgery* **40,** 433–441.

Allen, G. S., Gold, L. H. A., Chou, S. N., and French, L. A. (1974b). Cerebral arterial spasm. Part 3: In vivo intracisternal production of spasm by serotonin and blood and its reversal by phenoxybenzamine. *Journal of Neurosurgery* **40,** 451–8.

Allen, G. S., Gross, C. J., French, L. A., and Chou, S. N. (1976). Cerebral arterial spasm. Part 5: in vitro contractile activity of vasoactive agents including human CSF on human basilar and anterior cerebral arteries. *Journal of Neurosurgery* **44,** 594–600.

Anderson, J. V., Coupe, M. O., Morris, J. A., Hodgson, H. J. F., and Bloom, S. R. (1987). Remission of symptoms in carcinoid syndrome with a new 5-hydroxytryptamine M receptor antagonist. *British Medical Journal* **294,** 1129.

Anthony, M., Hinterberger, H., and Lance, J. W. (1967). Plasma serotonin in migraine and stress. *Archives of Neurology* **16,** 544–52.

Ashton, J. H., Benedict, C. R., Fitzgerald, C., Raheja, S., Taylor, A., Campbell, W. B., Buja, L. M., and Willerson, J. T. (1986). Serotonin as a mediator of cyclic flow variations in stenosed canine coronary arteries. *Circulation* **73,** 572–8.

Astrom, A., and Samelius, U. (1957). The action of 5-hydroxytryptamine and some of its antagonists on the umbilical vessels of the human placenta. *British Journal of Pharmacology* **12,** 410–14.

Ayuma, H., Ishikawa, M., and Sakiyaki, S. (1986). Endothelium-dependent inhibition of platelet aggregation. *British Journal of Pharmacology* **80,** 411–15.

Baele, G., Bogaerts, H., Clement, D. L., Pannier, R., and Barbier, F. (1981).

Platelet activation during treadmill exercise in patients with chronic peripheral arterial disease. *Thrombosis Research* **23,** 215–23.

Bing, R. J., Burger, W., Chemnitius, J. M., Saeed, M., and Metz, M. Z. (1985). Effect of intact endothelium against platelet-induced coronary artery spasm in isolated rabbit hearts. *American Journal of Cardiology* **55,** 1596–600.

Bounameaux, H., Devos, P., and Verhaeghe, R. (1985a). Acute effect of ketanserin on the leg perfusion assessed by Doppler pressure measurements and ^{201}Tl scintigraphy in patients with intermittent claudication. *Journal of Cardiovascular Pharmacology* **7,** 1083–6.

Bounameaux, H., Holditch, T., Hellemans, H., Berent, A., and Verhaeghe, R. (1985b). Placebo-controlled, double-blind, two-centre trial of ketanserin in intermitten claudication. *Lancet* **2,** 1268–71.

Bove, A. A. and Dewey, J. D. (1983). Effects of serotonin and histamine on proximal and distal coronary vasculature in dogs: comparison with alpha-adrenergic stimulation. *American Journal of Cardiology* **52,** 1333–9.

Brazenor, R. M. and Angus, J. A. (1981). Ergometrine contracts isolated canine coronary arteries by a serotonergic mechanism: no role for alpha adrenoceptors. *Journal of Pharmacology and Experimental Therapeutics* **218,** 530–6.

Brown, H. (1977). Serotonin-producing tumors. In *Serotonin in Health and Disease* (ed. W. Essman) pp. 393–423. Spectrum, New York.

Brum, J. M., Sufan, Q., Lane, G., and Bove, A. A. (1984). Increased vasoconstrictor activity of proximal coronary arteries with endothelial damage in intact dogs. *Circulation* **70,** 1066–73.

Bush, L. R. (1987). Effects of the serotonin antagonists, cyproheptadine, ketanserin and mianserin, on cyclic flow reductions in stenosed canine coronary arteries. *Journal of Pharmacology and Experimental Therapeutics* **240,** 674–82.

Chahine, R. A., Raizner, A. E., and Luchi, R. J. (1975). Coronary arterial spasm in classic angina pectoris. *Catheterization and Cardiovascular Diagnosis* **1,** 337–42.

Chand, N. and Altura, B. M. (1980). Serotonin receptors subserve only contraction in canine and rat pulmonary arteries and veins. *Artery* **7,** 232–45.

Cocks, T. M. and Angus, J. A. (1983). Endothelium-dependent relaxation of coronary arteries by noradrenaline and serotonin. *Nature (London)* **305,** 627–30.

Cohen, R. A. (1986). Contraction of isolated canine coronary arteries resistant to S_2-serotonergic blockade. *Journal of Pharmacology and Experimental Therapeutics* **237,** 548–52.

Cohen, R. A. and Coffman, J. D. (1984). Digital vasospasm: the pathophysiology of Raynaud's phenomenon. *International Angiology* **3,** 47–55.

Cohen, R. A. and Vanhoutte, P. M. (1985). Platelets, serotonin, and endothelial cells. In *Serotonin and the cardiovascular system* (ed. P. M. Vanhoutte) pp. 105–12. Raven Press, New York.

Cohen, R. A., Shepherd, J. T., and Vanhoutte, P. M. (1983a). Inhibitory role of the endothelium in the response of isolated coronary arteries to platelets. *Science* **221,** 273–4.

Cohen, R. A., Shepherd, J. T., and Vanhoutte, P. M. (1983b). 5-Hydroxytryptamine can mediate endothelium-dependent relaxation of coronary arteries. *American Journal of Physiology* **245,** H1077–80.

Cortellaro, M., Boschetti, C., Antoniazzi, V., Moreo, G., Repetto, S., Verna, E., Boscarini, M., Limido, A., Binaghi, G., and Polli, E. E. (1983). Transcoronary platelet thromboxane A2 formation without platelet trapping in patients with coronary stenosis—effect of sulphinpyrazone treatment. *Thrombosis and Haemostasis (Stuttgart)* **50,** 857–9.

Curry, R. C., Pepine, C. J., Sabom, M. B., and Conti, C. R. (1979). Similarities of egonovine-induced and spontaneous attacks of variant angina. *Circulation* **59,** 307–12.

Davies, M. J., and Thomas, A. (1984). Thrombosis and acute coronary-artery lesions in sudden cardiac ischaemic death. *New England Journal of Medicine* **310,** 1137–40.

De Caterina, R., Carpeggiani, C., and L'Abbate, A. (1984). A double-blind, placebo-controlled study of ketanserin in patients with Prinzmetal's angina. Evidence against a role for serotonin in the genesis of coronary vasospasm. *Circulation* **69,** 889–94.

De Clerck, F. and Van Nueten, J. M. (1982). Platelet-mediated vascular contractions: inhibition of the serotonergic component by ketanserin. *Thrombosis Research* **27,** 713–27.

De Clerck, F., David, J. L., and Janssen, P. A. J. (1982). Inhibition of 5-hydroxytryptamine-induced and -amplified human platelet aggregation by ketanserin (R41468), a selective 5-HT_2-receptor antagonist. *Agents and Actions* **12,** 388–97.

De Clerck, F., Somers, Y., and van Gorp, L. (1984). Platelet-vessel wall interactions in hemostasis: implication of 5-hydroxytryptamine. *Agents and Actions* **15,** 627–35.

De Cree, J., Leempoels, J., Geukens, H., De Cock, W., and Verhaegen, H. (1981). The antihypertensive effects of ketanserin (R41468), a novel 5-hydroxytryptamine-blocking agent, in patients with essential hypertension. *Clinical Science* **61,** 473S-6S.

De Cree, J., Leempoels, J., Geukens, H., and Verhaegen, H. (1984). Placebo-controlled double-blind trial of ketanserin in treatment of intermittent claudication. *Lancet* **2,** 775–9.

De Mey, J. G. and Vanhoutte, P. M. (1982). Heterogeneous behavior of the canine arterial and venous wall. Importance of the endothelium. *Circulation Research* **51,** 439–47.

De Mey, J. G., Claeys, M., and Vanhoutte, P. M. (1982). Endothelium-dependent inhibitory effects of acetylcholine, adenosine triphosphate, thrombin and arachidonic acid in the canine femoral artery. *Journal of Pharmacology and Experimental Therapeutics* **222,** 166–73.

Degli Uberti, E. C., Trasforini, G., Margutti, A. R., Rotola, C. A., and Pansini, R. (1982). Effect of ketanserin, a new inhibitor of 5-HT_2 receptors, on plasma renin activity and aldosterone levels in normal subjects. *Hormone Research* **16,** 244–8.

Demling, R. H., Wong, C., Fox, R., Hechtman, H., and Huval, W. (1985). Relationship of increased lung serotonin levels to endotoxin-induced pulmonary hypertension in sheep: effect of a serotonin antagonist. *American Review of Respiratory Disease* **132,** 1257–61.

Doenicke, A., Siegel, E., Hadoke, M., and Perrin, V. L. (1987). Initial clinical study of AH 25086B (5-HT_1-like agonist) in the acute treatment of migraine. *Cephalalgia* **7,** 438–9.

Dormandy, J. A. (1987). Serotonin and haemorrheology. *International Journal of Cardiology* **14,** 213–9.

Dzurik, R., Fetkovska, N., Brimichova, G., and Tison, P. (1985). Blood pressure, 5-OH indoleacetic acid, vanilmandelic acid excretion and blood platelet aggregation in hypertensive patients treated with ketanserin. *Journal of Cardiovascular Pharmacology* **7** (Suppl. 7), S29–31.

Edvinsson, L., Degueurce, A., Duverger, D., MacKenzie, E. T., and Scatton, B. (1983). Central serotonergic nerves project to the pial vessels of the brain. *Nature (London)* **306,** 55–7.

Epstein, S. and Hamilton, S. (1977). Cyproheptadine inhibition of stimulated plasma renin activity. *Journal of Clinical Endocrinology and Metabolism* **45,** 1235–7.

Fagard, R., Cattaert, A., Lijnen, P., Staessen, J., Vanhees, L., Moerman, E., and Amery, A. (1984). Responses of the systemic circulation and of the renin–angiotensin–aldosterone system to ketanserin at rest and exercise in normal man. *Clinical Science* **66,** 17–25.

Fanchamps, A. (1982). The evolution of thinking about the role and site of action of serotonin in migraine. In *Advances in neurology,* Vol. 33, *Headache: physiopathological and clinical concepts* (ed. M. Critchley) pp. 31–3. Raven Press, New York.

Feniuk, W., Humphrey, P. P. A., and Watts, A. D. (1983). 5-Hydroxytryptamine-induced relaxation of isolated mammalian smooth muscle. *European Journal of Pharmacology* **96,** 71–8.

Folts, J. D., Crowell, E. B., and Rowe, G. G. (1976). Platelet aggregation in partially obstructed vessels and its elimination with aspirin. *Circulation* **54,** 365–70.

Folts, J. D., Gallagher, K., and Rowe, G. G. (1982). Blood flow reductions in stenosed canine coronary arteries: vasospasm or platelet aggregation? *Circulation* **65,** 248–55.

Fozard, J. R. (1982). Mechanism of the antihypertensive effect of ketanserin. *Journal of Cardiovascular Pharmacology* **4,** 829–38.

Fozard, J. R. (1985). 5-Hydroxytryptamine in the pathophysiology of migraine. In *Vascular neuroeffector mechanisms*, Vol. 5 (eds. J. A. Bevan, T. Godfraind, R. A. Maxwell, J. C. Stoclet, and M. Worcel) pp. 321–28. Elsevier, Amsterdam.

Freedman, S. B., Chierchia, S., Rodriguez-Plaza, L., Bugiardini, R., Smith, G., and Maseri, A. (1984). Ergonovine-induced myocardial ischaemia: no role for serotonergic receptors? *Circulation* **70,** 178–83.

Freeman, W. K., Rorie, D. K., and Tyce, G. M. (1981). Effects of 5-hydroxytryptamine on neuroeffector junction in human pulmonary artery. *Journal of Applied Physiology* **51,** 693–8.

Furchgott, R. F. (1983). Role of the endothelium in responses of vascular smooth muscle. *Circulation Research* **53,** 557–73.

Furlong, B., Henderson, A. H., Lewis, M. J., and Smith, J. A. (1987). Endothelium-derived relaxing factor inhibits in vitro platelet aggregation. *British Journal of Pharmacology* **90,** 687–92.

Ganey, P. E., Sprugel, K. H., Hadley, K. B., and Roth, R. A. (1986). Monocrotaline pyrrole-induced cardiopulmonary toxicity is not altered by metergoline or ketanserin. *Journal of Pharmacology and Experimental Therapeutics* **237,** 226–31.

Gasic, S., Eichler, H. G., and Korn, A. (1985). Effect of ketanserin on phenylephrine-dependent changes in splanchnic hemodynamics and systemic blood pressure in healthy subjects. *Journal of Cardiovascular Pharmacology* **7,** 219–23.

Gertz, S. D., Uretsky, G., Wajnberg, R. S., Navot, N. and Gotsman, M. S. (1981). Endothelial cell damage and thrombus formation after partial arterial constriction: relevance to the role of coronary artery spasm in the pathogenesis of myocardial infarction. *Circulation* **63,** 476–86.

Graham, J. R. (1967). Cardiac and pulmonary fibrosis during methysergide therapy for headache. *American Journal of Medical Science* **254,** 1.

Gross, M. D., Grekin, R. J., Gniadek, T. C., and Villareal, J. Z. (1981). Suppression of aldosterone by cyproheptadine in idiopathic aldosteronism. *New England Journal of Medicine* **305,** 181–5.

Gryglewski, R. J., Dembinska-Kiec, A., Zmuda, A., and Gryglewska, T. (1978). Prostacyclin and thromboxane A_2 biosynthesis capacities of heart, arteries, and platelets at various stages of experimental atherosclerosis in rabbits. *Atherosclerosis* **31,** 385–94.

Gustafsen, J., Lendorf, A., Raskov, H., and Boesby, S. (1986). Ketanserin versus placebo in carcinoid syndrome. *Scandinavian Journal of Gastroenterology* **21,** 816–8.

Halpern, A., Kuhn, P. H., Shaftel, H. E., Samuels, S. S., Shaftel, N., Selman, D., and Birch, H. G. (1960). Raynaud's disease, Raynaud's phenomenon, and serotonin. *Angiology* **11,** 151–67.

Heistad, D. D., Armstrong, M. L., Marcus, M. L., Piegors, D. J., and Mark, A. L. (1984). Augmented responses to vasoconstrictor stimuli in hypercholesteolemic and atherosclerotic monkeys. *Circulation Research* **54,** 711–8.

Henry, P. D. and Yokoyama, M. (1980). Supersensitivity of atherosclerotic rabbit aorta to ergonovine. Mediation by a serotonergic mechanism. *Journal of Clinical Investigation* **66,** 306–13.

Heric, E. and Tackett, R. L. (1985). Altered vascular reactivity in the rabbit during hypercholesterolemia. *Pharmacology* **31,** 72–81.

Hirsh, P. D., Hillis, L. D., Campbell, W. B., Firth, B. G., and Willerson, J. T. (1981). Release of prostaglandins and thromboxane into the coronary circulation in patients with ischaemic heart disease. *New England Journal of Medicine* **304,** 685–91.

Holtz, J., Held, W., Sommer, O., Kuhne, G., and Bassenge, E. (1982). Ergonovine-induced constriction of epicardial coronary arteries in conscious dogs: alpha-adrenoceptors are not involved. *Basic Research in Cardiology* **77,** 278–91.

Houston, D. S. and Vanhoutte, P. M. (1986). Serotonin and the vascular system: role in health and disease, and implications for therapy. *Drugs* **31,** 149–63.

Houston, D. S. and Vanhoutte, P. M. (1988). Comparison of serotonergic receptor subtypes on the smooth muscle and endothelium of the canine coronary artery. *Journal of Pharmacology and Experimental Therapeutics* **244,** 1–10.

Houston, D. S., Shepherd, J. T., and Vanhoutte, P. M. (1985). Adenine nucleotides, serotonin, and endothelium-dependent relaxations to platelets. *American Journal of Physiology* **248,** H389–95.

Houston, D. S., Shepherd, J. T., and Vanhoutte, P. M. (1986). Aggregating human platelets cause direct contraction and endothelium-dependent relaxation of isolated canine coronary arteries. *Journal of Clinical Investigation* **78,** 539–44.

Houston, D. S., Burnstock, G., and Vanhoutte, P. M. (1987). Different P_2-purinergic receptor subtypes on endothelium and smooth muscle in canine blood vessels. *Journal of Pharmacology and Experimental Therapeutics* **241,** 501–6.

Humphrey, P. P. A., Feniuk, W., Perren, M. J., Oxford, A. W., Brittain, R. T., and Jack, D. (1987). The pharmacology of selective 5-HT_1-like receptor agonists for the acute treatment of migraine. *Cephalalgia* **7,** 400–1.

Humphrey, P. P. A., Feniuk, W., Perren, M. J., Oxford, A. W., Coates, I. H., and Butina, D. (1988). GR 43175, a selective agonist for the 5-HT_1-like receptor in dog isolated saphenous vein. *British Journal of Pharmacology* **94,** 1123–32.

Huttner, I. and Gabbiani, G. (1983). Vascular endothelium in hypertension. In *Hypertension: physiopathology and treatment*, 2nd ed. (eds. J. Genest *et al.*) pp. 473–88. McGraw Hill, New York.

Hutton, R. A., Mikhailidis, D. P., Bernstein, R. M., Jeremy, J. Y., Hughes, G. R. V. *et al.* (1984). Assessment of platelet function in patients with Raynaud's syndrome. *Journal of Clinical Pathology* **37,** 182–7.

Huval, W. V., Mathieson, M. A., Stemp, L. I., Dunham, B. M., Jones, A. G. *et al.* (1983). Therapeutic benefits of 5-hydroxytryptamine inhibition following pulmonary embolism. *Annals of Surgery* **197,** 220–5.

Huval, W. V., Lelcuk, S., Shepro, D., and Hechtman, H. B. (1984). Role of serotonin in patients with acute respiratory failure. *Annals of Surgery* **200,** 166–72.

Jauernig, R. A. and Moulds, R. F. W. (1978). A human arterial preparation for studying the effects of vasoactive agents. *Circulation Research* **42,** 363–8.

Kalsner, S. and Richards, R. (1984). Coronary arteries of cardiac patients are hyperreactive and contain stores of amines: a mechanism for coronary spasm. *Science* **223,** 1435–7.

Kamal, L., Quan-Bui, K. H. L., and Meyer, P. (1984). Decreased uptake of ^{3}H-serotonin and endogenous content of serotonin in blood platelet in hypertensive patients. *Hypertension* **6,** 568–73.

Kawachi, Y., Tomoike, H., Maruoka, Y., Kikuchi, Y., Araki, H., Ishii, Y., Tanaka, K., and Nakamura, M. (1984). Selective hypercontraction caused by ergonovine in the canine coronary artery under conditions of induced athersclerosis. *Circulation* **69,** 441–50.

Ku, D. D. (1982). Coronary vascular reactivity after acute myocardial ischaemia. *Science* **218,** 576.

Kuhn, D. M., Wolf, W. A., and Lovenberg, W. (1980). Review of the role of the central serotonergic neuronal system in blood pressure regulation. *Hypertension* **2,** 243–55.

Lamping, K. G., Marcus, M. L., and Dole, W. P. (1985). Removal of the endothelium potentiates canine large coronary artery constrictor responses to 5-hydroxytryptamine in vivo. *Circulation Research* **57,** 46–54.

Lewis, H. D., Davis, J. W., Archibald, D. G., Steinke, W. E., Smitherman, T. C., Doherty, J. E., Schnaper, W. H., LeWinte, M. M., Linares, E., Pouget, J. M., Sabharwal, S. C., Chesler, E., and DeMots, H. (1983). Protective effects of aspirin against acute myocardial infarction and death in men with unstable angina. Results of a vetarans administration cooperative study. *New England Journal of Medicine* **309,** 396–403.

Lewy, R. I., Wiener, L., Walinsky P., Lefer, A. M., Silver, M. J., and Smith, J. B. (1980). Thromboxane release during pacing-induced angina pectoris:

possible vasoconstrictor influence on the coronary vasculature. *Circulation* **61,** 1165–71.

Lindblad, L. E., Shepherd, J. T., and Vanhoutte, P. M. (1984). Cooling augments platelet-induced contraction of peripheral arteries of the dog. *Proceedings of the Society of Experimental Biology and Medicine* **176,** 119–22.

Loisy, C., Beorchia, S., Centonze, V., Fozard, J. R., Schechter, P. J., and Tell, G. P. (1985). Effect on migraine headache of MDL 72222, an antagonist at neuronal 5-HT receptors. Double-blind, placebo-controlled study. *Cephalgia* **5,** 79–82.

Longstaff, J., Gush, R., Williams, E. H., and Jayson, M. I. V. (1985). Effects of ketanserin on peripheral blood flow, haemorheology, and platelet function in patients with Raynaud's phenomenon. *Journal of Cardiovascular Pharmacology* **7** (Suppl. 7), S99–101.

Luchi, R. J., Chahine, R. A., and Raizner, A. E. (1979). Coronary artery spasm. *Annals of Internal Medicine* **91,** 441–9.

McGoon, M. D. and Vanhoutte, P. M. (1984). Aggregating platelets contract isolated canine pulmonary arteries by releasing 5-hydroxytryptamine. *Journal of Clinical Investigation* **74,** 828–33.

Marco, E. J., Balfagon, G., Salaices, M., Sanchez-Ferrer, C. F., and Marin, J. (1985). Serotonergic innervation of cat cerebral arteries. *Brain Research* **338,** 137–9.

Marwood, J. F. and Stokes, G. S. (1984). Studies on the mechanism of action of the hypotensive effect of ketanserin. *Clinical and Experimental Pharmacology and Physiology* **11,** 125–32.

Maseri, A., Pesola, A., Marzilli, M., Severi, S., Parodi, O., L'Abbate, A., Ballestra, A. M., Maltinti, G., De Nes, D. M., and Biagini, A. (1977). Coronary vasospasm in angina pectoris. *Lancet* **1,** 713–17.

Maseri, A., Severi, S., De Nes, M., L'Abbate, A., Chierchia, S., Marzilli, M., Ballestra, A. M., Parodi, O., Biagini, A., and Distante, A. (1978). 'Variant' angina: one aspect of a continuous spectrum of vasospastic myocardial ischaemia. *American Journal of Cardiology* **42,** 1019–35.

Maseri, A., Chierchia, A., and Davies, G. (1986). Pathophysiology of coronary occlusion in acute infarction. *Circulation* **73,** 233–9.

Mata-Bouchart, L. A., Waters, D. D., Bouchard, A., Miller, D. D., and Theroux, P. (1985). Failure of ketanserin, a serotonin inhibitor, to prevent spontaneous or ergonovine-induced attacks of variant angina. *Canadian Journal of Cardiology* **1,** 168–71.

Mehta, J. and Mehta, P. (1981). Platelet function in hypertension and effect of therapy. *American Journal of Cardiology* **47,** 331–4.

Meuleman, T. R., Hill, D. C., Port, J. D., Stanley, T. H., Pace, N. L., *et al.* (1983). Ketanserin prevents platelet aggregation and endotoxin-induced pulmonary vasoconstriction. *Critical Care Medicine* **11,** 606–11.

Middlemiss, D. N. (1984). Stereoselective blockade at [^{3}H]5-HT binding sites and at the 5-HT autoreceptor by propranolol. *European Journal of Pharmacology* **101,** 289–93.

Mohan, J. (1980). The neurosurgeon's view. In *Cerebral vasospasm* (ed. D. J. Boullin) pp. 15–35. John Wiley, Chichester.

Montenegro, R., Knuppel, R. A., Shah, D., and O'Brien, W. F. (1985). The effect of serotonergic blockade in postpartum preeclamptic patients. *American Journal of Obstetrics and Gynecology* **153,** 130–4.

Moulds, R. F. W., Iwanov, V., and Medcalf, R. (1984). The effects of platelet-derived contractile agents on human digital arteries. *Clinical Science* **66,** 443–51.

Muck-Seler, D., Deanovic, Z., and Dupelj, M. (1979). Platelet serotonin (5-HT) and 5-HT releasing factor in plasma of migrainous patients. *Headache* **19,** 14–17.

Mullane, K. M., Bradley, G., and Moncada, S. (1982). The interaction of platelet-derived mediators on isolated canine coronary arteries. *European Journal of Pharmacology* **84,** 115–8.

Nagai, H., Noda, S., and Mabe, H. (1975). Experimental cerebral vasospasm. Part 2: Effects of vasoactive drugs and sympathectomy on early and late spasm. *Journal of Neurosurgery* **42,** 420–8.

Nevelsteen, A., De Clerck, F., Loots, W., and De Gryse, A. (1984). Restoration of post-thrombotic peripheral collateral circulation in the cat by ketanserin, a selective 5-HT_2-receptor antagonist. *International Archives of Pharmacodynamics and Therapeutics* **270,** 268–79.

Nielsen, S. L., Vitting, K., and Rasmussen, K. (1983). Prazosin treatment of primary Raynaud's phenomenon. *European Journal of Clinical Pharmacology* **24,** 421–3.

Olson, N. C. (1987). Role of 5-hydroxytryptamine in endotoxin-induced respiratory failure of pigs. *American Review of Respiratory Disease* **135,** 93–9.

Pearson, J. G. and Gordon, T. L. (1985). Nucleotide metabolism by endothelium. *Annual Review of Physiology* **47,** 617–27.

Peroutka, S. J., Huang, S., and Allen, G. S. (1986). Canine basilar artery contractions mediated by 5-$hydroxytryptamine_{1A}$ receptors. *Journal of Pharmacology and Experimental Therapeutics* **237,** 901–6.

Preibisz, J. J., Sealey, J. E., Laragh, J. H., Cody, R. J., and Weksler, B. B. (1983). Plasma and platelet vasopressin in essential hypertension and congestive heart failure. *Hypertension* **5** (Suppl. I), I129–38.

Raskin, N. H. (1981). Pharmacology of migraine. *Annual Review of Pharmacology and Toxicology* **21,** 463–78.

Richter, J. A., Baransky, A., Gob, E., and Spath, P. (1986). Cardiovascular effects of ketanserin in the treatment of hypertension during coronary artery bypass surgery. *Journal of Cardiovascular Pharmacology* **8,** 108–15.

Robertson, R. M., Robertson, D., Friesinger, G. C., Timmons, S., and Hawiger, J. (1980). Platelet aggregates in peripheral and coronary-sinus blood in patients with coronary-artery spasm. *Lancet* **2,** 829–31.

Roddie, I. C., Shepherd, J. T., and Whelan, R. F. (1955). The action of 5-hydroxytryptamine on the blood vessels of the human hand and forearm. *British Journal of Pharmacology* **10,** 445–50.

Rubanyi, G. M., Frye, R. L., Holmes, D. R., and Vanhoutte, P. M. (1987). Vasoconstrictor activity of coronary sinus plasma from patients with coronary artery disease. *Journal of the American College of Cardiology* **9,** 1234–49.

Rubenstein, M. D., Wall, R. T., Baim, D. S., and Harisson, D. C. (1981). Platelet activation in clinical coronary artery disease and spasm. *American Heart Journal* **102,** 363–7.

Rusch, N. J., Chyatte, D., Sundt, T. M., and Vanhoutte, P. M. (1985). 5-Hydroxytryptamine: source of activator calcium in human basilar arteries. *Stroke* **16,** 718–20.

Rydzewski, W. (1976). Serotonin (5-HT) in migraine: levels in whole blood in and between attacks. *Headache* **16,** 16–19.

Salmon, S., Bonciani, M., Fanciullacci, M., Marianelli, L., Michelacci, S., *et al.* (1982). A putative 5-HT central feedback in migraine and cluster headache attacks. In *Advances in neurology,* Vol. 33, *Headache: physiopathological and clinical concepts* (eds. M. Critchley *et al.*) pp. 265–274. Raven Press, New York.

Saxena, P. R. (1974). Selective vasoconstriction in carotid vascular bed by methysergide: possible relevance to its antimigraine effect. *European Journal of Pharmacology* **27,** 99–105.

Schalekamp, M. A. D. H., Woittiez, A. J., Wenting, G. J., Van Den Meiracker, A. H., and Man In't Veld, A. J. (1986). Ketanserin: haemodynamic effects and mechanism of action. *Journal of Hypertension* (Suppl.) **4,** S7–12.

Schaub, R. G., Meyers, K. M., and Sande, R. D. (1977). Serotonin as a factor in depression of collateral blood flow following experimental arterial thrombosis. *Journal of Laboratory and Clinical Medicine* **90,** 645–53.

Schneider, S. H., Tendler, M., Apelian, A., Jageneau, A. H. M., and Khachadurian, A. K. (1985). Effects of ketanserin, a 5-HT_2-receptor antagonist, on the blood flow response to temperature changes in the diabetic foot. *Journal of Clinical Pharmacology* **25,** 413–17.

Schroeder, J. S., Bolen, J. L., Quint, R. A., Clark, D. A., Hayden, W. G., Higgens, C. B., and Wexler, L. (1977). Provocation of coronary spasm with ergonovine maleate. *American Journal of Cardiology* **40,** 487–91.

Seibold, J. R. (1985). Serotonin and Raynaud's phenomenon. *Journal of Cardiovascular Pharmacology* **7** (Suppl. 7), S95–8.

Seibold, J. R. and Jageneau, A. H. M. (1984). Treatment of Raynaud's phenomenon with ketanserin, a selective antagonist of the serotonin-2 (5-HT_2) receptor. *Arthritis and Rheumatism* **27,** 139–46.

Shimokawa, H., Aarhus, L. A., and Vanhoutte, P. M. (1987). Porcine coronary arteries with regenerated endothelium have a reduced endothelium-dependent responsiveness to aggregating platelets and serotonin. *Circulation Research* **61,** 256–70.

Sibbald, W., Peters, S., and Lindsay, R. M. (1980). Serotonin and pulmonary hypertension in human septic ARDS. *Critical Care Medicine* **8,** 490–4.

Sicuteri, F., Testi, A., and Anselmi, B. (1961). Biochemical investigations in headache: increase in the hydroxyindoleacetic acid excretion during migraine attacks. *International Archives in Allergy* **19,** 55–8.

Sjaastad, O. (1975). The significance of blood serotonin levels in migraine: a critical review. *Acta Neurologica Scandinavica* **51,** 200–10.

Somerville, B. W. (1976). Platelet-bound and free serotonin levels in jugular and forearm venous blood during migraine. *Neurology* **26,** 41–5.

Spatz, M. (1979). Serotonin in cardiac disease. In *Serotonin in health and disease,* Vol. V, *Clinical applications* (ed. J. Essman) pp. 389–401. Spectrum, New York.

Stranden, E., Roald, O. K., and Krohg, K. (1982). Treatment of Raynaud's phenomenon with the 5-HT_2-receptor antagonist ketanserin. *British Medical Journal* **285,** 1069–71.

Sweetman, H. E., Shepro, D., and Hechtman, H. B. (1981). Inhibition of thrombocytopenic petechiae by exogenous serotonin administration. *Haemostasis* **10,** 65–

Syvälahti, E., Kangasniemi, P., and Ross, S. B. (1979). Migraine headache and blood serotonin levels after administration of zimelidine, a selective inhibitor of serotonin uptake. *Current Therapeutic Research* **25,** 299–310.

Tada, M., Kuzuya, T., Inoue, M., Kodama, K., Mishima, M., Yamada, M., Inui, M., and Abe, H. (1981). Elevation of thromboxane B_2 levels in patients with classic and variant angina pectoris. *Circulation* **64,** 1107–15.

Taylor, E. E., Duckles, S. P., and Nelson, D. L. (1986). Dissociation constants of serotonin agonists in the canine basilar artery correlate to K_i values at the 5-HT_{1A} binding site. *Journal of Pharmacology and Experimental Therapeutics* **236,** 118–25.

Taylor, L. M. and Porter, J. M. (1986). Drug treatment of claudication: vasodilators, hemorrheologic agents, and antiserotonin drugs. *Journal of Vascular Surgery* **3,** 374–81.

Thompson, J. H. (1977). Serotonin (5-hydroxytryptamine) and the alimentary system. In *Serotonin in health and disease,* Vol. 4, (ed. J. Essman) pp. 201–392. Spectrum, New York.

Todd, M. H., Forrest, J. B., and Cragg, D. B. J. (1983). The effects of aspirin and methysergide on responses to clot-induced pulmonary embolism. *American Heart Journal* **105,** 769–76.

Utsunomiya, T., Krausz, M. M., Shepro, D., and Hechtman, H. B. (1981). Prostaglandin control of plasma and platelet 5-hydroxytryptamine in normal and embolized animals. *American Journal of Physiology* **241,** H766–71.

van der Starre, P. J. A., Scheijgrond, H. W., Reneman, R. S., and Kolling, J. B. (1983). The use of ketanserin, a 5-hydroxytryptamine receptor antagonist, for treatment of postoperative hypertension following coronary artery bypass surgery. *Anesthesia and Analgesia* **62,** 63–9.

Vanhoutte, P. M. (1985a). Serotonin and hypertension. In *Vascular neuroeffector mechanisms* (ed. J. A. Bevan, T. Godfraind, R. A. Maxwell, J. C. Stoclet, and M. Worcel) pp. 205–10. Elsevier, Amsterdam.

Vanhoutte, P. M. (1985b). Peripheral serotonergic receptors and hypertension. In *Serotonin and the cardiovascular system* (ed. P. M. Vanhoutte) pp. 123–33. Raven Press, New York.

Vanhoutte, P. M. (1986). Serotonin and the blood-vessel wall. *Journal of Hypertension* **4** (Suppl. 5), S112–15.

Vanhoutte, P. M. (1987a). Serotonin and the vascular wall. *International Journal of Cardiology* **14,** 189–203.

Vanhoutte, P. M. (1987b). Cardiovascular effects of serotonin. *Journal of Cardiovascular Pharmacology* **10** (Suppl. 3), S8–11.

Vanhoutte, P. M., and Cohen, R. A. (1983). The elusory role of serotonin in vascular function and disease. *Biochemical Pharmacology* **32,** 3671–4.

Vanhoutte, P. M., and Houston, D. S. (1988). Platelets, endothelium and vasospasm. *Circulation* **72,** 728–34.

Vanhoutte, P. M., and Houston, D. S. (1986). Serotonin, platelets, and vascular function. In *Central and peripheral mechanisms of cardiovascular regulation* (eds. J. Magro *et al.*) pp. 135–61. Plenum, New York.

Vanhoutte, P. M., and Shepherd, J. T. (1970). Effect of temperature on reactivity of isolated cutaneous veins of the dog. *American Journal of Physiology* **218,** 187–90.

Vanhoutte, P. M., and Shepherd, R. F. J. (1986). Serotonin, serotonergic antagonists, and hypertension. In *Primary hypertension* (eds. W. Kaufmann *et al.*) pp. 210–24. Springer-Verlag, New York.

Vanhoutte, P. M., Cohen, R. A., and Van Nueten, J. M. (1984). Serotonin and arterial vessels. *Journal of Cardiovascular Pharmacology* **6,** S421–8.

Vanhoutte, P. M., Rubanyi, G. M., Miller, V. M., and Houston, D. S. (1986). Modulation of vascular smooth muscle contraction by the endothelium. *Annual Review of Physiology* **48,** 307–20.

Van Nueten, J. M., Janssen, P. A. J., Van Beek, J., Xhonneux, R., Verbeuren, T. J., and Vanhoutte, P. M. (1981). Vascular effects of ketanserin (R41468), a novel antagonist of 5-HT_2 serotonergic receptors. *Journal of Pharmacology and Experimental Therapeutics* **218,** 217–30.

Vayssairat, M., Priollet, P., Hagege, A., Housset, E., and Foray, J. (1986). Does ketanserin relieve frostbite? *The Practitioner* **230,** 406–8.

Verheyen, A., Vlaminckx, E., Dauwers, F., Van Den Broeck, C., and Wouters, L. (1984). Serotinin-induced blood flow changes in the rat hind-legs after unilateral ligation of the femoral artery. Inhibition by the S_2 receptor antagonist ketanserin. *International Archives of Pharmacodynamics and Therapeutics* **210,** 280–94.

Voldby, B., Engbaek, F., and Enevoldsen, E. M. (1982). CSF serotonin concentrations and cerebral arterial spasm in patients with ruptured intracranial aneurysm. *Stroke* **13,** 184–9.

Webb, R. C. and Vanhoutte, P. M. (1985). Responsiveness of the hypertensive blood vessel wall to serotonergic stimuli. In *Serotonin and the cardiovascular system* (ed. P. M. Vanhoutte) pp. 113–22. Raven Press, New York.

Weiner, C. P., Socol, M. L., and Vaisrub, N. (1984). Control of pre-eclamptic hypertensive by ketanserin—a serotonin receptor antagonist. *American Journal of Obstetrics and Gynecology* **149,** 496–500.

Weksler, B. B., Ley, C. W., and Jaffe, E. A. (1978). Stimulation of endothelial cell prostacyclin production by thrombin, trypsin, and the ionophore A23187. *Journal of Clinical Investigation* **62,** 923–30.

Wenting, G. J., Woittiez, A. J. J., Man In't Veld, A. J., and Schalekamp, M. A. D. H. (1984). 5-HT, alpha-adrenoceptors, and blood pressure: effects of ketanserin in essential hypertension and autonomic insufficiency. *Hypertension* **6,** 100–9.

Whigham, K. A. E., Howie, P. W., Drummond, A. H., and Prentice, C. R. M. (1978). Abnormal platelet function in pre-eclampsia. *British Journal of Obstetrics and Gynaecology* **85,** 28–32.

Winkelmann, R. K., Goldyne, M. E. and Linscheid, R. L. (1976). Hypersensitivity of scleroderma cutaneous vascular smooth muscle to 5-hydroxytryptamine. *British Journal of Dermatology* **95,** 51–6.

Yokoyama, M., Akita, H., Mizutani, T., Fukuzaki, H., and Watanabe, Y. (1983). Hyperreactivity of coronary arterial smooth muscles in response to ergonovine from rabbits with hereditary hyperlipidemia. *Circulation Research* **53,** 63–71.

Young, H. A. Kolbeck, R. C., and Schmidek, H. H. (1986). Hemorrhage-induced alterations of rabbit basilar artery reactivity and sensitivity to serotonin. *Neurosurgery* **19,** 502–6.

Zahavi, I., Chagnac, A., Hering, R., Davidovich, S., and Kuritzky, A. (1984). Prevalence of Raynaud's phenomenon in patients with migraine. *Archives of Internal Medicine* **144,** 742–4.

Zervas, N. T., Candia, M., Candia, G., Kido, D., Pessin, M. S. *et al.* (1979). Reduced incidence of cerebral ischaemia following rupture of intracranial aneurysms. *Surgical Neurology* **11,** 339–44.

Index